Advanced Manufacturing Techniques for Engineering and Engineered Materials

R. Thanigaivelan
Muthayammal Engineering College, India

N. Rajan
Vinayaga Mission Kripananda Variyar Engineering College, India

T.G. Argul
St. Mary's Engineering College, India

A volume in the Advances in Chemical and Materials Engineering (ACME) Book Series

IGI Global
PUBLISHER of TIMELY KNOWLEDGE

Published in the United States of America by
IGI Global
Engineering Science Reference (an imprint of IGI Global)
701 E. Chocolate Avenue
Hershey PA, USA 17033
Tel: 717-533-8845
Fax: 717-533-8661
E-mail: cust@igi-global.com
Web site: http://www.igi-global.com

Library of Congress Cataloging-in-Publication Data

Names: Thanigaivelan, Rajasekaran, 1977- editor. | Rajan, N., 1977- editor. | Arul, T. G., 1976- editor.
Title: Advanced manufacturing techniques for engineering and engineered materials / Rajasekaran Thanigaivelan, N. Rajan, and T.G. Arul, editors.

Description: Hershey, PA : Information Science Reference, an imprint of IGI Global, [2022] | Includes bibliographical references and index. | Summary: "This book highlights the latest trends in manufacturing processes such as 3D Printing, Casting, Welding, Surface Modification, CNC, Non- Traditional, Industry 4.0 Ergonomics and Hybrid Machining Methods"-- Provided by publisher.
Identifiers: LCCN 2021036885 (print) | LCCN 2021036886 (ebook) | ISBN 9781799895749 (h/c) | ISBN 9781799895756 (s/c) | ISBN 9781799895763 (ebk)
Subjects: LCSH: Materials. | Manufacturing processes.
Classification: LCC TA403 .A2565 2022 (print) | LCC TA403 (ebook) | DDC 620.1/1--dc23
LC record available at https://lccn.loc.gov/2021036885
LC ebook record available at https://lccn.loc.gov/2021036886

This book is published in the IGI Global book series Advances in Chemical and Materials Engineering (ACME) (ISSN: 2327-5448; eISSN: 2327-5456)

British Cataloguing in Publication Data
A Cataloguing in Publication record for this book is available from the British Library.

All work contributed to this book is new, previously-unpublished material. The views expressed in this book are those of the authors, but not necessarily of the publisher.

For electronic access to this publication, please contact: eresources@igi-global.com.

Advances in Chemical and Materials Engineering (ACME) Book Series

J. Paulo Davim
University of Aveiro, Portugal

ISSN:2327-5448
EISSN:2327-5456

Mission

The cross disciplinary approach of chemical and materials engineering is rapidly growing as it applies to the study of educational, scientific and industrial research activities by solving complex chemical problems using computational techniques and statistical methods.

The **Advances in Chemical and Materials Engineering (ACME) Book Series** provides research on the recent advances throughout computational and statistical methods of analysis and modeling. This series brings together collaboration between chemists, engineers, statisticians, and computer scientists and offers a wealth of knowledge and useful tools to academics, practitioners, and professionals through high quality publications.

Coverage

- Computational methods
- Numerical Techniques
- Ductility and Crack-Resistance
- Toxicity of Materials
- Coatings and surface treatments
- Ceramics and glasses
- Statistical Methods
- Materials to Renewable Energies
- Heat Treatments
- Fracture Mechanics

IGI Global is currently accepting manuscripts for publication within this series. To submit a proposal for a volume in this series, please contact our Acquisition Editors at Acquisitions@igi-global.com or visit: http://www.igi-global.com/publish/.

The Advances in Chemical and Materials Engineering (ACME) Book Series (ISSN 2327-5448) is published by IGI Global, 701 E. Chocolate Avenue, Hershey, PA 17033-1240, USA, www.igi-global.com. This series is composed of titles available for purchase individually; each title is edited to be contextually exclusive from any other title within the series. For pricing and ordering information please visit http://www.igi-global.com/book-series/advances-chemical-materials-engineering/73687. Postmaster: Send all address changes to above address.

Titles in this Series

For a list of additional titles in this series, please visit: www.igi-global.com/book-series

Handbook of Research on Nanoemulsion Applications in Agriculture, Food, Health, and Biomedical Sciences
Karthikeyan Ramalingam (B.S. Abdur Rahman Crescent Institute of Science and Technology, ndia)
Engineering Science Reference • © 2022 • 671pp • H/C (ISBN: 9781799883784) • US $295.00

Handbook of Research on Advancements in the Processing, Characterization, and Application of Lightweight Materials
Kaushik Kumar (Birla Institute of Technology, India) B. Sridhar Babu (CMR Institute of Technology, India) and J. Paulo Davim (University of Aveiro, Porugal)
Engineering Science Reference • © 2022 • 458pp • H/C (ISBN: 9781799878643) • US $285.00

Emerging Applications and Implementations of Metal-Organic Frameworks
Shimaa Mohamed Elsaeed (Egyptian Petroleum Research Institute, Egypt) Elsayed Zaki (Egyptian Petroleum Research Institute, Egypt) and Abdel-Azim Abdel-Azim (Egyptian Petroleum Research Institute, Egypt)
Engineering Science Reference • © 2021 • 254pp • H/C (ISBN: 9781799847601) • US $225.00

Normal Partitions and Hierarchical Fillings of N-Dimensional Spaces
Gennadiy Vladimirovich Zhizhin (Russian Academy of Natural Sciences, Russia)
Engineering Science Reference • © 2021 • 280pp • H/C (ISBN: 9781799867685) • US $195.00

Applications of Nanomaterials in Agriculture, Food Science, and Medicine
Mohd Amin Bhat (Higher Education Department, Government of Jammu and Kashmir, India) Irshad Ahmad Wani (Government Degree College for Boys, Anantnag, India) and Shah Ashraf (National Institute of Technology Srinagar (NIT Srinagar), India)
Engineering Science Reference • © 2021 • 442pp • H/C (ISBN: 9781799855637) • US $215.00

Technology Development for Adsorptive Heat Energy Converters Emerging Research and Opportunities
Kostyantyn M. Sukhyy (Ukrainian State University of Chemical Engineering, Ukraine) Elena A. Belyanovskaya (Ukrainian State University of Chemical Engineering, Ukraine) and Mikhailo P. Sukhyy (Ukrainian State University of Chemical Engineering, Ukraine)
Engineering Science Reference • © 2021 • 305pp • H/C (ISBN: 9781799844327) • US $195.00

Nanotechnologies and Clusters in the Spaces of Higher Dimension Emerging Research and Opportunities
Gennadiy Vladimirovich Zhizhin (Independent Researcher, Russia)
Engineering Science Reference • © 2021 • 286pp • H/C (ISBN: 9781799837848) • US $195.00

IGI Global
PUBLISHER of TIMELY KNOWLEDGE

701 East Chocolate Avenue, Hershey, PA 17033, USA
Tel: 717-533-8845 x100 • Fax: 717-533-8661
E-Mail: cust@igi-global.com • www.igi-global.com

List of Reviewers

R. Boopathi, *Mahendra Engineering College, India*
S. Chandika, *Algonquin College of Applied Arts and Technology, Canada*
T. Geethapriyan, *SRM Institute of Science and Technology, India*
M. Girija, *VIT University, India*
J. M. Jafferson, *VIT University, India*
Pradeep Kumar Krishnan, *National University of Science and Technology, Oman*
A. Ranjith Kumar, *Sona College of Technology, India*
S. Maniraj, *Paavai College of Engineering, India*
T. Mothilal, *KCG College of Technology, India*
T. Mythili, *Mahendra Engineering College, India*
G. Suganya Priyadharshini, *Coimbatore Institute of Technology, India*
E. Rajkeerthi, *Anna University, India*
K. G. Saravanan, *Sona College of Technology, India*
N. Sivashankar, *Kongunadu College of Engineering and Technology, India*
M. Soundarrajan, *Muthayammal Engineering College, India*
S. Sudhagar, *Muthayammal Engineering College, India*
P. Suresh, *Muthayammal Engineering College, India*
S. Suresh, *Muthayammal Engineering College, India*
Praveen Kumar V., *VIT University, India*
J. R. Vinod Kumaar, *Mahendra Engineering College, India*
R. Viswanathan, *AVS College of Technology, India*
T. Yuvaraj, *Muthayammal Engineering College, India*

Table of Contents

Detailed Table of Contents

Surface modification techniques are mainly considered to change the surface of the material based on the application. The surface modified materials possess excellent mechanical, tribological, and biological characteristics. In this chapter, the principle of various surface modification techniques including their outcomes, advantages, and limitations are reviewed. Among the various surface coating techniques, the most frequent techniques such as chemical vapor deposition (CVD), physical vapor deposition (PVD), micro-arc oxidation (MAO) coating, and electro-deposition coating were discussed in detail. The effect of surface modification on mechanical and tribological characteristics along with recent applications are described in detail. Finally, the major surface characterization techniques and their outcomes are elaborated in this chapter.

Micromachined components are finding application in a variety of fields such as medical, semiconductor, and aerospace. Electrochemical micromachining (ECMM) is a technology used to realize the micro components with enhanced features and functionality. This chapter reviews the impact of cathode profile on the machining rate and accuracy. An indigenous developed ECMM setup is used with special attachment of rotating electrode. The control parameters such as machining voltage, duty cycle, electrolyte

modern manufacturing industries are posses the tough fight on current requirements. On the other hand, continuously developing new technology and trends of advanced machining methods on modern technology for industries and academicians are not well established in the open web sources. Hence, this chapter will be eye opener for both industrialists and researchers about the various novel techniques, trends, and developments in conventional machining.

In this experiment, the work piece is electrochemically machined which finds its applications in brine heaters, heat exchangers, propeller shafts, and pumps. The process is carried out using sodium chloride electrolyte and copper beryllium wire as tool electrode which is heat treated in three different methods which are annealing, quenching, and normalizing. The response parameters like theoretical and experimental metal removal rate have been measured and studied by varying machining parameters like voltage, frequency, concentration of electrolyte, and duty cycle. Based on the values obtained from the experiment, it is found that quenched tool electrodes have better machining capabilities than the other heat-treated tool electrodes and untreated tool electrodes.

Automated manufacturing is an advanced manufacturing process which is enabled to correct its parameters autonomously during the machining process to achieve an assured objective. It can be developed through the institution of interactions with various systems, including machine tools, sensors and controller networks, actuators and simulation-based designs, as well as control algorithms. The automated manufacturing process can be realized in order to optimize process parameters automatically in real time, obtaining optimum processing performance and product quality.

A substantial effort has recently been made to fabricate numerous combinations of metal matrix composites. Aluminium metal matrix composite (AlMMCs) has been extremely popular among MMCs and is best suited for lighter materials in various industries and domestic applications. AlMMCs have significant faults in terms of production costs. More research is needed to develop AlMMCs that are both economic and suitable for a wide range of industrial applications. AlMMCs are manufactured using the casting technique, solid-state processing, infiltration method, accumulative roll bonding, spray deposition, friction stir processing, and other technologies. This chapter will summarize the current state of the art in aluminum-based metal matrix composites, as well as recent developments in processing and application, all in one location, so that relevant stakeholders can benefit the most and new advancements on such themes can be fostered.

Clinical performances of chemotherapeutic drugs which are used to manage different stages of cancers are usually facing numerous pharmacological challenges such as tumor microenvironment, high dose requirements, poor selectivity towards cancer cells, life-threatening cytotoxicity, and frequent drug resistance incidences, in addition to pharmacotechnical issues such as poor aqueous solubility, uncontrolled drug-release, low stability, non-specific bio-distribution, and erratic bioavailability profiles. The chapter aims to provide a brief account of advancements made in nanotechnology-enabled manufacturing engineering tools for manipulating polymeric materials as efficient carriers so that loaded anti-cancer drugs would exhibit better therapeutic applications and optimized clinical significance in cancers.

Most of the manufacturers in the world are already being adopted to industrial digitization in the perspective of Industry 4.0. Every conventional manufacturing method is being converted into cyber-physical systems in order to govern the process digitally. Friction stir welding is an advanced solid state welding method that has been utilized in many industries, and the process is no exception to digital transformation. This chapter aims to discuss the various aspects of digitizing the friction stir welding process by the application of different sensors. Implementation of various sensors such as current, force, sound, and vision in friction stir welding machine are able to collect valuable data that can be used for adopting Industry 4.0.

The need of welding in the current scenario of manufacturing is increasing due to the application of various components. New innovations in welding are created in order to meet benefits such as high production rate, quality surface finish, high mechanical strength of joint, difficult geometry, new materials for new components. This chapter provides insight on the current advancements of welding to explore the aforementioned benefits. Welding process parameters, novel implements, research backgrounds, applications, and ability of welding process along with its merits and demerits are covered in this chapter. In addition, advanced welding techniques in the areas of similar and dissimilar welding are presented. Moreover, this chapter would unfold the latest trends in advanced welding techniques for current requirements in industries.

The additive manufacturing process denotes modern manufacturing technologies that create a practical model from digital data. These days, the 3D (three-dimensional) printing technology signifies a great prospect to support medicinal and healthcare firms to produce new definite medicines, allowing quick manufacture of medicinal transplants, and moving the approach that specializes surgeon and physician strategy measures. For example, currently, in the practice of modern medical treatment, patient-specific anatomical models (3D-printed) are used. Soon, functional implantable organs by 3D (three-dimensional) printed process will possibly be offered, decreasing the queue time and growing the total of lives protected. This modern manufacturing technology for healthcare and medical is still required to a great extent of work in development; however, it is applied in numerous dissimilar habits in a medicinal and therapeutic area that previously reeled below a huge burden concerning optimum presentation.

Agro waste polymer composite (AWPC) comprising agro waste as reinforcement and polymer as the matrix is known to contribute additional fuel load when exposed to heat sources owing to their constituent make up. This has restricted their use in areas where strict compliance to fire regulations does exist. GA was mixed with flame retardants (FR) such as aluminium tri-hydroxide, ammonium polyphosphate, and carbon black at 0%, 12%, 15%, and 18% loading proportion, and these were then used to fabricate the AWPC using hand lay-up and compression moulding. The AWPC were tested for flammability properties in the cone calorimeter apparatus at 50kW/m2. Flammability properties such as time to ignition, heat released rates, mass loss rates, and smoke production rates obtained showed the AWPC processed with GA hybridized FR were greatly improved when compared to those without FR. Therefore, GA addition in AWPC proved to be a good FR that could be used in building applications to meet required fire safety standards.

Agro-waste fibres (AWFs) are of very reasonable economic value and have attracted a lot of research interest globally. AWFs have significant potential in composites due to low cost, high strength, environmental benign nature, availability, and sustainability. Nevertheless, dozens of the investigations have failed to give the desired attention of the fire behaviour of AWF composite materials when exposed to heat atmosphere. In this chapter, a detailed account of AWF composite material combustion process was discussed to understand the reason for the high heat release rates (HRR) caused by cellulosic content. HRR is a critical component of the flammability properties of AWF composite materials as the fire behaviour largely depends on it. The prospective examination of AWF composite material fire behaviour gives the required direction about the limits of their utilization and developmental trends for future industrial applications in Nigeria.

Preface

The advent of new materials finds applications in various fields such as biomedical, aerospace and automobile. These value-added materials that behave better than conventional materials, which have characteristics like lighter, broader service temperature ranges, multifunctional, and have better life-cycle performance. The materials fabrication involves different processes such as reinforcement of new material and addition/growth/merging of materials. Such processing of materials must assure the demands of geometric and property of the application without lowering them. Therefore, control of processing particulars and related technologies become more and more important. Moreover the nonexistence of standard material in the market and for these materials the market is particularly specialized and inadequate, significant capital investment or flexible manufacturing schemes will be required to set up consistent, reproducible production potential. The machining technology adds a value to the piece of materials by removing some of its volume. The quantity of value added is a function of the manner in which the machined components interact or fit with other components and the design of the form itself. The progress of machining technologies and observation of recent years material development the production in the manufacturing industries are relatively easy. The most highly efficient machine tool working at close tolerances can now be attained by much more ubiquitous equipment. The accessibility of this advanced machining ability carry expands new forms of added value and to market products with high degree precision and accuracy, less constrained by manufacture realities. This book sets out to provide an overview of this area and is intended to be a lead-in to the subject; it will apprise the non-specialist reader of some of the key technologies available and will provide assistance in the selection of appropriate techniques.

New materials and manufacturing techniques opens up new market, new technology, and new methods for existing products. Researchers and engineers are still finding new ways to innovate and remain competitive in ever-changing markets and the areas of application by no means exhaustive. Additive manufacturing methods including 3-D printing, create highly complex assemblies from one continuous material. Through this technology, manufacturers trim out failure points in a system, industries that benefit from additive manufacturing include aerospace, medical, prototyping, automotive, consumer goods, and many more sectors, and will grow as the technology gets cheaper and more intuitive. Advanced materials have provides scope for the creation of highly precise blends of metals, plastics, glass, ceramics, etc. that serve definite applications. Advanced materials such as composite materials are accurately varied in physical and chemical properties, generating more performance breakthroughs. Some important composite materials in current markets include high strength alloys, eco-friendly plastics and advanced ceramics/glass. The automated systems allow for precision movement & joining, and get better the consistency of work across many production units. Robotics is also invaluable for tasks that are traditionally hazardous, limiting risk, faster, and cheaper products. Robotics is more common in automotive, aerospace, forging, consumer goods, and numerous other industries. Laser technology is applied for welding and surface modification and these processes are currently used for pressure vessels, proximity sensor welding, battery welding, sensitive electronics, and much more, providing safer and more accurate products across many generations.

The advanced manufacturing automation process market, by application, is segmented into healthcare, transportation, aerospace and defense, food and beverages, and electronics and semiconductors. The advanced manufacturing automation process market, by geography, is segmented into four regions, namely, North America, Europe, Asia-Pacific (APAC), and the Rest of the world (RoW). The employment demand for highly skilled labour relevant to advanced manufacturing is increasing. The training for developing a skilled labour pool in advanced manufacturing will be improved through various program such as vocation training and education programmes, as well as continuing professional development schemes. Various universities and Institutes are teaching advanced materials and manufacturing and many students, industrialist, professors and research scholars will be interested in it.

ORGANIZATION OF THE BOOK

The book is organized into 16 chapters. A brief description of each of the chapters follows:

Chapter 1

Surface modification techniques are mainly considered to change the surface of the material based on the application. Among the various surface coating techniques the most frequent techniques such as chemical vapor deposition (CVD), physical vapor deposition (PVD), micro-arc oxidation (MAO) coating and electro-deposition coating were discussed in detail. The effect of surface modification on mechanical and tribological characteristics along with recent application is described in detail.

Chapter 2

This chapter reviews the interaction between the input parameters and machining speed & overcut were studied. The input parameters considered are machining voltage, duty cycle, electrolyte concentration and cathode tool rotation. In this chapter various multi-objective optimization techniques applied in optimization of ECMM process were reviewed.

Chapter 3

This chapter discusses the different technology used for surface modification of material for biomedical applications. The surface modification process is generally used to obtain the antibacterial properties, antifouling, enhanced mechanical properties, reduced cytotoxicity, anti-inflammatory potential, improved hemocompatibility and osseointegration capabilities. Moreover surface modification of dental implants and polymers were reviewed and recent progress and challenges in application of biomaterials are reported.

Chapter 4

Various methods are available for machining of composites materials and alloys are Electrochemical Micro-machining and Wire Electrical Discharge Machining. These methods are the apt methods for micron level machining of these materials by complex shapes with good surface quality. This chapter provide the insight information into the recent developments in machining of this alloys and composite materials by ECMM and WEDM.

Chapter 5

In this chapter, continuous developing new technology and trends of advanced machining methods on modern technology for industries and academicians are discussed. This book chapter will be eye opener for both industrialists and researchers to know the various novel techniques, trends, and developments in conventional machining.

Chapter 6

In this chapter electrochemical machining process detailed and effect of tool electrode which is heat treated in three different methods which are annealing, quenching, and normalizing on the output parameters are discussed.

Chapter 7

In this chapter the automated manufacturing process can be realized in order to optimize process parameters automatically in real time, obtaining optimum processing performance and product quality.

Chapter 8

This chapter will summarize the current state of the art in aluminum-based metal matrix composites, as well as recent developments in processing and application, all in one location, so that relevant stakeholders can benefit the most and new advancements on such themes can be fostered.

Chapter 9

The current chapter aims to provide a brief account of advancements made in nanotechnology-enabled manufacturing engineering tools for manipulating polymeric materials as efficient carriers so that loaded anti-cancer drugs would exhibit better therapeutic applications and optimized clinical significances in cancers.

Chapter 10

This chapter discuss the various aspects of digitizing friction stir welding process by the application of different sensors. Implementation of various sensors such as current, force, sound and vision in friction stir welding machine, able to collect valuable data which can be used for adopting industry 4.0.

Chapter 11

This chapter provides insight on the current advancements of welding to explore the aforementioned benefits. Welding process parameters, novel implements, research backgrounds, applications and ability of welding process along with its merits and demerits are covered in this chapter.

Chapter 12

This chapter covers the recent developments in the engineering industry require joining of like and unlike materials with different properties such as melting point, coefficient of thermal conductivity, solubility and difference in electrochemistry. Weldments testing, including destructive and non-destructive examinations, are crucial during procedure qualification, welder qualification, and the production welding process are discussed in detail.

Chapter 13

This chapter discuss the Additive manufacturing technique used in the reentry launch vehicle. Space Industry has been progressively utilizing additive manufacturing in their production which is elaborated in the chapter.

Chapter 14

This chapter discussed and concluded on the 3D (Three-dimensional) printing technology which signifies a great prospect to support medicinal and healthcare firms to produce new definite medicines, allowing quick manufacture of medicinal transplants, and moving the approach that specialists' surgeon and physician strategy measures.

Chapter 15

This chapter provides insight about the Agro waste polymer composite (AWPC) comprising agro waste as reinforcement and polymer as the matrix is known to contribute additional fuel load when exposed to heat sources owing to their constituents make up. AWPC production and its applications to meet required fire safety standards are discussed.

Chapter 16

This chapter discuss the behaviour of Agro-waste fibres (AWF) composite material on combustion process to understand the reason for the high heat release rates (HRR) caused by cellulosic content. The prospective examination of AWF composite material fire behaviour gives the required direction about the limits of their utilization and developmental trends for future industrial applications in Nigeria.

This book acts as a bridge between this high-level research and the large number of academics and practitioners looking to advanced manufacturing processes for innovative solutions, providing them with practical and approachable information. Applications in aerospace, automotive, medical, and food industries are addressed, featuring technical details that will help successful implementation. This unique book also provides broad coverage of the theory behind this surface modification technology, including material development, as well as the technical details required for readers to investigate the novel applications of the involved methods for themselves.

First and foremost, the Editors would like to thank the Almighty for His blessings for completing the work to their satisfaction. The Editors would also like to thank all the Chapter Contributors, the Reviewers, Book Development Editor, and the team of IGI global Personnel for their availability for work on this editorial project. Last, but positively not least, the editors would like to thank all the well wishers, colleagues, students, and all who were directly or indirectly involved in this project.

R. Thanigaivelan
Muthayammal Engineering College, India

N. Rajan
Vinayaka Mission Kirupananda Variyar Engineering College, India

T. G. Arul
St. Mary's Engineering College, India

Acknowledgment

The editors would like to recognize the help of all the people concerned in this project and, more particularly, to the authors and reviewers that took part in the review process. Without their support, this book would not have become a reality.

First, the editors would like to thank each one of the authors for their contributions. Our sincere gratitude goes to the authors of chapter's who contributed their time and expertise to this book.

Second, the editors wish to acknowledge the valuable contributions of the reviewers regarding the improvement of quality, coherence, and content presentation of chapters. Most of the authors also served as referees; we highly appreciate their double task.

R. Thanigaivelan
Muthayammal Engineering College, India

N. Rajan
Vinayaka Mission Kirupananda Variyar Engineering College, India

T. G. Arul
St. Mary's Engineering College, India

Section 1
Machining and Surface Modification

Chapter 1
Surface Modification Techniques for Improving the Material Performance

Thanigaivelan R.
Muthayammal Engineering College, India

Saravanan K. G.
Sona College of Technology, India

Makesh M.
Paavai Engineering College, India

ABSTRACT

Surface modification techniques are mainly considered to change the surface of the material based on the application. The surface modified materials possess excellent mechanical, tribological, and biological characteristics. In this chapter, the principle of various surface modification techniques including their outcomes, advantages, and limitations are reviewed. Among the various surface coating techniques, the most frequent techniques such as chemical vapor deposition (CVD), physical vapor deposition (PVD), micro-arc oxidation (MAO) coating, and electro-deposition coating were discussed in detail. The effect of surface modification on mechanical and tribological characteristics along with recent applications are described in detail. Finally, the major surface characterization techniques and their outcomes are elaborated in this chapter.

INTRODUCTION

Surface engineering deals with various fields such as engineering, materials science and physics and the current developments in coating and treatment technologies solves the complex issues in the surface engineering. By understanding, surface is the most important part of any engineering component. Most of the components fail due to the ill effects of the surface condition which often constraints the technological

DOI: 10.4018/978-1-7998-9574-9.ch001

development. In general, components with characteristics of, resistance to friction, wear and corrosion are required and to meet these requirements, development of new surface with different modification techniques is need. In practice, surface modification process in materials brings modifications in micro structural and compositions. Micro structural modification process such as surface transformation hardening, surface melting, shot peening and compositional modification process attributes to coating and deposition, Electro-deposition, conversion coating, thermal coating, hot dipping, chemical vapour deposition and physical vapour deposition are currently applied to improve the surface property. The surface modified difficult to cut materials machining efficiency is improved by using hybrid surface modification process. This chapter discusses the most recent technologies for surface modification for industrial components.

METHODS OF SURFACE MODIFICATIONS

Surface Transformation Hardening

Transformation hardening is the one of the hardening process for steels. During heat treatment process the components are heated to the required temperature in bulk and then quenched in water or oil to achieve the preferred hardness on the surface. In the majority of engineering applications, the wear in components arise only in specific areas hence, it is adequate to harden areas which to susceptible to wear for enhancing the performance of the component. Surface hardening is performed through nitridation, oxidation and boronizing as shown in the figure1.The laser gun emits laser energy on the surface to be hardened which moves in both the direction horizontally. In recent years laser has be extensively applied for surface processing due to its advantages such as monochromatic nature and controlled amounts of energy to preferred regions (Grum and Sturm, 1997). The limitation of this process is that irregular hardness distribution and a non-uniform hardened depth and width.

Figure 1. Schematic of hardening process

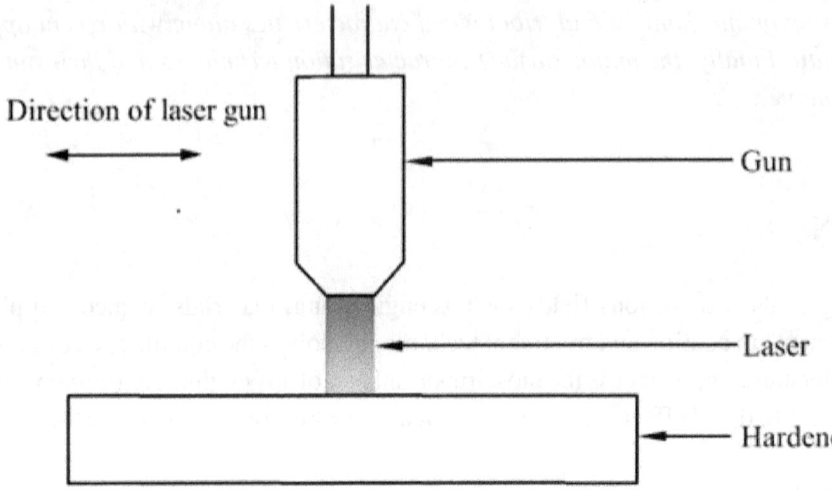

Surface Melting

Surface melting is applied to achieve heat-resistant, wear-resistant, corrosion-resistant and porous bio-compatible coatings on materials. Surface melting involves heat sources, which produce high power densities at the surface with the help of laser and electron beam processes as shown in figure 2. Laser melting is extensively applied for surface modification of complex geometries and Bruzzo et al. (2021) have deposited the 316L stainless steel powder on the tubular specimens, and re-melted the surfaces using the same laser machine. The laser re-melting develops new texture on the surface and a correct combination of laser parameters prevents the creation of an anisotropic surface. Similarly, electron beam energy densities play a prominent role on surface micro structuring, phase composition, elements distribution and tensile property of samples. The nanosecond small diameter pseudo spark electron beam radiation is directed on the 304 stainless steel surface. After thousands of microspark treatment, uniform submicron grains with slip bands spreaded across the modified layer. This treatment induces dislocations, and nano-twins with the lamella thickness < 8 nm in the adapted layer. The number of the microspark is directly proportional to the surface micro-hardness and corrosion resistance. The micro-structure of the treated surface is improved through this process (Cao et al..2021). This method has an inherent disadvantage that the heat produced by the laser beam should be quickly transferred from the molten bath to avoid the development of residual stresses. Due to this, the thermal conductivity of the materials should be higher to reduce the quantity of unwanted heat, resulting in reduced residual stress (Voznesenskaya et al., 2020)

Figure 2. Laser melting process

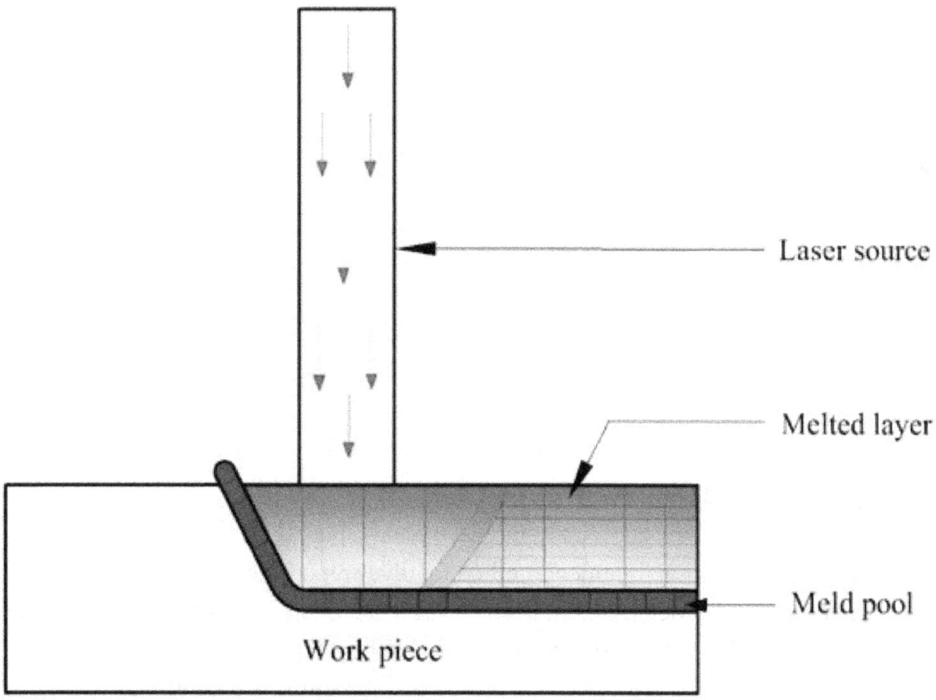

Shot Peening

Shot peening is one of the cold working process. It creates a layer of compressive residual stress and alters the mechanical characteristics of metals and composites. Figure 3 shows the schematic of shot peening process. It is nothing but hitting a surface with round ceramic or metallic particles with sufficient force to develop plastic deformation. This technique develops irregular surface with micro and nanometer features. You et al. (2021) have used shot peening process to modify the surface of the heat sink for electronic applications. A shot of diameter 0.6 mm, made of 410 stainless steel, ejection speed was about 60 m/s is considered for the surface modification and irregular microscopic defects, and increase in surface roughness upto 5.63 μm is witnessed. The thermal emissivity of the shot peened heat sink increase to 2.5 times with increase in surface roughness.

Figure 3. Schematic of shot peening process

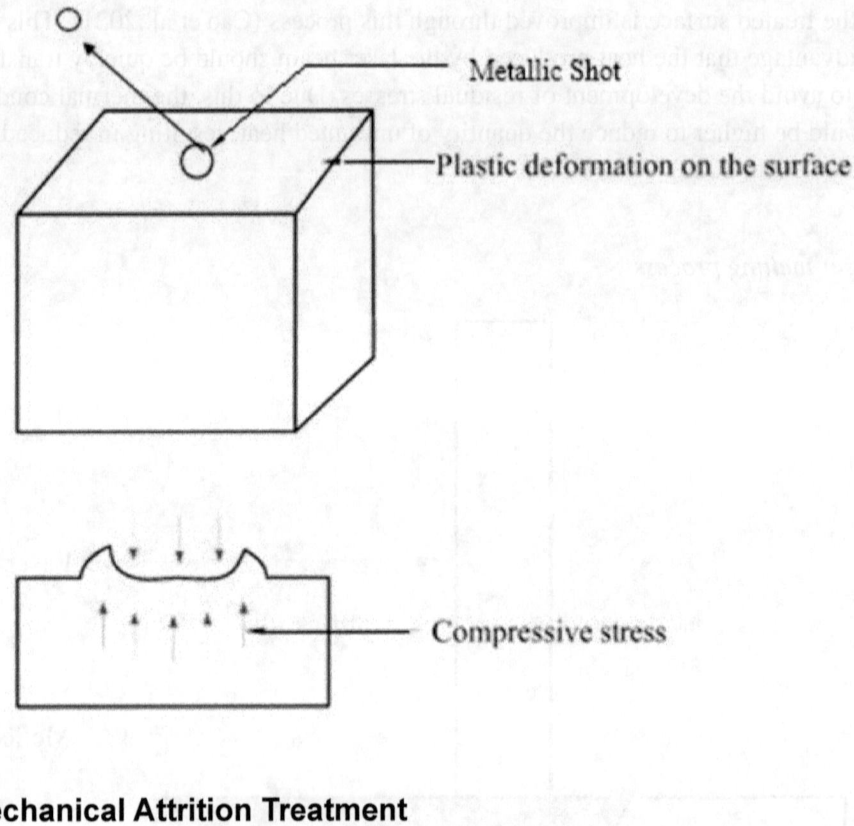

Surface Mechanical Attrition Treatment

The surface mechanical attrition treatment method involves repetitive multidirectional impacts by speeding balls to make surface hardening of bulk material as shown in figure 4. This impact creates severe plastic deformation on the surface of the material, leading to grain refinement without changing in composition. This method produces severe surface nanocrystallization and brings out the slow change in base material ranging from nano-meter to regular size. Elastic deformation is seen in the area in close vicinity of the base material and extreme plastic deformation occurs at the uppermost surface. Hence the modified surface shows favourable surface hardness, tensile strength, resistance to fatigue and wear resistance due to the formation of nano-crystallites at their surface and subsurface layer (Arifvianto, & Mahardika, 2018).

Figure 4. Surface mechanical attrition process

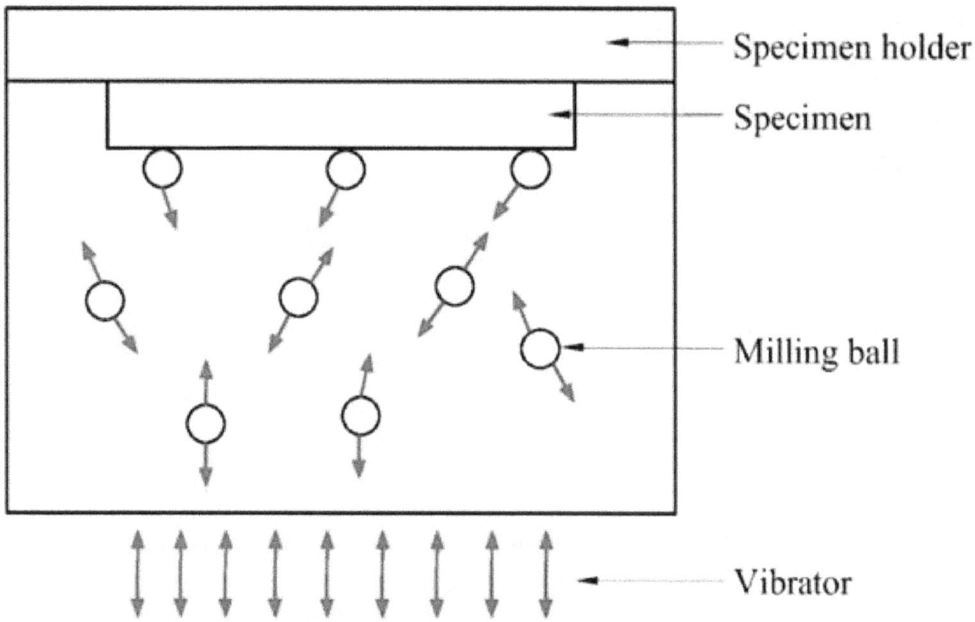

FREQUENT METHODS USED FOR SURFACE COATING

Chemical Vapour Deposition (CVD)

Chemical vapour deposition (CVD) is a vacuum deposition method applied to fabricate superior quality, high-performance, solid materials. The method is frequently used in the semiconductor industry to fabricate thin films. Figure 5 shows the schematic of the CVD process. CVD is a coating process which includes thermally started chemical reactions at the parent material with the assistance of vaporous or fluid reactant containing the favoured component within the reaction chamber. During this process the chosen material will be deposited on the surface by heating the solid substrate and particular chemical reaction with mixed vapour (Carlsson & Martin 2010). CVD is used for developing 2D materials and polymeric thin films which find applications in electronics and medical devices (Sun et al. 2021).

Figure 5. Schematic of CVD setup

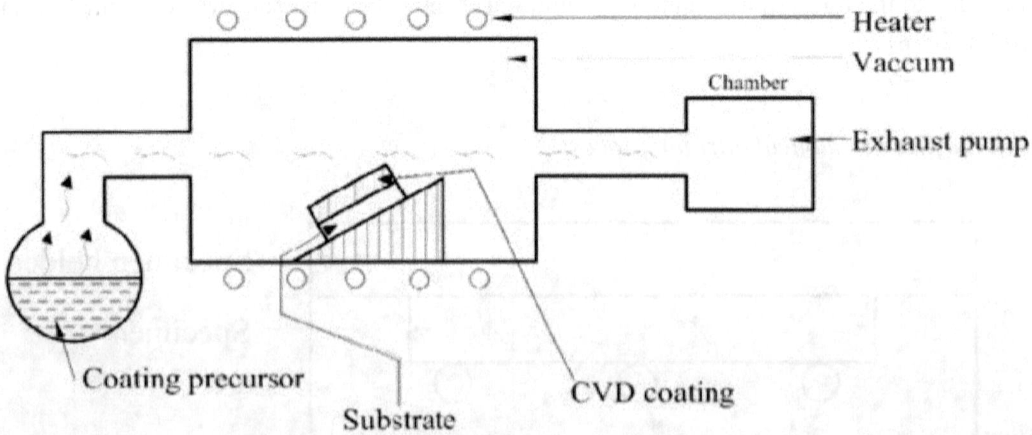

Physical Vapour Deposition (PVD)

Physical vapour deposition (PVD) process is well know process for developing corrosion and wear resistance surface with thin protective films. It is applied to materials extending from fancy objects to mechanical components/parts. The preferences of this method are that the mechanical, erosion, and creative properties of the coating layers can be modified on necessity (Baptista et al. 2018). In common, PVD happens in a elevated vacuum environment where solid/liquid is vanished to vapour state taken after by metal vapor condensation as shown on the schematic figure 6. The thermal energy needed for the evaporation was provided by different units, namely electron beam, heating wire, laser beam, molecular beam, etc. The vaporized atoms pass through the vacuum chamber and deposited onto the substrate. Physical vapour deposition coated parts find application where the components encounters high wear rate. Physical vapour deposition posses some drawback namely coating of complex shapes, high operation cost and reduced output, and process complexity.

Micro-Arc Oxidation (MAO) Coating

The figure 7 shows the setup for micro-arc oxidation coating in which cathode is source and anode is the substrate. A high voltage distinction is kept up between the cathode and anode to create micro-arcs as plasma channels. The created smaller scale arcs hit the substrate and liquefy a fraction of the surface, based on the concentration of the micro-arcs. During micro arc oxidation coating process a plasma channel created induces the micro-arcs between the cathode and anode which melts the surface. The generations of plasma channels release the pressure to carry out the deposition process on the substrate surface using coating materials in the electrolyte. The oxygen inside the electrolyte actuates a chemical response of oxidation and stores the oxides on the surface of the substrate materials. The efficiency coating is with the selection of coating materials in the electrolyte. Aluminium, magnesium, titanium, and their alloys are coated using the micro-arc oxidation. Micro-arc oxidation provides high corrosion resistance, being a porous structure; this coating layer finds application in biomedical implants and fixations (Zhao et al., 2010).

Figure 6. Physical vapour deposition setup

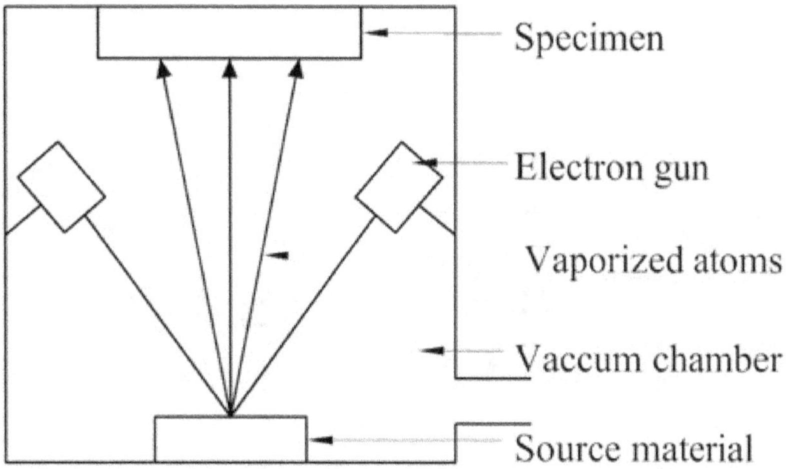

Figure 7. MAO coating setup

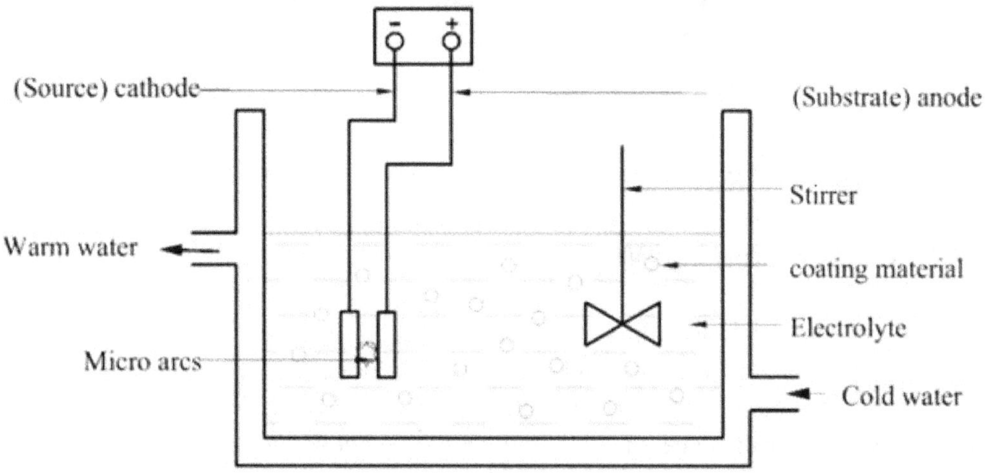

Electro Deposition Coating

Electro deposition is a coating process in which metallic ions are coated on a substrate. In this method of coating, ions transfer takes due the applied voltage between anode and cathode in the electrochemical cell. During this process, a layer of coating is formed on the substrate immersed in the electrolyte, by getting ions from the anode. Broad studies have been conducted on electro deposition of materials. With respect to existing literature, the electro assisted coatings notably improve the corrosion properties of the substrate. In addition, this type of coating method has been used for producing superhydrophobic polymeric coatings such as polythiophene (Darmanin et al., 2010). Figure 8 shows the electro deposition coating HAP powder on titanium substrate (Rakesh Kumar et al. 2021).

Figure 8. Electro deposition coating
(Rakesh kumar et al. 2021)

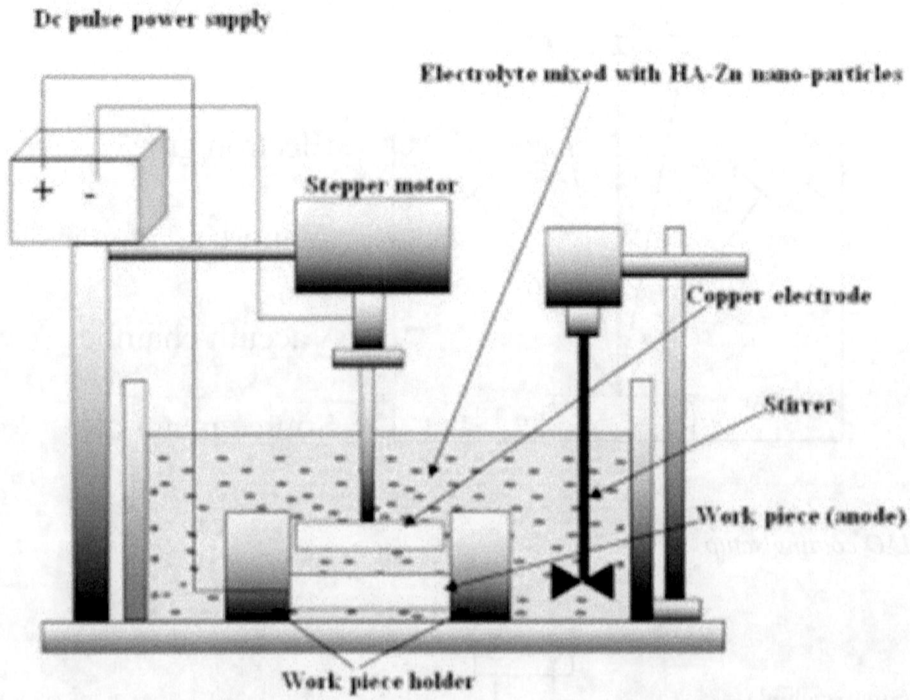

SURFACE MODIFICATION FOR IMPROVING THE PROPERTY OF MATERIALS

Resistance to Corrosion and Wear

Corrosion and wear in materials, both directly and indirectly, costs huge to the industries such as chemical, paper and mining. Corrosion always accompanies the wear process in all environments, except vacuum and inert atmospheres. The materials are protected from corrosion by coating using above discussed techniques. The MAO coating process is used to form Al_2O_3 coating layers on high-strength aluminum alloys to enhance corrosion resistance properties. The shaped coating layers were inspected and thick oxide layer is strongly adhered into the aluminum substrate increasing the corrosion resistance properties of MAO coating substrate. (Dzhurinskiy et al. 2021). In other research porous(nano) oxide layer was developed on the titanium alloy(Ti6Al4V) surface using ethylene glycol electrolyte and lactic acid solution through electro deposition process. Based on the analytical instrument studies the nanopore of size φ 89nm and ~ 5.412 µm thickness of oxide layer were formed on the surface. Potentiodynamic polarization study of nanoporous oxide layer in an 8 g/l NaCl medium shows that the nanoporous oxide layer fabricated on Ti6Al4V surfaces shows good corrosion resistance property than that of uncoated Ti6Al4V (Saha et al. 2021). A special additive manufacturing technology namely selective laser melting is a laser based additive manufacturing technology, used in the manufacturing of 316L stainless steel components. The mechanical and tribological properties of 316L stainless steel is improved by this technique and plasma oxidation treatment process. The stainless steel surface were coated with Ti6Al4V layers through oxidization at temperature range of 650 °C and 750 °C for the duration of 1 to 4 hr in the plasma environment. The wear tests were conducted to understand the wear characteristics of the Ti6Al4V layers using Al_2O_3 balls at dry load of 10 N, with pin-on-disk tribometer. The tribometer results affirm that the hardness and wear resistance of Ti6Al4V coated surface and plasma-oxidized surfaces appears superior comes about than the un-coated 316L surface.The development of titanium oxide and diffusion zone profundity attributes for best wear resistance and elevated hardness value (Tekdir & Yetim. 2021). Laser cladding technology has been used for surface modification of titanium alloys.

Ceruim oxide is coated on titanium alloys (Ti811) through coaxial powder feeding strategy. The coated surface basically contained TiC, TiB2, Ti2Ni, and α-Ti. This method of coating develops a surface with distribution of homogenous microstructure and fine grain size. Laser power is considered for the coating and 45 J/mm^2 of laser power were used to develop the ceruim oxide of 2 wt%. The process enhance the micro- hardness and wear resistance due to fine grain fortifying and scattering reinforcing impacts of CeO2, contributing straightforwardly to produce a high normal hardness of 811.67 $HV_{0.5}$ with a lower friction (Yang et al. 2021). The sequence applications of surface modification process have fetched good results on high strength low alloy automotive steel. This automotive steel experiences intense surface plastic deformation through controlled groove pressing, intense shot peening and ultrasonic process for developing nano-crystal surface. This process creates a nano-crystal layer < 50–100 μm with superior resistance to wear and friction in particular at small loads compared to untreated specimens (Karademir et al. 2021).

Resistance to Fatigue

The fabricated parts undergo static loading, including tensile, compressive, hardness, etc., A major challenges and concerns for metallic parts in application is their behaviour under cyclic loading, i.e. their fatigue resistance which is a familiar mode for mechanical failure in many engineering structures. Hence research on enhancement of fatigue resistance is performed worldwide. Ultrasonic nanocrystal surface alteration is a recently developed strategy which employs low amplitude ultrasonic frequency vibrations superimposed on inactive stack to form severe plastic deformation leading surface nanocrystallization. (Ye, et al. 2021). In this process the tungsten carbide ball hits the specimen surface at frequency range of 10-30 kHz. This strategy is utilized to improvise the fatigue properties of A100 steel. During this process the martensite microstructure is surprisingly refined; moreover surface compressive residual stress is improved to 1700 MPa due to the severe lattice distortion. The upgraded surface hardness and the compressive residual stresses notably improve the fatigue resistance of A100 steel by bringing down the fatigue crack nucleation and propagation (Zhao et al. 2021). Aluminium alloy (AA7075-T651) fatigue property is improved using laser shock peening (LSP) process. During this process a deep compressive residual stresses above 300MPa were developed and fatigue test shows a two times increase in overall life, due to the mechanism of crack initiation (Sanchez and coworker 2021). Selective laser melting (SLM) technique is considered for preparing aluminium-Ni-Cu alloy. The nickel and copper of 3.8–4.2 and 1.9–2.2 wt.% respectively is deposited on the aluminium to improve the fatigue resistance of the specimen. The SLM Al-Ni-Cu alloy with high fracture resistance has potential in the aerospace field (Chang et al. 2021). The impact of laser process parameters and the kinetic energy on the microstructure, mechanical characteristics and fatigue performance of nickel-based super-alloy Inconel 718 is analysed. The increment in kinetic energy successfully advances the surface grain refinement and develops a compressive residual stress in Inconel 718. Among the various methods, ultrasonic nanocrystal surface modification is propounded to be the foremost appropriate strategy for improving the fatigue performance.

Enhancement of Thermal Property

In recent decade boiling has got significance for its viability in cooling of micro-electronic gadgets due to its higher heat extraction capacity as compared to air or single-phase liquid cooling. Micro- and nano-surfaces have been created for this reason either by surface coating or by micro-machining. With the advent of micromachining technology surfaces with ribs, dimples, arrays of protrusions and roughness have been developed on heat sinks for enhancing heat dissipation. The thermal radiation coating paint was sprayed onto the heat sink surfaces by a spray gun along with compressed air. The thickness of the coating was managed by the number of layers. For each layer has a thickness of about 0.8 μm (You et al.. 2021). Thanigaivelan and Deepa (2018) have coated the aluminium and copper micro-fin with aluminum paint using spray paint technology to understand the effect of paint coating on the convective and radiative heat transfer. The paint coated aluminum test pieces of fin spacing 5 mm shows highest value for convective heat transfer coefficient ie 24.54 W/m^2 K.

APPLICATION OF SURFACE MODIFICATION

Surface Modified Cutting Tools

The application of cutting tools such as drill bits, turning tool and high speed tools are enormous and at the same time it suffers severe surface wear, during cutting of difficult-to-cut materials. Amid machining process the friction between the tool-work and tool-chip interfacing produces ceaseless heat and abrasion. To improve the device life span and quality of machined components, different methods have been applied to diminish external loading, improve tribological performance and/or superior chemical steadiness. In order to improve the drill bits life and corrosive resistance used for wood industries, a joint technique of surface hardening process is considered in the drill bit i.e., diffusing the nitrogen the stage 1 using ammonia and oxidising the nitrided surface at the stage 2 using water vapour. This process develops the surface layer with oxide, nitride, carbonitride, and oxycarbonitride elements which attributes for improved surface properties of high speed tool (Kh Eshkabilov et.al.2021). The plasma hardening method is used to improve the soil cultivating instruments made up of 65 G steel. The plasma quenching improves the wear resistance of normalized steel by four times. The resistance to wear for the hardened steel is studied using the corundum and steel indenter and former shows 15.3 times more wear resistance than with a steel indenter (Mashrabov, et al 2021).

The wear resistance of milling tool fabricated through three distinctive strategies ie two casting methods namely gravity and vacuum and one conventional metallurgical course) with same chemical composition. It is apparent that the quality of eutectic carbide spatial dispersion in the milling tool plays a critical part in wear performance, and, at the same hardness, it is autonomous of twisting quality strength and impact durability. Hence, the end mills made from the casting process appears to have best wear resistance to abrasive and oxidation wear (Chaus et al 2021). The surface textured inserts are used for machining hard materials such as titanium alloy. Figure 9 shows the tool insert with micro slots and micro drilled holes. Polycrystalline cubic boron nitride (PCBN) and polycrystalline diamond (PCD) inserts are modified using micro holes surface textures on rake and flank face for high-speed machining of Ti6Al4V alloy along with minimum quantity lubrication. The polycrystalline diamond with micro holes surface textures showed best performance characteristics (Rao et al.2021). Laser finished tool also well known in which micro-scale dimple is produced on HSS tool surface using pulsed Nd:YAG laser. The laser finished tool resulted in lower cutting forces and reduced tool – chip contact (Sasi et al. 2017). Micro-EDM technology is also considered for developing micro-textures on high-speed steel and this technology ensures the creation of uniform textures (Bhim Singh et al. 2019). Moreover more number of micro holes in different diameter, depth and pitch were fabricated on the tool inserts and its influence is studied on aluminium metal matrix composites. Textured inserts with solid lubrication enhances the cooling and lubrication functions and there is 21% and 19% reduction in surface roughness and power consumption respectively (Sasi et al., 2017). Amoung the diameter, depth and pitch, the pitch shows less impact on output responses. The regular and uniform textures are prepared on PVD coated surface of WC/Co substrate through plasma-assisted laser machining. The regular and uniform micro-scale textures greatly reduced the cutting force generated during turning of SS316L. Ion implantation technology is also used for improving the performance of many drill tools; moreover its durability of the guide pads was not addressed properly. The coating of tools with TiN has some disadvantages, namely dimensional change and delaminated layer during the work (Meng et al., 2021).

Figure 9. Surface textured tool insert

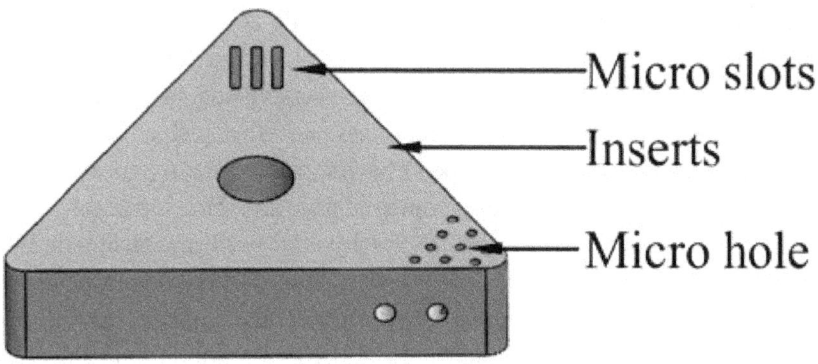

Surface Modified Workpiece

Most problems encountered during machining of titanium are heat generation and friction at the tool–chip and tool–workpiece interfaces. Several strategies have been used with some success in the development of machinability of titanium alloys. However, novel applications and machining strategies for hard-to-cut alloys are ceaselessly being developed by worldwide researchers. It is obvious that amid the electrochemical process, a lean electrolytic inactive film (oxide layer) develop on the workpiece surface for the particular range of potential and electrolyte pH. This layer has properties that are distinctive than the base material that can be separated with the decreased mechanical cutting force. Skoczypiec et al. (2016) have applied this technique for machining of stainless steel and studied the cutting force by varying the cutting speed and the feed rate. This electrochemical-assisted mechanical machining is more suitable for passivating materials such as stainless steel, titanium alloys and aluminum. Since, titanium alloy finds application in various fields the research on machining of titanium alloys is given due importance. In case of Ti4Al6V alloy the temperature of machining increases with the cutting speed. The high cutting temperature worsens tool life, surface quality, and cutting accuracy. Moreover, the micromachining of Ti4Al6V alloy is generally accompanied by the formation of burrs. It is also vital to contain these burrs during micromachining because it is not like conventional machining, a post finishing process is not possible. Hence in this electrochemical assisted hybrid micro machining process will be suitable process to machine Ti4Al6V alloy. During electrochemical process the development of passive layer on the surface of the specimen can be machined with less cutting force thereby reduces the machining temperature. The problems of machining titanium are many folds which depend on types of titanium alloy.

There have been many surface hardening technologies which involve heating system for quenching of steels. The heating system requires dedicated thermal equipment for obtaining required structural and phase transformations on the surfaces. Here a novel surface hardening which directly hardens the surface of the workpiece. The friction and plastic deformation between the tool and workpiece during cutting process is advantageous for creating the heat. This heat generation hardens the workpiece surface continuously. The joint effect of intense deformation, confined heat, and fast quenching speed makes the machined surface to go through both physical and metallurgical changes. Cylindrical samples with a diameter of 60 to 80 mm made of steel after normalization is machined in screw-cutting lathe using oxide ceramic, cemented carbide, and SiAlON tools. The multi-component dynamometers were generally used to obtain the cutting force, the piezoelectric based dynamometer is preferred over the strain gauge type for its accuracy and reliability. The temperatures in the cutting zone were measured by thermocouple. Based on the experiments the following inference is made by authors: Phase transformations happen at higher temperature and speed of deformation could not be established during laser quenching. The normalized workpiece surface depicts homogeneous hardness over other surface hardening process (Nikolay Zubkov et al., 2017). Gunasekaran and co workers have modified the magnesium alloy surface using cryogenics. The specimen is treated for different range of time and the temperature is reduced at the rate of 20°C/min from room temperature. The cryogenic treatment is performed using liquid nitrogen and the treated specimen is turned using Kirloskar variable speed lathe. The input parameters such as cryogenic soaking time, feed, speed and depth of cut on cutting temperature, surface roughness and cutting force were examined. Orthogonal cutting forces such as Fx, Fy, Fz were measured using piezoelectric dynamometer kept under the tool holder along with the suitable load amplifier. The roughness measurement is done using a stylus based surface roughness tester (Make: Taylor Hobson, Sutronic 25). The cutting temperature can be measured by non-contact IR Thermometer.

In manufacturing industries, laser surface alloying shown in figure 10 is recently considered as an important candidate for enhancing the surface properties of metals. The light alloys namely aluminium and magnesium are the key engineering materials used extensively in the manufacturing sector especially automobile industries. These materials have good strength, corrosion resistance, weldability and cost effective which favours their application in several engineering industries. In recent study, the surface alloying of pure aluminum was done through CO_2 laser. Researchers have considered alloying powders in the ratio of 2(copper):1(magnesium):1(manganese) on the aluminium surface. With the help of laser surface alloying the hardness of the Al-Cu-Mg-Mn increased by two to seven times at a 1.7 kW laser power (34). Wear rate of laser alloyed surface is measured using a modified Lancaster wear coefficient and pin-on-disc device. At 10 N and 20 N loads with diverse sliding speeds show diminishment of wear by 30–50% due to surface alloying. Moreover, microstructure and surface characterization studies depict good metallurgical bonding without defects. At outset the laser alloyed surface has more wear resistance and can be recommended for the moving parts.

Figure 10. Schematic of surface alloying

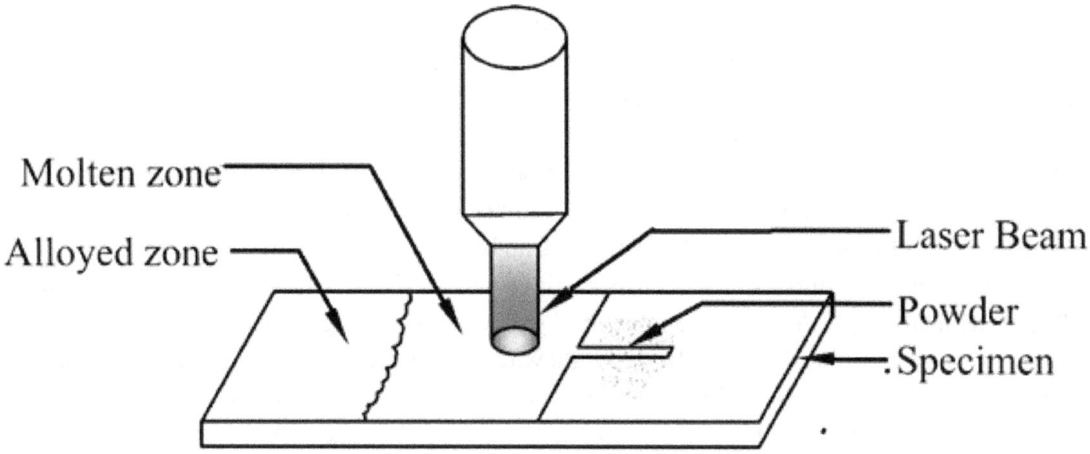

Surface Modification of Electrodes

The surface structure of cathodes within the Li-ion batteries and super capacitors plays a overwhelming part and influence the system working such as structural solidness and material's conductivity (35). Different strategies are being used to enhance the electrochemical execution of these components to be specific doping of move metal ion, generation of composite materials, generation of nano materials, and surface alteration of materials. Among these methods, surface alteration of the terminal materials is broadly utilized since of ease of generation and financial viability.

The surface coating is one of the surface modification process in which thick coating layer is generated on the surface to provide superior protection and to limit the ion and electron transport. For high performance cathode materials, homogenous thin surface coating is preferred and generation of this type of coating is challenging. The surface coating by sintering and dissolving the surface wipe out the surface structure of the cathode materials and hence it is vital to make use of suitable coating material and consider a suitable coating technique to generate high-performance cathode materials.

SURFACE CHARACTERIZATION TECHNIQUES

To understand the surface properties such as feedstock powder, porosity, thickness, roughness, hardness, and phase composition of the modified surface the following various surface characterization methods are used:

- Microscopy (optical),
- Scanning Electronic Microscopy (SEM),
- Energy dispersive
- X-ray spectroscopy (EDS), and
- X-ray Diffraction (XRD),

Surface Analysers

Metallographic analysis can be obtained using Olympus GX51 microscope with magnification of 1000x on the polished surface. In many research article surface characterizations were carried out using SEM (Make:EVO 18, Carl Zeiss, Germany) and X-ray spectroscopy to obtain the surface structure and elements compositions and distributions(24). Moreover, a white light interferometer (Make: Wyko NT9300, Veeco Inc., USA) is used to analysis the surface structure. For structural data and residual stress studies X-ray diffraction and focus variation microscopy are used (Bruzzo et al., 2021). At the same time, the broadly used sin2ψ method and electron backscatter diffraction analysis were conducted for micro-structural characterization and cross section analysis (Gunasekaran et al., 2021; Karademir et al., 2021). The features of additively manufactured surfaces such as high slopes, undercuts and varying optical properties are analyzed using polarized light with 10X magnification lens. The details of residual stress are calculated using CrKα radiation, upto 2 µm below the substrate surface.

The SEM with different magnifications such as x1000, x5000 and x10,000 were used for detailed surface analysis. X-ray photoelectron spectroscopy was used to analyze the chemical composition and chemical bonding of the working surface. The surfaces of ion implantation were examined using a photoelectron spectroscope (Devaraj et al., 2021). The chemical conditions of surface elements were also determined using high-energy resolution spherical sector spectrometer. In some studies the depth report of implanted tungsten, carbon, cobalt, oxygen and nitrogen were determined using the secondary ion mass spectrometry analysis. This analysis is performed on an SAJW-05 analyzer outfitted with a mass spectrometer and physical electronics ion gun. Energy dispersive X-ray spectroscopy were used to make analysis of the composition of aluminum substrate containing 0.6 wt.% Si, 0.3 wt.% Fe, 0.1 wt.% Cu and balanced aluminum.

Moreover the oxide layers composition obtained at various oxidation temperatures on pre-nitrided substrate is analysed using X-ray method. The nitride-oxide layer surface and microstructures are analysed using a SEM, and structure compositions were determined using an energy-dispersive X-ray spectrometer (Eshkabilov et al., 2021). During characterization of modified surface the working distance is maintained at 8.5 mm.

Surface Roughness, Wear and Friction Analysers

The surface roughness is measured using Surftester (Make: Mitutoyo, Japan) and high-precision portable roughness tester. The high-precision surface tester can measure 8200 points upto 4mm of test length with speed of 0.20 mm/s. Using the instruments average roughness, Root Mean Square (RMS) and maximum peak to valley were determined. The abrasive wear resistance of the coated surface is analyzed using rubber wheel test (Meng et al., 2021). Moreover the analysis of wear and friction were conducted on pin-on-disc tribometer (Make: DUCOM instruments) at parameter setting of 30 min at room temperature for dry condition, at ASTM G99-95a standard (30). SEM/EDS were also used for characterization of wear track of various textured tools occurred during machining. Amsler apparatus were considered for tribological studies to measure the friction at glide velocity of 0.83 m/s, the load of 25 N and the friction distance of 2500 m. Friction torque is also noted at the time of tests. Different testing condition is adapted to study the wear and frictional performance of treated samples using reciprocating tribometer. At dry sliding condition, 6 mm diameter tungsten carbide ball at force of 5N, 10N and 15 N were considered for the study. The wear probe slides at a speed of 8cm/s in a linear path of 12 mm for a total sliding distance of 500m. Volume loss is calculated approximately by multiplying the wear probe stroke length, track width and track depth. The wear stroke is kept constant as 12 mm for the whole specimens in this experiment. The optical profilometer were used to measure the wear track width and depth and the arithmetic mean of the width and depth were obtained.

Mechanical Strength Analysis

To test the hardness of the material, the material surface is touched with a rigid indenter and the imprint left is measured to determine the hardness value. The rigid indenter is of different shapes such as spherical, conical or pyramidal. Micro-hardness tester (Make: EMCOTEST DuraScan 70, Germany) having an indenter of diamond pyramid with regular quadrihedral shape with a load of 100 kg is used for the evaluation of micro hardness (Meng et al., 2021). Vickers method is an another testing method uses Wilson Wolpert 401 MWD microhardness tester (Make:Buehler, Lake Bluff, IL, USA) for measuring the micro hardness. Before carrying out the test the specimen is sectioned and rough polished along the thickness. To obtain accurate results, all the specimens were subjected to minimum of 6 times repeated test, and the average of those values can be used for the final value. The surface hardness of materials are measured using micro combi tester with a spherical indenter of radius 5µm. At maximum loading condition of 5000 mN, the loading and unloading were done for time period of 30 s and 10 s and the load displacement indentation curves were documented. The hardness was estimated from these curves based on the Oliver and Pharr (1992) method. Micro-hardness measurements are also performed by hardness tester for make Qness Q10M (Tekdir & Yetim, 2021).

Tensile tests were conducted using universal tension and compression testing machine functioning at an invariable test speed of 5 mm/min at room temperature. Tensile test specimens were fabricated according to ISO 6892 standard (17). A rotating fatigue testing machine were considerd to test the strength of the Al-Ni-Cu alloy in some studies. The fatigue characteristics of the Al-Ni-Cu alloy were compared with the stress–number of cycles to failure curve. The number of cycles to failure is calculated by the average of 3–5 rotating fatigue tests. Moreover, the fatigue fracture surface can be observed through SEM to confirm the fracture occurrence and determine the fracture mechanism (Chang et al., 2021).

Scratch Analyser

The coating adhesion of the coated samples can be tested using nano-scratch tester with a diamond stylus of 100 µm radius (30). In some studies, scratch test is conducted at maximum load of 180 N with a loading speed of 100 N/min along with scratch length of 4.5mm. The measured data are related to coefficient of friction and acoustic emission signal. Coating thickness was measured using eddy current probe, and each thickness is measured 3 to 6 times and final measurements are the average of all measured values. Microstructure and surface morphology studies of coatings was performed by field emission scanning electron microscopy according to ASTM E1588 (36). The surface elements namely oxygen and nitrogen, in the surface layer of hundred nanometers thick, were determined with use of it. Other instruments such as energy dispersive spectroscopy and dedicated Quantax 200, Esprit 1.9 code are also used as a scratch analyser (Dzhurinskiy et al., 2021).

Cutting Surface Analyser

Cutting force of surface modified tools and workpiece were studied using, a KISTLER 9275A piezo-electric ceramic and strain gauge dynamometer (Meng et al., 2021). Temperatures of deformed cutting surface were measured using thermocouple method. Online temperature of the modified surface can be measured by indirect method using IR thermometer (Make: IRTIS-2000S thermograph, Russia).

Corrosion Analyser

The corrosion resistance of the specimen was analysed using electrochemical measurements. The galva-nostat/potentiostat are considered for the analyses (Dzhurinskiy et al., 2021). The open circuit potential measurements and DC polarization tests are used to determine the passive characteristics of the surface based on the range of potentials and surface oxide film. Tafel extrapolation, is used to find the values of corrosion current density and corrosion potential. The surface oxide layer behavior can also be studied by electrochemical impedance spectroscopy (EIS).

SUMMARY

The surface modification process is inevitable when it comes to real time application. There various types of surface modification process are considered and one over other the significance process depends on the applications and cost. To understand the surface properties variety of techniques were considered for evaluating the mechanical properties, surface topography and surface composition. A combined utiliza-tion of different methods also fulfill the needs, hence it is essential to completely realize the principles and effects of many modification techniques, and their newest advances proceeding to the processing.

REFERENCES

Arifvianto, B., & Mahardika, M. (2018). Surface morphology, roughness and wettability of the medical grade 316L stainless steel processed with surface mechanical attrition treatment and electropolishing for the preparation of osteosynthesis plate. *The Journal of Physiological Sciences; JPS*, *29*(3), 83–94.

Baptista, A., Silva, F., Porteiro, J., Míguez, J., & Pinto, G. (2018). Sputtering physical vapour deposition (PVD) coatings: A critical review on process improvement and market trend demands. *Coatings*, *8*(11), 402. doi:10.3390/coatings8110402

Bruzzo, F., Catalano, G., Demir, A. G., & Previtali, B. (2021). Surface finishing by laser re-melting applied to robotized laser metal deposition. *Optics and Lasers in Engineering*, *137*, 106391. doi:10.1016/j.optlaseng.2020.106391

Cao, X., Hu, J., Huo, W., Xi, X., & Zhao, W. (2021). Surface microstructure and property modifications in AISI 304 stainless steel induced by pseudospark pulsed electron beam treatments. *Vacuum*, *184*, 109914. doi:10.1016/j.vacuum.2020.109914

Carlsson, J. O., & Martin, P. M. (2010). Chemical vapor deposition. In *Handbook of Deposition Technologies for films and coatings* (pp. 314–363). William Andrew Publishing. doi:10.1016/B978-0-8155-2031-3.00007-7

Chang, K. C., Zhao, J. R., & Hung, F. Y. (2021). Microstructure, Mechanical Properties, and Fatigue Fracture Characteristics of High-Fracture-Resistance Selective Laser Melting Al-Ni-Cu Alloys. *Metals*, *11*(1), 87. doi:10.3390/met11010087

Chaus, A. S., Sahul, M., Moravčík, R., & Sobota, R. (2021). Role of microstructural factor in wear resistance and cutting performance of high-speed steel end mills. *Wear*, *474*, 203865. doi:10.1016/j.wear.2021.203865

Darmanin, T., Taffin de Givenchy, E., Amigoni, S., & Guittard, F. (2010). Hydrocarbon versus fluorocarbon in the electrodeposition of superhydrophobic polymer films. *Langmuir*, *26*(22), 17596–17602. doi:10.1021/la103310m PMID:20879773

Devaraj, S., Malkapuram, R., & Singaravel, B. (2021). Performance analysis of micro textured cutting insert design parameters on machining of Al-MMC in turning process. *International Journal of Lightweight Materials and Manufacture*, *4*(2), 210–217. doi:10.1016/j.ijlmm.2020.11.003

Dzhurinskiy, D. V., Dautov, S. S., Shornikov, P. G., & Akhatov, I. S. (2021). Surface Modification of Aluminum 6061-O Alloy by Plasma Electrolytic Oxidation to Improve Corrosion Resistance Properties. *Coatings*, *11*(1), 4. doi:10.3390/coatings11010004

Eshkabilov, K., Berdiyev, S., & Kamolov, B. (2021, January). Hardening of cutting tools by combined gas nitriding method. *IOP Conference Series. Materials Science and Engineering*, *1030*(1), 012019. doi:10.1088/1757-899X/1030/1/012019

Grum, J., & Šturm, R. (1997). Laser surface melt-hardening of gray and nodular irons. *Applied Surface Science*, *109*, 128–132. doi:10.1016/S0169-4332(96)00648-4

Guan, P., Zhou, L., Yu, Z., Sun, Y., Liu, Y., Wu, F., & Chu, D. (2020). Recent progress of surface coating on cathode materials for high-performance lithium-ion batteries. *Journal of Energy Chemistry*, *43*, 220–235. doi:10.1016/j.jechem.2019.08.022

Gunasekaran, K., Kumar, P. G., Thanigaivelan, R., Arunachalam, R., & Shanmugam, V. (2021). Optimization of Turning Parameters of Cryogenic Soaked AZ91 Magnesium Alloy using TOPSIS coupled Taguchi Technique. *Journal of New Materials for Electrochemical Systems*, *24*(1), 49–54. doi:10.14447/jnmes.v24i1.a09

Karademir, I., Celik, M. B., Husem, F., Maleki, E., Amanov, A., & Unal, O. (2021). Effects of constrained groove pressing, severe shot peening and ultrasonic nanocrystal surface modification on microstructure and mechanical behavior of S500MC high strength low alloy automotive steel. *Applied Surface Science*, *538*, 147935. doi:10.1016/j.apsusc.2020.147935

Kumar, R., Thanigaivelan, R., Rajanikant, G. K., Jagadeesha, T., & Das, J. (2021). Evaluation of hydroxyapatite-and zinc-coated Ti-6Al-4V surface for biomedical application using electrochemical process. *Journal of the Australian Ceramic Society*, *57*(1), 107–116. doi:10.100741779-020-00517-6

Mashrabov, N., Erofeev, V. V., & Korotkov, V. A. (2021). Plasma hardening of elements for soil cultivating instruments. *Welding International*, 1-5.

Meng, X., Zhang, K., Guo, X., Wang, C., & Sun, L. (2021). Preparation of micro-textures on cemented carbide substrate surface by plasma-assisted laser machining to enhance the PVD tool coatings adhesion. *Journal of Materials Processing Technology*, *288*, 116870. doi:10.1016/j.jmatprotec.2020.116870

Morozow, D., Barlak, M., Werner, Z., Pisarek, M., Konarski, P., Zagórski, J., & Caban, J. (2021). Wear Resistance Improvement of Cemented Tungsten Carbide Deep-Hole Drills after Ion Implantation. *Materials (Basel)*, *14*(2), 239. doi:10.3390/ma14020239 PMID:33418938

Rao, C. M., Sachin, B., Rao, S. S., & Herbert, M. A. (2021). Minimum Quantity Lubrication through the micro-hole textured PCD and PCBN inserts in the machining of the Ti–6Al–4V alloy. *Tribology International*, *153*, 106619. doi:10.1016/j.triboint.2020.106619

Saha, S. K., Park, Y. J., & Cho, S. O. (2021). Fabrication of highly ordered nanoporous oxide layer on Ti6Al4V surfaces for improved corrosion resistance property. *Journal of Molecular Structure*, *1223*, 129244. doi:10.1016/j.molstruc.2020.129244

Sanchez, A. G., You, C., Leering, M., Glaser, D., Furfari, D., Fitzpatrick, M. E., Wharton, J., & Reed, P. A. S. (2021). Effects of laser shock peening on the mechanisms of fatigue short crack initiation and propagation of AA7075-T651. *International Journal of Fatigue*, *143*, 106025. doi:10.1016/j.ijfatigue.2020.106025

Sasi, R., Subbu, S. K., & Palani, I. A. (2017). Performance of laser surface textured high speed steel cutting tool in machining of Al7075-T6 aerospace alloy. *Surface and Coatings Technology*, *313*, 337–346. doi:10.1016/j.surfcoat.2017.01.118

Singh, B., Sasi, R., Subbu, S. K., & Muralidharan, B. (2019). Electric discharge texturing of HSS cutting tool and its performance in dry machining of aerospace alloy. *Journal of the Brazilian Society of Mechanical Sciences and Engineering*, *41*(3), 1–11. doi:10.100740430-019-1654-6

Skoczypiec, S., Grabowski, M., & Ruszaj, A. (2016). The impact of electrochemical assistance on the microturning process. *International Journal of Advanced Manufacturing Technology*, *86*(5), 1873–1880. doi:10.100700170-015-8320-7

Sun, L., Yuan, G., Gao, L., Yang, J., Chhowalla, M., Gharahcheshmeh, M. H., Gleason, K. K., Choi, Y. S., Hong, B. H., & Liu, Z. (2021). Chemical vapour deposition. *Nature Reviews Methods Primers*, *1*(1), 1–20. doi:10.103843586-020-00005-y

Tekdir, H., & Yetim, A. F. (2021). Additive manufacturing of multiple layered materials (Ti6Al4V/316L) and improving their tribological properties with glow discharge surface modification. *Vacuum*, *184*, 109893. doi:10.1016/j.vacuum.2020.109893

Thanigaivelan, R., & Deepa, D. (2018). Heat transfer enhancement by coated fins in the microscale domain. *Thermal Science*, *22*(6 Part B), 2783–2789. doi:10.2298/TSCI160901089T

Tolcha, M. A., Gutu Jiru, M., & Lemu, H. G. (2021). Experimental Investigation of Al-CuMgMn Alloy Developed by Laser Beam Surface Modification for Wear Resistance. *Metals*, *11*(5), 712. doi:10.3390/met11050712

Voznesenskaya, A. A., Kochuev, D. A., Raznoschikov, A. S., Zhdanov, A. V., & Morozov, V. V. (2020, July). Effect of part orientation on temperature stresses in the structure of a material during selective laser melting. *IOP Conference Series. Materials Science and Engineering*, *896*(1), 012131. doi:10.1088/1757-899X/896/1/012131

Yang, L., Yang, X., Zhang, T., & Sun, R. (2021). Optimization of microstructure and properties of composite coatings by laser cladding on titanium alloy. *Ceramics International*, *47*(2), 2230–2243. doi:10.1016/j.ceramint.2020.09.063

Ye, C., Telang, A., Gill, A. S., Suslov, S., Idell, Y., Zweiacker, K., Wiezorek, J. M. K., Zhou, Z., Qian, D., Mannava, S. R., & Vasudevan, V. K. (2014). Gradient nanostructure and residual stresses induced by Ultrasonic Nano-crystal Surface Modification in 304 austenitic stainless steel for high strength and high ductility. *Materials Science and Engineering A*, *613*, 274–288. doi:10.1016/j.msea.2014.06.114

You, Y., Zhang, B., Tao, S., Liang, Z., Tang, B., Zhou, R., & Yuan, D. (2021). Effect of Surface Microstructure on the Heat Dissipation Performance of Heat Sinks Used in Electronic Devices. *Micromachines*, *12*(3), 265. doi:10.3390/mi12030265 PMID:33806561

Zhao, L., Cui, C., Wang, Q., & Bu, S. (2010). Growth characteristics and corrosion resistance of micro-arc oxidation coating on pure magnesium for biomedical applications. *Corrosion Science*, *52*(7), 2228–2234. doi:10.1016/j.corsci.2010.03.008

Zhao, W., Liu, D., Yang, J., Zhang, H., Ma, A., Zhang, X., Ren, Z., Zhang, R., Dong, Y., & Ye, C. (2021). Improving plain and fretting fatigue resistance of A100 steel using ultrasonic nanocrystal surface modification. *International Journal of Fatigue*, *148*, 106204. doi:10.1016/j.ijfatigue.2021.106204

Zubkov, N., Poptsov, V., & Vasiliev, S. (2017). Surface Hardening by Turning without Chip Formation. *Jordan Journal of Mechanical & Industrial Engineering*, *11*(1).

Chapter 2
Electrochemical Micromachining Performance Optimization:
Impact of Cathode Profile and Rotation on Machining Speed and Accuracy

Vempannan C.
Muthayammal Engineering College, India

Kanish T. C.
Vellore Institute of Technology, India

Thirumanikandan D.
Muthayammal Engineering College, India

Aswin B.
Muthayammal Engineering College, India

ABSTRACT

Micromachined components are finding application in a variety of fields such as medical, semiconductor, and aerospace. Electrochemical micromachining (ECMM) is a technology used to realize the micro components with enhanced features and functionality. This chapter reviews the impact of cathode profile on the machining rate and accuracy. An indigenous developed ECMM setup is used with special attachment of rotating electrode. The control parameters such as machining voltage, duty cycle, electrolyte temperature, and rotating RPM were considered on improving the output performance. The interaction between the input parameters and machining speed and overcut was studied. The input parameters considered are machining voltage, duty cycle, electrolyte concentration, and cathode tool rotation. There are various optimization processes considered to evaluate the working range of control factors for the machine on output performance. In this chapter, various multi-objective optimization techniques applied in optimization of ECMM process were reviewed.

DOI: 10.4018/978-1-7998-9574-9.ch002

INTRODUCTION

Micromachining is a technology deals with machining of conducting materials in micron range. The micro machined components finds application in biomedical, automobile, defence and ornamental industries. Among various micromachining techniques electrochemical micromachining (ECMM) is found to be superior for its advantages of no tool wear, higher machining speed, good surface finish and no residual stress. ECMM works on a principal of Faradays law of electrolysis in which cathode act as tool electrode and workpiece is considered as an anode (Jain et al. 2012). The material removal on the workpiece takes place due to ions dissociation with the help of applied voltage. In ECM process a cathode material is chosen based on the ability to permits the electrolytic breakdown of water and liberation of hydrogen. The hydrogen bubbles evolution and the electrolysis of water produces hydroxide ions. The ions from the anode combine with hydroxide ions formed from the electrolysis of water to produce metal hydroxides. These precipitate away of the electrolyte, eliminating the ions from the electrolyte and avoiding them from cathode deposition (Leese & Ivanov 2016). In ECMM process the anodic dissolution takes place in the range of size 1-999μm with applied voltage in the range of 0-10 Volts (Bhattacharyya et al.2004). The narrow gap between the cathode and anode place an important role in material removal and it is called inter electrode gap (IEG). The optimized gap is essential for achieving higher material removal rate (MRR), by increasing the IEG the MRR is significantly affected. The electrolyte namely active and passive electrolyte is considered in the ECMM process (Shunmugam & Kanthababu, 2019). Passive electrolytes such as sodium nitrate, sodium chlorate, weak acid were considered for ECMM apart from active electrolytes (strong acids such as sulphuric acid, hydrochloric acid and sodium chloride). In case of ECMM the stagnant electrolyte is found to show more efficacy compared to flow electrolyte. Sometimes pulsating electrolyte flow is also preferred over the continuous flow of electrolyte. The electrolysis is initiated by potential difference between the electrodes, there are two types of power supply considered for the ECMM namely direct current and pulse power supply. The pulse power supply is found to be suitable for micromachining which offer higher MRR and good surface quality. The pulse power supply in the range of millisecond, microsecond and nanosecond were used for the machining based on the application. The various factors that affect the performance of ECMM are machining voltage, current, pulsed power supply, electrolyte type, temperature and concentration, cathode profile, IEG gap width and workpiece material. In this chapter the effect of cathode profile and rotation is discussed, since in ECMM process cathode shape, temperature, coating and rotation has shown the significant improvement in the MRR and accuracy. The fishbone diagram shown in figure 1, shows the input parameters for improving the ECMM performance.

EFFECT OF CATHODE

Cathode Geometry

The shape created on the workpiece depends on the profile of the cathode. Normally cylindrical shaped electrodes were considered in the ECMM process for making circular holes. In ECMM process the accuracy of the micro-hole depends on the size of the cylindrical electrode. The size of the micro-hole machined by ECMM process is normally 2-3 times the cathode. Hence the proper choice of electrode plays a vital role in ECMM process. Thanigaivelan et al. (2017) have introduced the different types of tool electrode tips in ECMM process in the view of improving the machining speed and accuracy. They used flat electrode, conical electrode, wedged electrode, truncated electrode tips and found that conical electrode tip is suitable for producing higher accuracy and wedged electrode for higher machining speed. Figure 2 shows the schematic of cathode tool with insulated and uninsulated area along with current density. It can be understood that from the figure that in conical tool tip shape the scope for stray current is less which is attributing lower overcut. Yao 2019 et al have developed high strength thin hollow cathode to improve the machining efficiency and developed different structure on TB6 using this electrode. Liu et al (2021) have used glass tube electrode coated with silver in which the glass tube remains as an insulated substrate, and the coated silver layer function as the cathode. They developed micro holes with aspect ratio of 3 and no accrual of debris products and short circuits are witnessed. Soundarrajan and Thanigaivelan (2020) have used short length solid tool and obtained 74.3% higher MR over normal tool electrode at 7 V, 31 g/L electrolyte concentration, and 90% duty cycle. Rajkeerthi et al 2020 have used hollow tapered of diameter 0.3 mm with reverse potential drop off approach and compared the performance with hollow cylindrical tool (CHT) of same diameter. The use of hollow tapered tool enhances the machining rate, geometrical accuracy and surface finish.

Figure 1. Fishbone diagram for ECMM
(Source: Reprinted from Arunchalam et al 2018)

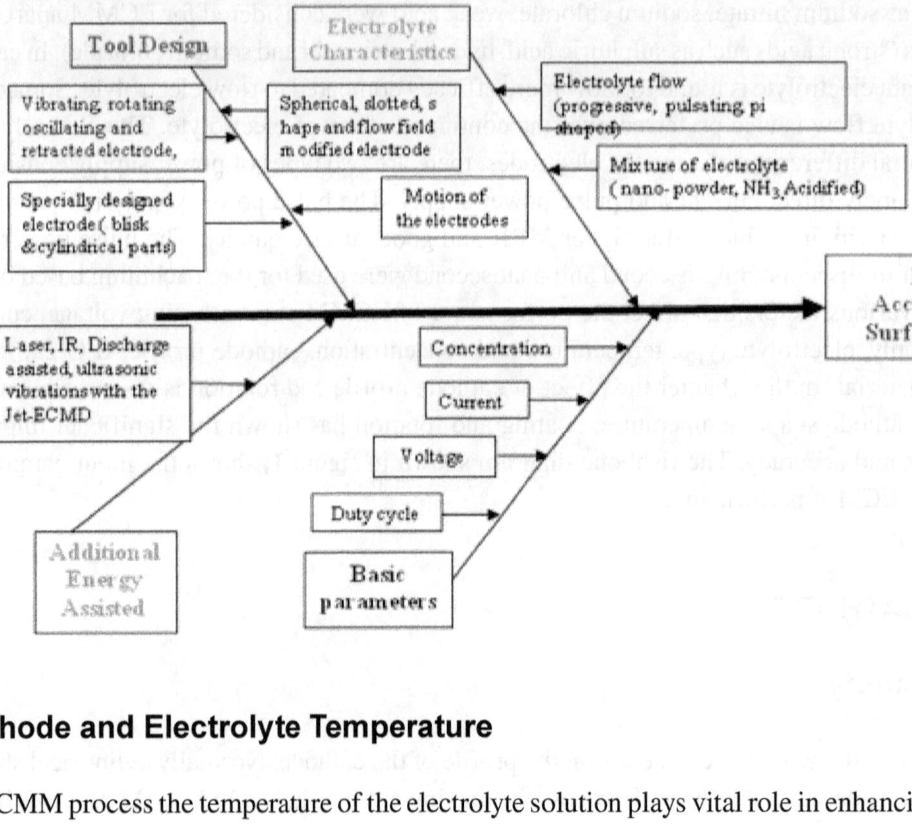

Cathode and Electrolyte Temperature

In ECMM process the temperature of the electrolyte solution plays vital role in enhancing the machining speed and accuracy. Maniraj and Thanigaivelan (2019) have used heated electrode for machining the aluminium metal matrix composites. The use of heated electrode have improved the machining speed by

88.37% and reduction of radial overcut and conicity factor is seen at 8 V, 90% duty cycle, 35 g/l electrolyte concentration and 60°C electrode temperature. Tang et al 2014 have considered electrolyte temperature for electrochemical machining of S-03 special stainless steel. At parameter combination of machining voltage 24 V, electrolyte temperature at 30 °C and 0.8 MPa electrolyte pressure, the surface roughness is Ra 0.08 µm and the MRR is 411.4 mm^3/min. Soundarrajan and Thanigaivelan 2021 have used used hot air assited electrolyte to improve the machining performance. The hot air is mixed electrolyte produces 2.87 times increased MRR. This research group has also used IR and UV rays to heat the electrolyte and found that the significant increase in machining speed. The use of heated electrolyte has negative effect on the surface of the microhole. Rajan et al (2019) have considered heated electrolyte for machining B_4C reinforced aluminium composites. They concluded that machining speed increases with the electrolyte temperature. The reinforcement of 5% mass fraction of B_4C in Al7075 shows a higher machining speed at 30°C and increases further with electrolyte temperature up to 39 °C.

Figure 2. Cathode with insulation and stray current
(Source: Reprinted from Thanigaivelan et al 2018)

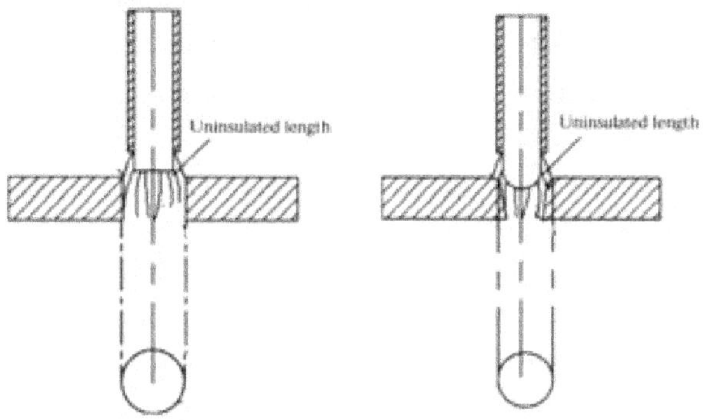

Cathode Coating

In ECMM process the cathode when supplied with electric power supply experience the stray current around the surface of the electrode. This occurrence of stray current produces overcut on the workpiece. The overcut can be measured by finding the diametrical/radial/ area difference between the cathode and finished micro hole. The stray current can be prevented by coating cathode with insulating material. There are different techniques are available for insulating the tool electrode namely chemical vapour disposition, physical vapour deposition and epoxy coating. Coating of micro tool posses various technical challenges like selecting the suitable coating material for a specific application, attachment of coating material to the base metal, homogenous of coating and the thickness of coating which decides the coating strength, porosity, and resistance to corrosion. Recently soundarrajan and thanigaivelan (2020) have used white putty for the tool insulation, they have also considered ethylene-vinyl acetate hot melt adhesive glue as an insulation material. This putty coated tool produces 43.1% lesser overcut over plain tool at the parameter combination of 23 g/L, 15 V, and 85%. The hot melt adhesive coated tool provides 29.5% lesser over plain tool electrode. The hollow wedge tool shows the subsequent highest MR compared to the plain tool electrode. Swain et al. 2012 have electroplated the nickel on tungsten electrode DC and pulse power supply. The electroplated electrode show a correlation between the input process parameters and the nickel coated tungsten microelectrode.

Cathode Rotation

In micromachining any forms or features developed on the substrate needs high quality and accuracy, especially in electrochemical micro drilling the use of thin tool electrode provides very narrow flow in IEG. This narrow IEG gets obstructed due to debris created during machining and hence proper flushing of debris is essential to prevent undue effects. Since in ECMM slow moving or stagnant electrolyte is preferred, the rapid carrying away of the reaction products, namely hydrogen gas, sludge and Joule heat, from the machining area in essential (Thanigaivelan et al 2018). The ions dissociated during pulse-on period cannot be simply rejuvenated by convection and diffusion of the electrolyte. The improper flushing of electrolyte in IEG results in short-circuit, attributing for damage of tool and workpiece surface. To overcome the ill effect of reaction products Tsui et al 2008 have proposed the rotating tool electrode. The tool is a micro helical drill bit which solves the problem of renewing of electrolyte. The rotation of micro-helical electrode swirl the fresh electrolyte into the machining area. This phenomenon discard the debris through the helical groove of the electrode out of the machining zone, contributing for renewal of the electrolyte. In line with this study a fresh study is planned with rotating electrode for improving the machining speed and accuracy.

EXPERIMENTAL DETAILS

The developed EMM machine set-up shown in figure 3 consists of machine structure, pulsed power supply unit and IEG control system. The machine structure consists of one base plate, one vertical column, three angle plates made up of mild steel. The angle plates house the stepper driver and mechanism for tool feed. The machining enclosure is placed on the base plate with electrolyte pump and filter (Thanigaivelan et al 2010). A pulsed power supply of 20V and 30A with the provision for controlling the voltage, current and pulse width were used. The IEG is maintained with the help of the microprocessor. In tool feeding arrangement, threads of the lead screw is 40 threads per inch. When the microprocessor sends the pulse to the stepper motor it creates a linear movement of 4μm. Hence by using the press switch in the control kit the required IEG is set. Initial the tool is made to touch the workpiece and multimeter connected to the anode and cathode were short circuited to produce the beep sound and slowly the cathode tool is reversed to set the required IEG. Figure 4 shows the attachment of rotating electrode to the tool feeding system. The attachment consists of dc motor, speed controller, tool holder and tiny micro 0.3mm twist drill bit. The commercially available dc motor (775) with voltage of 12V and speed of 6000rpm is considered for the setup. The speed control system consists of MOSFET, 10K pot and resistors. Using the speed control circuit the speed of the high speed dc motor is controlled.

The nickel based alloys are finding application in various fields and inconel 718 is popular for its application as a turbine material. Hence in this research the commercially available inconel 718 is used as a workpiece. As received inconel 718 plate is cut into suitable dimension using Wire EDM. The dimension of the work piece used for the study is 50x 50x1 mm. This dimension will have suitable fit in the workpiece holder.

Figure 3. EMM setup

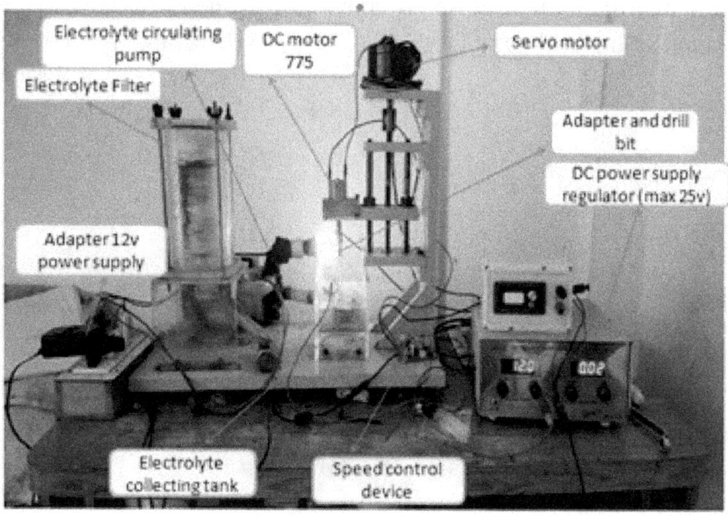

Figure 4. Tool feeding system with rotation assembly

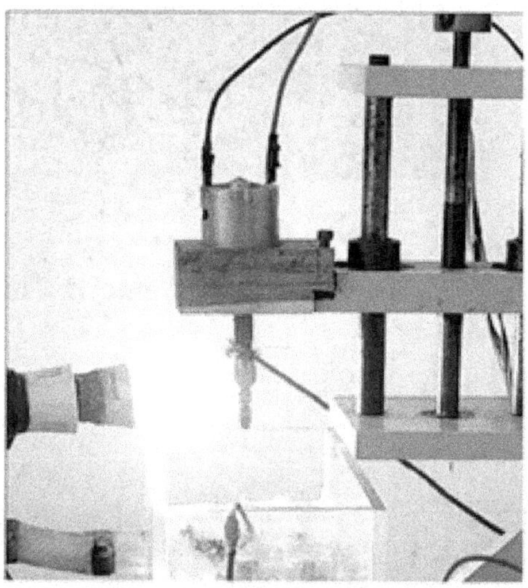

The workpiece holder is made of acrylic material which consists of two blocks and workpiece is held between the blocks and fastened with screws. The power supply is given to the workpiece through the screws. Tiny micro 0.3mm twist drill bit is used a cathode. The circumference of the cathode tool is coated with coated with adhesive glue for insulating purpose to avoid stray current effect. The sodium nitrate brine solution is prepared for various concentrations with distilled water. The brine solution is prepared by mixing the weighed sodium nitrate salts in one liter of distilled water. The solution is mixed thoroughly with magnetic stirrer. For each experiments fresh electrolyte are prepared and used. The interaction between the input parameters and machining speed & overcut were studied. The figure 1.5 shows the sample of the ECMMed specimen for micro hole. The input parameters considered are machining voltage (V), duty cycle (%), electrolyte concentration (gm/L) and cathode tool rotation (rpm). The duty cycle is defined as the ratio of pulse on time to total time for fixed frequency of 50 Hz. Table 1 shows the experimental values and output response. The experimental design follows the varying one parameter at a time.

Figure 5. Machined specimen

Sample calculation of overcut and machining speed (Maniraj & Thanigaivelan 2019)

Overcut = Electrode diameter/radius (r) – Hole diameter/radius (R)

Where,
r is a electrode radius in micrometer.
R is a hole radius in micro meter.

Overcut = 300-730 = 430μm.

MRR= thickness of work piece / time taken for machining complete hole

Where,
T is a thickness of work piece in micrometer. (1mm= 1000 micron)
t is a time taken for machining in sec. (1 min = 60 sec)

MRR = 1000/780 = 1.282 μm/sec.

Table 1. Experimental combinations

S.No	Machining voltage (V)	Duty cycle (%)	Electrolyte concentration (gm/L)	Cathode tool rotation (rpm)	Machining Rate (µm/s)	Radial Overcut (ROC) (µm)
1	6	90	35	3500	0.544	375
2	7	90	35	3500	0.577	422
3	8	90	35	3500	0.595	514
4	9	90	35	3500	0.667	586
5	10	90	35	3500	0.904	592
6	10	50	35	3500	0.586	360
7	10	60	35	3500	0.642	436
8	10	70	35	3500	0.756	486
9	10	80	35	3500	0.854	542
10	10	90	35	3500	0.904	594
11	10	90	15	3500	0.397	386
12	10	90	20	3500	0.424	452
13	10	90	25	3500	0.526	498
14	10	90	30	3500	0.725	512
15	10	90	35	3500	0.910	590
16	10	90	35	2500	0.522	366
17	10	90	35	2750	0.636	450
18	10	90	35	3000	0.689	494
19	10	90	35	3250	0.746	526
20	10	90	35	3500	0.900	590

RESULTS AND DISCUSSION

Effect of Machining Voltage on Machining Rate and ROC

From the figure 6, machining rate increases with raise in machining voltage, the raise in machining voltage increases the current value required for machining. Machining voltage application in the range of 9-10 V causes steep increase in the machining rate. At higher voltage the more current density is contributing for higher material removal since the ECM principal the material removal is proportional to the applied voltage. The trends for ROC are found to increase with the voltage. The linear increase in ROC is seen for the voltage range of 7-9V and the curve flattens for higher voltage application. At higher voltage, due rotation of cathode tool, the reaction products which interrupts the machining were dispelled from the machining zone attributing for minimization of micro sparks and stray currents.

Figure 6. Machining voltage vs. machining rate and ROC

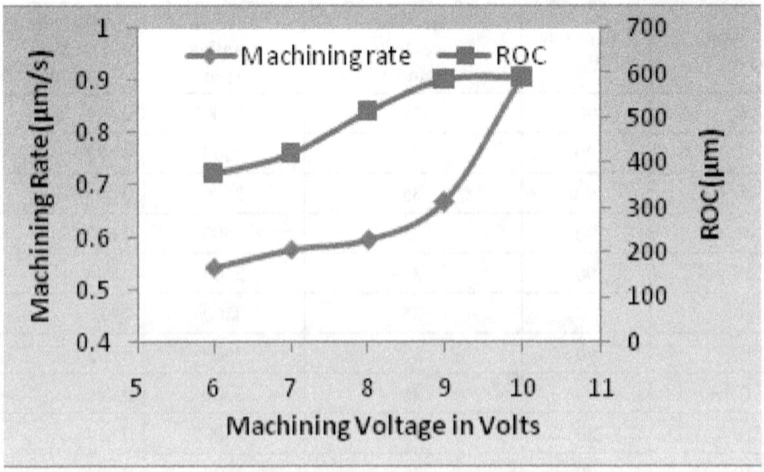

Effect of Duty Cycle on Machining Rate and ROC

The increase in duty cycle increases the machining rate as shown in figure 7. The increase in duty cycle for the fixed frequency the pulse on time increases. In ECMM the material removal mechanism takes place during pulse on time and evacuation of debris takes during pulse off time. Higher duty cycle means more pulse on time which attributes for longer exposure of current on the workpiece. With increase in pulse on time the time required for evacuation get reduces which results in short circuit and microsparks. The generation on inconsistent microsparks removes extra material from the machining area attributing for higher overcut. In figure 6 machining rate increases more linearly in the duty cycle range of 60 to 90%. ROC shows a linear trend in the duty cycle range of 60 to 90%. The rotation of electrode attributes lesser overcut due to proper washing out reaction products.

Figure 7. Duty cycle vs. machining rate and ROC

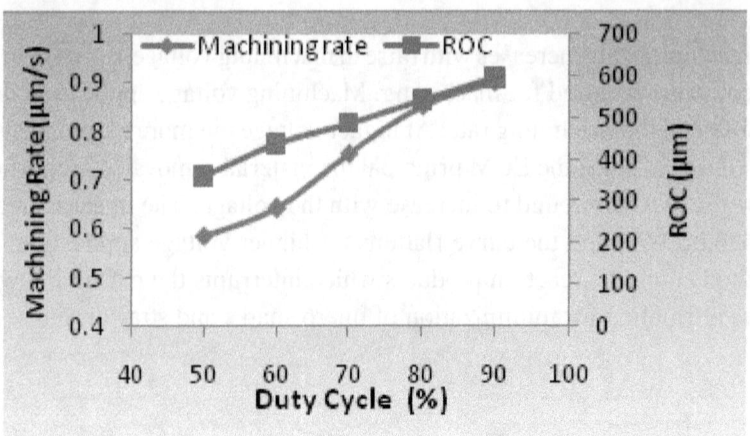

Effect of Electrolyte Concentration on Machining Rate and ROC

The increase in electrolyte concentration increases the no. of ions required for exchange during the machining process. From the figure 8 it is evident that the electrolyte concentration in the range of 20 to 35gm/L increases the machining rate drastically due to more availability of ions. The overcut increases less linearly for the electrolyte concentration of 15 to 30 gm/L, it is due to fact that the introduction of tool rotation enhances micro- hole quality.

Figure 8. Electrolyte concentration vs. machining rate and ROC

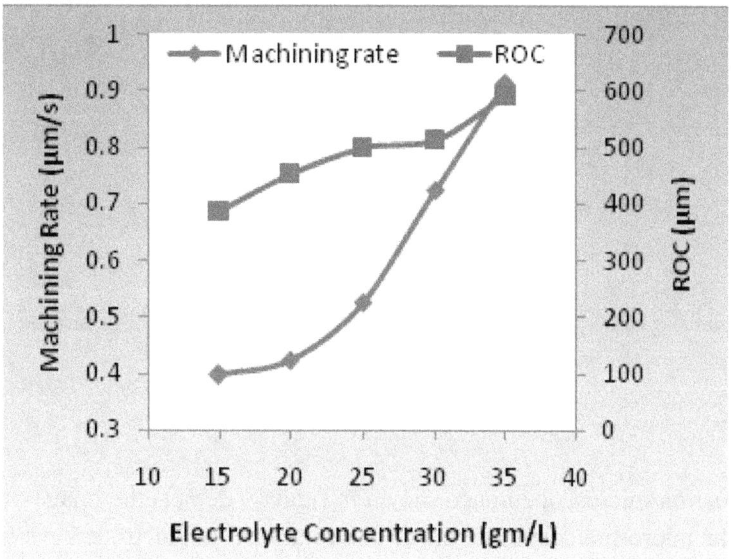

Effect of Rotational Speed of Electrode on Machining Rate and ROC

The rotational speed of electrode is varied between 2500 to 3500 rpm and graph shows the relationship between the machining rate and ROC. The increase in rotational speed the machining rate increases more linearly for the rpm of 2500 to 2750 and less linearly for the rpm range of 2750 to 3250 and further increases drastically to higher machining rate (figure 9). The rotational electrode at lower rpm mixes the fresh electrolyte and renews the electrolyte in the IEG. At the initial stage of machining the available of ions in the electrolyte is more contributing for linear increase in machining rate. Moreover the available of reaction products in the electrolyte is less hence the machining efficiency is better. By increasing the rpm further the reaction products (sludge) in the electrolyte increases the density of the electrolyte. The heating effect also attributes for increase in density of the electrolyte which reduces the electrode speed by creating the resistance. Further increase in electrode rpm overcomes the resistance caused by sludge results in proper evacuation of reaction products from the machining area attributing for higher machining rate. The increase in rotational speed has significant effect on ROC. The ROC increases gradually with rotational speed, the increase in rotational speed efficiently renews the electrolyte and clears the sludge's. The breaking of hydrogen bubbles evolved from cathode develops micro stirring effect in the machining zone.

Figure 9. Cathode tool rotation vs. machining rate and ROC

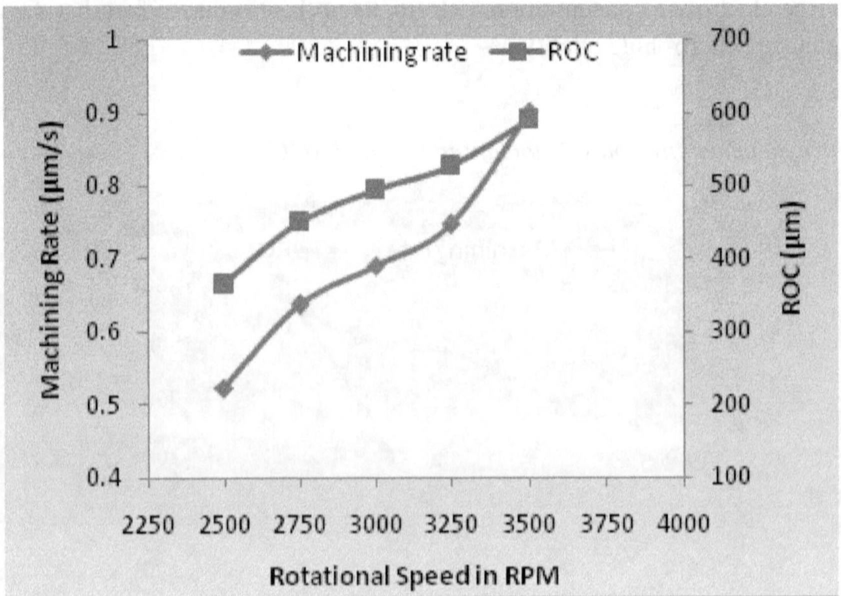

The scanning electron microscope image shown in figure 9 depicts the image of the micro-hole and expanded view of the micro-hole circumference. The rotational of electrode have significant influence on the generation of micro hole. The micro-hole generated at electrode rotational speed of 3000rpm with 10V, 35gm/L electrolyte concentration and 90% duty cycle shows good circular micro-hole with better surface quality without any stray current and micro spark effect in the SEM picture.

Figure 10. Pictures of micro-hole

MULTIOBJECTIVE OPTIMIZATION TECHNIQUES IN ECMM

When it comes to productivity and quality of the components, many manufacturing industries foresee the application of optimization in manufacturing process. Researchers have considered single objective and multi-objective optimization for optimization of ECMM parameters. Since in ECMM the major output responses are material removal rate, surface quality and accuracy, and industries also look for optimized range of input parameters for type of material and features machined. Compared to single objective, multi-objective optimization techniques were found to be suitable for providing solutions for real time problems. For eg in manufacturing industry the press worked components shows burrs and other sheet metals oriented defects. Among the defective components some requires rework to meet the demands of original equipment manufacturer (vendor). The cost of extra man/machine hours on the components raises the cost of production and this problem is the exact application of multi-objective optimization technique. Here in this case the productivity has to be improved and cost of production has to be reduced. Hence the possible input factors are to be controlled to obtain better output characteristics. These complications are known as multi-objective optimization problems. Maximizing profit and minimizing the cost of a component; is the typical example of multi-objective optimization problems (Cong et al. 2020). Figure 11 fish bone diagram depicts various optimization techniques in ECMM

Figure 11. Fish bone diagram depicting various optimization techniques in ECMM

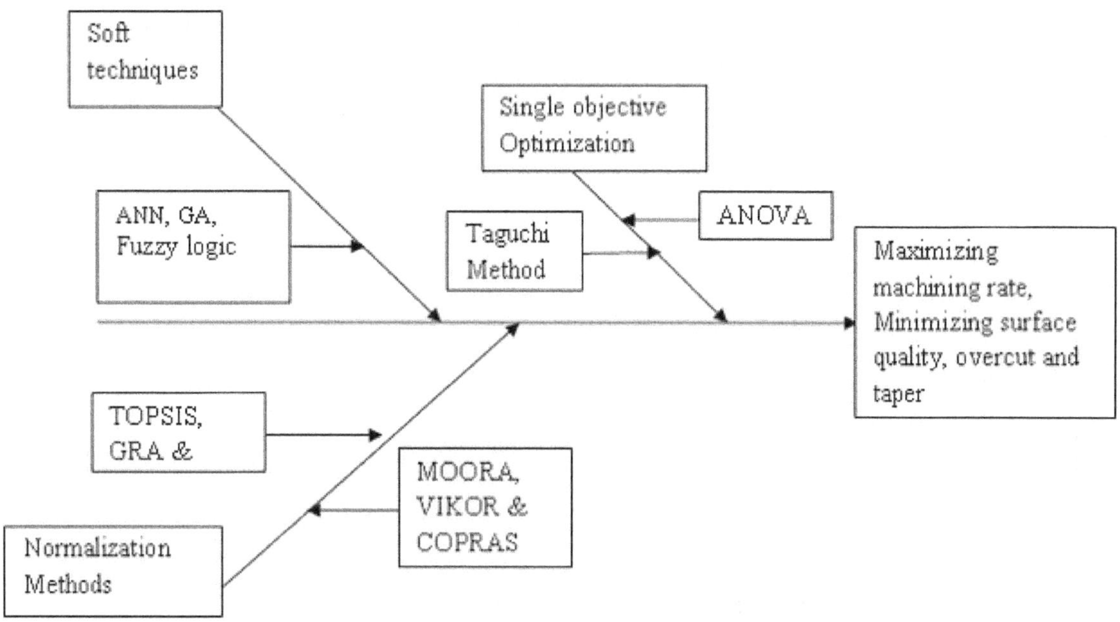

The major components of multi-objective optimization are discussed here:

Determination of Signal-to-Noise (S/N) Ratio for Individual Performances

The output characteristics are initially determined using Signal to Noise (S/N) ratios. It represents the wanted and unwanted values for every output quantity. As per the objectives of the response, the S/N ratio is classified as Lower-the-better, Higher-the-better, Nominal-the-best.

Lower-the-better characteristics is evaluated using

$$S / N = -10\log\left(\frac{1}{n}\sum_i^n Y_i^2\right) \tag{1}$$

Higher-the-better characteristics is calculated using

$$S / N = -10\log\left(\frac{1}{n}\sum_i^n Y_i^{-2}\right) \tag{2}$$

Where n is the total number of tests and *i* indicate the number of experiments.

In multi objective optimization techniques the assigning weight to the responses needs a complete insight about the process otherwise it will definitely result in vain. The various methods are used by researchers for assessing the weight. The following sections discuss the effective techniques used in ECMM for assessing the weight of the responses.

Entropy Weight Calculation

The entropy weight method is one of the most effective methods of evaluating the weight of responses. The following procedures are followed here.

$$Z = \begin{bmatrix} Z_{11} & Z_{12} & Z_{13} & \cdots & Z_{1n} \\ Z_{21} & Z_{22} & Z_{23} & \cdots & Z_{2n} \\ Z_{31} & Z_{32} & Z_{33} & \cdots & Z_{3n} \\ \vdots & \vdots & \vdots & \ddots & \vdots \\ Z_{m1} & Z_{m2} & Z_{m3} & \cdots & Z_{mn} \end{bmatrix} \tag{3}$$

The responses are organized in the decision matrix according to equation 3 that contains n attributes and m alternatives for the decision. The responses are normalized using equation

$$y_{ij} = \frac{k_{ij}}{\sum_{i=1}^m k_{ij}^2} \; j = 1,2,\ldots,.n \tag{4}$$

Where y_{ij} is a normalized value between the interval [0,1] for i^{th} alternative and j^{th} attribute.

The Entropy value E_j is evaluated using equation 5.

$$E_j = -k \sum_{i=1}^{m} y_{ij} \ln(y_{ij}) \quad j = 1,2,\ldots\ldots,n \tag{5}$$

Where $= \frac{1}{ln}$, m is constant, m is the no of alternatives

The degree of divergence can be calculated using equation

$$d_j = 1 - E_j \tag{6}$$

The result of this weight of jth criteria could be calculated using equation

$$W_{ij} = \frac{d_j}{\sum_{j=1}^{n} d_j} \tag{7}$$

Equal Weights

Equal weights can be evaluated from the equation given below (Dawes & dan Corrigan, 1974):

$$A_i = \frac{1}{n} \tag{8}$$

Where i=1,2,3,…,n and n is the number of objective functions.

Rank Order Centroid (ROC) Weights

In this method the weights put every criterion in a proportional position. The weights can be obtained using the following equation (Einhorn & McCoach, 1977):

$$A_i = \frac{2\left(n+1-i\right)}{n\left(n+1\right)} \tag{9}$$

The weight determined can be used in following multi criterion method to obtain the optimal solution. In the scalarization method, the minimizing function is noted as negative, whereas the maximizing function is noted as positive. To offer logic of equality among each objective function, it is essential to normalize using root mean square (Gunantara, Sastra, & Hendrantoro, 2014). The following is an example of a scalarization from the 3 objective functions:

$$F(x) = -\frac{w_1 f_1(x)}{\sqrt{E f_1^2(x)}} + \frac{w_2 f_2(x)}{\sqrt{E f_2^2(x)}} - \frac{w_3 f_3(x)}{\sqrt{E f_3^2(x)}} \tag{10}$$

where $F(x)$ is the fitness functions, $f_1(x)$, $f_2(x)$, $f_3(x)$ are objective functions 1, 2, 3, and $w_1 w_2 w_3$ is the weight of 1, 2, 3.

Technique for Order of Preference by Similarity to Ideal Solution (TOPSIS)

TOPSIS is a multi-criteria decision analysis method is based on the conception that the selected alternative must have the smallest geometric distance from the positive ideal solution (PIS) and the biggest geometric distance from the negative ideal solution (NIS).

Step-by-step procedure for TOPSIS analysis is expressed in a series of following steps.

1. Decision matrix consisting of m alternatives and n criteria,

$$\text{Decision Matrix} = \begin{bmatrix} x_{11} & x_{12} & \ldots\ldots & x_{1n} \\ x_{21} & x_{22} & \ldots. & x_{2n} \\ \ldots\ldots & \ldots\ldots & \ldots\ldots & \ldots. \\ x_{m1} & x_{m2} & \ldots. & x_{mn} \end{bmatrix} \tag{11}$$

2. The normalized value N_{ij} is calculated as

$$N_{ij} = \frac{x_{ij}}{\sqrt{\sum_{j=1}^{m} x_{ij}^2}} \ldots\ldots \quad j = 1,2,3\ldots,m \quad i = 1,2,3\ldots n \tag{12}$$

3. Weighted normalized decision matrix is determined as

$$M_{ij} = W_i N_{ij} \ldots\ldots \; j=1,2,3,\ldots,m \;\; i=1,2,3,\ldots,n \tag{13}$$

4. Positive-ideal and negative-ideal solutions were obtained from the following expressions

$$V^+ = \left(M_1^+, M_2^+, \ldots\ldots M_n^+\right) \text{ for max values} \tag{14}$$

$$V^- = \left(M_1^-, M_2^-, \ldots\ldots, M_n^-\right) \text{for min values} \tag{15}$$

5. Separation measures are obtained by using the following expression

$$d_j^+ = \sqrt{\sum_{i=1}^{n}\left(M_{ij} - M_i^+\right)^2} \quad j = 1,2\ldots m \tag{16}$$

$$d_j^- = \sqrt{\sum_{i=1}^{n}\left(M_{ij} - M_i^-\right)^2} \quad j = 1,2,\ldots m \tag{17}$$

6. Calculate the relative closeness coefficient by using the following equation

$$C_i = \frac{d_j^-}{d_j^+ + d_j^-} \tag{18}$$

7. The C_i values are arranged in the descending order for finding most and least preferred values.

MOORA

Multi-objective optimization on the basis of ratio analysis (MOORA) method is used to solve different decision-making problems. As in TOPSIS the decision matrix is are developed as

$$Y = \begin{bmatrix} y_{11} & y_{12} & y_{13} & \cdots & y_{1n} \\ y_{21} & y_{22} & y_{23} & \cdots & y_{2n} \\ y_{31} & y_{32} & y_{33} & \cdots & y_{3n} \\ \vdots & \vdots & \vdots & \ddots & \vdots \\ y_{m1} & y_{m2} & y_{m3} & \cdots & y_{mn} \end{bmatrix} \tag{19}$$

The matrix responses are normalized using equation

$$k_{ij} = \frac{y_{ij}}{\sum_{i=1}^{m} y_{ij}^2} \quad j = 1,2\ldots,n \tag{20}$$

Where k_{ij} is a dimension less number which belongs to the interval [0, 1] for i^{th} alternative and j^{th} attribute which denotes the normalized performance.

The maximum normalized performance values are should added and minimum normalized values are should subtracted as in equation.

$$q_i = \sum_{j=1}^{g} w_j k_{ij} - \sum_{j=g+1}^{n} w_j k_{ij} \quad (j = 1,2,\ldots n) \tag{21}$$

Where g is the number of attributes to be maximized, (n-g) is the number of attributes to be minimized, and q_i is the normalized assessment value.

The importance of responses to the attributes w_j should be multiplied with corresponding weight as in equation.

$$q_i = \sum_{j=1}^{g} w_j k_{ij} - \sum_{j=g+1}^{n} w_j k_{ij} \quad (j = 1,2,\ldots n) \tag{22}$$

Where, w_j is the weighted value obtained using entropy method which is discussed in previous section. Then these q_i values are ranked as per the preference values, which the higher value deserves the optimal combination.

Grey Relational Analysis (GRA)

In GRA, is one of the effective multi objective optimization technique. Here the values are normalized to the range of 0 to 1. The values are normalized using the following equations (Thanigaivelan et al.2013). The maximization and minimization of responses has been calculated using below relation 23 and 24.

$$X_i^*(P) = \frac{X_i(P) - MinX_i(P)}{MaxX_i(P) - MinX_i(P)} \quad \text{For } (i = 1,2\ldots m, \quad P = 1,2\ldots n) \tag{23}$$

$$X_i^*(P) = \frac{\text{Max } x_i(P) - x_i(P)}{\text{Max } x_i(P) - \text{Min } x_i(P)} \quad \text{For } (i = 1,2\ldots m, \quad P = 1,2\ldots n) \tag{24}$$

Where m is the total number of experiments, n is the total number of observed data. The normalized values are used to calculate the Grey relational coefficient (GRC)

$$\xi_i(P) = \frac{\Delta Min + \zeta \Delta Max}{\Delta_{oi}(P) + \zeta \Delta Max} \tag{25}$$

Here $\Delta_{oi}(P)$ deviation sequence is calculated from reference sequence $X_0^*(P)$ and comparability sequence $X_i^*(P)$. The range $0 \leq \xi \leq 1$ have used for the distinguishing coefficient

$$\gamma_i = \frac{1}{n} \sum_{P=1}^{n} \xi_i(P) \tag{26}$$

A Grey relational grade (γi_j is a weighted summation of grey relational coefficients which is represented in the equation. γi_is helps to find the correlation of reference and comparability values.

Finally, the database in ECMM shows that many authors have handled problems for improving MRR, surface quality and reducing the overcut. Multiobjective genetic algorithm (MOGA) optimization technique is applied for enhancing material removal and in EMM. They considered high carbon high chromium die tool steel (HCHCr die tool steel) with the use of copper nano particles mixed sodium nitrate electrolyte (Sathyamoorthy et al.2015). Apart from, mixed electrolyte and hybrid techniques, tool design optimization plays a key role in EMM process. To attain the accurate shape of tool, various optimization techniques such as grey relational analysis, NSGA –II, PSO-DF and RSM are used in this process (Thanigaivalean et al.2013). Also simple optimization techniques such as TOPSIS, GRA, MOORA, VIKOR and COPRAS attract the researchers.

Krishnan et al.(2020) optimized the process parameters of ECMM using TOPSIS and VlseKriterijumska Optimizacija I Kompromisno Resenje (VIKOR) method. SS 304 work material is machined using $NaNO_3$is used as electrolyte. The experiments are planned based on Taguchi L_{16} OA. TOPSIS and VIKOR methods are used to find the performance scores and the optimal combination is found to be 21V, 80% duty cycle, 600 rpm rotating speed and 0.8 μm s^{-1} feed rate. Mouliprasanth et al.(2019) used the TOPSIS technique in ECMM to find the optimal process variables with variety of electrolytes on machining rate and accuracy. They used passivating, non-passivating and composite electrolytes in the experiment. The input parameters are connected with the output responses of ECMM and the optimal combinations for better response is 7 V, 0.5mm/min feed rate, 0.7 duty cycle with composite electrolyte. Pradeep et al. (2019) studied the ECMM performance with polymer graphite cathode with $NaNO_3$ electrolyte on stainless steel material. The performance of ECMM is evaluated in terms on machining rate and accuracy. They determined the optimal process values using TOPSIS method. As per the ANOVA results, machining voltage plays important role in ECMM which is takes 52.29% of all input variables for higher machining rate. Maniraj et al.(2019) optimized the process parameters of ECMM on ground-granulated blast furnace slag (GGBS)-reinforced aluminium metal matrix composites using TOPSIS method. L_{18} orthogonal array (OA) experimental design is used with electrical parameters and percentage of reinforcement composition on machining rate, ROC. Using regression analysis, a model equation is created, as per ANOVA, percentage of reinforcement composition plays a essential role machining performance. The optimal combination found to be 10 V, 50% duty cycle, 35 g/l electrolyte concentration and 12% of GGBS composition. Geethapriyan et al. (2019) coupled the Taguchi method with grey relational analysis method to study the ECMM performance on stainless steel. Two different tool electrodes such as copper and brass are used in the experiment. The output responses like machining rate, surface roughness and ROC are analyzed using coupled GRA method. The copper electrode develops 20.91% and 29.12% of higher machining rate and ROC respectively compared to brass electrode. Brass electrode produces 23.62% higher surface roughness over copper electrode. Subburam et al.(2018) adopted the Taguchi's L_9 OA to conduct the experiments on inconel 625 alloy with EMM process using acidified $NaNO_3$ electrolyte. The conical shape stainless steel needle is used as electrode and insulated with epoxy resin. They suggest the optimal parameter combination that16 V, 30 g/l electrolyte concentration and 45% duty cycle for higher machining rate and lower overcut. Cole et al. (2017) optimized the ECMM performance using Taguchi method. Zirconium based Bulk metallic glass (BMG) material and $NaNO_3$ is used as work and electrolyte respectively. Taguchi method used to evaluate the effect of pulsed voltage range, duty cycle and electrolyte flow rate. ANOVA and signal-noise ratios applied to optimize ECMM process parameters. Using this technique, 0.4 l/min of electrolyte flow rate, 1:10 of duty cycle and 2.235–3V machining voltage range expected to machine the quality micro holes. Mehrvar et al. (2017) used the differential evolution algorithm to optimize the process parameters of ECM with voltage, feed rate, electrolyte concentration and flow rate on machining rate, accuracy and surface roughness. The mathematical model developed for machining rate, accuracy and surface roughness using response surface methodology. Nayak et al.(2016) used the computational fluid dynamics (CFD) simulation to study the effect of electrolyte changes happening in IEG namely pressure variations, passivation changes, velocity variations, turbulence kinetic energy and temperature changes on machining responses. I shaped tool electrode and nickel based inconel 825 work materials are considered for the experiment. Based on CFD results, the pointed corner of electrode diminish the velocity distribution when the turbulence is highest. Also, effect of process parameters such voltage, electrolyte concentration and feed rate are studied and optimized using GRA method. Rao et al.(2015) ECM ed Al/B_4C composites and optimized its factors using Taguchi method with utility concept. L_{27} OA is planned using voltage, feed rate, concentration and % of reinforcement and output response measured in terms of machining rate, surface roughness and ROC. ANOVA results shows that the feed rate contributes more on responses. Das, et al. (2014) applied the GRA-Taguchi method to optimize the process variables such electrolyte concentration, voltage, feed rate and inter electrode gap on machining rate and surface roughness. Experiments are conducted with L_{27} OA on of EN31 tool steel using $NaNO_3$electrolyte. Electrolyte concentration is plays notable role on output responses as per the ANOVA. Also from the confirmation test, 48% improvement is noticed in ECMM performance. Mukherjee et al.(2013) have used the biogeography-based optimization (BBO) algorithm to optimize the factors of ECM. In this, both single and multi-response optimization models are used. The results of optimization were compared with the other algorithms based optimization techniques such as population-based algorithms i.e., genetic algorithm and artificial bee colony algorithm. Also, the authors reported that BBO algorithm is mainly suitable for ECMM than other technique because those methods consume more time to find the optimal combination.

SUMMARY

In ECMM process cathode profile plays the major role which decides the accuracy of the features machined on it. The literatures have shown many developments related to electrode such as coating, shape modification, oscillating electrodes and heated electrodes for improving the machining rate and accuracy. Still the development of cathode design for complex shapes and high aspect ratio holes are challenging due the size and handling. In ECMM the renewal of electrolyte is another focus area the researchers are working and use of rotational electrode alleviates the problems of sludge accumulation in the machining zone. The increase in rotational speed has significant effect on ROC. The increase in rotational speed efficiently renews the electrolyte and clears the sludge's. The breaking of hydrogen bubbles evolved from cathode build up micro stirring mechanism in the machining zone. The micro-hole generated at electrode rotational speed of 3000rpm with 10V, 35gm/L electrolyte concentration and 90% duty cycle shows good circular micro-hole with better surface quality without any stray current and micro spark effect in the SEM picture. Moreover it is evident from the above discussion that researchers have conducted machinability studies on alloys and AMMCs through ECMM and studied the effect of various process parameters on machining rate, surface topography and accuracy. Even though considerable contribution on ECMM machining of metal matrix composites is made by researchers there is still the gap prevails due to the advancement of new materials. Hence it is essential to explore more on machining of metal matrix composites and nickel based alloys. The optimization of process parameter is essential for the commercialization of the ECMM, hence the multi - objective optimization techniques like TOPSIS, GRA, VIKOR, MOORA, AHP, Fuzzy and ELECTRE, PROMTHEE methods can be explored. Additionally, the most significant process parameter that affects the machining is identified using ANOVA.

REFERENCES

Arunachalam, R., Thanigaivelan, R., & Devadula, S. (n.d.). *21 Advances and Applications of Nontraditional Machining Practices for Metals and Composite Materials*. Academic Press.

Bhattacharyya, B., Munda, J., & Malapati, M. (2004). Advancement in electrochemical micro-machining. *International Journal of Machine Tools & Manufacture*, *44*(15), 1577–1589. doi:10.1016/j.ijmachtools.2004.06.006

Cole, K., Kirk, D., Singh, C., & Thorpe, S. (2017). Optimizing electrochemical micromachining parameters for Zr-based bulk metallic glass. *Journal of Manufacturing Processes*, *25*, 227–234. doi:10.1016/j.jmapro.2016.11.015

Cong, P., Xu, G., Wei, T., & Li, K. (2020). A survey of profit optimization techniques for cloud providers. *ACM Computing Surveys*, *53*(2), 1–35.

Das, M.K., Kumar, K., Barman, T.K., &Sahoo, P. (2014). Optimization of surface roughness and MRR in electrochemical machining of EN31 tool steel using grey-Taguchi approach. *Procedia Materials Science, 6*, 729-740.

Einhorn, H. J., & McCoach, W. (1977). A simple multiattribute utility procedure for evaluation. *Behavioral Science*, *22*(4), 270–282. doi:10.1002/bs.3830220405

Geethapriyan, T., Lakshmanan, P., Prakash, M., Iqbal, U. M., & Suraj, S. (2019). Influence of Tool Electrodes on Machinability of Stainless Steel 420 Using Electrochemical Micromachining Process. In *Advances in Manufacturing Processes* (pp. 441–456). Springer. doi:10.1007/978-981-13-1724-8_43

Gunantara, N., Sastra, N. P., & Hendrantoro, G. (2014). Cooperative Diversity Paths selection Protocol With multi-Objective Criterion In Wireless Ad-Hoc Networks. *International Journal of Applied Engineering Research: IJAER*, *9*(23), 22395–22407.

Jain, V. K., Kalia, S., Sidpara, A., & Kulkarni, V. N. (2012). Fabrication of micro-features and micro-tools using electrochemical micromachining. *International Journal of Advanced Manufacturing Technology*, *61*(9), 1175–1183. doi:10.100700170-012-4088-1

Krishnan, N., Deepak, J., & Hariharan, P. (2020). Multi-response optimization of electrochemical micromachining on masked SS304. *Engineering Research Express*, *2*(1), 015041. doi:10.1088/2631-8695/ab5eb9

Leese, R. J., & Ivanov, A. (2016). Electrochemical micromachining: An introduction. *Advances in Mechanical Engineering*, *8*(1), 1687814015626860. doi:10.1177/1687814015626860

Liu, G., Tong, H., Li, Y., & Zhong, H. (2021). Novel structure of a sidewall-insulated hollow electrode for micro electrochemical machining. *Precision Engineering*, *72*, 356–369. doi:10.1016/j.precisioneng.2021.05.009

Maniraj, S., & Thanigaivelan, R. (2019). Effect of electrode heating on performance of electrochemical micromachining. *Materials and Manufacturing Processes*, *34*(13), 1494–1501. doi:10.1080/10426914.2019.1655153

Maniraj, S., & Thanigaivelan, R. (2019). Optimization of electrochemical micromachining process parameters for machining of AMCs with different% compositions of GGBS using taguchi and TOPSIS methods. *Transactions of the Indian Institute of Metals*, *72*(12), 3057–3066. doi:10.100712666-019-01772-3

Mehrvar, A., Basti, A., & Jamali, A. (2017). Optimization of electrochemical machining process parameters: Combining response surface methodology and differential evolution algorithm. *Proceedings of the Institution of Mechanical Engineers. Part E, Journal of Process Mechanical Engineering*, *231*(6), 1114–1126. doi:10.1177/0954408916656387

Mouliprasanth, B., & Hariharan, P. (2019). Measurement of Performance and Geometrical Features in Electrochemical Micromachining of SS304 Alloy. *Experimental Techniques*, 1–15.

Mukherjee, R., & Chakraborty, S. (2013). Selection of the optimal electrochemical machining process parameters using biogeography-based optimization algorithm. *International Journal of Advanced Manufacturing Technology*, *64*(5-8), 781–791. doi:10.100700170-012-4060-0

Pradeep, N., Sundaram, K. S., & Kumar, M. P. (2019). Multi-response optimization of electrochemical micromachining parameters for SS304 using polymer graphite electrode with $NaNO_3$ electrolyte based on TOPSIS technique. *Journal of the Brazilian Society of Mechanical Sciences and Engineering*, *41*(8), 323. doi:10.100740430-019-1823-7

Rajan, N., Thanigaivelan, R., & Muthurajan, K. G. (2019). Machinability studies on an a17075 composite with varying amounts of b4c using an induction-heated electrolyte in electrochemical machining. *Materiali in Tehnologije*, *53*(6), 873–880. doi:10.17222/mit.2019.077

Rajkeerthi, E., Hariharan, P., & Pradeep, N. (2021). Enhancing machining rate and geometrical accuracy in electrochemical micromachining of Co-Ni-Cr-W superalloy. *Materials and Manufacturing Processes*, *36*(4), 488–500. doi:10.1080/10426914.2020.1843672

Rao, S. R., & Padmanabhan, G. (2015). Parametric optimization in electrochemical machining using utility based Taguchi method. *Journal of Engineering Science and Technology*, *10*(1), 81–96.

Robyn, M. (1974). Dawes and Bernard Corrigan. Linear models in decision making. *Psychological Bulletin*, *81*(2), 95–106. doi:10.1037/h0037613

Sathiyamoorthy, V., Sekar, T., & Elango, N. (2015). Optimization of processing parameters in ECM of die tool steel using nanofluid by multiobjective genetic algorithm. *TheScientificWorldJournal*, *2015*, 2356–6140. doi:10.1155/2015/895696 PMID:26167538

Shunmugam, M. S., & Kanthababu, M. (2019). *Advances in Micro and Nano Manufacturing and Surface Engineering*. Academic Press.

Soundarrajan, M., & Thanigaivelan, R. (2020). Effect of coated and geometrically modified tools on performance of electrochemical micromachining. *Materials and Manufacturing Processes*, *35*(7), 775–782. doi:10.1080/10426914.2020.1740252

Soundarrajan, M., & Thanigaivelan, R. (2021). Electrochemical Micromachining of Copper Alloy through Hot Air Assisted Electrolyte Approach. *Russian Journal of Electrochemistry*, *57*(2), 172–182. doi:10.1134/S1023193521020117

Subburam, V., Ramesh, S., Kumar, P. M., & Srinivasan, A. (2018). Performance optimisation of electrochemical micromachining of micro-holes on Inconel 625 alloy. *International Journal of Precision Technology*, *8*(1), 66–84. doi:10.1504/IJPTECH.2018.094083

Swain, A. K., Sundaram, M. M., & Rajurkar, K. P. (2012). Use of coated microtools in advanced manufacturing: An exploratory study in electrochemical machining (ECM) context. *Journal of Manufacturing Processes*, *14*(2), 150–159. doi:10.1016/j.jmapro.2011.11.005

Tang, L., Li, B., Yang, S., Duan, Q., & Kang, B. (2014). The effect of electrolyte current density on the electrochemical machining S-03 material. *International Journal of Advanced Manufacturing Technology*, *71*(9-12), 1825–1833. doi:10.100700170-014-5617-x

Thanigaivelan, R., Arunachalam, R. M., Kumar, M., & Dheeraj, B. P. (2018). Performance of electrochemical micromachining of copper through infrared heated electrolyte. *Materials and Manufacturing Processes*, *33*(4), 383–389. doi:10.1080/10426914.2017.1279304

Thanigaivelan, R., Arunachalam, R. M., & Subburam, V. (2010). Design and Development of a Table Top Electrochemical Micromachining Tool. *Proceedings of the 16th International Symposium on Electromachining*.

Thanigaivelan, R., Senthilkumar, R., Arunachalam, R. M., & Natarajan, N. (2017). Impact of the Shape of Electrode-tool on Radical Overcut of Micro-hole in Electrochemical Micromachining. *Surface Engineering and Applied Electrochemistry*, *53*(5), 486–492. doi:10.3103/S1068375517050143

Tsui, H. P., Hung, J. C., You, J. C., & Yan, B. H. (2008). Improvement of electrochemical microdrilling accuracy using helical tool. *Materials and Manufacturing Processes*, *23*(5), 499–505. doi:10.1080/10426910802104237

Yao, J., Chen, Z. T., Nie, Y. J., & Wang, F. (2019). Flow field design and experimental study of electrochemical machining using thin hollow cathodes. *Proceedings of the Institution of Mechanical Engineers. Part B, Journal of Engineering Manufacture*, *233*(3), 766–775. doi:10.1177/0954405417752511

Chapter 3
Surface Modification Techniques for Bio-Materials:
An Overview

Thanigaivelan R.
Muthayammal Engineering College, India

Satya Prakash
https://orcid.org/0000-0001-6303-5970
Muthayammal Engineering College, India

Maniraj S.
Paavai Engineering College, India

ABSTRACT

The increasing population demands materials with higher expectations associated to quality of life cropping up from an aging population. Recently there has been an ever-increasing search for novel biomaterials as the material needs for complex biomedical devices shoots up with time. This chapter discusses the different technology used for surface modification of material for biomedical applications. The surface modification process is generally used to obtain the antibacterial properties, antifouling, enhanced mechanical properties, reduced cytotoxicity, anti-inflammatory potential, improved hemocompatibility, and osseointegration capabilities. Various metallic materials are considered as biocompatible materials such as titanium alloys, stainless steel alloys, and magnesium alloys. Moreover, surface modification of dental implants and polymers were reviewed, and recent progress and challenges in application of biomaterials are reported related to the surface modification process.

DOI: 10.4018/978-1-7998-9574-9.ch003

INTRODUCTION

Bio-materials as a whole possesses unique property which may not be exhibited by other materials and hence in order to make them to conveniently applied for specific application the surface modification come to the existence. By using various physical and chemical means, the surface compositions and structures can be adjusted to assist seamless adaptation to the physiological surroundings and at the same time perform the necessary functions. There various techniques for surface modification of biomaterials are depicted in the figure 1. The surface modified surfaces can prevent corrosion, enhance biocompatibility, and osseointegration without sacrificing the bulk properties. The following sections discuss the different approach of surface modification for improving the responses such as antibacterial, antifouling, mechanical, corrosion and tribological, osseointegration, cytotoxicity, anti-inflammatory and hemocompatibility, moreover this chapter touches upon the surface modification of dental implants and polymers in dental application.

Figure 1. Surface modification techniques

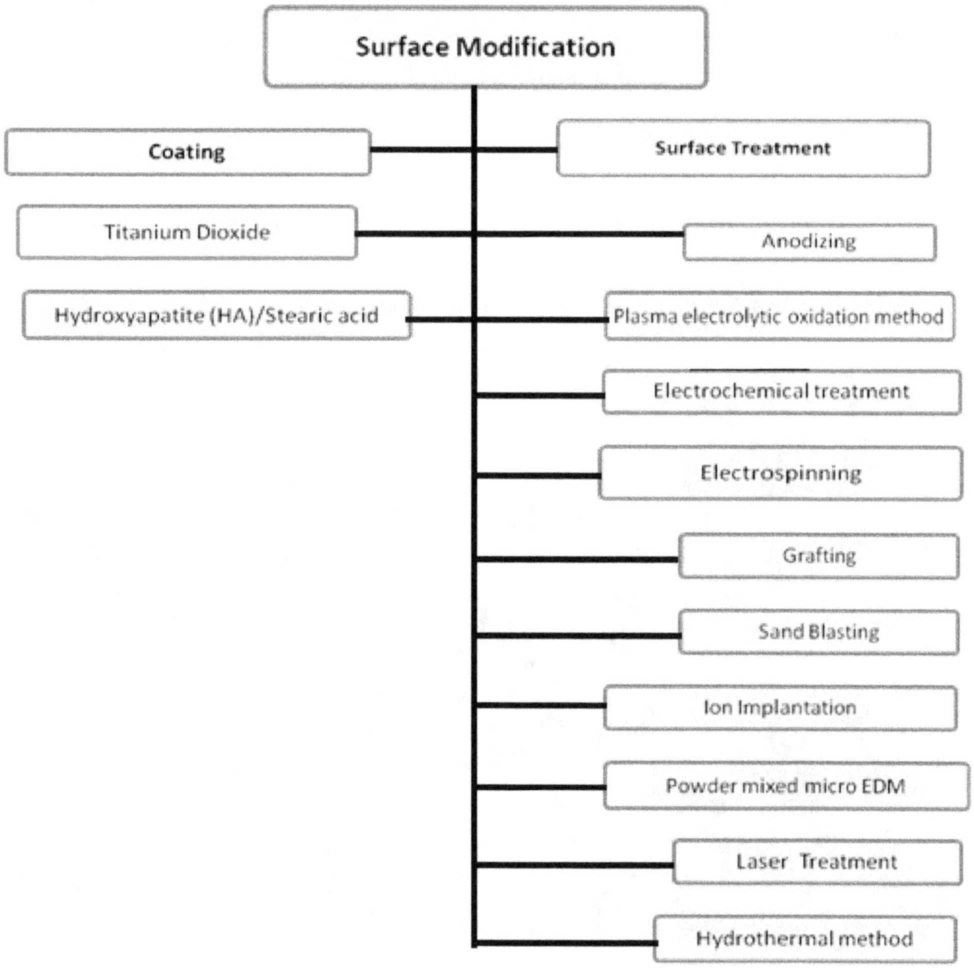

Antibacterial Response

Implants sometimes develop infection which impairs host defense mechanisms and, result in resistant chronic infection reducing the extended usage. Normally metallic implants come with the risk of bacterial colonization and implant associated infection. The common bacteria such as staphylococci, and staphylococcus aureus, have a strong affinity to colonize foreign bodies and cause implant associated infection. It is evident that surface property of the implants plays the significant role and hence modified surfaces alleviate the problems of infection.

The various suitable surface modification processes is in progress to prevent the bacterial colonization and one such process is anodizing. Anodizing is a process of creating the thickness of natural oxide layer on the surface of metals especially titanium, stainless steel, aluminium and magnesium alloys through electrolytic passivation process in which the part to be treated act as the anode electrode of an electrolytic cell. This method is found to be effective for antibacterial response. The electrochemical etching of the surface has double advantages ie., creating passive layer and development of micro and nano-structure with surface roughness.

Jang et al. (2018) have electrochemically etched the stainless steel 316L surface and generated nanopores with pore diameters of 20–25 nm and sharp nano-protrusions. Figure 2 shows the electrochemically hydroxyapatite nano-powder coated titanium sample which exhibits micro/nanopore surface. The electrochemically etched surface exhibits a superior passive layer, attributing for improved antibacterial characteristics.

Figure 2. Hydroxyapatite coated titanium surface

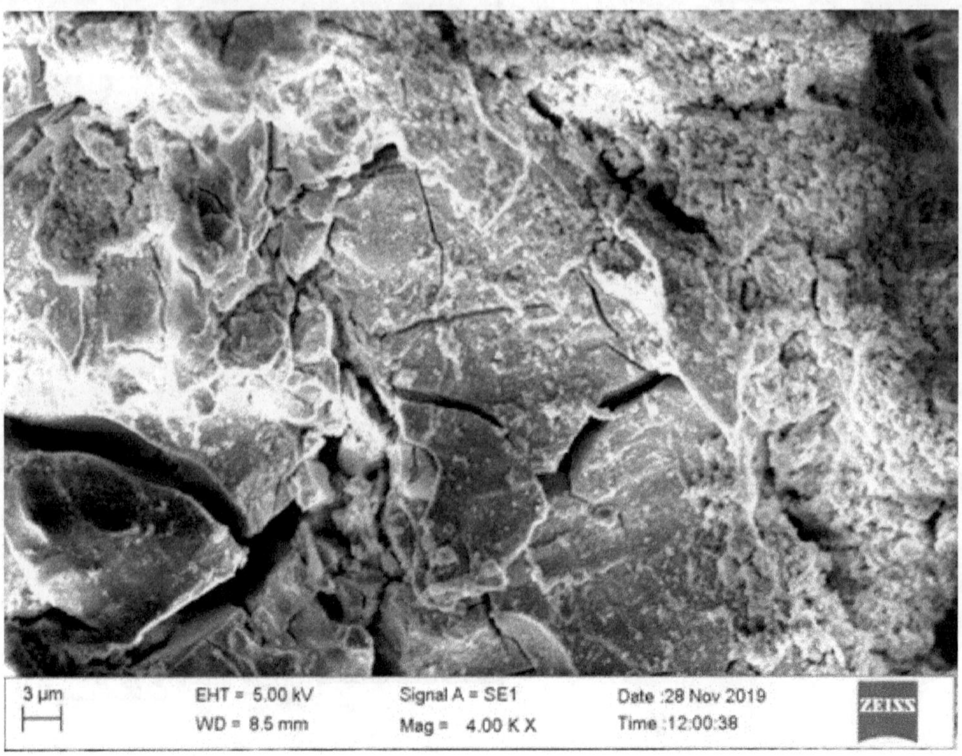

Recently hydroxyapatite coating attracts many researchers for developing the antibacterial coating. Li et al. (2021) have developed super-hydrophobic composite coating of hydroxyapatite /stearic acid using hydrothermal method. Hydrothermal coating is also a proven method to prevent the infection. It develops nano-particles on the surface involving high vapour pressure and high-temperature aqueous solution. This process enhances corrosion resistance and antibacterial adhesion by developing hydroxyapatite micro-/nano-scale structure surface. These structures induce the quick deposition of the mineralized apatite layer contributing for the short-term antibacterial adhesion property in magnesium alloy implants. Another method is layer-by-layer self-assembly method in which a thin-film of opposite charges particles is formed in alternating layers. Researchers have also considered hybrid techniques, Wang et al. (2021) have surface modified the surgical grade titanium alloy using silk fibroin, Ag and strontium for supporting bone formation and hindering osteoclasts. They considered technologies namely anodizing, hydrothermal synthesis, and layered self-assembly to convert the titanium surface as antibacterial surface. The antibacterial analysis shows that the functional coating has excellent antibacterial activity due to its inhibiting property of gram negative and gram positive bacteria at the same time (Wei et al. 2019).

Other methods used for making super-hydrophobic surfaces are templating, lithography, plasma treatment, electrochemical deposition, sol-gel deposition, wet chemical reaction, chemical vapor deposition, and electrospinning (Chakraborty et al. (2021). Li et al 2021 have developed superhydrophobic composite coating of hydroxyapatite /stearic acid on magnesium alloy (AZ31B) using hydrothermal method. The wettability, corrosion resistance and antibacterial adhesion capacity of the composite coating shows excellent super-hydrophobicity with a contact angle about 152.52° and a sliding angle about 2° (Jiang et al. 2015).

Using plasma electrolytic oxidation method the Ca, P and I doped TiO_2 coatings were developed to improve antibacterial efficiency (de Viteri, 2016). This method is used in treatment of Ti-15Mo alloy surfaces in baths containing 0.1 M calcium hypophosphite with copper(II) phosphate. The oxide coatings were developed to improve the bacteriostatic and cytocompatible properties. The surface morphology, roughness, and wettability of the oxide coatings were studied. The spectroscopic and crystallographic analysis techniques shows that the presence of copper compounds and bioactive calcium and phosphorus and the highest percentage of copper concentration (3.5 at.%) was detected in the oxide layer. The response of bacteriostatic characteristics of Cu-contained oxide layers were tested for gram-positive staphylococcus aureus and gram-negative escherichia coli. The Cu-incorporated surfaces, show bacterial colonies of ~10–100 times lower than on the reference surface (Leśniak-Ziółkowska et al 2021).

Polymers like polyethylene and polyurethane are used as medical polymers. Low temperature plasma treatment is done to activate the polymeric surface to enhance surface and adhesion properties. Using electro-spinning technique porous microstructures were developed with poly dimethylsiloxane and polyamide. Finally, natural black seed oil is sprayed on the microstructure to produce a lubricating layer. The modified PE and PU substrates shows slippery behavior moreover, the antimicrobial effects of the produced slippery liquid-infused porous surfaces using black seed oil were proven against gram-positive staphylococcus aureus and gram-negative escherichia coli (Habib et al. 2021).

Dry processes such as ion implantation (Ca^+, N^+, F^+), oxidation (anode oxidation, titania spraying), ion plating (TiN, alumina), and ion beam mixing (Ag, Sn, Zn, Pt) with Ar^+ on polished pure titanium plates is considered to study its antibacterial effect (Yoshinari et al. 2001). F^+ implanted specimens notably reduced the growth of both P. gingivalis and A. actinomycetemcomitans. The other surface-modified specimens did not show useful antibacterial activity against both bacteria.

Antifouling Response

Antimicrobial materials act as an inhibitory or lethal to encountering microorganisms, while antifouling materials prevent the adhesion of microbes and proteins. Bio-fouling occurs on surgical equipment, medical implants, and prothetic devices. Microbial contamination, and the consequent threat of infection, and implant non-acceptance, is a main motive in developing capable antifouling strategies. While bio-fouling is related to proteins absorption, inhibition of protein adsorption avoids the reason of infection at the source. The film of adsorbed protein on surfaces provides a platform for cell attachment, and subsequent bacterial formation leads to the creation of bacterial films. There are techniques to reduce the bio-fouling first one is use of antibacterial coating which kills the bacteria and other micro-organisms. Second is to keep away the proteins, and then the cells, and thus prevent bio-fouling. Third is to develop surfaces that auto-clean so that the organisms do not get attached. Nature has developed a blending of these approaches for antifouling. The figure 2 shows the prosthetic knew with and without bio-film.

Figure 3. Prosthetic knew with and without bioflim
(Schoenmakers et al. 2021)

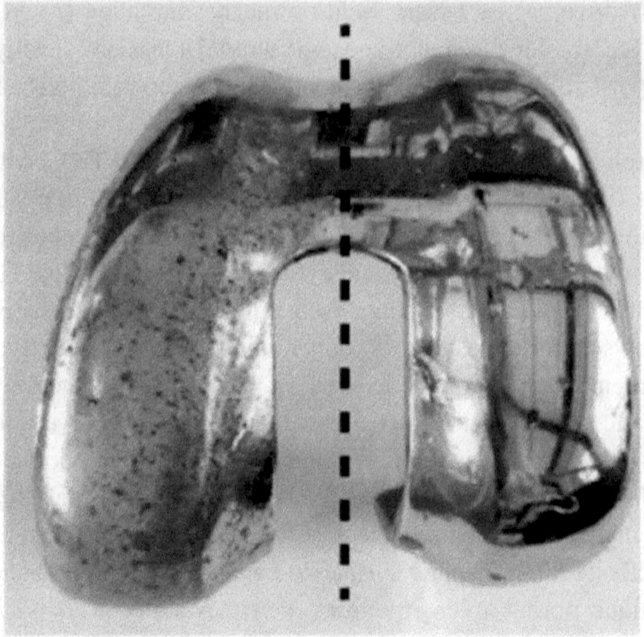

Coating of surface with nitrogen and silver ions can dissolve biofilm through a number of mechanisms together with the production of oxidative or nitrosative stress within the biofilm structure. Another method is coat the surface with heparin, albumin (hydrophilic protein) and the presence of these proteins develops the hydrophobic surfaces between the surface and bacterial cell wall, thereby preventing the initial bacterial adhesion (Harris and Richards, 2006). Surface Modification of silicone is done by developing a cross-linked poly (ethylene glycol) dimethacrylate) (P(PEGDMA)) polymer layer was covalently grafted on medical-grade silicone surface to improve its antibacterial and antifouling properties (Li et al 2012). The amalgamation of silk sericin and natural tannic acid are deposited on the titanium surface to develop eco-friendly antifouling coatings. The silk sericin which is hydrophilic nature is sericulture by product is degummed from raw silk and mostly disposed in silk processing wastewater, causing water pollution. The natural tannic acid and silk sericin were joined together via hydrogen bonding interactions, and deposited on the titanium surfaces through tri-hydroxyphenyl present in tannic acid. It is an eco-friendly way to fabricate antifouling coatings. The tannic acid / silk sericin co-deposited titanium surfaces show good protein repellent as well as platelet and bacterial anti-adhesive properties (Cheng et al., 2020).

Antifouling approach is based on the alteration of basic polymers to generate polymers with new surface characteristics. Poly-ethylene glycol is considered as an antifouling polymers but, reports show that it is not suitable for long-term applications. Therefore, several other materials such as poly vinylpyrrolidone, poly 2-oxazolines, polybetaines, poly hydroxyfunctional acrylates, polyacrylamide, heparin and phosphorylcholine have been examined for anti-adhesive applications which show the properties of decreased protein adsorption, cell adhesion/bacterial attachment and as a result reduce biofilm formation (Marques et al. 2017). A variety of antifouling polymers are being used in medical devices namely catheters, orthopedic/dental implants, vascular grafts and contact lenses. The limitation of these polymers is due to reduce/delay bio-film development on medical device surfaces, a hardly any microorganisms can in time adhere and form a bio-film. Hence anti-fouling coatings are unlike from each other; therefore the kind of coating should be personalized for a precise application.

Mechanical, Corrosion and Tribological Response

The investigation in tribology, corrosion and tribocorrosion field has been developing in the recent years due to their clinical significance in orthopedic areas. The synergistic effect of wear and corrosion on orthopedic and dental implants occurs once it is implanted in the body due to load and body fluids on the implant-bone interface (Villanueva and coauthors 2017). Researchers are improving the efficacy of the implants through different approach. Tricalcium phosphate and titanium dioxide are mixed with alumina to produce bioactive coating for improving the mechanical and tribological properties of orthopedic implants (Barkallah et al. 2021). A pin-on-disk tribometer is used to measure friction and wear studies against zirconia ball. A 2D profilometer and SEM are used to evaluate the wear mechanism. Another method, upset forging which plastically deforms metal under high pressure into high strength components of varying size. This method is considered for creating severe plastic deformation technique to produce fine-grained thick plates. The repeated upsetting process evolves the ZK60 Mg surface with microstructure, texture, bio mechanical properties, and bio-corrosion behavior after applying 3 and 5 passes of repeated upsetting process (Fakhar et al., 2021). However, the 3-passed specimen can be considered as biodegradable implant due to its good combination of acceptable strength, ductility, and corrosion resistance.

Chrome nitride medical grade 316 low carbon vaccum melt stainless steel is coated with AlCrN and Ag alloy using cathodic arc deposition method. The morphological, structural, nano-mechanical and tribological study on AlCrN/CrN coating show good results in nano-scratch and confirmed lower wear in simulated tribo-test in physiological simulated body fluid condition (Patnaik et al. 2021). Stainless steel is considered as biocompatible materials and its variety of composition shows possible application in biomedical application. A 316L stainless steel is laser dimpled and electrochemically treated for improved surface characteristics such as surface roughness, hardness and contact angle (Thanigaivelan et al., 2020). A L_9 orthogonal array experiments are designed using the dimple distance, laser power, inter-electrode gap (IEG) voltage and electrolyte concentration. The technique for order of preference by similarity to ideal solution (TOPSIS) is used to identify the best parameter combination, based on the TOPSIS analyse, the optimal combination of lowest dimple distance of 150 µm, laser power of 14 W, supply electrode voltage of 4 V and electrolyte concentration of 10 g/l of $NaNO_3$ shows surface roughness value of 2.64 µm before electrochemical treatment and 1.08 µm after electrochemical treatment, surface hardness of 538.21 N/mm^2 and contact angle of 111.1°. Figure 4 shows the micro textured surface using laser scribing. Mechanical properties such as hardness and fracture toughness were estimated using semi-circular bending and indentation tests. The study proposes that, the best mechanical and fracture properties, pertaining to micro-hardness and fracture toughness, were attained for Al_2O_3-10 wt% TCP –5 wt% TiO_2 composition.

Figure 4. Micro textured 316L stainless surface

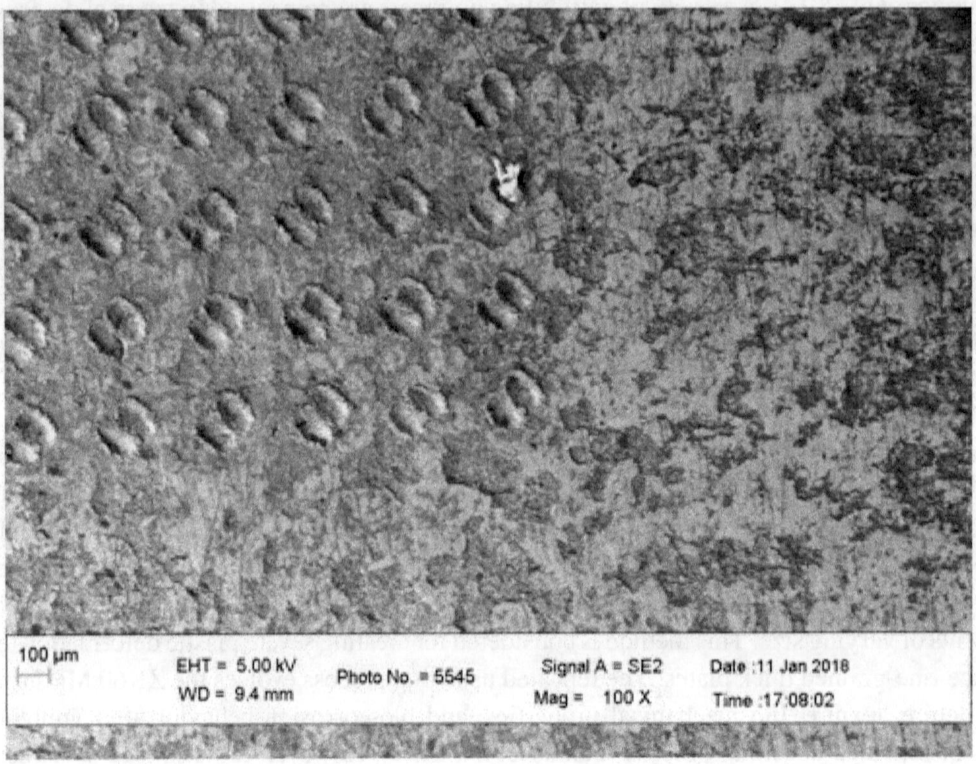

In recent days novel metals finds application in biomedical field namely gum metal with the non-toxic composition Ti–36Nb–2Ta–3Zr–0.3O (wt. %) is a comparatively new alloy which fit in to the group of metastable beta Ti alloys. Gum metal performance has been analysed in terms of its mechanical properties, corrosion resistance and cell culture response. The advantageous mechanical characteristics of gum metal are low Young's modulus, high strength and a large range of reversible deformation, which is much required for implants (Golasiński et al. 2021). The long and short duration of electrochemical treatment of gum metal showed high corrosion resistance. Evaluation of the in vitro biocompatibility of the alloys shows Increased in vitro MC3T3-E1 osteoblast viability and proliferation.

Osseointegration Response

Osseointegration is defined as a structural and functional connection between living bone and the surface of an implant. The implant tissue connection is a dynamic region of interaction. This intricate interaction not only includes biomaterial and biocompatibility issues but it also depends on the mechanical environment. To enhance osseointegration, the implant surface modification is done at nano and micro scale between the body tissue and implant. The titanium medical grade alloy is surface modified using powder-mixed electric discharge machining (Mughal et al., 2021). Pulse current, on/off time, and various silicon carbide powder concentrations are used as input parameters to modify the surfaces. Powder concentration plays an important role to control surface roughness and recast layer depth. Oxides and carbides supplemented surface improves the micro-hardness of the re-solidified layer from 320 HV to 727 HV. Moreover surface topology study shows the presences of nano-porosity (50–200 nm) which contributes for osseointegration.

Titanium samples are also roughened through grit-blasting which is a familiar process for surface treatment. In this process hard ceramic particles are casted out by compressed air at a high velocity through a nozzle. The size of the ceramic particles attributes for the surface roughness which will be ranging from 110 to 250 μm. The ceramic particles should posses properties namely stability and biocompatibility, and should not affect the in growth of bone cells. These ceramics abrasive particles after grit blasting sticks to the surfaces and cannot be removed fully after cleaning, contributes for osseointegration and reduction of corrosion resistance of implant surfaces in the body fluid environment (Wang et al.. 2020). Moreover microstructure surface of ceramic blasted implants shows smooth surface as shown in figure 5 and also the surface chemistry of ceramic blasted implants shows different characteristics compared to other coating techniques. Beta titanium alloys namely Ti10Mo8Nb alloy, are considered to be bio-inert and modifying the surface characteristics is essential to improve cell adhesion and, osseointegration. Anodic oxidation method is used to alter the alloy surface with an organic electrolyte under 20 V for 10.8 kilosecond at room temperature. The modified surface exhibits hydrophilic nano-porous layer of TiO_2 contributing for osseointegration (Carobolante et al..2020)

Biological modified surface such as cell seeding and biological coatings, promotes the osteoblast attachment, propagation, and differentiation. Diverse cells and proteins is seeded onto porous implants surface ie., bone marrow derived stem cells, embryonic stem cells and vascular endothelial growth factor. The consequence of biological techniques varies with respect to density, position, differentiation potential of the seeded cells, and the design of the substrates (Liu et al. 2020). Figure 6 presents the anodic oxidized TiO_2 surface with nano-pores which is suitable for osseointegration. The modified surface improves the cell adhesion which are mainly depends on protein adsorption and expansion of filopodium. The surface characteristics like skewness, kurtosis and spatial distribution of peaks play the important role on cell adhesion. Surfaces with negative skewness depict broader peaks (Figure.7), which facilitates cell adhesion (Rakesh Kumar et al. 2020).

Figure 5. Titanium sample blasted by grit
(Łukaszewska-Kuska et al. 2019)

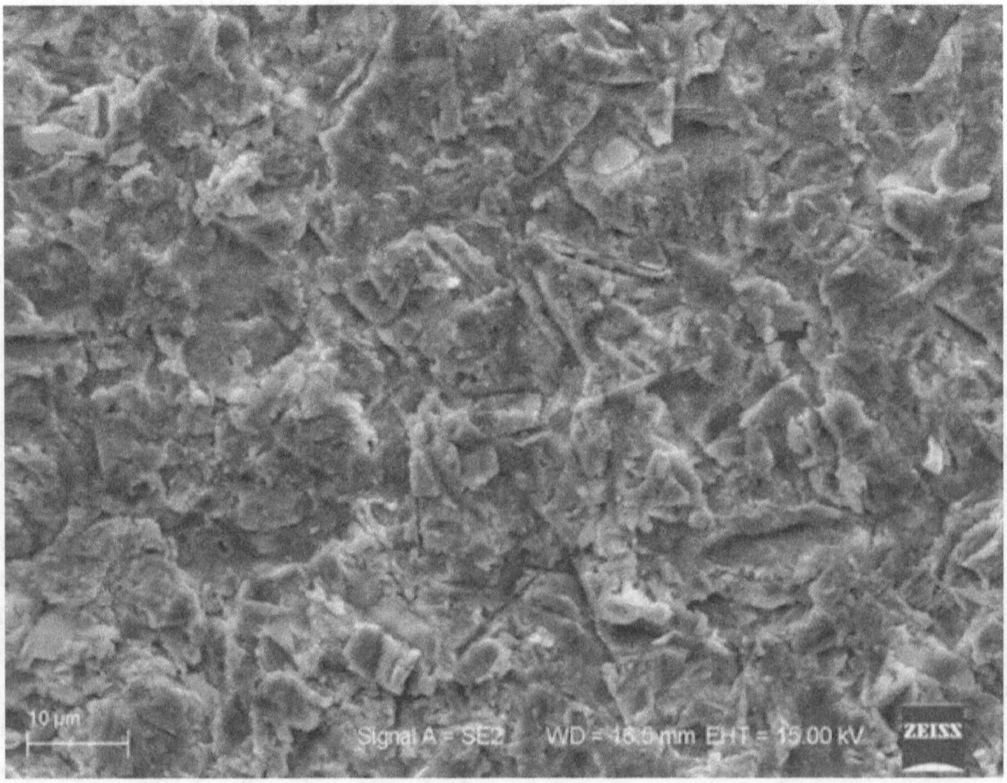

Cytotoxicity Response

Cytotoxicity refers to the ability of certain chemicals/elements to destroy living cells. Cytocompatibility can be improved through surface modification process. AZ91 magnesium alloy is coated with bioactive glass nanoparticles (~30 nm) fibre through electrospinning process (Panahi et al. 2020). The nanocomposite film thickness of 30 µm ± 5 is developed on the alloy. The surface characterization studies after 7 days of incubation show the creation and precipitation of hydroxyapatite and magnesium hydroxides on the surface of fibrous film. In addition in vitro bioactivity of coated surface, revealed better cell adhesion and cytocompatibility. Plasma assisted microwave chemical vapor deposition method is applied for surface modification of Ti-6Al-4V and γ-Ti Al alloys (Kyzioł et al. 2014). This technique improves hardness, roughness, surface energy and contact angle of the surface. Deposition of SiCNH coating on the γ-TiAl alloy surface, without plasma nitriding process, resulted in the most hydrophobic structure. The selected deposition parameters help to develop the attachment of CT26 cells and promoted their growth. The cell viability is improved on the surface-modified implants using electrochemical coating of hydroxyapatite powder and nano zinc particle on the Ti-6Al-4V alloy (Rakesh Kumar et al. 2020). The electrolyte concentration is prepared by mixing the hydroxyapatite and nano zinc particle of varying quantity and the effect of electrochemical parameter on surface coating thickness, surface quality and cell viability were studied. The study reveals that the cell viability is maximum at hydroxyapatite and nano zinc particle concentration of 6g/L and 0.8g/L respectively. The nano-zinc particle concentration in the range of 0.8 to 1.2g/L and voltage of 12-13V are appropriate for getting the controlled coating thickness. During the electrochemical process, the nano-pores hydroxyapatite structures with a pore size of 215-786 nm are obtained supports for better cells attachment on the surface. The cell viability is found to high (0.366) at 6g/L hydroxyapatite and 1.6g/L of nano zinc particle concentration at 14V.

Figure 6. Electrochemically modified nanoporous surface
(Fadl-allah et al., 2012)

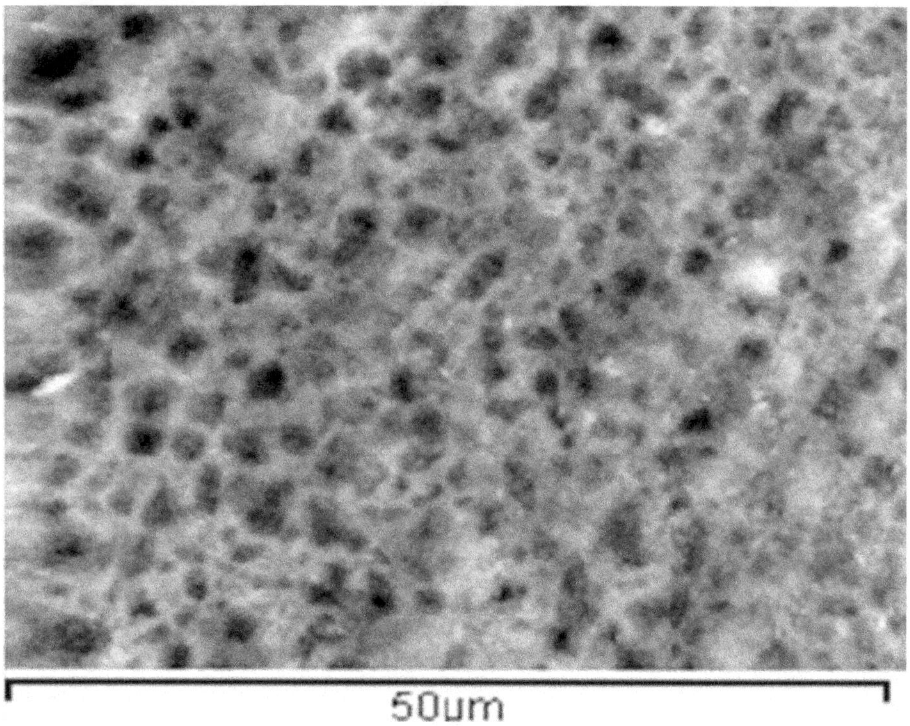

Anti-Inflammatory Response

However, achieving good biocompatibility in vivo is not an easy deal. Besides the risk of bacterial or fungal infection, an inflammatory response always occurs. In its acute phase, the inflammatory response recruits immune cells to the diseased site, resulting in the elimination of the pathogens or the damaged tissue. In its resolution phase, inflammation is helpful for the healing and the tissue regeneration. However, a severe acute or strong chronic inflammation can lead to foreign body response (FBR) and collateral tissue damage. That is why, nowadays, biomaterials researchers are developing new systems to decrease the inflammatory response to the implanted materials. A diversity of materials in diferent forms (coatings, hydrogels, nanoparticles etc.) has been produced for this purpose. Of note, interaction between diferent materials and the biological environment seems to be the key to create new systems that enable the management of the inflammatory response and lead to a decreased risk of chronic inflammation However, achieving good biocompatibility in vivo is not an easy deal. Besides the risk of bacterial or fungal infection, an inflammatory response always occurs. In its acute phase, the inflammatory response recruits immune cells to the diseased site, resulting in the elimination of the pathogens or the dam-aged tissue. In its resolution phase, inflammation is helpful for the healing and the tissue regeneration. However, a severe acute or strong chronic inflammation can lead to foreign body response (FBR) and collateral tissue damage.

Figure 7. Graph of skewness and amplitude distribution curve
(Rakesh Kumar et al., 2020)

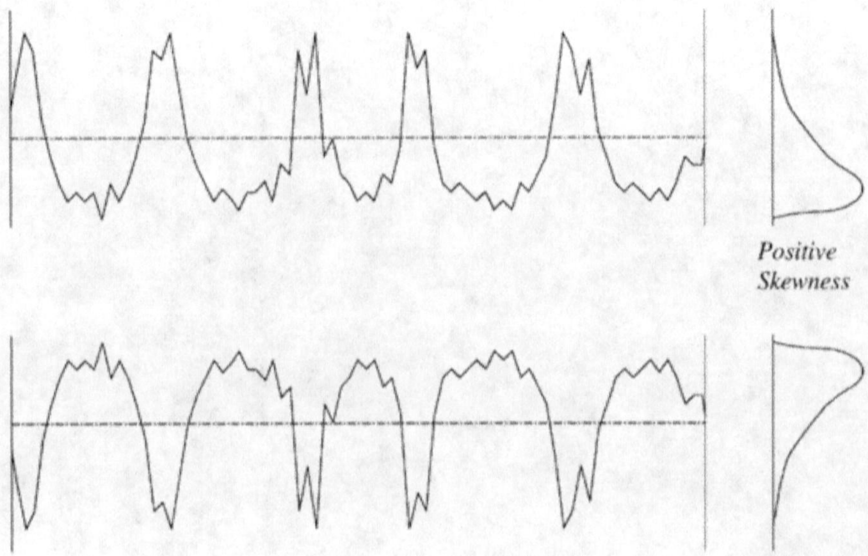

That is why, nowadays, biomaterials researchers are developing new systems to decrease the inflammatory response to the implanted materials. A diversity of materials in diferent forms (coatings, hydrogels, nanoparticles etc.) has been produced for this purpose. Of note, interaction between diferent materials and the biological environment seems to be the key to create new systems that enable the management of the inflammatory response and lead to a decreased risk of chronic inflammation However, achieving good biocompatibility in vivo is not an easy deal. Besides the risk of bacterial or fungal infection, an inflammatory response always occurs. In its acute phase, the inflammatory response recruits immune cells to the diseased site, resulting in the elimination of the pathogens or the dam- aged tissue. In its resolution phase, inflammation is helpful for the healing and the tissue regeneration. However, a severe acute or strong chronic inflammation can lead to foreign body response (FBR) and collateral tissue damage. That is why, nowadays, biomaterials researchers are developing new systems to decrease the inflammatory response to the implanted materials.

A diversity of materials in diferent forms (coatings, hydrogels, nanoparticles etc.) has been produced for this purpose. Of note, interaction between diferent materials and the biological environment seems to be the key to create new systems that enable the management of the inflammatory response and lead to a decreased risk of chronic inflammation. To achieve good biocompatibility in vivo is not an easy deal and there always inflammatory response occurs due to bacterial or fungal infection. In its severe phase, the inflammatory response sends immune cells to the diseased site, resulting in the exclusion of the pathogens or the damaged tissue resulting in healing and the tissue regeneration. On other hand, a strong chronic inflammation can cause foreign body response and tissue damage. Hence, nowadays, biomaterials researchers are evolving new systems to decrease the inflammatory response to the implanted materials. A variety of materials in different forms (coatings, hydrogels, nanoparticles etc.) has been developed for this purpose. A new modified titanium material with anti-inflammatory properties is developed using low-temperature plasma jet technology (Wang et al..2020). The surface roughness and corrosion resistance have got increased. The bacterial adhesion experiment established that the sample suppress the bacterial adhesion after plasma treatment. Other surface treatment methods commonly used in order to improve the roughness and the hydrophilicity of the titanium surface is a sandblast/acid etching. The formation of new structure limits adhesion of macrophages onto the surface, reducing the inflammatory reaction and enhancing the integration of the implant. Sometimes chemical composition of the materials also contributes for inflammatory response (Lebaudy et al. 2021).

Synthetic polymers like polyethylene, polystyrene, polyetheretherketone, poly (methyl methacrylate), etc. are also subjected to the anti inflammatory studies. In a study by Rostam et al..2016, polystyrene sample is surface treated using O_2 plasma in order to oxidize the surface and make it hydrophilic. The untreated and the treated surface with similar roughness showed that the hydrophobic surface of untreated surface results with higher anti-inflammatory response. These results disagree with the results obtained for the titanium rough material. The studies show that, the highest anti-inflammatory response was obtained for rough hydrophilic titanium surface.

Techniques for Improving Hemocompatibility

Hemocompatibility is essential for medical devices or materials contacting blood. There are different techniques are used to improve the hemocompatibility characteristics of biomaterial surfaces. Researchers have used micro-arc oxidation technique and following super-hydrophobic treatment for developing super hydrophobic TiO_2 coatings on Ti-6Al-4V alloys. The modified surface is studied for surface morphology, surface roughness, water contact angle, corrosion resistance and hemocompatibility. The rough and porous micrometer-size structure TiO_2 coating on Ti-6Al-4V alloy is developed. Further during hydrophobic treatment process the TiO_2 is grafted with low surface energy film resulting in the formation of super-hydrophobic surfaces with the water contact angle of 153.39° (Jiang et al. 2015). The hemolysis ratio and platelets attachment characteristics of the Ti-6Al-4V alloys were enhanced through the micro-arc oxidation treatment and subsequent super-hydrophobic treatment. This super-hydrophobic surface repels all the platelet and hence, the TiO_2 coatings of Ti-6Al-4V alloys with higher hemocompatibility show good scope for blood-contacting applications. In case of cardiovascular medical devices which routinely come into contact with blood, hence hemocompatibility has been considered as an important candidate. Various surface modification techniques namely heparin modification, albumin coating, surface anodization, plasma etching, and hydrothermal treatments have been investigate to improve the hemocompatibility of titanium-based materials. Still, there are several limitations related to the robustness of the surfaces and long-term efficacy in vivo. Titanium and its alloy Ti–6Al–4V were hydrothermally treated to develop nano-structured surfaces to enhance their hemocompatibility (Manivasagam et al. 2020). These modified surfaces were evaluated for their wettability, surface morphology, surface chemistry, and crystallinity. The hemocompatibility of these surfaces was characterized by evaluating blood plasma protein adsorption, platelet adhesion and activation, platelet–leukocyte complex formation, and whole blood clotting. The studies indicate that lower fibrinogen adsorption, cell adhesion, platelet activation, and whole blood clotting is seen on hydrothermally treated surfaces. Hence hydrothermal technique may be a capable approach to avoid thrombosis for several titanium blood-contacting medical devices. Presently in order to overcome the prevailing problems of blood-friendly coating materials, graphene has been proposed which is a multifunctional and find wide range of biomedical applications. The chemical unresponsiveness, atomic smoothness, and high robustness make graphene suitable for surface coating material for implantable devices. The CVD coated graphene, which is free of polymer contamination, considerably reduced platelet adhesion and activation, extended coagulation time, and decreased ex vivo thrombosis development (Meng et al. 2021). The graphene coated surface shows excellent hemocompatible properties which can be considered for next-generation cardiovascular devices.

In case of mechanical valve prosthesis, wettability gradient surface and low flow resistance, exhibit good hemocompatibility. The wettability gradient surface on a nickel-titanium (NiTi) alloy was prepared via combined laser microfabrication and surface stearic acid self-assembly approaches (Zhang et al. 2020). Scanning electron microscopy (SEM) confirms that the wettability gradient surface consist of a smooth NiTi region with different pore sizes and distances. Contact angle measurement exposed that, along with a low surface energy layer, the structural topography gradient create a wettability gradient surface on NiTi alloy which drive droplet motion. On compared with bare NiTi, such wettability gradient surface show anti-adhesion property, which is helpful to improve hemocompatibility. In another research super-hydrophobic titania nanoflower surfaces were fabricated on a titanium alloy Ti-6Al-4V substrate with hydrothermal synthesis and vapor-phase silanization (Montgomerie and Popat 2021). Surface hemocompatibility was examined through lactate dehydrogenase cytotoxicity analysis, blood-plasma protein adsorption, platelet and leukocyte adhesion and activation, and whole blood clotting analysis. The results indicated a reduction of protein adsorption, platelet and leukocyte adhesion and activation, bacterial adhesion, and biofilm formation as well as improved contact angle stability compared to control surfaces.

Surface Modification of Dental Implants

The dental implants may suffer various complications such as mechanical, biological, or technical complications. There are three categories of implant materials namely metals, ceramics and polymers are considered for dental applications. The some of the desirable properties of the dental implants are biotolerant, bioinert and bioactive. The materials possess these characteristics is considered for the dental implants otherwise there surface properties are varied to the requirements. There is various surface modification methods are considered to make the materials biocompatible. In dental implants loosening of abutment screws is one of the most common problems and it's countered by plasma nitriding treatment (Sun et al. 2020).

Laser has been applied on the surface of alumina toughened zirconia to improve the cellular response of this ceramic. Micro-grooves and grid-like geometries were prepared on this ceramics which show improved cell response, with increased metabolic activity (Carvalho et al., 2020). Nano zirconia coatings possess excellent bio-compatibility properties which makes it a suitable candidate for dental applications. A nano-zirconia coating was performed on 316L stainless steel using electrophoretic deposition with 2% suspension in Isopropanol. The electrochemical performance of nano-zirconia on artificial saliva shows corrosion-resistant behaviour (Mohandoss and cowokers.2020).

Recently, the γ-TiAl alloy was considered as the most reliable candidates for the preparation of dental implants because of its excellent mechanical strength, chemical stability and biocompatibility. Coarse and ultra-fine grained Ti grade 4 shows significant grain size refinement which increase the proof stress by 65% and fatigue endurance limit increased from 523 MPa to 698 MPa (Fintová et al., 2020). The influence of sandblasting combined with acid etching and results confirm better mechanical resistance of ultra-fine grained materials for dental implants in the comparison with coarse-grain one. Researchers have considered the effects of high-speed milling assisted the minimum quantity of lubrication, micro wire electrical discharge machining, laser and sandblasting/large grit /acid-etching treatments on surface morphology, topography, chemical composition, wettability and biofilm-associated infections on the surface. These processes create different effect (more irregular surfaces) on the surface namely micro wire electrical discharge and laser machining. Results show that micro wire electric discharge machining and laser-treated surfaces revealed hydrophobic behavior. A significant decrease of biofilm generation is seen on micro wire electric discharge machined surface due to the hydrophobicity and existence of the copper element in the recast layer. Surface analysis confirms that the zirconium, silicon, and fluorine elements on γ-TiAl surface have a direct effect on the anti-bacterial activity (Koopaie et al., 2020).

Surface Modification of Polymers for Dental Application

The polymers used in dental application should have higher surface hydrophilicity and wettability and these properties can be developed by chemical polar groups on surfaces. Introduction of carboxyl, hydroxyl, or amine groups on a surface results in higher wettability, and more hydrophilic surface. The plasma surface modification including sputtering, deposition, and implantation are also considered for this purpose. Grafting of polymers onto the surfaces is achieved by graft polymerization. Grafting can be divided into chemical grafting, mechano-chemical grafting and radiation induced-grafting. Radiation-induced grafting can be obtained by using different kinds of radiation. High-energy radiation (γ-rays, X-rays and electron beams), mid energy radiation (UV rays, laser or plasma sources) and low-energy radiation (infrared, ultrasonic, microwave (MW) and visible sources) which brings desired changes in the polymers, based on the irradiation time and energy of radiation (Koodaryan and Hafezeqoran 2016). It is evident that research in the field combined methods or used methods that are not always straight forward to classify. Wet-chemical methods are acceptable for surface modification which involves the reaction between a chemical compound in solution and a surface. Treatment with hydrogen peroxide is considered a wet chemical method. The disadvantage of this method is that reactions are non-specific, degree of surface modification may not be repeatable and irregular surface etching. Another method is oxidization of the surface of a biomaterial by exposing it to ozone. UV-ozone treatment of different materials such as polyethylene (PE), poly(etheretherketone) and poly (vinyl fluoride) were studied. Polymers used in prosthetic materials are consisting of polymethacrylate and polydimethylsiloxane which are considered as denture base material and soft liners (Delfi et al. 2020). To overcome bonding challenges between these polymeric materials, several methods have been developed with the aim of surface alteration. Different mechanical and chemical surface treatments have been applied to improve the bond strength of silicone soft liner to polymethacrylate. Plasma surface treatment can considered to improve the bond strength between two different types of resins.

SUMMARY

Among the various surface modification techniques the oxide layer generation through anodizing improves the antibacterial response and hydroxyapatite coating and electrochemical method improves the cell adhesion property. Coating of surface with nitrogen and silver ions, natural tannic acid and silk sericin improves antifouling property. Grit blasting method found suitable for osseointegration of implants. Bioactive glass nano-particles support the anti-inflammatory properties. In a nut shell application of surface modification techniques sole depends on the material property, strength adaptability, reliability and cost. In vivo environment the implants behaviour changes with the biochemical reactions. Hence it is important to entirely realize the principles and effects of many modification techniques in vivo environment, and their modern advances proceeding to the application.

REFERENCES

Barkallah, R., Taktak, R., Guermazi, N., Elleuch, K., & bouaziz, J. (2021). Mechanical properties and wear behaviour of alumina/tricalcium phosphate/titania ceramics as coating for orthopedic implant. *Engineering Fracture Mechanics*, *241*, 107399. doi:10.1016/j.engfracmech.2020.107399

Carbolated, J. P. A., da Silva, K. B., Chaves, J. A. M., Netipanyj, M. F. D., Popat, K. C., & Claro, A. P. R. A. (2020). Nanoporous layer formation on the Ti10Mo8Nb alloy surface using anodic oxidation. *Surface and Coatings Technology*, *386*, 125467. doi:10.1016/j.surfcoat.2020.125467

Carvalho, A., Grenho, L., Fernandes, M. H., Daskalova, A., Trifonov, A., Buchvarov, I., & Monteiro, F. J. (2020). Femtosecond laser microstructuring of alumina toughened zirconia for surface functionalization of dental implants. *Ceramics International*, *46*(2), 1383–1389. doi:10.1016/j.ceramint.2019.09.101

Chakraborty, A., Mulroney, A. T., & Gupta, M. C. (2021). Superhydrophobic Surfaces by Microtexturing: A Critical Review. *Reviews of Adhesion and Adhesives*, *9*(1), 35–64.

Cheng, Y. F., Mei, Y. H., Sathishkumar, G., Lu, Z. S., Li, C. M., Wang, F., Xia, Q. Y., & Xu, L. Q. (2020). Tannic acid-assisted deposition of silk sericin on the titanium surfaces for antifouling application. *Colloid and Interface Science Communications*, *35*, 100241. doi:10.1016/j.colcom.2020.100241

de Viteri, V. S., Bayón, R., Igartua, A., Barandika, G., Moreno, J. E., Peremarch, C. P. J., & Pérez, M. M. (2016). Structure, tribocorrosion and biocide characterization of Ca, P and I containing TiO2 coatings developed by plasma electrolytic oxidation. *Applied Surface Science*, *367*, 1–10. doi:10.1016/j.apsusc.2016.01.145

Delfi, M., Ghomi, M., Zarrabi, A., Mohammadinejad, R., Taraghdari, Z. B., Ashrafizadeh, M., Zare, E. N., Agarwal, T., Padil, V. V. T., Mokhtari, B., Rossi, F., Perale, G., Sillanpaa, M., Borzacchiello, A., Kumar Maiti, T., & Makvandi, P. (2020). Functionalization of polymers and nanomaterials for biomedical applications: Antimicrobial platforms and drug carriers. *Prosthesis*, *2*(2), 117–139. doi:10.3390/prosthesis2020012

Fadl-allah, S. A., Quahtany, M., & El-Shenawy, N. S. (2013). *Surface modification of titanium plate with anodic oxidation and its application in bone growth*. Academic Press.

Fakhar, N., & Sabbaghian, M. (2021). A good combination of ductility, strength, and corrosion resistance of fine-grained ZK60 magnesium alloy produced by repeated upsetting process for biodegradable applications. *Journal of Alloys and Compounds*, *862*, 158334. doi:10.1016/j.jallcom.2020.158334

Fintová, S., Kuběna, I., Palán, J., Mertová, K., Duchek, M., Hutař, P., Pastorek, F., & Kunz, L. (2020). Influence of sandblasting and acid etching on fatigue properties of ultra-fine grained Ti grade 4 for dental implants. *Journal of the Mechanical Behavior of Biomedical Materials*, *111*, 104016. doi:10.1016/j.jmbbm.2020.104016 PMID:32799132

Golasiński, K. M., Detsch, R., Szklarska, M., Łosiewicz, B., Zubko, M., Mackiewicz, S., Pieczyska, E. A., & Boccaccini, A. R. (2021). Evaluation of mechanical properties, in vitro corrosion resistance and biocompatibility of Gum Metal in the context of implant applications. *Journal of the Mechanical Behavior of Biomedical Materials*, *115*, 104289. doi:10.1016/j.jmbbm.2020.104289 PMID:33388535

Habib, S., Zavahir, S., Abusrafa, A. E., Abdulkareem, A., Sobolčiak, P., Lehocky, M., Vesela, D., Humpolíček, P., & Popelka, A. (2021). Slippery Liquid-Infused Porous Polymeric Surfaces Based on Natural Oil with Antimicrobial Effect. *Polymers*, *13*(2), 206. doi:10.3390/polym13020206 PMID:33430082

Harris, L. G., & Richards, R. G. (2006). Staphylococci and implant surfaces: A review. *Injury*, *37*(2), S3–S14. doi:10.1016/j.injury.2006.04.003 PMID:16651069

Jang, Y., Choi, W. T., Johnson, C. T., García, A. J., Singh, P. M., Breedveld, V., & Champion, J. A. (2018). Inhibition of bacterial adhesion on nanotextured stainless steel 316L by electrochemical etching. *ACS Biomaterials Science & Engineering*, *4*(1), 90–97. doi:10.1021/acsbiomaterials.7b00544 PMID:29333490

Jiang, J. Y., Xu, J. L., Liu, Z. H., Deng, L., Sun, B., Liu, S. D., Wang, L., & Liu, H. Y. (2015). Preparation, corrosion resistance and hemocompatibility of the superhydrophobic TiO_2 coatings on biomedical Ti-6Al-4V alloys. *Applied Surface Science*, *347*, 591–595. doi:10.1016/j.apsusc.2015.04.075

Jiang, J. Y., Xu, J. L., Liu, Z. H., Deng, L., Sun, B., Liu, S. D., Wang, L., & Liu, H. Y. (2015). Preparation, corrosion resistance and hemocompatibility of the superhydrophobic TiO2 coatings on biomedical Ti-6Al-4V alloys. *Applied Surface Science*, *347*, 591–595. doi:10.1016/j.apsusc.2015.04.075

Koodaryan, R., & Hafezeqoran, A. (2016). Surface modification of dental polymers by plasma treatment: A review. *Biomedical & Pharmacology Journal*, *9*(1), 317–321. doi:10.13005/bpj/942

Koopaie, M., Bordbar-Khiabani, A., Kolahdooz, S., Darbandsari, A. K., & Mozafari, M. (2020). Advanced surface treatment techniques counteract biofilm-associated infections on dental implants. *Materials Research Express*, *7*(1), 015417. doi:10.1088/2053-1591/ab6a57

Kumar, R., Thanigaivelan, R., Rajanikant, G. K., Jagadeesha, T., & Das, J. (2021). Evaluation of hydroxyapatite-and zinc-coated Ti-6Al-4V surface for biomedical application using electrochemical process. *Journal of the Australian Ceramic Society*, *57*(1), 107–116. doi:10.100741779-020-00517-6

Kyzioł, K., Kaczmarek, Ł., Brzezinka, G., & Kyzioł, A. (2014). Structure, characterization and cytotoxicity study on plasma surface modified Ti–6Al–4V and γ-TiAl alloys. *Chemical Engineering Journal*, *240*, 516–526. doi:10.1016/j.cej.2013.10.091

Lebaudy, E., Fournel, S., Lavalle, P., Vrana, N. E., & Gribova, V. (2021). Recent Advances in Antiinflammatory Material Design. *Advanced Healthcare Materials*, *10*(1), 2001373. doi:10.1002/adhm.202001373 PMID:33052031

Leśniak-Ziółkowska, K., Kazek-Kęsik, A., Rokosz, K., Raaen, S., Stolarczyk, A., Krok-Borkowicz, M., Pamuła, E., & Simka, W. (2021). Plasma electrolytic oxidation as an effective tool for production of copper incorporated bacteriostatic coatings on Ti-15Mo alloy. *Applied Surface Science*, *563*, 150284. doi:10.1016/j.apsusc.2021.150284

Li, M., Neoh, K. G., Xu, L. Q., Wang, R., Kang, E. T., Lau, T., Olszyna, D. P., & Chiong, E. (2012). Surface modification of silicone for biomedical applications requiring long-term antibacterial, antifouling, and hemocompatible properties. *Langmuir*, *28*(47), 16408–16422. doi:10.1021/la303438t PMID:23121175

Li, Q., Bao, X., Cai, S., Xie, Y., Liu, Y., Liu, J., & Xu, G. (2021). Fabrication of superhydrophobic composite coating of hydroxyapatite/stearic acid on magnesium alloy and its corrosion resistance, antibacterial adhesion. *Journal of Materials Science*, *56*(8), 5233–5249. doi:10.100710853-020-05592-5

Li, Q., Bao, X., Sun, J., Cai, S., Xie, Y., Liu, Y., Liu, J., & Xu, G. (2021). Fabrication of superhydrophobic composite coating of hydroxyapatite/stearic acid on magnesium alloy and its corrosion resistance, antibacterial adhesion. *Journal of Materials Science*, *56*(8), 5233–5249. doi:10.100710853-020-05592-5

Liu, Y., Rath, B., Tingart, M., & Eschweiler, J. (2020). Role of implants surface modification in osseointegration: A systematic review. *Journal of Biomedical Materials Research*, *108A*(3), 470–484. doi:10.1002/jbm.a.36829 PMID:31664764

Łukaszewska-Kuska, M., Krawczyk, P., Martyla, A., Hędzelek, W., & Dorocka-Bobkowska, B. (2019). Effects of a hydroxyapatite coating on the stability of endosseous implants in rabbit tibiae. *Dental and Medical Problems*, *56*(2), 123–129. doi:10.17219/dmp/103851 PMID:31274249

Manivasagam, V. K., & Popat, K. C. (2020). In Vitro Investigation of Hemocompatibility of Hydrothermally Treated Titanium and Titanium Alloy Surfaces. *ACS Omega*, *5*(14), 8108–8120. doi:10.1021/acsomega.0c00281 PMID:32309720

Marques, A. B. L. R. D. S. (2017). *Antifouling surfaces as a strategy to prevent microorganism's adhesion on medical devices* (Doctoral dissertation).

Meng, X., Cheng, Y., Wang, P., Chen, K., Chen, Z., Liu, X., Fu, X., Wang, K., Liu, K., Liu, Z., & Duan, X. (2021). Enhanced Hemocompatibility of a Direct Chemical Vapor Deposition-Derived Graphene Film. *ACS Applied Materials & Interfaces*, *13*(4), 4835–4843. doi:10.1021/acsami.0c19790 PMID:33474941

Mohandoss, S., Ganeasan, M. S., Manovasuki, J., & Sridhar, T. M. (2020). Surface Modification of 316L Stainless Steel dental implants by Novel Bioinert Nano Zirconia coating. *Trends in Biomaterials & Artificial Organs*, *34*(3).

Montgomerie, Z., & Popat, K. C. (2021). Improved hemocompatibility and reduced bacterial adhesion on superhydrophobic titania nanoflower surfaces. *Materials Science and Engineering C*, *119*, 111503. doi:10.1016/j.msec.2020.111503 PMID:33321602

Mughal, M. P., Farooq, M. U., Mumtaz, J., Mia, M., Shareef, M., Javed, M., ... Pruncu, C. I. (2021). Surface modification for osseointegration of Ti6Al4V ELI using powder mixed sinking EDM. *Journal of the Mechanical Behavior of Biomedical Materials, 113*, 104145.

Panahi, Z., Tamjid, E., & Rezaei, M. (2020). Surface modification of biodegradable AZ91 magnesium alloy by electrospun polymer nanocomposite: Evaluation of in vitro degradation and cytocompatibility. *Surface and Coatings Technology*, *386*, 125461. doi:10.1016/j.surfcoat.2020.125461

Patnaik, L., Maity, S. R., & Kumar, S. (2021). Mechanical and tribological assessment of composite AlCrN or aC: Ag-based thin films for implant application. *Ceramics International*, *47*(5), 6736–6752. doi:10.1016/j.ceramint.2020.11.016

Rostam, H. M., Singh, S., Salazar, F., Magennis, P., Hook, A., Singh, T., Vrana, N. E., Alexander, M. R., & Ghaemmaghami, A. M. (2016). The impact of surface chemistry modification on macrophage polarisation. *Immunobiology*, *221*(11), 1237–1246. doi:10.1016/j.imbio.2016.06.010 PMID:27349596

Schoenmakers, J. W., Heuker, M., López-Álvarez, M., Nagengast, W. B., van Dam, G. M., van Dijl, J. M., Jutte, P. C., & van Oosten, M. (2021). Image-guided in situ detection of bacterial biofilms in a human prosthetic knee infection model: A feasibility study for clinical diagnosis of prosthetic joint infections. *European Journal of Nuclear Medicine and Molecular Imaging*, *48*(3), 757–767. doi:10.100700259-020-04982-w PMID:32901352

Sun, F., Wang, L., Li, X. C., Cheng, W., Lin, Z., Ba, D. C., Song, G.-Q., & Sun, C. S. (2020). Effect of surface modification on the long-term stability of dental implant abutment screws by plasma nitriding treatment. *Surface and Coatings Technology*, *399*, 126089. doi:10.1016/j.surfcoat.2020.126089

Thanigaivelan, R., Arunachalam, R. M., Nithish, A., Venkatesh, S., Naveenkumar, P., & Selvaganapathy, S., & Aravind, A. S. (2020). Optimization of Laser and Electrochemical Process Parameters for Surface Modification of Hardness and Hydrophobicity on 316L Steel. *Lasers in Engineering*, 45.

Villanueva, J., Trino, L., Thomas, J., Bijukumar, D., Royhman, D., Stack, M. M., & Mathew, M. T. (2017). Corrosion, tribology, and tribocorrosion research in biomedical implants: Progressive trend in the published literature. *Journal of Bio- and Tribo-Corrosion*, *3*(1), 1–8. doi:10.100740735-016-0060-1

Wang, B., Wu, Z., Lan, J., Li, Y., Xie, L., Huang, X., Zhang, A., Qiao, H., Chang, X., Lin, H., Zhang, H., Li, T., & Huang, Y. (2021). Surface modification of titanium implants by silk fibroin/Ag co-functionalized strontium titanate nanotubes for inhibition of bacterial-associated infection and enhancement of in vivo osseointegration. *Surface and Coatings Technology*, *405*, 126700. doi:10.1016/j.surfcoat.2020.126700

Wang, L., Wang, W., Zhao, H., Liu, Y., Liu, J., & Bai, N. (2020). Bioactive effects of low-temperature argon–oxygen plasma on a titanium implant surface. *ACS Omega*, *5*(8), 3996–4003. doi:10.1021/acsomega.9b03504 PMID:32149226

Wang, Q., Zhou, P., Liu, S., Attarilar, S., Ma, R. L. W., Zhong, Y., & Wang, L. (2020). Multi-scale surface treatments of titanium implants for rapid osseointegration: A review. *Nanomaterials (Basel, Switzerland)*, *10*(6), 1244. doi:10.3390/nano10061244 PMID:32604854

Wei, T., Yu, Q., & Chen, H. (2019). Responsive and synergistic antibacterial coatings: Fighting against bacteria in a smart and effective way. *Advanced Healthcare Materials*, *8*(3), 1801381. doi:10.1002/adhm.201801381 PMID:30609261

Yoshinari, M., Oda, Y., Kato, T., & Okuda, K. (2001). Influence of surface modifications to titanium on antibacterial activity in vitro. *Biomaterials*, *22*(14), 2043–2048. doi:10.1016/S0142-9612(00)00392-6 PMID:11426884

Zhang, Q., Dong, J., Peng, M., Yang, Z., Wan, Y., Yao, F., Zhou, J., Ouyang, C., Deng, X., & Luo, H. (2020). Laser-induced wettability gradient surface on NiTi alloy for improved hemocompatibility and flow resistance. *Materials Science and Engineering C*, *111*, 110847. doi:10.1016/j.msec.2020.110847 PMID:32279801

Chapter 4
Recent Trends in Non-Traditional Machining of Alloys and Composites

Vinod Kumaar J. R.
Mahendra Engineering College, India

Mythili T.
Mahendra Engineering College, India

Suganya Priyadharshini G.
Coimbatore Institute of Technology, India

ABSTRACT

In the modern manufacturing era, machining alloys and composite materials enable proper machining methods to get the required shape and dimensions. The usage of alloys and composite materials is increasing in several industries including aerospace, automobile, MEMS, electronics, medical, biomedical, pharmaceutical industries, and so on because of less weight and more strength. Though various methods are available for machining of composites materials and alloys, only electrochemical micro-machining (ECMM) and wire electrical discharge machining (WEDM) are apt for micron-level machining of these materials by complex shapes with good surface quality. This chapter attempts to provide insight into the recent developments in machining of these alloys and composite materials by ECMM and WEDM.

INTRODUCTION

In recent times, the miniaturization of micro-components/products is continuously becoming more popular nowadays. An alloys and composites usually cover wide range of materials. A combination of one or more metals with other elements is known as alloy and whereas the combination of two or more elements but usually it won't contain metals is termed as composites. Manufacturing of such micro-products using alloys and composites with complex shapes, accuracy and good surface finish has turned out to be

DOI: 10.4018/978-1-7998-9574-9.ch004

challenging for the industries as well as for the researcher. To mention a few of the micro-products and its practical applications in the electrical and electronics industries are LED TV, mobile phones, computers, many electronic gadgets like iPod, iPad, tabs and electrical accessories of the home appliances etc. In the aerospace industries the usage of micro-components includes satellites, navigational systems, nozzles, diffusers, sensors such as gas sensor, temperature sensor, pressure sensor and sound sensor and actuators like magnetic, hydraulic, pneumatic, thermal actuators and electrostatic actuator are used to control and monitor the various functions of the system. In automobile industries we find many micro sized-components which includes nozzles, gears, timing belts, valves, engine parts etc. Likewise, in the medical field the surgical instruments, medical implants, medical measuring devices, medical monitoring devices. Also, in the pharmaceutical industry preparation of drugs, handling, filling and transporting of drugs needs precised micro-components. To manufacture the micro-components/products of alloys and composite material, micro-machining is one of the leading technologies to fulfill the aforementioned needs. The classification of machining process is shown in the figure.1.

Figure 1. Various machining processes of alloys and composites
(Hofy. 2018)

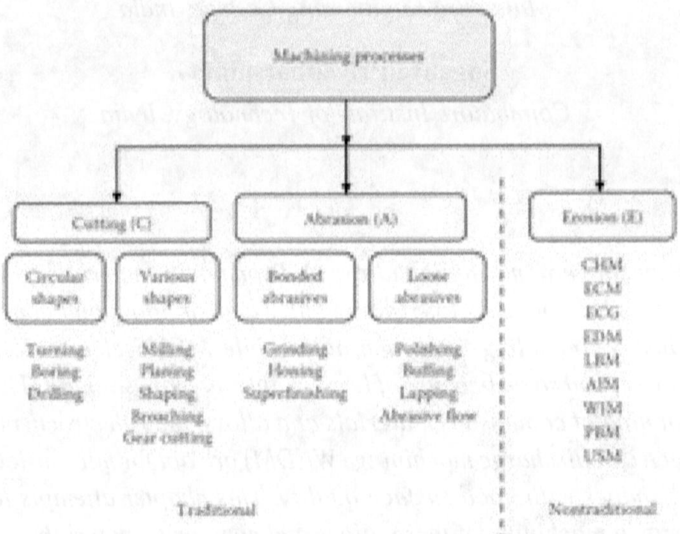

These micro-components of alloys and composite materials can be machined with traditional machining processes and non-traditional machining processes but machining of micro-components through traditional methods results in poor surface finish and less accuracy. In non-traditional machining methods, micro-machining of alloys and composites can be machined effortlessly using ECMM and WEDM processes owing to its favorable advantages like difficult to cut materials, machining complex and small intricate shapes, high precision and accuracy with close tolerance of machining is possible unlike other non-traditional machining process. (Bagotsky. 2005)

Origin of ECMM

In the 19th century, the concept of electrochemical dissolution was discovered by Michael Faraday. Faraday formed the law of electrolysis which is based on the concept of electrochemical dissolution. In the year 1929, the Russian investigator Gusseff first introduced the concept of machining the metal with the help of an electrolytic method. Electrochemical micro-machining works on the principle of electrolysis where reverse electroplating method is followed to remove the material instead of adding the material like electro deposition and it is governed by the Faraday's law of electrolysis. (Bhattacharyya.2015)

Basic Principle of ECMM

In ECMM, anodic dissolution mechanism is followed for material removal during the process of electrolysis. In the ECMM process, pulsed power supply is given in between cathode and anode. The basic working principle of ECMM is as shown in the figure 2. Here in ECMM workpiece acts as anode and it is connected to positive terminal and micro tool acts as a cathode and it is connected to the negative terminal. Both the terminals are submerged in the electrolyte and the gap maintained between the two terminals is known as inter-electrode gap. Once the current is passed through the electrolyte there will be exchange of ions where positive ions move towards the cathode and negative ions move towards the anode. By anodic dissolution the material is dissolved into metallic ions through electro chemical reactions and thus the generation of hydrogen gas bubbles are obtained on the surface of the cathode. Finally, the mirror image of the shape of the tool is obtained as the shape of the work material. (Rajurkar et al. 2013)

Figure 2. Basic principle of ECMM
(Anasane and Bhattacharyya. 2016)

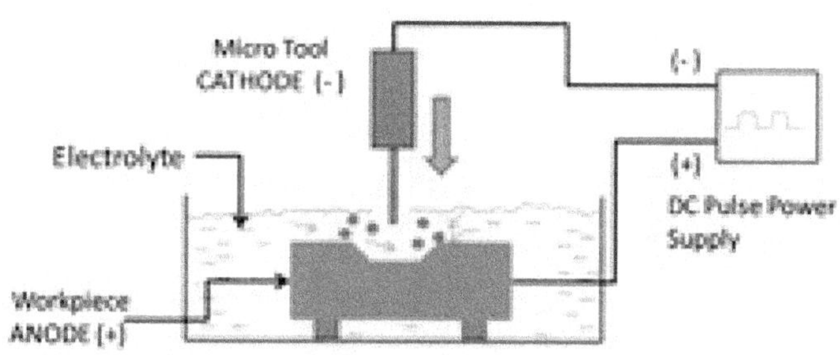

The salt solutions like H_2SO_4, $NaNO_3$ and NaCl are most commonly used an electrolyte material for machining the material. While machining the material, sludges are usually formed and deposited in the electrolyte and these sludges are filtered from the electrolyte and sometimes the purified electrolyte is recirculated again for further machining process. The sludges contain metal oxide ions and harmful substances which are to be treated properly before it is directly disposed into the environment as it causes environmental pollution and becomes hazard to the environment. (Anasane and Bhattacharyya. 2016). ECMM is capable of machining wide range of materials like super alloys, magnesium alloys, titanium alloys, stainless steel alloys, copper alloys, aluminium metal matrix composites and hybrid metal matrix composites which are used for various purposes in aerospace, automobile, electronics, MEMS, medical, biomedical, and pharmaceutical applications etc.

Process Characteristics of ECMM

Process characteristics of ECMM decides the accuracy and precision of machining. The performance of ECMM can be enhanced by choosing the right process variables. The main process characteristics that control the process behavior are shown in the Figure.3. The process characteristics such as machining voltage, type of electrolyte, electrolyte concentration, temperature, flow rate, machining gap or inter electrode gap (IEG) directly affect the material removal rate, surface quality, accuracy and precision. Maintaining the proper machining gap and current density enhances better material removal rate, good accuracy and surface finish. The output response parameters such as MRR, Overcut, circularity, conicity and surface roughness are usually calculated in ECMM.

Figure 3. Process characteristics of ECMM
(Hofy. 2018)

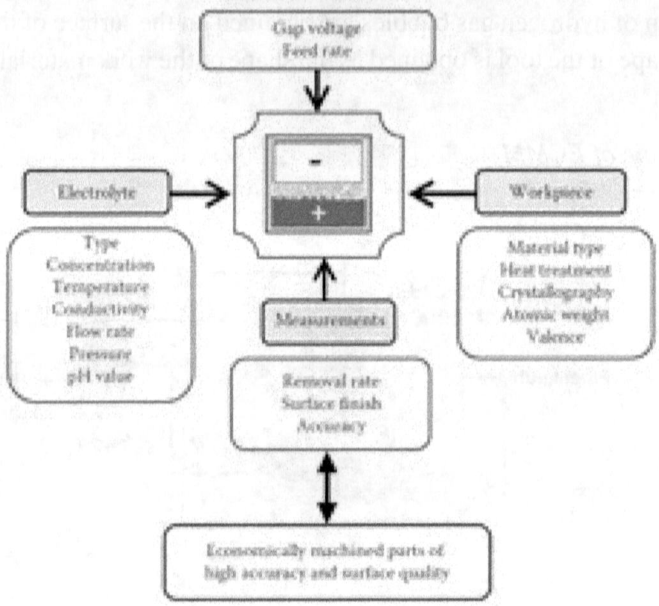

Influence of Process Characteristics on Material Removal Rate

Figure 4 shows the process relations which affects the MRR with the various variables. material removal rate is greatly influenced by the current density, electrolyte type, concentration of the electrolyte, temperature of the electrolyte and machining gap. MRR is directly proportional to its process variables.

Figure 4. Process characteristics affects the MRR
(Hofy. 2018)

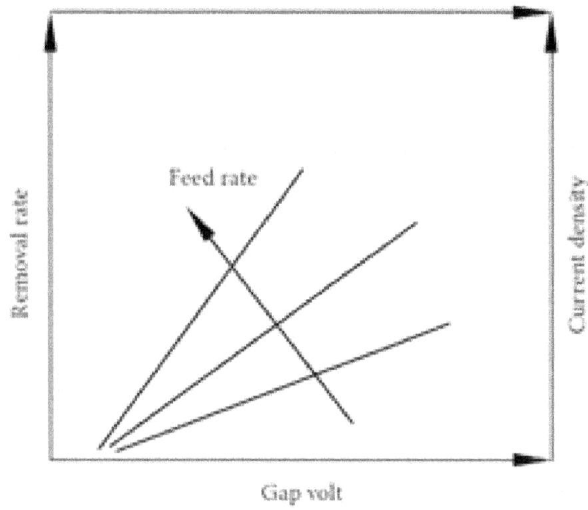

Influence of Process Characteristics on Electrode Gap Width

Figure 5 shows the process relations which affects the electrode gap width with the various variables. The accuracy of the machining process is inversely proportional to the size of the machining gap, feed rate and gap voltage. (Hofy. 2018)

Figure 5. Process characteristics affects the electrode gap width
(Hofy. 2018)

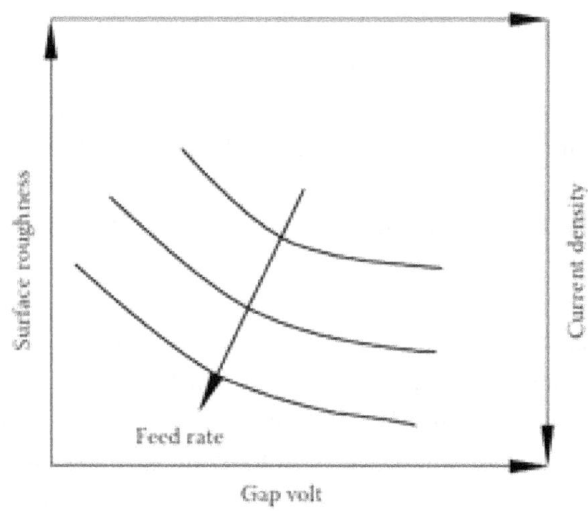

Influence of Process Characteristics on Surface Roughness

Figure 6 shows the process relations which affects the surface roughness with the various variables. The surface roughness is directly proportional to its current density, feed rate, gap voltage.

Figure 6. Process characteristics affecting the surface roughness
(Hofy. 2018)

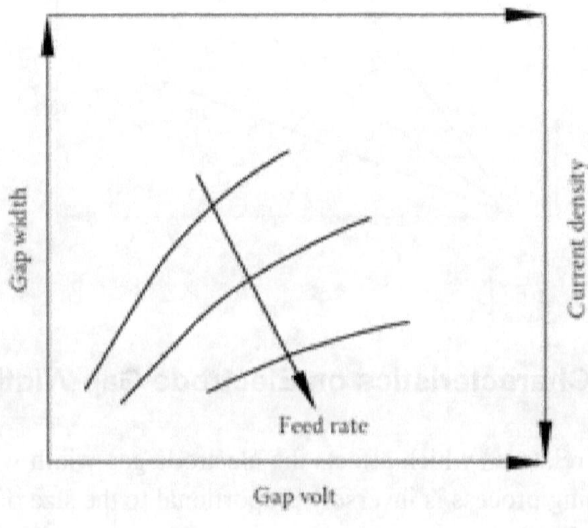

Manufacturing Cost of ECMM

Figure 7 shows the manufacturing cost of ECM. It is clearly understood from the figure.7 that at the optimal feed rate the manufacturing cost of ECMM gets minimized. Figure 8. Shows the variation of manufacturing cost of ECMM with the effect of feed rate. At lower feed rates time of the machining gets increased and also it increases surface finish and accuracy. At higher feed rates the machining time gets reduced and worsens the accuracy and surface finish of the machined surface.

Machining of Alloys and Composites through ECMM

ECMM has more potential to machine wide range of materials like the combinations of various alloys and composites. More research studies have been conducted in ECMM with recent trends for machining the wide range of materials. In this pipeline, titanium alloy (Ti6A14V) is known for the excellent corrosive resistance, strength to weight ratio, toughness and is very difficult to machine the material due to self-passivation. Titanium alloy has been performed in electrochemical machining operation using NaCl electrolyte, $NaNO_3$ electrolyte and the combination of NaCl and $NaNO_3$ electrolyte, the result indicates that mixed electrolyte produces reduced pitting corrosion even in high current density and the surface quality of the mixed electrolyte was better than the NaCl electrolyte and $NaNO_3$electrolyte. (He et al. 2018).

Figure 7. Manufacturing cost of ECMM
(Hofy. 2018)

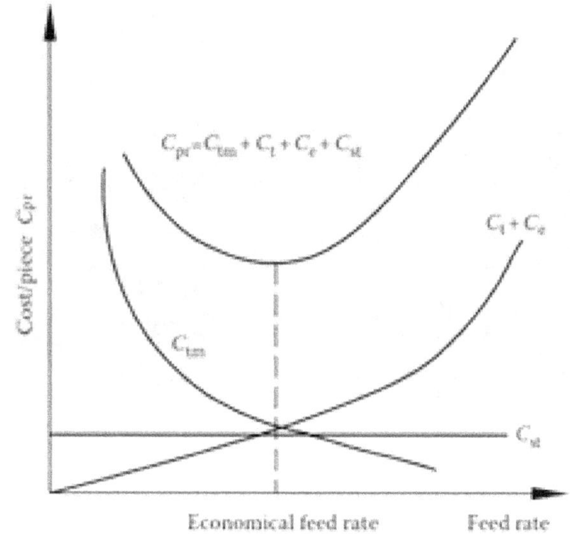

Figure 8. Variation of manufacturing cost of ECMM with the effect of feed rate
(Hofy. 2018)

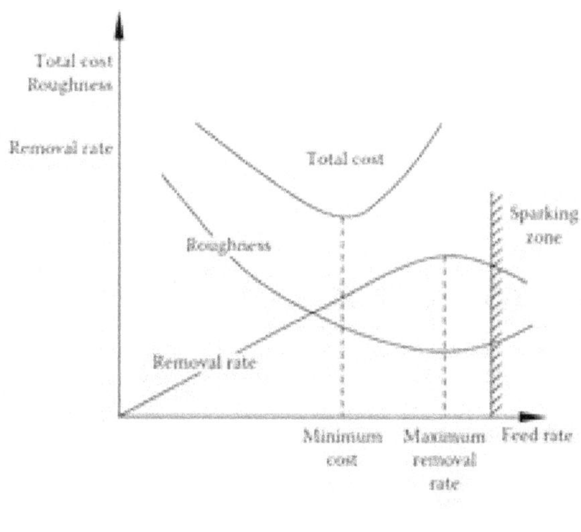

Machining of titanium alloys have been investigated in ECMM to study the various process parameters such as MRR, Radial overcut, conicity using the seven different types of electrolytes. The mixture of ethylene glycol and sodium bromide shows remarkable results compared with other hydrochloric acid, sodium perchlorate, ethylene glycol, sodium bromide and ethylene glycol with sodium bromide and NaCl (Anasane and Bhattacharyya. 2016).

Figure 9. Sample SEM image of the micro-hole hole using EG+Br electrolyte with titanium material
(Anasane and Bhattacharyya. 2016)

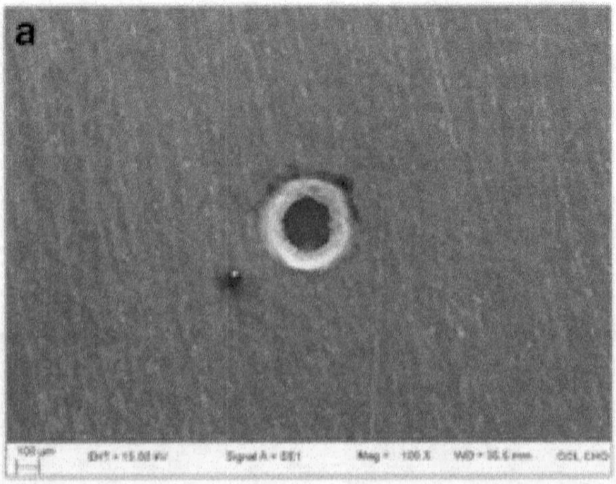

In aircraft engines, the high temperature materials like titanium alloy of Ti 60 is importantly used. By using ECMM process the orthogonal experiments were planned and conducted on Ti 60 using NaCl electrolyte to determine the surface roughness. Frequency and electrolyte temperature influences much on the surface roughness. The surface roughness was obtained as 0.912 μm for the best optimal parameters shows with 13wt% of electrolyte concentration, 20 V, 0.4 kHz, duty cycle of 0.3, temperature of 23 °C, and anode feed rate of 0.5 mm/min. (Chen et al. 2016).

Figure 10. Sample specimen machined using ECMM
(Chen et al. 2016).

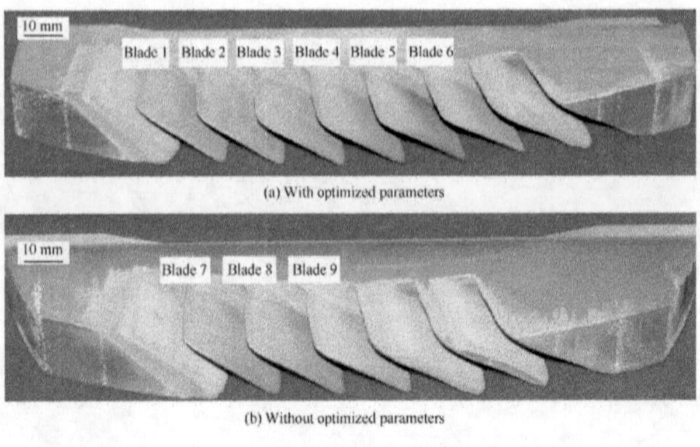

ECMM process has been effectively used for fabricating the complex micro shapes on nickel and nickel based super alloys. The effects of machining parameters such as pulse on time, applied voltage, pulse period, electrode diameters and side gap were considered for the study. The study suggested that, to achieve the best machining accuracy the machining operation has to be conducted on lower machining voltage, smaller pulse on time and pulse period, and minor diameter electrode. (Huang and Liu. 2014)

Figure 11. Sem micro- image of the nurbs curve
(Huang and Liu. 2014)

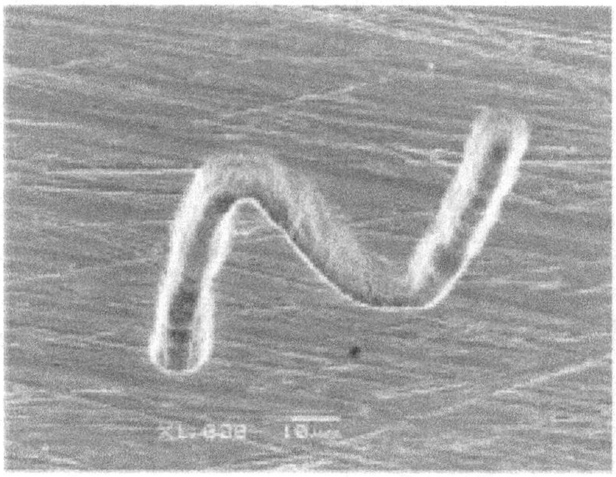

Figure 12. Sem micro- image of the heart shape
(Huang and Liu. 2014)

Aluminium metal matrix composites (AMMC) are mainly used in aerospace industries, automobile industries rather than unreinforced alloys are due to high hardness, good strength to weight ratio, lower coefficient of thermal expansion, high thermal shock, better wear and lower density.AMMC are of less weight and as well as low cost that are the certain reasons for increasing demand in various industries. (Koli at al. 2015, Sharma et al. 2020).

Figure 13. Various applications of aluminium metal matrix composites
(Sharma et al. 2020)

Machining of aluminium metal matrix composites through ECMM was conducted along with abrasive assisted electro chemical machining, electrochemical grinding and electrochemical drilling. The study revealed that variation of applied voltage, Electrolyte concentration, flow rate and tool feed rate simultaneously increase the value of MRR. The ROC was increased by electrolyte concentration and higher voltage. Moreover, ROC is decreased with increase of tool feed rate and the surface roughness got affected by increase of machining voltage and tool feed rate. (Hynes et al. 2018)

Figure 14. Sem image of the aluminium metal matrix composite machined using ECMM
(Hynes et al. 2018)

Hard metals like TiC, TiN, pure Ti and the binder metals like Fe, Co and Ni are machined through ECM with the help of cyclovoltammograms using $NaNO_3$ solution. Hydrogen evolution takes place for all the materials at the cathode surface and at the cathode all the samples were conductive. Anodically, all the combinations of TiC and TiN were dissolved. They created a surface model in which Fe seems to be valid for all the materials and except Ti. (Walther et al. 2007). Nickel aluminium bronze alloy is specifically used in marine water propellers, land gear bearings, pressure sensor, pharmaceutical and biomedical industries. Nickel aluminium bronze alloy mainly contains copper as a major proportion and the remaining it is followed by nickel, aluminium and iron. Machining of nickel aluminium bronze alloy through traditional technique produces higher circularity and irregular surfaces and it can be best machined through ECMM. In ECMM for machining of Nickel aluminium bronze alloy the three electrodes were utilized namely brass, copper and tungsten carbide and NaCl was used as an electrolyte medium. To find out the best parameter RSM was used to find out the optimal parameters. The result indicates that by using copper electrode more amount of materials was removed and the circular micro-holes were produced by tungsten carbide electrode with reduced overcut and whereas the brass material shows the intermediate performance than the other two electrodes. (Palani et al. 2020).

Haste alloy C 276 comes under nickel based super alloy, nickel is well known for good corrosive resistance. Haste alloy is most commonly used in chemical/food processing, pharmaceutical and petrochemical industries. A micro-hole on haste alloy C 276 material were carried out using ECMM process with $NaNO_3$, NaCl and a combination of citric acid with $NaNO_3$and NaCl as hybrid electrolyte of (20,25 and 30 g/l) concentrations were utilized to conduct the experiments. They used L_9 orthogonal array and Taguchi, grey relational analysis to find out the optimum levels. Among the electrolyte's hybrid electrolytes produces higher MRR, reduced overcut, lower conicity. The optimum values of hybrid electrolyte using taguchi grey were at 10V of machining voltage, 30 g/l of electrolyte concentration, 33% of duty cycle and feed rate of 1mm/min. (Kumarasamy et al. 2020)

Figure 15. Sem image of the Nickel aluminium bronze alloy machined using ECMM a) brass electrode b) tungsten carbide electrode c) copper electrode
(Palani et al. 2020)

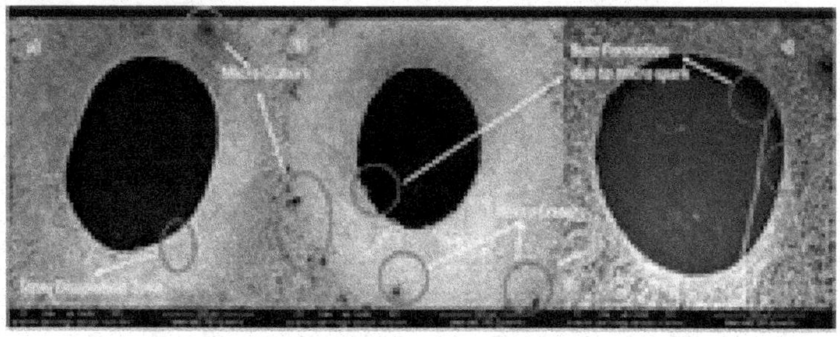

Figure 16. Sem image of the micro-hole machined on haste alloy C 276 with a) NaCl electrolyte b) $NaNO_3$ electrolyte c) hybrid electrolyte using ECMM
(Kumarasamy et al. 2020) (Better)

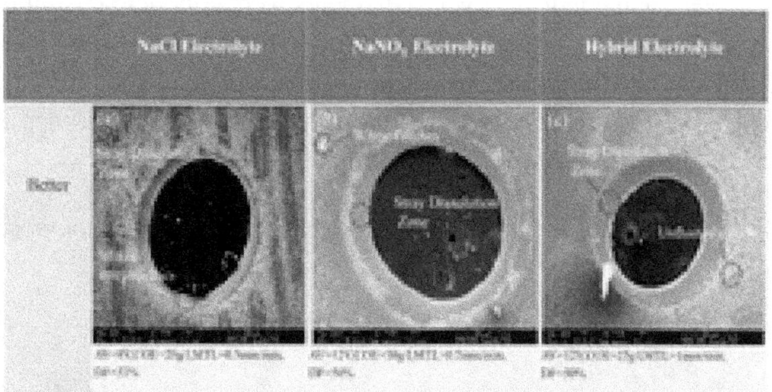

Figure 17. Sem image of the micro-hole machined on haste alloy C 276 with d) NaCl electrolyte e) $NaNO_3$ electrolyte f) hybrid electrolyte using ECMM
(Kumarasamy et al. 2020) (Moderate)

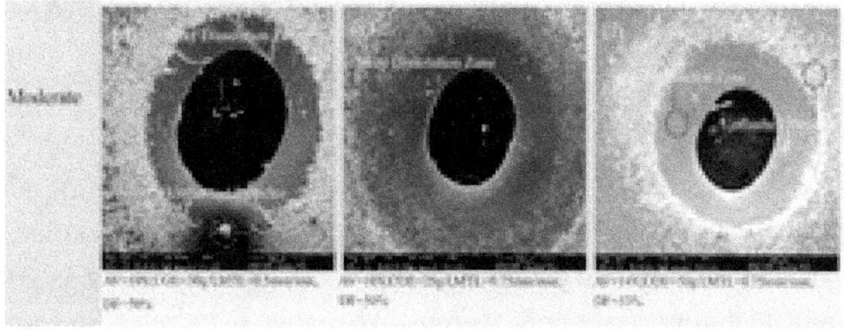

Figure 18. Sem image of the micro-hole machined on haste alloy C 276 with g) NaCl electrolyte h) $NaNO_3$ electrolyte i) hybrid electrolyte using ECMM
(Kumarasamy et al. 2020) (Poor)

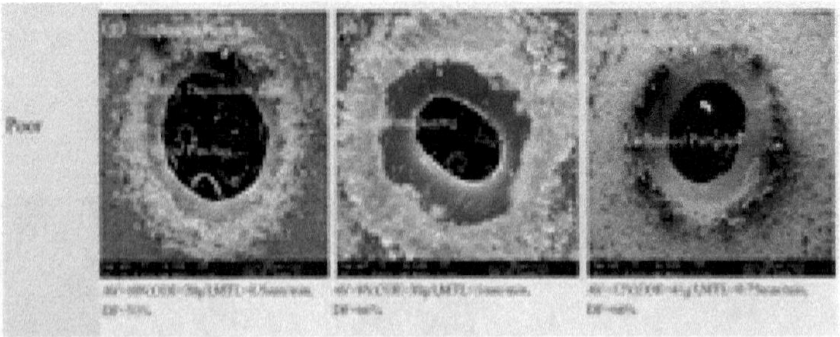

Aluminium alloy (AA) 7075 reinforced with nano silicon carbide particles with 1.5 wt.% was machined with ECMM. The confirmation experiments were performed with both RSM and TLBO based optimum conditions and the result shows that, values had coincided very well with the RSM result considering a targeted surface roughness value of 0.4 µm which, shows TLBO was greater than RSM. (Prakash and Gopalakannan. 2021). In general, aluminium is a light weight material and it is reinforced with some composites to have additional benefits. In AA6061 alloy along with Mg_2Si has moderate strength and this material is especially used in light weight structures and ship building industry. TiB_2 has greater hardness and excellent electrical conductivity. Experiments were conducted using ECMM to produce a drilled micro-hole on AA6061 - TiB_2 in situ composites and it was validated with Entropy-VIKOR method to find out the best optimal conditions showed 2 mol of electrolyte concentration, 12 v of machining voltage and 4 A of current for greater MRR and reduced overcut. (Chandrasekhar and Prasad. 2020)

Figure 19. SEM image of the machined micro-hole on AA6061-TiB_2 in situ composites a) exterior portion b) interior portion
(Chandrasekhar and Prasad. 2020)

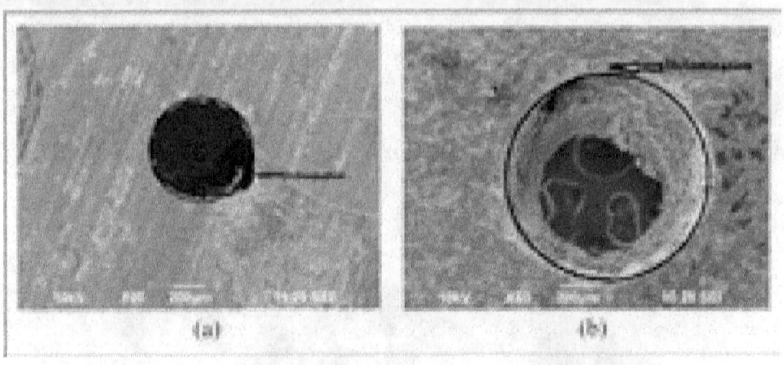

Experimental studies were conducted in ECMM to produce micro holes on aluminium composite material (Al 7075-87%, Al_2O_3-6% and borosilicate powder- 7%) as a workpiece with hollow brass electrode using NaCl as an electrolyte. By varying the process variables such as voltage, pulse on time, current they calculated Machining Rate and overcut. At higher parameter levels it affects the surface finish and where as they achieved good accuracy at the lower level of the input parameters. (Ramesh and Subburam. 2019)

Figure 20. SEM image of the machined micro-hole on aluminium composite
(Ramesh and Subburam. 2019)

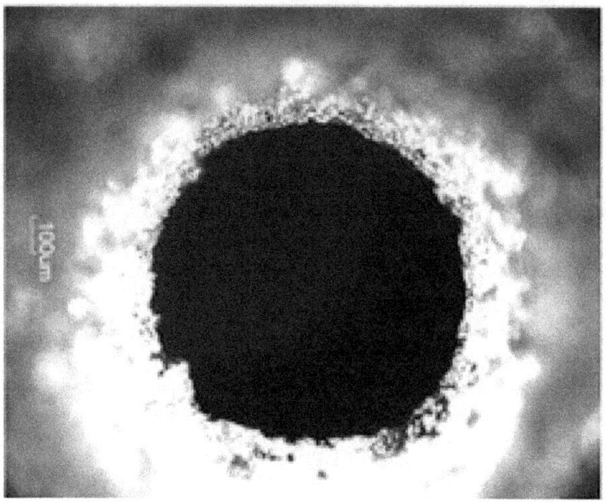

A study was conducted on metal matrix composites using ECMM process can be efficiently used for machining complicated micro holes because of its favorable merits when compared to traditional methods. (Arumugam and Rajalingam. 2020). Magnesium and its alloys are used in the human body specifically because of better bio degradability rather than materials like titanium and stainless steel. Magnesium and its alloys have wide applications in armaments, textile, sports and building industries. Magnesium and its alloys reduce further surgery cost. The usage of titanium and stainless steel in the human body causes some allergic issues. Reviewed the machining of magnesium and its alloy by the various machining methods. Magnesium and its alloys have the demerits of poor mechanical strength and more corrosive behavior. (Chalisgaonkar. 2020)

Figure 21. Image of the magnesium and its alloy implanted in the arm
(Chalisgaonkar. 2020)

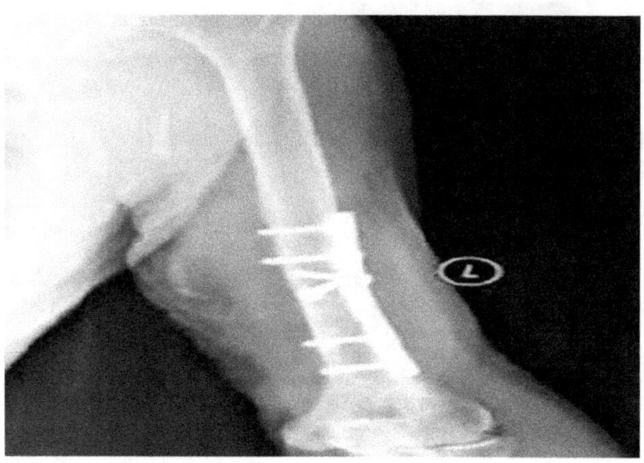

Figure 22. Fixation of various medical implants in the human body
(Chalisgaonkar. 2020)

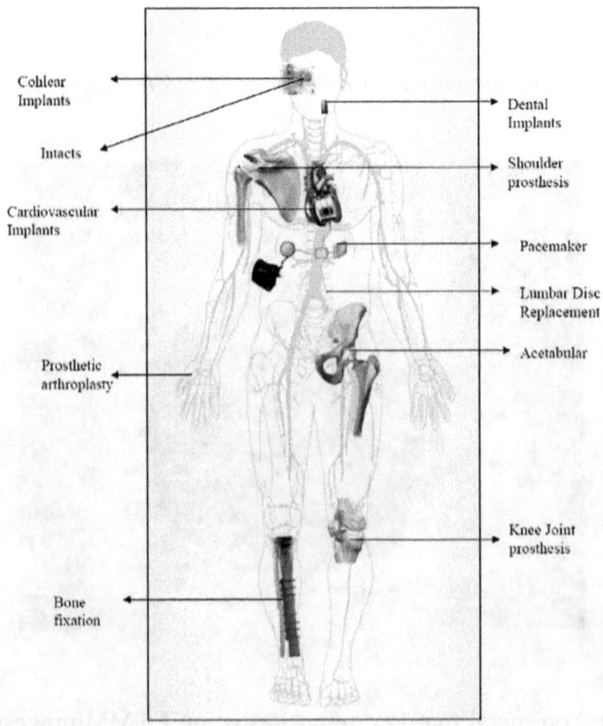

Machining of Titanium alloy (Ti-6Al-4V) was carried out in ECMM using different electrolytes with varying concentrations such as NaCl+$NaNO_3$, NaCl+ Glycerol ($C_3H_8O_3$) and NaCl +citric acid. The Taguchi's design of experiments helps to find out the best optimal parameters. The result revealed the best MRR was found for NaCl+$NaNO_{3.}$ Highest overcut and circularity is found for the electrolyte combination of NaCl +citric acid. However, the optimal overcut is found for the parameter combination of NaCl+ Glycerol (Thangamani et al. 2021)

Figure 23. SEM image of the machined micro-hole on Ti-6Al-4V (Nacl +$NaNo_3$)
(Thangamani et al. 2021)

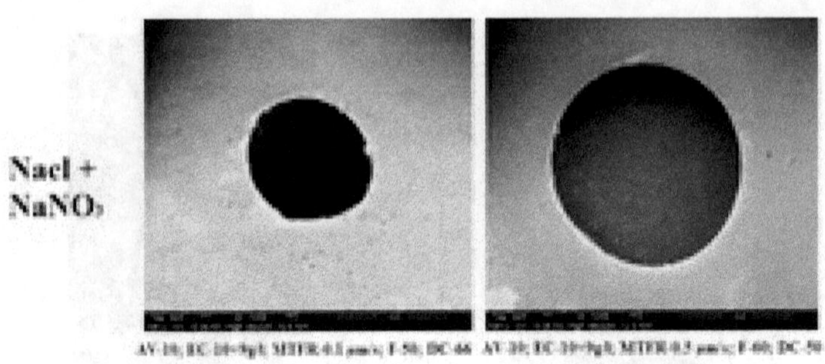

Figure 24. SEM image of the machined micro-hole on Ti-6Al-4V (Nacl+Glycerol)
(Thangamani et al. 2021)

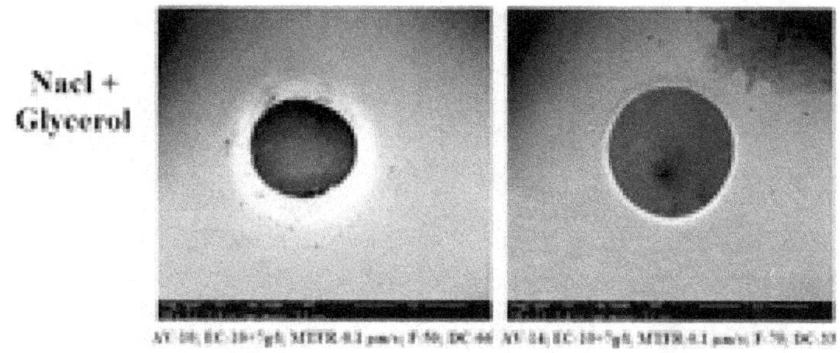

Figure 25. SEM image of the machined micro-hole on Ti-6Al-4V (Nacl+ citric acid)
(Thangamani et al. 2021)

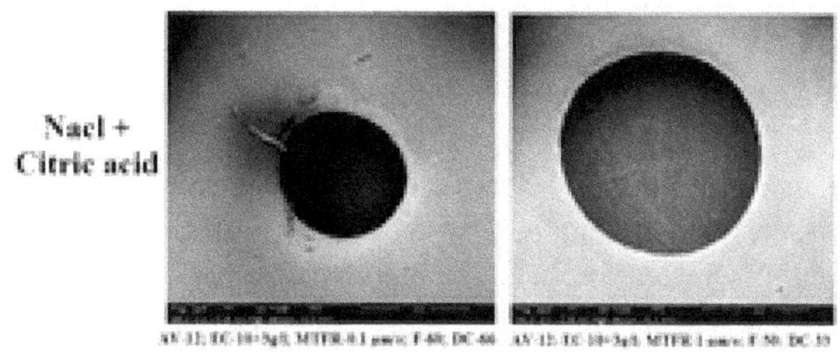

WIRE ELECTRICAL DISCHARGE MACHINE

Wire electric discharge machining technology has grown since last 30 years. This wire cutting method was introduced in the industries around 1960. For produce the required shape of the component, the optical lie follower system was used. This follower system was introduced by D.H. Dulebohn. Due to its flexible capabilities WEDM is used to make components for the automotive, aerospace, and tool and die manufacturing industries. Other applications are biomedical and jewellery R&D areas. The WEDM is only process for machining intricate shapes in conductive parts, exotic materials, difficult to cut materials and conductive ceramics (Ho et al.2004).

Wire EDM Mechanism

Wire EDM process is a thermo electric process which is used to remove the material from the specimen by the repetitive electric spark (Li and Laghari. 2019). In this process wire is used as electrode, it made of copper, brass or tungsten of diameter 0.05 to 0.30mm. This wire is capable for achieving very small corner radii also (Praveen et al. 2020). The gap between wire (electrode) and work piece is called as spark gap. The dielectric fluid is constantly served in the machining zone for flush out the removed debris from the specimen. Dielectric fluid should be carefully chosen to reduce the spark gap and allow for precision machining. Normally kerosene and deionized water used as dielectric fluid. Kerf width is width of material removed from the specimen through WEDM process as shown in figure 26. (Karmakar and Maji. 2021)

Figure 26. Wire EDM process and machined surface

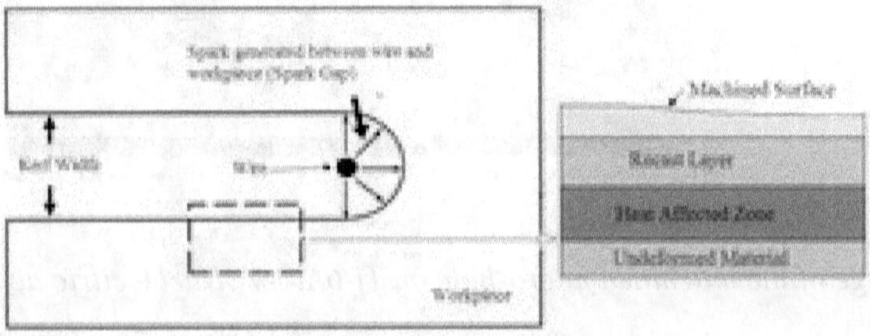

WEDM Process Parameter

Input process parameters can be generally classified as electrical parameters and non electrical parameters. Electrical parameters are pulse on time, pulse off time, peak current, etc, Non electrical parameters such as wire material, wire tension, etc., The input process parameters that affects the machining characteristics of WEDM process is shown in Ishikawa cause and effect diagram (Figure 27). (Praveen et al. 2020, Patel and Maniya. 2018).

Figure 27. Ishikawa cause and effect diagram

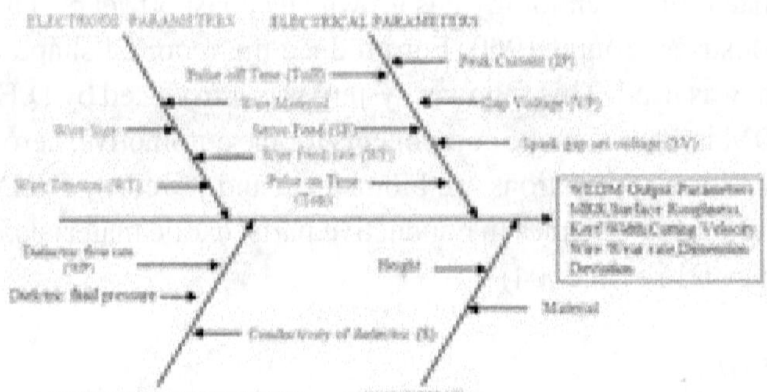

WEDM Machining Features

Generally metal removal rate, surface roughness, kerfwidth cutting velocity wire wear rate and dimension deviation are the machining characteristics or output responses of the WEDM process. Among all the characteristics Metal removal rate and surface roughness are the most important characteristics for getting better performance. Metal removal rate was calculated by the ratio between volume of material removal to time taken for the machining of specimen. The surface roughness (Ra) is obtained by using a surface roughness tester, which is most important parameter to be minimized. (Praveen et al. 2020)

WEDM Performance Characteristics on Alloys and Metal Matrix Composites

Now a day's hybrid nanometal matrix composites (HNMMCs) are used in the aerospace, defense and automotive industries due its outstanding mechanical properties. Al6082/Graphene/carbon nanotubes composite was prepared by using stir casting process. Because of the reinforcement the hardness of the HNMMC was increased compared to base alloy. While machining the composite, pulse on time was the most significant parameter for good metal removal rate. And also, poor surface finish was obtained due to higher voltage. (Perla et al.2021). When machining Aluminium alloy (LM25), Graphite (Gr) and Boron carbide (B_4C) hybrid composites using WEDM current acted as a most important parameter for metal removal rate. Current and pulse on time are the two influential parameters that affect the surface roughness. (Suresh et al. 2020). By using squeeze casting LM13/silicon carbide reinforced composite material was fabricated and machined. In this study reveals that, metal removal rate increased with increase of pulse on time & current and also it decreases with increasing the weight percentage of reinforcement particles in the base metal. (Vellingiri et al. 2021).

In Al-SiC-B_4C composites the kerf width slowly increased while increasing pulse current and severe change in kerf width occurred while increasing pulse on time. During the time of machining process heat energy was produced and it melts the materials broadly. So, the kerf width was increased while increasing pulse current and pulse on time. (Kumar et al. 2020). It cleared from analysis while increasing pulse on time the cutting speed also increased. A higher pulse on time and lower pulse off time generated high heat energy which results in produced big, deeper and overlapped crater on the machined surface of the work piece. Due to this reason larger volume of material removed from the work piece which results in higher cutting speed. At the same time higher cutting speed obtained at lower spark gap voltage. At lower pulse on time and lower peak current combination input parameters produced very small crater on the machined surface. Wire tension was another one parameter to produce a better surface finish on the machined surface. While increasing wire tension the wire deflection and vibrations are reduced. So, at stable machining condition good surface finish was obtained (Kumar 2018).

Aluminum metal matrix composites (AMMCs) consist of aluminium alloy (LM6) as matrix and 3, 6 and 9 percentage of particulate fly ash as reinforcement material was fabricated through the Stir casting process. Gap voltage is the most significant factor affect the metal removal rate. (Ananth et al. 2021). Straight and triangular profile cuts are done through WEDM on the self-lubricating composite. Here aluminum 6061 series self-lubricity composite fabricated with 5%, 10%, and 15 vol% boron carbide (B_4C) and hexagonal boron nitride (HBN) of 10 vol% particles. Self-lubricity composite material with 5 percent B_4C exposed larger kerf dimensional deviation due to fast melting and evaporation of specimen. Further, kerf dimensional deviation reduced while increasing B_4C particles. The increase in wire feed rate reduces the corner area inaccuracy to large extent.MMC with 15% B_4C composite shows unusual erosion and formation of large craters and pits are witnessed on the machined surface resulting in higher surface roughness. (Gnanavelbabu and Rajkumar. 2020). Composite materials are fabricated with AZ31/ Silicon-r-GO/ Magnesium by solvent based powder metallurgy process. The process parameters namely reinforcement proportion, Doping, electrical parameters and feed rate is analysed for MRR and surface roughness. ANOVA results describe that pulse on time and weight percentage of reinforcement wt.% are the major influencing parameter for metal removal rate. Pulse off time and weight percentage of reinforcement are the significant role factor in leading the Ra value. (Kavimani, et al. 2020).316L stainless steel is commonly used in engineering, marine and medical applications. The machined specimens are undergone the corrosion tests at different conditions. Pulses on time and peak current are the most important factor that affects the corrosion resistance of the machined surface. Due to the formation of thin and steady protective layer the corrosion resistance was increased at low level of pulse on time and peak current. While increasing peak current and pulse on time the surface roughness also increased. (Prathipati and Dora. 2019).

CONCLUSION

In this present work, the basic principle of ECMM and WEDM are discussed. The process parameters which governs the machining operation are discussed in detail. The recent methods of machining of various materials like alloys and composites with is characteristics and behavior are reviewed and discussed. Considering the emerging applications of materials in wide places ECMM and WEDM are the most appropriate methods for its remarkable performances.

REFERENCES

Ananth, S., Prakash, J. U., Juliyana, S. J., Rubi, C. S., & Sadhana, A. D. (2021). Effect of process parameters on WEDM of Al–Fly ash composites using Taguchi Technique. *Materials Today: Proceedings*, *39*, 1786–1790. doi:10.1016/j.matpr.2020.07.615

Anasane, S. S., & Bhattacharyya, B. (2016). Experimental investigation on suitability of electrolytes for electrochemical micromachining of titanium. *International Journal of Advanced Manufacturing Technology*, *86*(5), 2147–2160. doi:10.100700170-015-8309-2

Arumugam, R., & Rajalingam, M. (2020). A review on the effect of electro chemical machining on metal matrix composite materials. *Materials Today: Proceedings*, *33*, 3854–3857. doi:10.1016/j.matpr.2020.06.229

Bagotsky, V. S. (Ed.). (2005). *Fundamentals of electrochemistry* (Vol. 44). John Wiley & Sons. doi:10.1002/047174199X

Bhattacharyya, B. (2015). *Electrochemical micromachining for nanofabrication, MEMS and nanotechnology*. William Andrew.

Chalisgaonkar, R. (2020). Insight in applications, manufacturing and corrosion behaviour of magnesium and its alloys–A review. *Materials Today: Proceedings*, *26*, 1060–1071. doi:10.1016/j.matpr.2020.02.211

Chandrasekhar, S., & Prasad, N. B. V. (2020). Multi-response optimization of electrochemical machining parameters in the micro-drilling of AA6061-TiB2in situ composites using the Entropy–VIKOR method. *Proceedings of the Institution of Mechanical Engineers. Part B, Journal of Engineering Manufacture*, *234*(10), 1311–1322. doi:10.1177/0954405420911539

Chen, X., Xu, Z., Zhu, D., Fang, Z., & Zhu, D. (2016). Experimental research on electrochemical machining of titanium alloy Ti60 for a blisk. *Chinese Journal of Aeronautics*, *29*(1), 274–282. doi:10.1016/j.cja.2015.09.010

El-Hofy, H. (2018). *Fundamentals of machining processes: conventional and nonconventional processes*. CRC Press. doi:10.1201/9780429443329

Gnanavelbabu, A., & Rajkumar, K. (2020). Experimental characterization of dimensional and surface alternation of straight and angular cutting on self-lubricating composite: A wire EDM approach. *Arabian Journal for Science and Engineering*, *45*(7), 5859–5872. doi:10.100713369-020-04596-2

He, Y., Xiao, H., Gan, W., Yu, Q., & Yin, F. (2018). The Electrochemical Dissolution Behavior Research of Titanium Alloy under Different Electrolyte. *Procedia CIRP*, *68*, 751–756. doi:10.1016/j.procir.2017.12.132

Ho, K. H., Newman, S. T., Rahimifard, S., & Allen, R. D. (2004). State of the art in wire electrical discharge machining (WEDM). *International Journal of Machine Tools & Manufacture*, *44*(12-13), 1247–1259. doi:10.1016/j.ijmachtools.2004.04.017

Huang, S. F., & Liu, Y. (2014). Electrochemical micromachining of complex shapes on nickel and nickel-based superalloys. *Materials and Manufacturing Processes*, *29*(11-12), 1483–1487. doi:10.1080/10426914.2014.930897

Hynes, N. R. J., & Kumar, R. (2018). Electrochemical Machining of Aluminium Metal Matrix Composites. *Surface Engineering and Applied Electrochemistry*, *54*(4), 367–373. doi:10.3103/S1068375518040087

Karmakar, R., & Maji, P. (2021). Research advancements in machining of metal matrix composites through wire electro discharge machining technique. *Materials Today: Proceedings*, *44*, 4457–4461. doi:10.1016/j.matpr.2020.10.717

Kavimani, V., Prakash, K. S., Thankachan, T., Nagaraja, S., Jeevanantham, A. K., & Jhon, J. P. (2020). WEDM parameter optimization for silicon@ r-GO/magneisum composite using taguchi based GRA coupled PCA. *Silicon*, *12*(5), 1161–1175. doi:10.100712633-019-00205-6

Koli, D. K., Agnihotri, G., & Purohit, R. (2015). Advanced aluminium matrix composites: The critical need of automotive and aerospace engineering fields. *Materials Today: Proceedings*, *2*(4-5), 3032–3041. doi:10.1016/j.matpr.2015.07.290

Kumar, H., Manna, A., & Kumar, R. (2018). Modeling and desirability approach-based multi-response optimization of WEDM parameters in machining of aluminum metal matrix composite. *Journal of the Brazilian Society of Mechanical Sciences and Engineering*, *40*(9), 1–19. doi:10.100740430-018-1368-1

Kumar, S. S., Erdemir, F., Varol, T., Kumaran, S. T., Uthayakumar, M., & Canakci, A. (2020). Investigation of WEDM process parameters of Al–SiC–B4C composites using response surface methodology. *International Journal of Lightweight Materials and Manufacture*, *3*(2), 127–135. doi:10.1016/j.ijlmm.2019.09.003

Kumarasamy, G., Lakshmanan, P., & Thangamani, G. (2021). Electrochemical micromachining of hastelloy C276 by different electrolyte solutions. *Arabian Journal for Science and Engineering*, *46*(3), 2243–2259. doi:10.100713369-020-05032-1

Li, J., & Laghari, R. A. (2019). A review on machining and optimization of particle-reinforced metal matrix composites. *International Journal of Advanced Manufacturing Technology*, *100*(9), 2929–2943. doi:10.100700170-018-2837-5

Palani, S., Lakshmanan, P., & Kaliyamurthy, R. (2020). Experimental investigations of electrochemical micromachining of nickel aluminum bronze alloy. *Materials and Manufacturing Processes*, *35*(16), 1860–1869. doi:10.1080/10426914.2020.1813888

Patel, J. D., & Maniya, K. D. (2018). A Review on: Wire cut electrical discharge machining process for metal matrix composite. *Procedia Manufacturing*, *20*, 253–258. doi:10.1016/j.promfg.2018.02.037

Perla, V. R., Suraparaju, S. K., Thimmarayappa, M., & Pathi, V. R. (2021). Effect of wire electrical discharge machining process parameters of Al-6082 hybrid nano metal matrix composites. *Materials Today: Proceedings*, *43*, 551–556. doi:10.1016/j.matpr.2020.12.047

Prakash, J., & Gopalakannan, S. (2021). Teaching—Learning-Based Optimization Coupled with Response Surface Methodology for Micro Electrochemical Machining of ALUMINIUM NANOCOMPOSITE. *Silicon*, *13*(2), 409–432. doi:10.100712633-020-00434-0

Prathipati, R. Ch. R., & Dora, S. P. (2019). Corrosion behavior of surface induced by wire EDM on 316L stainless steel: An experimental investigation. *SN Applied Sciences*, *1*(9), 1–11. doi:10.100742452-019-0992-4

Praveen, D. V., Raju, D. R., & Raju, M. J. (2020). Optimization of machining parameters of wire-cut EDM on ceramic particles reinforced Al-metal matrix composites–A review. *Materials Today: Proceedings*, *23*, 495–498. doi:10.1016/j.matpr.2019.05.392

Rajurkar, K. P., Sundaram, M. M., & Malshe, A. P. (2013). Review of electrochemical and electro discharge machining. *Procedia CIRP*, *6*, 13–26. doi:10.1016/j.procir.2013.03.002

Ramesh, S., & Subburam, V. (2019). Electrochemical micromachining of aluminium alloy composite. In *Advances in Manufacturing Technology* (pp. 309–317). Springer. doi:10.1007/978-981-13-6374-0_36

Sharma, A. K., Bhandari, R., Aherwar, A., Rimašauskienė, R., & Pinca-Bretotean, C. (2020). A study of advancement in application opportunities of aluminum metal matrix composites. *Materials Today: Proceedings*, *26*, 2419–2424. doi:10.1016/j.matpr.2020.02.516

Suresh, V., Praneet, A. D., & Anoop, J. (2020). Ingenious analysis on machining parameters of aluminium alloy (LM25)/graphite (Gr)/boron carbide (B4C) hybrid composites using wire electrical discharge machining (WEDM). *Materials Today: Proceedings*.

Thangamani, G., Thangaraj, M., Moiduddin, K., Mian, S. H., Alkhalefah, H., & Umer, U. (2021). Performance Analysis of Electrochemical Micro Machining of Titanium (Ti-6Al-4V) Alloy under Different Electrolytes Concentrations. *Metals*, *11*(2), 247. doi:10.3390/met11020247

Vellingiri, S., Soundararajan, R., Mohankumar, N., Nithyananthakumar, K., & Muthuselvam, K. (2021). Exploration on WEDM process parameters effect on LM13 alloy and LM13/SiC composites using Taguchi method. *Materials Today: Proceedings*, *45*, 997–1003. doi:10.1016/j.matpr.2020.03.050

Walther, B., Schilm, J., Michaelis, A., & Lohrengel, M. M. (2007). Electrochemical dissolution of hard metal alloys. *Electrochimica Acta*, *52*(27), 7732–7737. doi:10.1016/j.electacta.2006.12.038

Chapter 5
Recent Developments in Conventional Machining for Metals and Composite Materials

Soundarrajan Madesh
https://orcid.org/0000-0002-7100-7192
Muthayammal Engineering College, India

Clement Christy Deepak Charles
Eppinger Tooling Asia Pvt Ltd, India

Sathishkumar D.
Nehru Institute of Engineering and Technology, India

ABSTRACT

Application of novel engineering materials and advanced technology has been increasing tremendously in many industrial needs. Among those challenging manufacturing methods, conventional machining is one of the proven and prominent processes. Although, fast growing and newer developments in modern manufacturing industries are posses the tough fight on current requirements. On the other hand, continuously developing new technology and trends of advanced machining methods on modern technology for industries and academicians are not well established in the open web sources. Hence, this chapter will be eye opener for both industrialists and researchers about the various novel techniques, trends, and developments in conventional machining.

INTRODUCTION

Manufacturing is the backbone of the industry and financial position of the nation is depends upon the range of manufacturing movement. It is apparent that the urge of global contest provide as a mechanism for amend in manufacturing techniques and structure (De Weck, 2014). The manufacturing technologies such as automated systems and various innovations are executed to enhance the CNC machine func-

DOI: 10.4018/978-1-7998-9574-9.ch005

tions which hinder the manufacturing expenditures. Also, this sector faces the various challenges due to continues growth of machine tools with high accuracy, high speed, digitalization and highly developed intelligence (Schiller, 2015). The components manufactured using the different machining techniques are required further machining to done for the final output. Figure 1 represents the advanced machining process of metals and composite materials with better machining quality.

Figure 1. Machining process of metals and composite materials

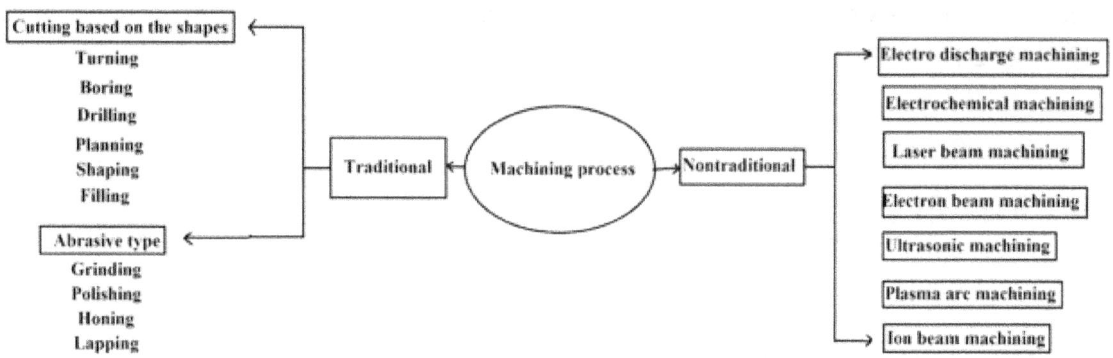

Moreover, machining is the major extensive metal seminal process in manufacturing industry. The rapid change and tough in machining creates the need in replacement of commercial metal using composite materials in consumer products. Also, machining process is a versatile nature due to the complication in shapes, range of scale (Micro and nano), cost effectiveness and time taken for machining. But, all those points satisfied with simple CNC machine. Components might have various profiles such as sharp corners, flat, interior and exterior features whereas these shapes can't produce using common machining process (Naves et al., 2013). The special machining features such as mirror exterior nature and high reflectivity are produced using only with diamond tools. However, every machining techniques, whether conventional or nonconventional owing its merits and demerits. There are many techniques exist in traditional or notational processes to machining the tough and complex materials for producing the quality component. But the user should select the suitable method for machining the same component with quality. Therefore, setoff guidelines are framed for the machining various metals and composite materials (Childs et al., 2000). Hence, in current scenario of machining is indeed a large section of manufacturing activities. Among those, metals and composite materials plays vital role for having right component with high precision (Byrne et al., 2003). On the other hand selection of machine perceived the machining characteristics. Although non traditional method posses good sign in machining nature, it is affected due the reasons such as machining cost, operator's skill, space and most importantly thermal ablations etc,. Moreover, the ability of traditional machining is explored lot by the researchers to enhance the machining performance. Therefore, in this chapter focus on the scope and process in conventional machining of metals and composite materials. In addition, advanced engineering materials, tool materials and machining process are discussed in this chapter to extent knowledge on conventional machining process.

CLASSIFICATION OF MACHINING OPERATIONS

In recent manufacturing incorporates the diversity of processes and engineers are impulse to select the peaceful process combination to produce a high accuracy component with suitable cost. Hence, the engineers should have a wide knowledge in materials processing and handling.

The machining process deals with material removal process from the workpiece which including with tools shapes, type of tool, materials of the tool (Boothroyd, 1988). Also, removal of material from the metal or compositecould be four basic methods which are based on the information related to the machining and energy applications on the workpiece. Table 1 represents the type of energy utilization and machining process.

Table 1. Classification of machining and type of energy utilization (König, 2008)

Group	Type of Energy	Machining process
A	Mechanical	Turning, Milling, Drilling, Grinding etc
		Water jet cutting, Abrasive jet machining, Blanking Punching, Shearing, Sand blasting, Ultrasonic machining, etc.
B	Thermal	Thermal cutting (melting) Electron beam machining Laser machining
		Electro discharge machining (EDM)
C	Chemical	Etching Thermal cutting (combustion)
		Electrochemical machining (ECM)

Group "A" induced with various common conventional machining processes with specific cutting edges. Other machining categories like B & C are comes under the thermal and chemical machining process and termed as non-conventional machining process. Also, these groups of process are not requiring the sharp cutting tools whereas removes the materials from work using energy beams or chemical reactions or evaporating. Recent scenario, hybrid machining processes are evolving significantly in order to create components in large scale (Knight & Boothroyd, 2019). From the table 1 groups of B & C are hybrid together and forms a novel machining process to machine tough materials in the industry. However, those methods are accompanied with high cost for initiation. In addition, manufacturing process is fluctuating rapidly in worldwide because of various factors such as increasing of consumer prospects, world markets and struggles, increasing knowledge intensity in products and technology etc. Also, current advances in market practice, engineering methodologies and manufacturing science enables the industries to produces novel and good components quickly in lower cost than ever before. Currently, machining process must compete with various manufacturing requirements and surface quality in addition with reliability and others. Moreover, the efficiency of production might be achieved based on the optimization of process parameters, tool dimensions and tool design. Therefore, the new concepts are implemented in the current machining technologies such as high-speed machining (HSM) (Jawahir & Van Luttervelt, 1993), high performance machining (HPM), high velocity machining and general sustainable machining.

COMMON FACTORS INFLUENCING THE MACHINING PROCESS

Chip Formation

The formation of chip plays major role on the surface finish of work material and optimization of machining process (Turley & Doyle, 1982). Chip formation occurred based on the physical nature of work material, tool shape and machining speed. In conventional machining using flat end tool liberates thechips such as continuous, elemental, segmental and discontinuous chip can be obtained(Rahman et al., 2005). The discontinuous chips are formed while machining the graphite, manganese sulphide containing materials such as cast ion or brittle materials.In case of ductile materials, lower machining speed and high feed causes for discontinues chips. The non-plastic deformation nature of the work material induces these kinds of chipsusing the cutting edge to the work surface. This factor resulting the chip distractions into small pieces in periodically. Figure 2 (a) presents the discontinuous chip image for brittle material.

Figure 2. Types of chip (a) discontinuous (b) elemental (c) segmented (d) continues

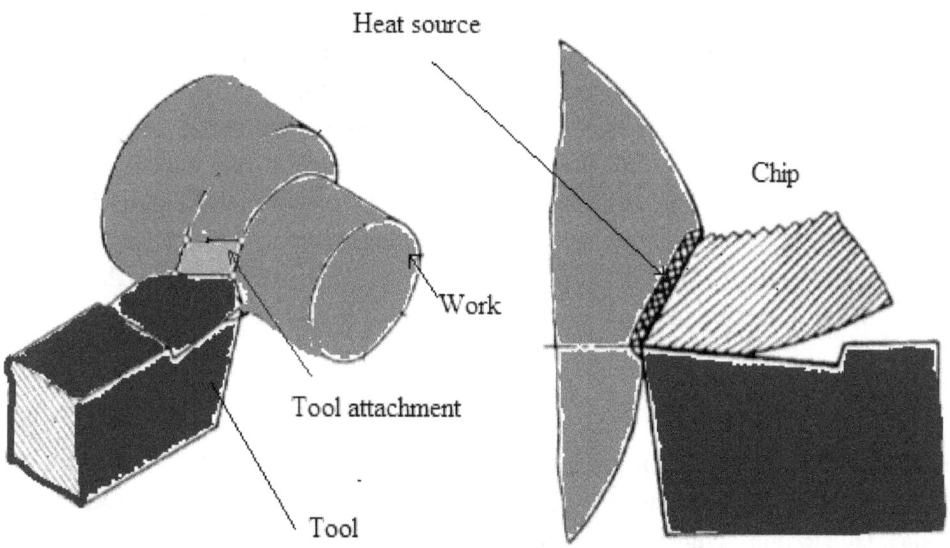

Figure 2 (d) shows continues chip formation which are occurred mostly with ductile materials such as mild steel, aluminum, copper and pure irons when machining in high cutting speed (Vyas & Shaw, 1999). The ductile nature of work materials does not require the high cutting force which resulting the good surface finish and produces continues chips. The shape of chip might have various types such as straight and helix due to the rigorous plastic deformation. This type of continues chip formation one of the major issues to the researchers and tool engineers for obtaining quality finishing of work material (Manyindo & Oxley, 1986). The segmented and elemental chips images are presented in figure 2 (c & b) which are considered through areas of concentrated shear detached from the area of material with lower deformation. The segmental chips are produced with most materials at high cutting speeds whereas the low thermal conductive materials such as titanium, stainless steel and alloys are takes place at lower machining speed (Prieto et al., 2018). Also, heat production at the machine tool when not distributed well narrow segments formed with work material. Therefore, controlling of chip is significant in conventional machining to explore the consistent machining conditions and monitoring of chip formation is the essential task in machining setups. To identify the chip formations in CNC machines various devices has been developed in last three decades (Kovacic et al., 2005). These devices functions based on the CNC machines nature which can alarming the machinist during cutting force exist beyond the limit, variations take place in length between tool and work, chip size variations, chip flow directions and variations in temperature of chip to bring the quality machining surface.

Cutting Vibrations

In machining process, machine tool and workpiece are fixed in the fixture and intricate with dynamic characteristics. During the machining process certain unwanted vibration occurred due the dynamic nature of machine. The consequences of tool vibration affect the machining surface intensively. Therefore, enhancement of machine tool is essential for obtaining better machining quality in conventional machining (Teti, 2002). In way various researchers are made attempt to control vibrations raised from the CNC machine. Some major suggestion from the literatures is followed. Structural alterations are added with machine setup to damping the vibrations which hinders the magnitude of harmonic reacceptance on the operations and cutting tool directions can be altered from the principal directions. A mass spring system is affixed with the machine to absorb the vibration. Use of high young's modulus materials such as tungsten, silica carbide, 2 to 4% Copper and nickel mixed alloys and phase sintered tungsten carbide metals are absorbing vibrations significantly. Cutting force control using feedback of machine which providing damping force trough piezoelectric actuator with internally controlled machining tool. Moreover, controlling of cutting force in conventional machining irrespective of all dynamic machineries are an essential requirement (Thandra & Choudhury, 2010). Although, various techniques followed in industries it posses significant modification in the machine tool adoption in setup.

Production of Heat

In machining process, due to the plastic deformation of metal produces 100 to 1000 Celsius degree generated at the machining zone and those heat energy distributed to the cutting tool edges and chips (Valiev, 2004). Due to high temperatures create impact on tool wear and friction at the chip and work interfaces which is presented in the figure 3.

According to this fact, the production of heat at the tool has been measured in various techniques such as thermoelectric effect, compositional effects, point measurement, field measurement. Based on the need, thermocouples, infrared, fiber optical sensor etc., are employed in measurement of heat generation which is present in figure 4.

The tool and work are electrically insulated from the machine fixtures. The electromagnetic field signals are collected from the cutting edge and converted into the temperature. Therefore, tool and work standardized with same material as in the machining test and thermocouple. In conduction methods, various novel techniques were created such as thin film sensors (TF), TF thermocouples (TFT), TF sensors to measure the heat in the tool and work interfaces area (Jiang et al., 2013). However, tool and work interference zone heat measuring techniques are employed to asses temperature which is need to control in the conventional machining. Hence, monitoring and controlling of temperature at the machining zone is very essential process in the conventional machining.

Figure 3. Tool and work interfaces

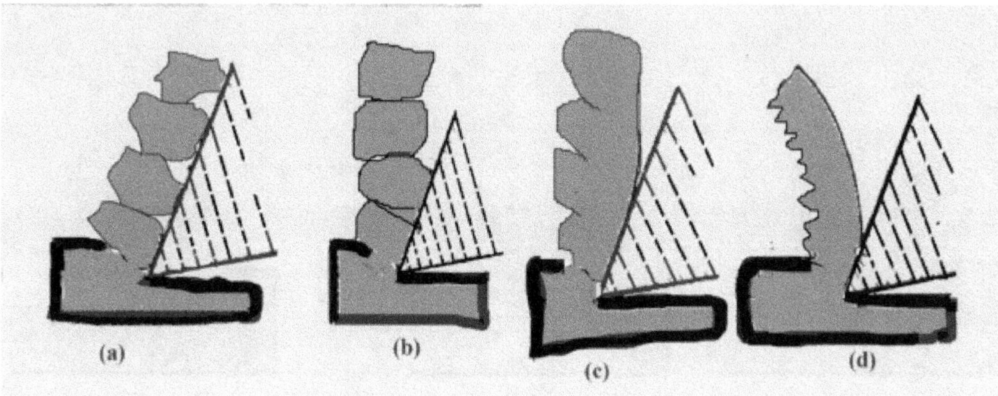

Figure 4. Tool and work thermocouple inserted

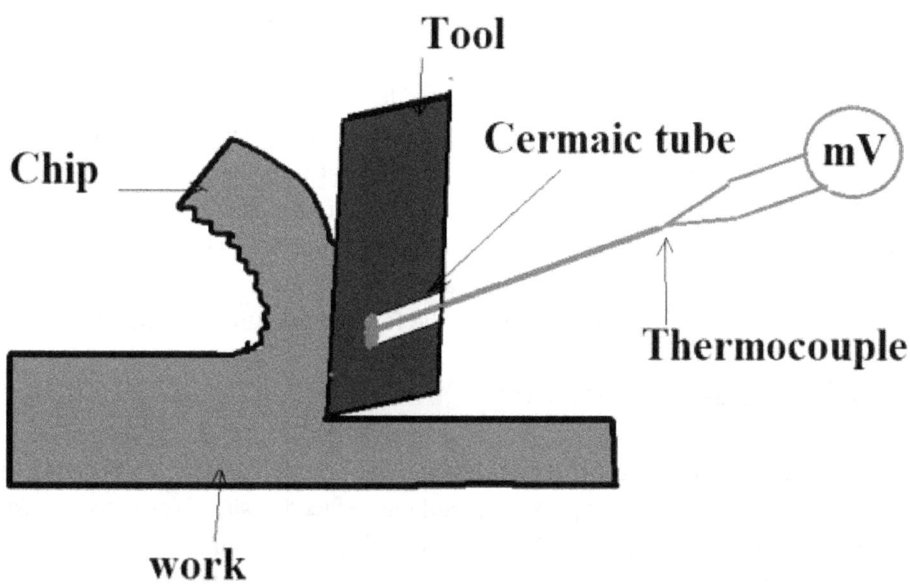

Cutting Fluids

Heat generation at the machining zone plays prominent role in conventional machining process for obtaining in good surface finish. The main purpose of cutting fluids is reducing heat generation on the machining zone which is more significant than cutting speed. Cooling is obtained through distributing and conducting segment of the heat generations. The cutting fluids are classified major categories such as non-water mixable cutting fluid (oil based), water mixable cutting fluid and water based cutting fluid (Lawal et al., 2012). The supply of cutting fluid to the machining zone using different techniques is presented in the figure 5.

Figure 5. Cooling techniques

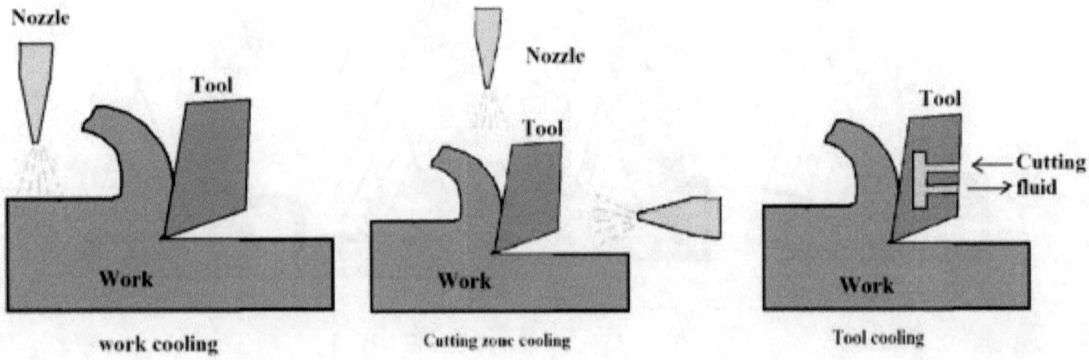

Non-water mixable cutting fluid - Purely oils are employed in machining zone with any dilution which are commonly mineral oils. Although, in view of improving thermal conductivity of cutting fluid other oils such as vegetable oils, easters, fats, chlorine, Sulphur and phosphorus are mixed with plain oil and utilized in machining zone. Also, researchers noted less lubricity with pure oils and employed more than 90% of oils in conventional machining. Water mixable cutting fluid – soluble minerals with help of emulsifiers are mixed with water and employed as cutting fluid. The emulsifiers categorize the size of oil droplet and conduct the heat significantly. Water based cutting fluid – plain water used as coolant and researchers mixed some organic compounds to improve lubricity. However, this lubricant exposed the corrosion on the machining surface due to atmospheric oxidization. Also, due to the microbial contaminations in water based cutting fluid leading to the formation of unpleasant odour.

Progress in Conventional Machining

In general, conventional machining process evolved significantly in last two decades for machining super alloys and composites which can provide excellent surface finish, lower power consumptions and higher tool life (M'Saoubi et al., 2015). Many factors affect the machinability nature of super alloys. Still composites or super alloys are withstanding in high yield strength and also exposes the ductile nature while machining, thereby needful higher forces to be cut. Moreover, inclusion of subsystem with conventional machining is an essential to improvise the machining quality. Therefore, the researchers are incorporating many techniques in conventional machining process. Those methods are provided below.

Ultrasonic Vibration Aided Machining

Ultrasonic vibration aided machining (UVAM) is a metal removal operation through employing vibrations on the cutting tool at a certain frequency. It enhances the machining performance of conventional machining technique for tough metals and alloys. Based on the authors (O'Toole et al., 2020) report, this vibration aided tool reduces the excessive noise and unwanted tool wear in addition with improvised surface finish. They used 0 to 40 μm amplitude and 0 to 500 Hz frequency vibration using a separate transducer. The vibrations an accomplishment diagram shown in figure 6.

Figure 6. Ultrasonic vibration aided machining system
(Xu & Zhang, 2015)

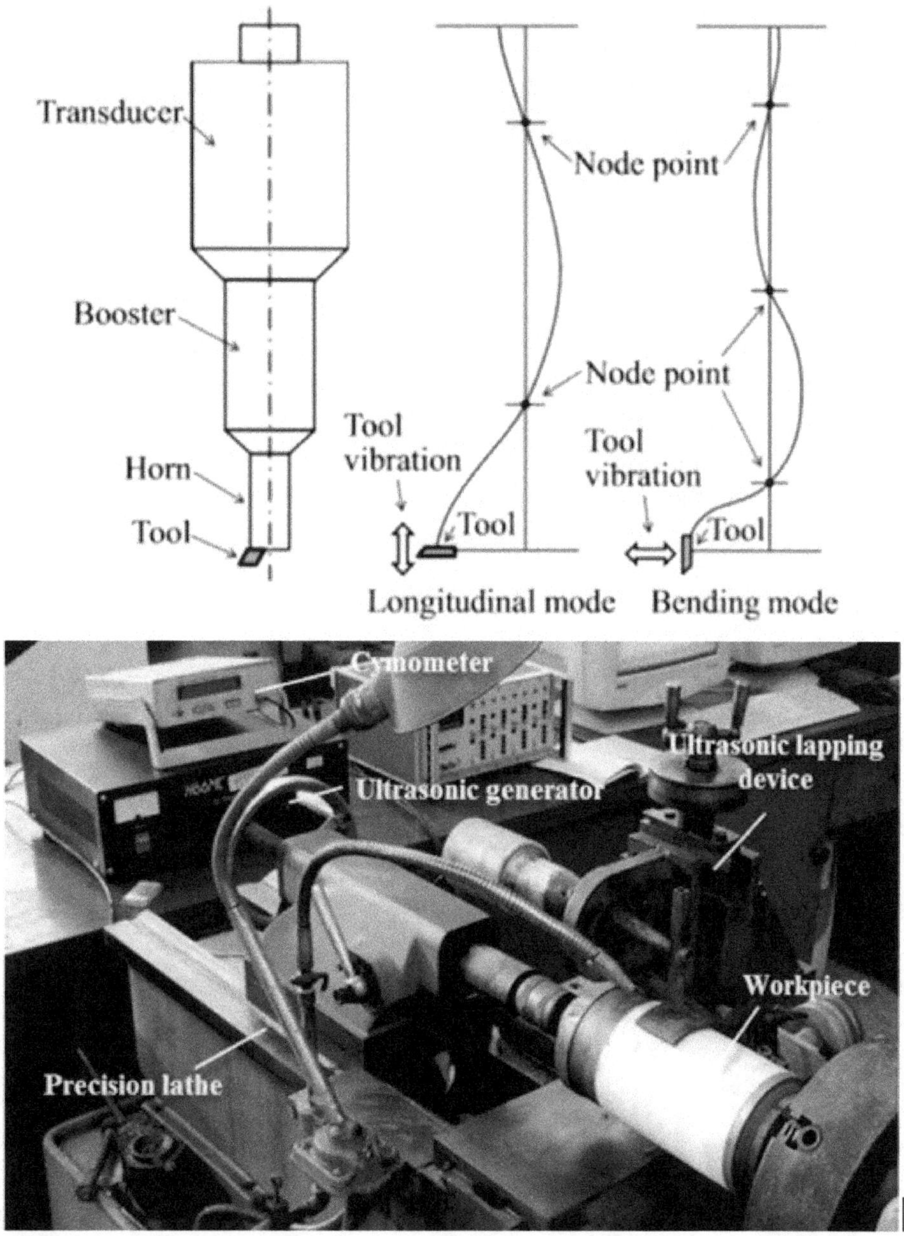

Tool has been fixed with end of the transducers which connected with horn. Although narrow frequency of vibrations are required for horn design, which can produce high accuracy machining. Due to the reasons such as horn cracking, over tightening, and sliding develops the ovel tool motion is presented in figure.

Heat-Assisted Machining

Heat assisted machining (HAM) is familiar method for machining aerospace materials such as nickel-based alloys and titanium alloys. During the machining various heating techniques are adopted with machining operations. The heat adoption in machining increases the softening of work material which can increases the tool life as well as hinders the cutting force and chatter instabilities. Moreover, heat adoption increases the plasticity of work material. This factor helps the tool to remove material easy and resulting the better surface finish. There are many heating techniques are endeavored by the researchers in conventional machining which are elaborated below.

Laser Assisted Machining

In this method heat energy in form of laser has been applied on the work piece surface which avoiding the materials deformation and this factor is significant for accuracy of metal cut. Also, laser heating method is one comfortable technique to direct the heat spread over the work surface and laser diameter can be tuned up to lower than 3 mm. Laser can be locate anywhere in the work surface till the range of accessible. In turning operation laser fixed over the machining area whereas in milling operation laser need to be move along with the work surface which is presented in figure 7 (a) & (b). During milling operations intensity of laser is increases than the turning operations. Since, the cyclic rotation in turning operations work piece rotated which causes the temperature rise (Jeon & Lee, 2012).

Figure 7. Laser assisted machining (a) Turning (b) Milling
(Sun et al., 2011)

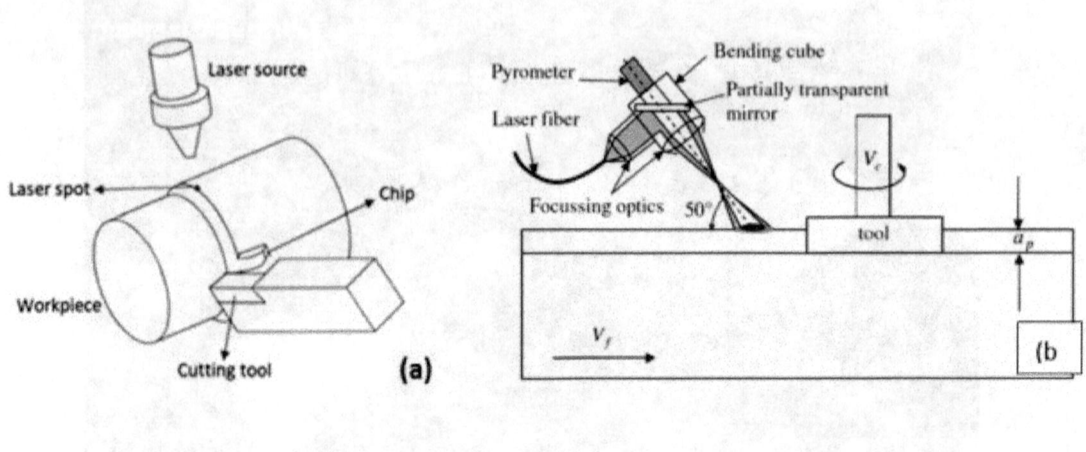

Moreover, single laser beam is not sufficient for heating the work piece for milling operations due to large area of work piece. Therefore multiple lasers can be included with milling operations. Location of laser to be placed on the work piece is very important, since the heated area must be cut immediately to avoid sudden cooling. The sudden cooling of work piece required higher cutting forces which resulting the improper machining such as excess tool wear, chatter and high surface roughness due to the increasing cutting forces (Kibria et al., 2013).

Plasma-Assisted Machining

Plasma heating is one of the prominent techniques before to machining of tough to cut materials or alloys. Plasma torch is placed over the work surface like as laser heating and its similar to the laser heating method. Direct current is applied to the tool and work piece which produces the arc and along with this plasma gas is supplied. Thereby high velocity plasma generated and supplied using nozzle over the work surface. Along with this inert gas is supplied around the plasma which protects the metal surface from the excess heat. This method heating temperature maintained between 400 to 1000 o C using indirect (convective) heating method (Lopez de Lacalle et al., 2004). Generally, this method applied in turning operations rather than milling operations due to its nature of functions. Because in milling, more complicated motions and less feed rates were applied. This leads to the more improper heating spread on the work surface which causes the less machining quality. The illustration of plasma heating method is shown in figure 8.

Figure 8. Plasma assisted machining
(Lopez de Lacalle et al., 2004)

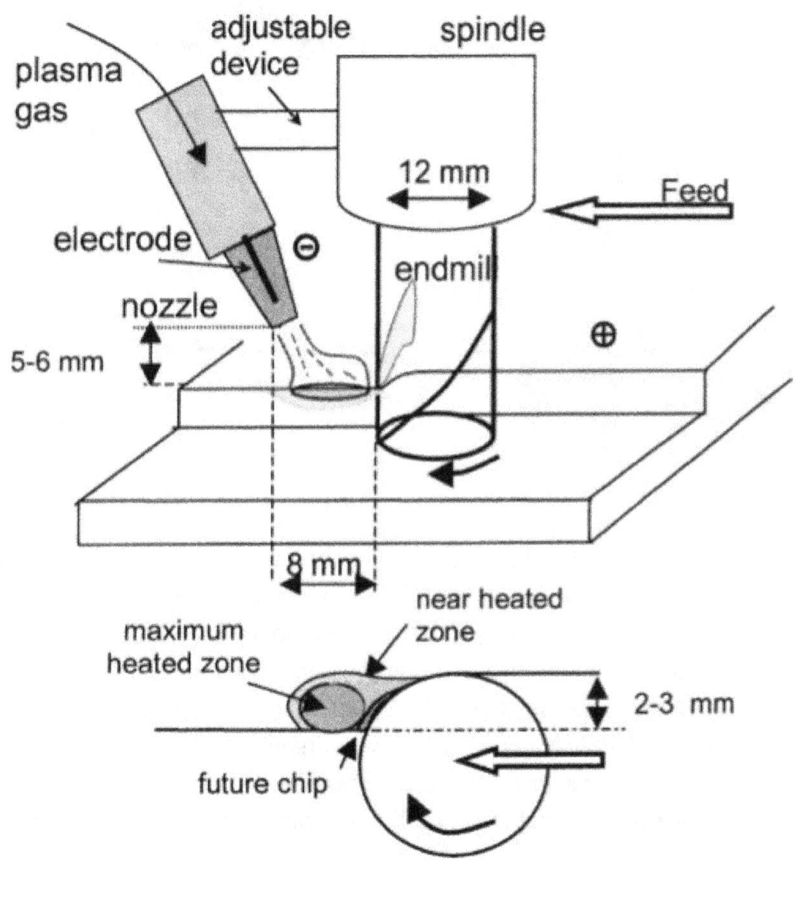

Induction-Assisted Machining

Induction electric coils are employed to heat the workpiece using high voltage electric current. This type of heating covers larger area of work piece than laser and plasma heating methods. The basic principle that magnetic force induced by the application of reverse direction of current in the electric coil. This current intensity produces the resistance when pass through the work piece which induce the heat energy. The rate of heat energy is influenced through various factors such as current voltage, frequency and type of work material. However, induction heating delivers uniform heating to the workpiece and system contains of transformer, condenser and cooling setup [28]. Based on the application and requirement different type of coil has been used for the heating which is present in figure 9.

Figure 9. Types of heating coil models
(Gürgen & Sofuoğlu, 2021; Pandey & Datta, 2021)

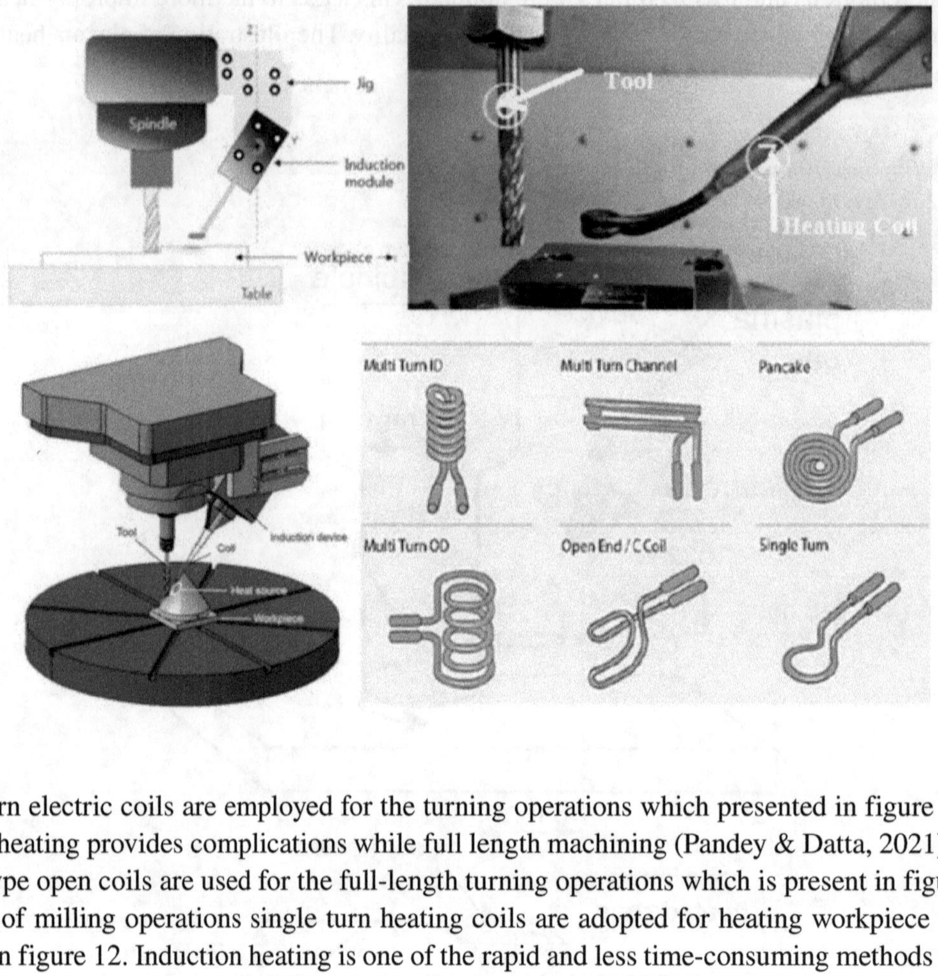

Multiturn electric coils are employed for the turning operations which presented in figure 10. This method of heating provides complications while full length machining (Pandey & Datta, 2021). Therefore, C – type open coils are used for the full-length turning operations which is present in figure 11.

In case of milling operations single turn heating coils are adopted for heating workpiece which is presented in figure 12. Induction heating is one of the rapid and less time-consuming methods for heating. Also, this method distributed the heat ununiformly (Gürgen & Sofuoğlu, 2021). Since,eddy current passed beneath the work surface for an extended area. This phenomenon much needed in milling operation; therefore, induction heating is suitable for milling operations than turning operation.

Figure 10. Multiturn induction heated turning and drilling
(Gürgen & Sofuoğlu, 2021)

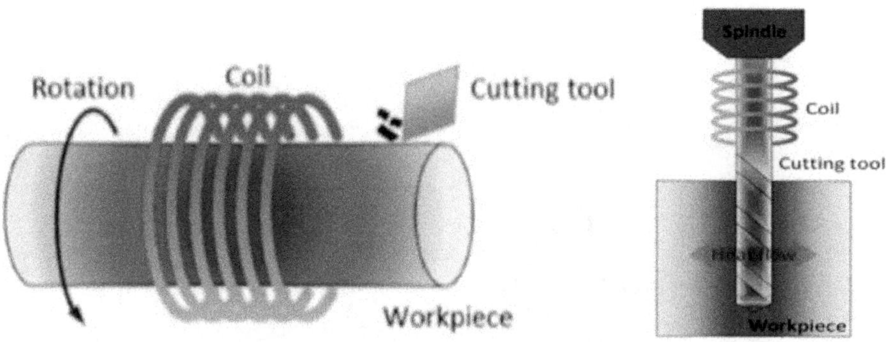

Figure 11. C-Type open heating coil
(Pandey & Datta, 2021)

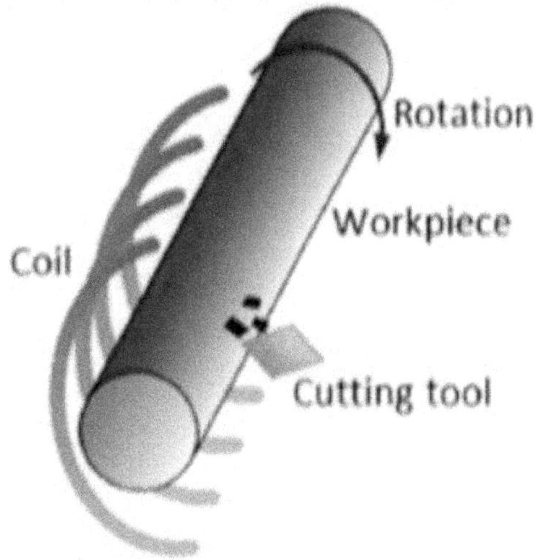

Infrared Assisted Machining

In order to reducesthe tool force and increase the machining rate for tough to cut materials infrared assistance has been used in conventional machining (Sanchez et al., 2014). In this method, due the work piece heat shear stress of material hindered for great level thereby cutting force reduced significantly.

Quartz electrical resistance was fixed near the work piece for heating while machining which is presented in figure 13. The experiment result predicts that life of cutting tool and surface roughness improved around 340% and 205% respectively. Moreover, continuous chip formation noticed on the machining which indicates that less energy consumption was paid in the machining process.

Figure 12. Single turn heating in milling operation
(Vieira et al., 2001)

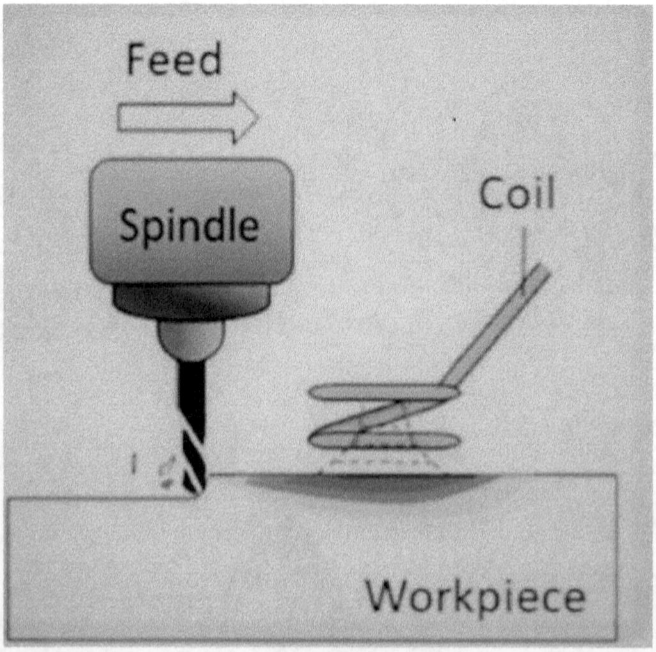

Figure 13. Infrared assisted machining
(Anderson et al., 2006)

Tunnel Furnace Heated Machining

The aforementioned techniques are some of the common heating techniques for heat aided machining. In the same way few more techniques are considered for heating the workpiece before machining operations. Therefore, tunnel furnace is used to heat the work piece only for turningoperations. Tunnel furnace image is shown in figure 14.

Figure 14. Tunnel furnace
(Witczak et al., 2015)

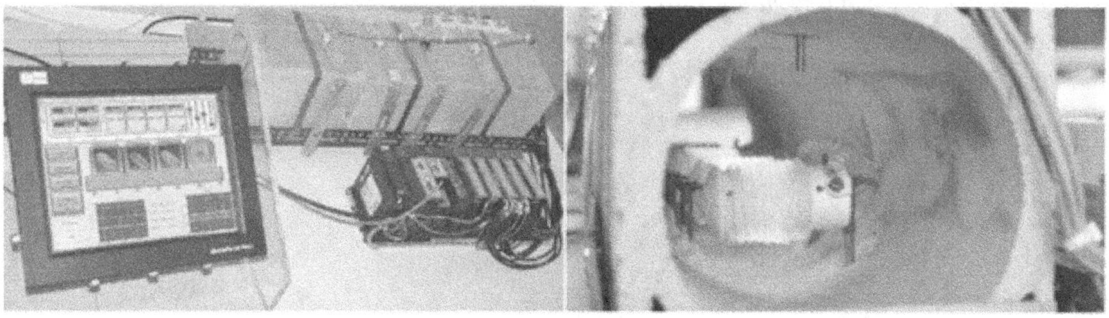

In this method, work piece heated for certain temperature using tunnel furnace and fixed in the fixer for turning operation. Also, excess number of thermal strains in the work materialleads to the higher unwanted material removal from the work piece. Moreover, oxidization take place on work material due to the large heat distributions on the material and work piece cool down quickly for the same reason (Witczak et al., 2015). Therefore, accuracy of machining diminished significantly. However, this method of heating is cost effective and can be used where the accuracy is not a subject of matter.

Flame Assisted Machining

In this method, commercially available gases such as liquid petroleum gas, acetylene, propane and natural gases are used for the production of flames and with the aim of increasing heat intensity higher oxygen ratio is used. Flame assisted machining is shown in figure 15. This method exhibits non controllable heat distribution among the work surface which leads to the uncertainty in cutting force. Also, micro structure of work material changes very suddenly due to the uncontrollable temperature (Parida & Maity, 2021). Thereby surface finishing is affected significantly and due to the open flame heating causes for the oxidization effect. Hence, flame heating method is not suitable for the aerospace materials.

Recent Development in Conventional Machining

High speed machining (HSM) is also termed as high-speed cutting (HSC) has provided a significant contribution to the manufacturing sector for last three decades. The research activities on HSM were mainly focused on turning and milling of various tough materials such as ferrous alloys, titanium, tungsten, composites materials and nickel-oriented alloys (Byrne et al., 2003). There are numerous developments were carried on the HSM process such as spindles, feed control systems, machining centers and digitalization of HSM which is presented in figure 16.

This kind of advancement in machining technology are very important for the manufacturing industries to obtain less production cost, short machining time, high accuracy and increasing of production. HSM process is a complex term to explain, since it has been controlled by many factors including the work and tool material properties, machining speed and feed. In general, HSM can be distinguished based on the cutting operation, work material to be machined, cutting force and process stability. The difference of HSM and conventional machining has presented using figure 17.

Figure 15. Flame assisted machining [34]

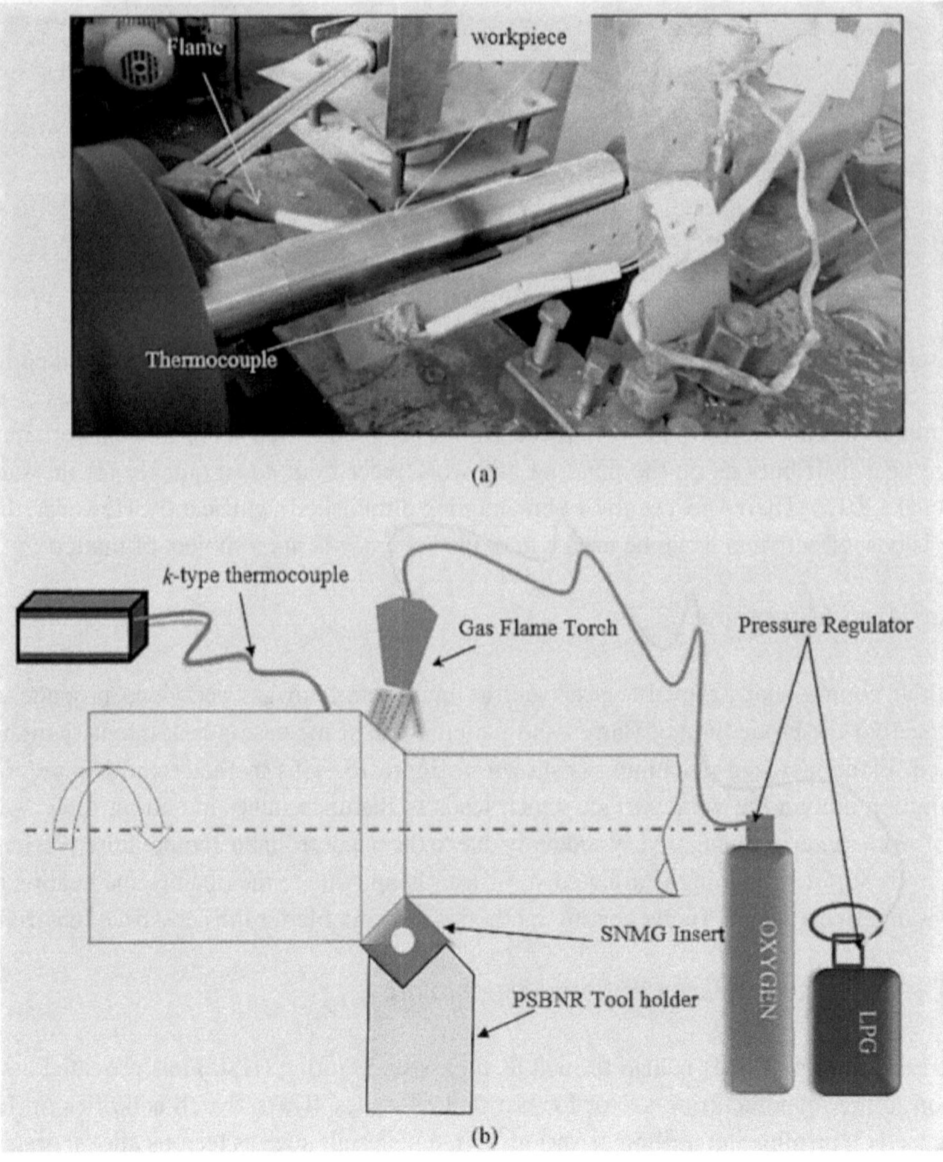

Figure 16. High speed machining machine
(Komanduri et al., 1985)

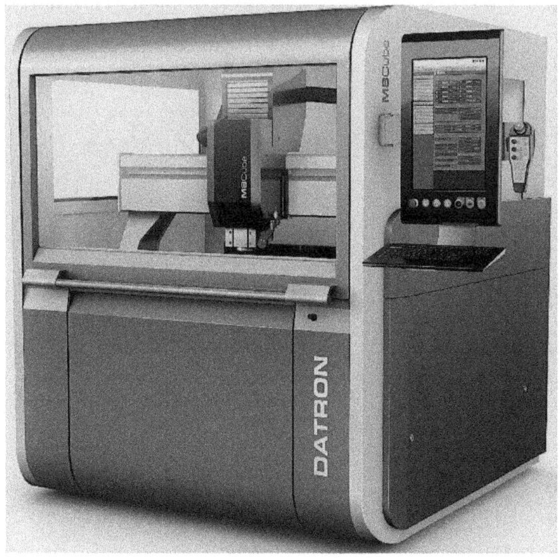

Figure 17. Turning (a) High speed machining (b) Conventional
(Schulz, 2001)

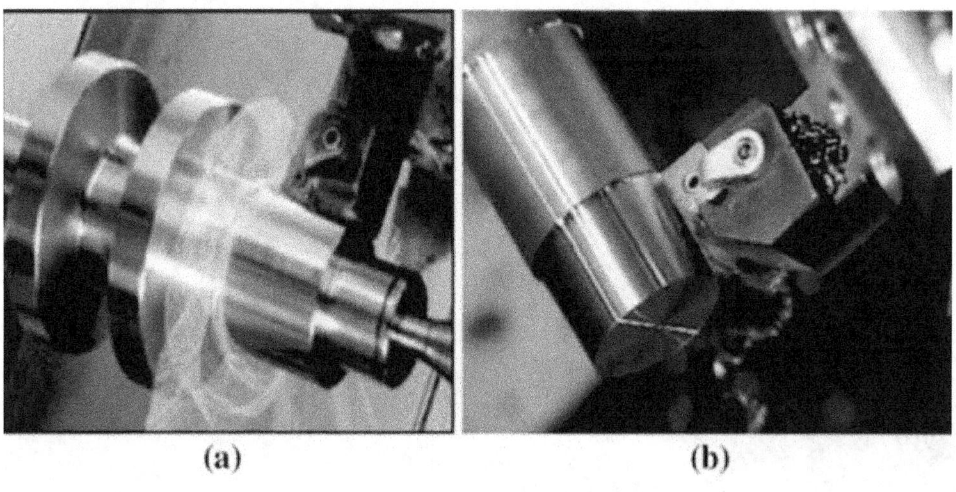

The speed of spindle produced from 200 to 300 RPM for machining of tough materials. This range of spindle speed employed almost all kind of conventional machining such as drilling, milling etc., in terms of materials removal process. The experiments carried out the experiments on conventional micro drilling on titanium alloy with the spindle speed of 10000 to 90000 RPM (Smith & Tlusty, 1997). But, high speed of tool exposes the aggressive tool wear which promotes the chatter surface on the work material. Therefore, various novel techniques are evolved recently to understand and improve the machining nature in HSM. Although various techniques upgraded for the enhancement, HSM is one of the prominent and significant development in conventional machining. In line up with that ultrasonic vibration is employed in HSM process which is displayed in figure 18.

Figure 18. Ultrasonic vibration assisted HSM process
(Toenshoff & Denkena, 2013)

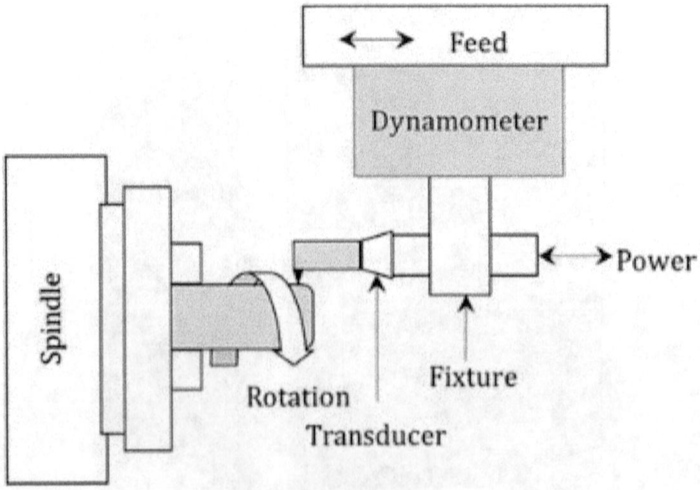

Figure 19. Coolant supplying Strategies(a) LN_2jet (b) MQL mist (c) Flood (d) dry machining
(Ye et al., 2012)

Using this method, burr free and increased tool life obtained by the researchers on machining of tough material such as titanium alloy. The vibration of tool arises in the route of cutting accomplishment and speed of cutting permits parting of tool tip from work material irregularlythough vibration surpasses the serious range. Sui et al. (Toenshoff & Denkena, 2013) studied the viability of ultrasonic vibration assisted turning at high speed on titanium. They noted 20 to 50% decreased cutting force when compared to conventional machining. Also, HSM process efficiency improved around 90% in material removal and surface finish. Dry machining is another one of novel approach with the conventional machining process. It is not possible technique due the technological limitations in HSM. But minimum quantity lubrication (MQL) might offer a better alternative to dry machining.

In general machining process 20 to 100 l/h coolant lubricants are employed whereas in MQL technique 0.03 to 0.2 l/h coolants are employed (Coromant, 1999). The researchers attempt with various forms of coolant oils such as in MQL technique. Therefore, cryogenic coolant liquid nitrogen (LN2) has been used in conventional turning with different strategies such as LN2 jet, MQL mist, Flood form which are represented in figure 19. LN2jet coolant produces better machining performance in terms of surface finish and materials removal rate. However, this method could not employ for all areas due to the cost.

CONCLUDING REMARKS

The aforementioned discussion that conventional machining processes are exposes the better machining ability with the novel upgraded techniques, especially for metals and composite materials. Conventional machining process still has potential to cutting the metals/difficult to cut metals and composite with the certain specifications. But the upgrading of technological limits is increasing the challenges for conventional machining. Hence, in order to meet the new challenges with manufacturing industry many outcomes and modifications are carried out such as vibration, heat, coolant and tool assistance. Although, due to the numerous applications still more modifications can be experiment with the conventional machining process. This chapter brings an outline of the different applications, issues and developments that are carried out in last few decades and lead the manufactures and researchers in selecting suitable conventional machining practices according to the requirement. Based on the discussion on the development on conventional machining practices are majorly focused on formation of chips which is very an essential for obtaining the quality surface finish on the work piece. Tool material and its geometry plays significant role for the higher machining rate. The adoption of new mechanism such as vibration, heat with machining is providing an excellent contribution on the surface finish. Whereas, controlling of temperature need to be modify for the betterment of machining performance. Findings of composite material machining with conventional is not well organized, hence there are various openings are found related to the current industry requirement. Moreover, due to digitalization everywhere machining and their process parameters need to be modified.

REFERENCES

Anderson, M., Patwa, R., & Shin, Y. C. (2006). Laser-assisted machining of Inconel 718 with an economic analysis. *International Journal of Machine Tools & Manufacture*, *46*(14), 1879–1891. doi:10.1016/j.ijmachtools.2005.11.005

Boothroyd, G. (1988). *Fundamentals of metal machining and machine tools* (Vol. 28). CRC Press.

Byrne, G., Dornfeld, D., & Denkena, B. (2003). Advancing cutting technology. *CIRP Annals*, *52*(2), 483–507. doi:10.1016/S0007-8506(07)60200-5

Childs, T. H. C., Maekawa, K., Obikawa, T., & Yamane, Y. (2000). *Metal cutting. In Theory and Applications*. Arnold.

Coromant, S. (1999). Application Guide: Die &Mould Making. *Sandvik Coromant*, 41-42.

De Weck, O. (2014). Trends in advanced manufacturing technology innovation. *Production in the Innovation Economy*, 235.

Gürgen, S., & Sofuoğlu, M. A. (2021). Advancements in conventional machining: a case of vibration and heat-assisted machining of aerospace alloys. In *Advanced Machining and Finishing* (pp. 143–175). Elsevier. doi:10.1016/B978-0-12-817452-4.00015-4

Jawahir, I. S., & Van Luttervelt, C. A. (1993). Recent developments in chip control research and applications. *CIRP Annals*, *42*(2), 659–693. doi:10.1016/S0007-8506(07)62531-1

Jeon, Y., & Lee, C. M. (2012). Current research trend on laser assisted machining. *International Journal of Precision Engineering and Manufacturing*, *13*(2), 311–317. doi:10.100712541-012-0040-4

Jiang, F., Liu, Z., Wan, Y., & Shi, Z. (2013). Analytical modeling and experimental investigation of tool and workpiece temperatures for interrupted cutting 1045 steel by inverse heat conduction method. *Journal of Materials Processing Technology*, *213*(6), 887–894. doi:10.1016/j.jmatprotec.2013.01.004

Kibria, G., Doloi, B., & Bhattacharyya, B. (2013). Experimental investigation and multi-objective optimization of Nd: YAG laser micro-turning process of alumina ceramic using orthogonal array and grey relational analysis. *Optics & Laser Technology*, *48*, 16–27. doi:10.1016/j.optlastec.2012.09.036

Knight, W. A., & Boothroyd, G. (2019). *Fundamentals of metal machining and machine tools*. CRC Press. doi:10.1201/9780429114243

Komanduri, R., Flom, D. G., & Lee, M. (1985). *Highlights of the DARPA advanced machining research program*. Academic Press.

König, W. (2008). *Fertigungsverfahren 1: Drehen, Fräsen, Bohren*. Springer-Verlag.

Kovacic, M., Brezocnik, M., Pahole, I., Balic, J., & Kecelj, B. (2005). Evolutionary programming of CNC machines. *Journal of Materials Processing Technology*, *164*, 1379–1387. doi:10.1016/j.jmatprotec.2005.02.047

Lawal, S. A., Choudhury, I. A., & Nukman, Y. (2012). Application of vegetable oil-based metalworking fluids in machining ferrous metals—A review. *International Journal of Machine Tools & Manufacture*, *52*(1), 1–12. doi:10.1016/j.ijmachtools.2011.09.003

Lopez de Lacalle, L. N., Sanchez, J. A., Lamikiz, A., & Celaya, A. (2004). Plasma assisted milling of heat-resistant superalloys. *J. Manuf. Sci. Eng.*, *126*(2), 274-285.

M'Saoubi, R., Axinte, D., Soo, S. L., Nobel, C., Attia, H., Kappmeyer, G., Engin, S., & Sim, W. M. (2015). High performance cutting of advanced aerospace alloys and composite materials. *CIRP Annals*, *64*(2), 557–580. doi:10.1016/j.cirp.2015.05.002

Manyindo, B. M., & Oxley, P. L. B. (1986). Modelling the catastrophic shear type of chip when machining stainless steel. *Proceedings of the Institution of Mechanical Engineers. Part C, Journal of Mechanical Engineering Science*, *200*(5), 349–358. doi:10.1243/PIME_PROC_1986_200_138_02

Naves, V. T. G., Da Silva, M. B., & Da Silva, F. J. (2013). Evaluation of the effect of application of cutting fluid at high pressure on tool wear during turning operation of AISI 316 austenitic stainless steel. *Wear*, *302*(1-2), 1201–1208. doi:10.1016/j.wear.2013.03.016

O'Toole, L., Kang, C., & Fang, F. (2020). Advances in rotary ultrasonic-assisted machining. *Nanomanufacturing and Metrology*, *3*(1), 1–25. doi:10.100741871-019-00053-3

Pandey, K., & Datta, S. (2021). Hot machining of difficult-to-cut materials: A review. *Materials Today: Proceedings*, *44*, 2710–2715. doi:10.1016/j.matpr.2020.12.687

Parida, A. K., & Maity, K. (2021). Study of machinability in heat-assisted machining of nickel-base alloy. *Measurement*, *170*, 108682. doi:10.1016/j.measurement.2020.108682

Prieto, J. R., Carbonell, J. M., Cante, J. C., Oliver, J., & Jonsen, P. (2018). Generation of segmental chips in metal cutting modeled with the PFEM. *Computational Mechanics*, *61*(6), 639–655. doi:10.100700466-017-1442-z PMID:31007328

Rahman, M. A., Rahman, M., Kumar, A. S., & Lim, H. S. (2005). CNC microturning: An application to miniaturization. *International Journal of Machine Tools & Manufacture*, *45*(6), 631–639. doi:10.1016/j.ijmachtools.2004.10.003

Sanchez, L. E., Mello, H. J., Neto, R. R. I., & Davim, J. P. (2014). Hot turning of a difficult-to-machine steel (saexev-f) aided by infrared radiation. *International Journal of Advanced Manufacturing Technology*, *73*(5-8), 887–898. doi:10.100700170-014-5879-3

Schiller, G. J. (2015, March). *Additive manufacturing for Aerospace. In 2015 IEEE Aerospace Conference*. IEEE.

Schulz, H. (Ed.). (2001). *Scientific fundamentals of HSC*. Hanser.

Smith, S., & Tlusty, J. (1997). *Current trends in high-speed machining*. Academic Press.

Sun, S., Brandt, M., Barnes, J. E., & Dargusch, M. S. (2011). Experimental investigation of cutting forces and tool wear during laser-assisted milling of Ti-6Al-4V alloy. *Proceedings of the Institution of Mechanical Engineers. Part B, Journal of Engineering Manufacture*, *225*(9), 1512–1527. doi:10.1177/0954405411411608

Teti, R. (2002). Machining of composite materials. *CIRP Annals*, *51*(2), 611–634. doi:10.1016/S0007-8506(07)61703-X

Thandra, S. K., & Choudhury, S. K. (2010). Effect of cutting parameters on cutting force, surface finish and tool wear in hot machining. *International Journal of Machining and Machinability of Materials*, *7*(3-4), 260–273. doi:10.1504/IJMMM.2010.033070

Toenshoff, H. K., &Denkena, B. (2013). *Basics of cutting and abrasive processes*. Academic Press.

Turley, D. M., & Doyle, E. D. (1982). Microstructural behavior--Its influence on machining. *On the Art of Cutting Metals—75 Years Later. A Tribute to FW Taylor*, 14-19.

Valiev, R. (2004). Nanostructuring of metals by severe plastic deformation for advanced properties. *Nature Materials*, *3*(8), 511–516. doi:10.1038/nmat1180 PMID:15286754

Vieira, J. M., Machado, A. R., & Ezugwu, E. O. (2001). Performance of cutting fluids during face milling of steels. *Journal of Materials Processing Technology*, *116*(2-3), 244–251. doi:10.1016/S0924-0136(01)01010-X

Vyas, A., & Shaw, M. C. (1999). *Mechanics of saw-tooth chip formation in metal cutting*. Academic Press.

Witczak, M., Mrugalski, M., & Korbicz, J. (2015). Towards robust neural-network-based sensor and actuator fault diagnosis: Application to a tunnel furnace. *Neural Processing Letters*, *42*(1), 71–87. doi:10.100711063-014-9387-0

Xu, W. X., & Zhang, L. C. (2015). Ultrasonic vibration-assisted machining: Principle, design and application. *Advances in Manufacturing*, *3*(3), 173–192. doi:10.100740436-015-0115-4

Ye, G. G., Xue, S. F., Ma, W., Jiang, M. Q., Ling, Z., Tong, X. H., & Dai, L. H. (2012). Cutting AISI 1045 steel at very high speeds. *International Journal of Machine Tools & Manufacture*, *56*, 1–9. doi:10.1016/j.ijmachtools.2011.12.009

Chapter 6
Effect of Tool Electrodes on Electrochemical Micromachining Processes

Geethapriyan T.
SRM Institute of Science and Technology, India

ABSTRACT

In this experiment, the work piece is electrochemically machined which finds its applications in brine heaters, heat exchangers, propeller shafts, and pumps. The process is carried out using sodium chloride electrolyte and copper beryllium wire as tool electrode which is heat treated in three different methods which are annealing, quenching, and normalizing. The response parameters like theoretical and experimental metal removal rate have been measured and studied by varying machining parameters like voltage, frequency, concentration of electrolyte, and duty cycle. Based on the values obtained from the experiment, it is found that quenched tool electrodes have better machining capabilities than the other heat-treated tool electrodes and untreated tool electrodes.

INTRODUCTION

Electrochemical Micromachining (EMM) finds application in various fields such as aerospace, medical and automobile (Soundarrajan & Thanigaivelan, 2020). There are various factors that influence the EMM performance namely current, voltage, electrolyte concentration, temperature and tool profile. Pang et al. 2019 proposed a new idea in electrochemical method with a floating cathode to regulate the inter electrode gap. The modelling and experimental research was conducted considering the electrochemical machining with a large span floating cathode. The study shows that a stable inter-electrode gap can be formed in ECM through the liquid electrolyte to support the cathode, and the inter-electrode gap. Bhattacharyya and Mitra (2002) states that electrochemical micromachining is preferred to other non-conventional machining techniques like laser beam machining, ultrasonic machining, electro discharge machining are thermal processes and causes excessive deformation on heat affected zones and micro

DOI: 10.4018/978-1-7998-9574-9.ch006

cracks on the work piece. It is inferred that machining rate depends on the atomic weight, current and time for dissolution and it does not depends on the hardness and other characteristics of the material. Uttarwar & Chopde (2013) discussed about the influencing parameters like voltage, current, electrolyte concentration, feed rate, flow rate, MRR and surface roughness. Machining was performed on SS304 using a copper tool and NaCl electrolyte. Irregular MRR was observed at high current and increase in current results in decrease of surface roughness. As electrolyte flow rate increases, the ion mobility between work piece and electrolyte increases thus increasing the chemical reaction speed and results in increased MRR.

Figure 1. Simplified illustration of ECMM Setup

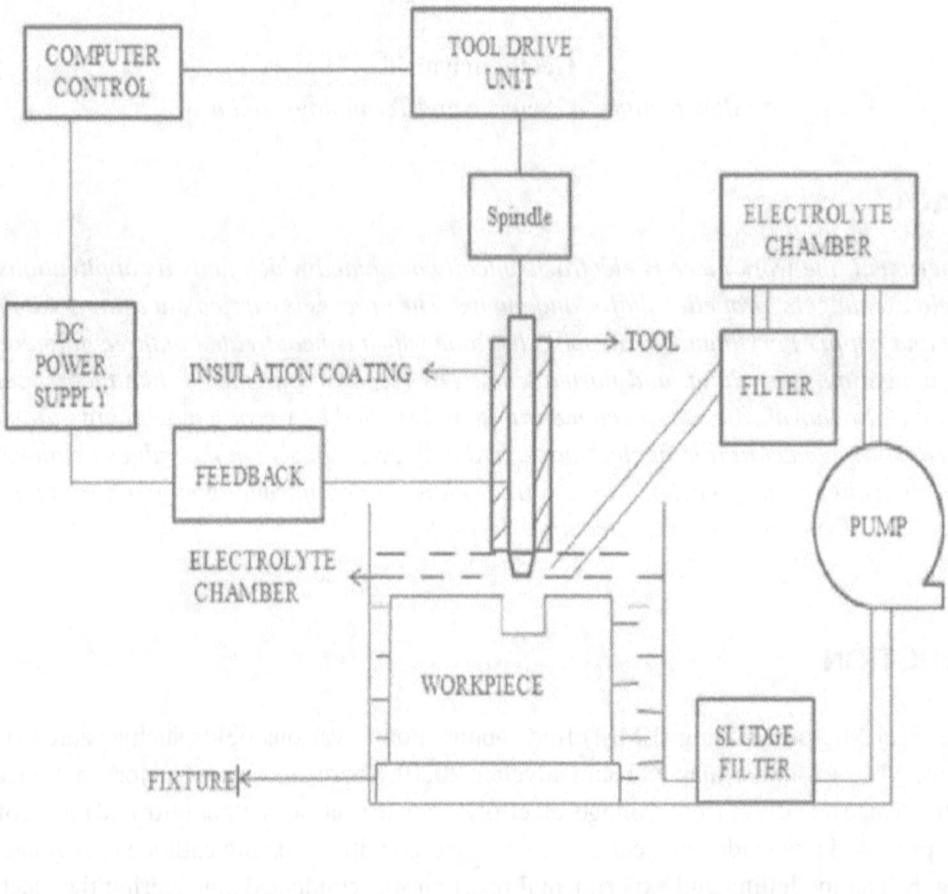

Bhattacharyya & Munda (2003) states that the electrochemical micro machining is considered to be among the few more effective advanced micromachining process using which a wide range of applications are calculated. Investigation indicates most predominant parameters like frequency, voltage, duty cycle and electrolyte concentration provide better MRR with low overcut. Kozak et al., (2004) states that the electrochemical machining is used in making advanced and complex shapes in aeronautical industries and its capability to machine macro and micro features for biomedical applications. Research and development in the process capabilities, tool, electrolytes and hybrid processes were presented. It is noted that ECM is used to fabricate micro electrodes and these tools have been used in ECM or EDM processes. Florence et al (2015) studied the technological advances in electrochemical machining processes. He concluded that electrolyte concentration and voltage are the most influencing parameter on overcut and material removal rate using grey relational analysis. It was noted that rate of machining and overcut were directly proportional to electrolyte concentration and applied voltage. Moon et al (2002). conducted experiment in micro ECM setup for fabrication of micro channel in hydrodynamic bearing. They found that 2 electrolytes, water mixed sodium nitrate and NaOH are used in this experiment to find better results. Also they highlighted about current identification method to find inter electrode gap characteristics, electrolyte concentration, pulse current and voltage based on model and experiments.

Raja et al. (2016) performed a study on the effect of electrochemical micro machining process parameters on material removal rate of the beryllium copper C17200 (work piece). The process parameters considered were electrolyte concentration, machining voltage and feed rate. To optimize the response measures, Grey Relational Analysis was used and concluded that the most influencing parameter on metal removal rate is feed rate. Wang et al., (2020) developed electro-chemical framework with the advancement of smaller scale, miniaturised scaling methods have turned into a hot issue in current industries.Smaller scale metal opening with high calibre are generally utilized in pieces of centre types gear in fields of vitality control, conveying and therapeutic apparatus. In the miniaturised scale openings, a couple of innovations, for example, smaller scale mechanical processing, small scale electro release machining and can be utilized to machine small machining are accessible. Miniaturised scale processing scale structures at rapid and minimal effort. The micron level holes were made on stainless steel by using ECM. To fulfil the requirements of the process electrochemical machining was build up now using the desired electrode made by block electrical discharge grinding a of analysis were done to influence the particular shape of electrode used in the machining efficiency and effectiveness when compared with a regular cylindrical electrode. Finally after the investigation it was concluded that the particular slot that were built by shape electrode were having a smaller side gap and smaller frontal gap as compared to the cylindrical electrode. Liu et al.(2018) demonstrated that stray current and insulation of tool are responsible for unacceptable ion dissolution in electro chemical micro machining (ECMM). They suggested that the insulated film on silicon-based tool micro electrodes is a critical process to restrict the stray current and improve the machining localization. Their experimental results show that they found reliable tool which is silicon based material and a insulation work on tool was better in order to prevent corrosion. Senthilkumar et al (2011) states that increase in applied voltage and tool feed rate causes a high current density at the inter electrode gap, and thus results in high material removal rate. As electrolyte flow rate and electrolyte concentration increases, the chemical reaction at the machining area gets speed up and result high material removal rate. From the experiment it was also observed that the surface roughness and electrolyte concentration are directly proportional to each other. With increase electrolyte flow rate the reaction products and bubbles generated in the area of machining are flushed away properly which leads to decrease surface roughness. Thanigaivelan et al (2010) studied about the performance of different sized tool, shape and design in ECMM. Flat, conical with rounded and wedged tool electrodes were used to determine the overcut and material removal rate by various parameters of electrolyte concentration. From the experimental result it is inferred that wedge tool produces higher MRR, whereas conical with rounded tool has better accuracy than flat tool electrode.

Millet et al., (2019) studied about the effect on heat treatment of copper beryllium upon shock response. In this study he carried out heat treatment by solution treating the copper beryllium (C17200) at 790 degree Celsius. Millet observed that, tensile strength, shear strength and elastic limit of the aged condition copper beryllium was significantly stronger than the solution treated copper beryllium. On heat treating the grains of the material gets altered and thus cause the material to restrict the range of elastic limit. Zhu et al (2006) studied about the change of mechanical properties and electrical conductivity of Cu-Te alloys on the effects of annealing process. The study shows that during the annealing of copper tellurium alloy, the re-crystallization and precipitation occur simultaneously and also it is found to be increase in grain with respect to the soaking time and temperature. The electrical conductivity increases due to increase in grain boundaries that restrict the dispersion of ions.

The marginal researches have been conducted to see the effect of change in tool electrode in electrochemical micro machining. In electrochemical micro machining, the bare tool electrode has less life duration. The tool must possess better electrochemical conductivity and stability in order to obtain better machining values. In this present work, to study the influence of heat treated copper beryllium tool electrode on ECMM process and to analyse various parameter such as theoretical and experimental material removal rate by varying the input parameters such as applied voltage, electrolyte concentration, frequency and duty cycle.

EXPERIMENTAL DETAILS

The work piece is fixed using a holder in a machining chamber. Tool which is fixed in a tool holder is made to go near the job with the help of control panel equipped with press buttons which helps to maintain the inter electrode gap. A micro controller (Arduino) governs the progress and man oeuvres the tool vertically using the help of NEMA 17 stepper motor. The input parameters like rate of feed, voltage, etc. are set. The machining is initiated by the electrolyte that flows between work piece metal and tool electrode. The machining is achieved when the tool creates a hole through the work piece due to dissolution

Figure 2. Experimental Setup of ECmM

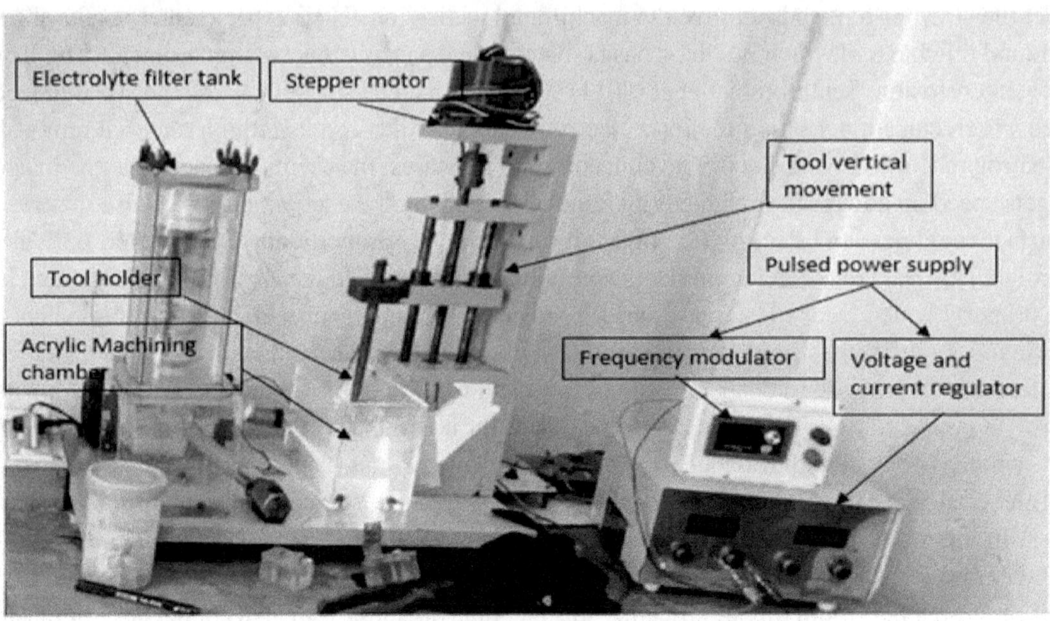

The ECmM setup consists of several components such as frequency modulator, voltage and current regulator, tool holder, electrolyte filter tank, filter pump and an machining chambaer made up of acrylic. The frequency modulator and volatage regulator together makes up the pulsed power supply unit. The frequency and the duty cycle can be varied using the frequency modulator.

MATERIALS AND METHODS

The work piece used in this study is Monel 400 alloy, whose composition consists of Iron, copper (cu), Nickel (Ni), Manganese (Mn), carbon, Sulphur and Silicon at different percentages by weight. The work piece Monel alloy is selected based on several factors such as hardness, conductivity etc., and Monel 400 alloy is stronger than pure nickel. Monel 400 alloy finds its applications in brine heaters, pumps, propeller shafts, and in other marine applications due to their high corrosion. Monel 400 alloy has high repetitive life cycle and relatively low cost. Monel 400 alloy is resistant to acids and alkalis.

Table 1. Chemical composition of Monel400 alloy

Material	Nickel	Copper	Iron	Manganese	Carbon	Silicon	Sulphur
MONEL 400	64.95%	32.37%	1.42%	0.71%	0.0001%	0.15%	0.010%

The tool electrode is selected based upon certain requirements such as easy fabrication, high electrical and thermal conductivity, electrochemical stability, stiffness to withstand electrolyte pressure, better machinability and high corrosion resistance. The tool electrode must possess high wear resistance and have high repetitive cycles. Based on these criteria, copper beryllium of 0.5mm diameter has been selected as the tool electrode.

Table 2. Chemical composition of copper beryllium tool electrode

Material	Beryllium	Copper	Iron	Aluminium	Silicon
Copper Beryllium	1.92%	97.36%	0.008%	0.010%	0.03%

The electrolytes used in Electro chemical micro machining process is of two types namely passivating and non-passivating electrolytes. Electrolytes are used in ECMM process to fill the inter electrode gap between the tool and the work piece by proving free ions to carry the electric current for anodic dissolution. Non-passivating electrolyte is chosen for their aggressive anions. The electrolyte used in the process is Sodium chloride (NaCl) - Analytical reagent grade and the sodium chloride (NaCl) is selected based on the aggressive nature to react with anode and remove more material from workpiece to improve productivity.

The copper beryllium tool electrode of 0.5mm diameter is cut into 27 pieces of 40mm length each. The copper beryllium tool electrode pieces are placed in three different containers for three methods of heat treatment (annealing, quenching and normalizing). The heat treatment process takes place in three stages such as heating period, soaking period and cooling period. In quenching, the heated tool is rapidly quenched by dipping it in water. The tool is kept at room temperature after heating for normalizing process. Annealing is done by furnace cooling for a time period of 24 hours.

RESULTS AND DISCUSSION

Influence of Process Parameters on Actual MRR

From the figure 3 it is inferred that the influencing parameter for the annealed tool is applied voltage and electrode where the MRR increases with the increase in the applied voltage. This is because the applied voltage linearly increases the erosion rate of the material from the work piece with increase in current. And due to the higher current available at frontal gap between work piece and electrode thereby reducing the machining time and increasing the material removal rate.

Figure 3. Main effect plot for actual MRR

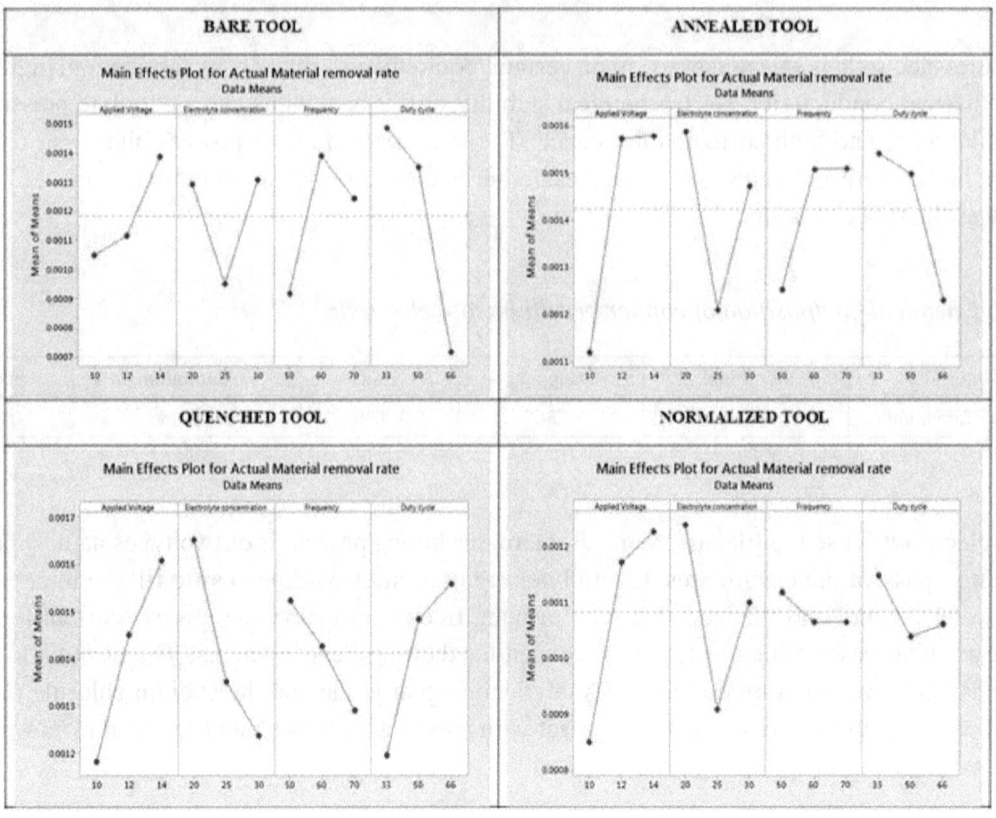

In the case of normalized tool electrode, it is found that the most influencing parameter is the duty cycle. From graph it can be observed that the MRR decreases with the increase in the duty cycle. At 33% duty cycle the material removal rate is high where the machining time less than the flushing time. And at 66% duty cycle the material removal rate is less even if the machining time is more than the flushing time. This is because the removed material or sludge which formed are not flushed away fully which creates spark as the tool touches the work piece thus resulting in decreased rate of material removal.

Effect of Heat Treated Tool on Actual MRR

From the obtained graph figure 4, we can infer that the quenched tool has highest MRR (actual). This is because the quenched tool electrode has fine grain structure compared to other tool electrodes. This facilitates the higher dissolution and removal of particles from the work piece thereby increasing the material removal rate. Also it was found that the MRR of quenched tool is 20.17% higher than untreated tool and normalized tool is 19.36% higher than untreated tool as they have higher electrical conductivity compared to other tool electrodes. Annealed tool has the least MRR having 8.57% lesser than untreated tool electrode due to the less electrical conductivity compared to other tool electrodes.

Figure 4. Comparison of actual MRR for the tool electrodes

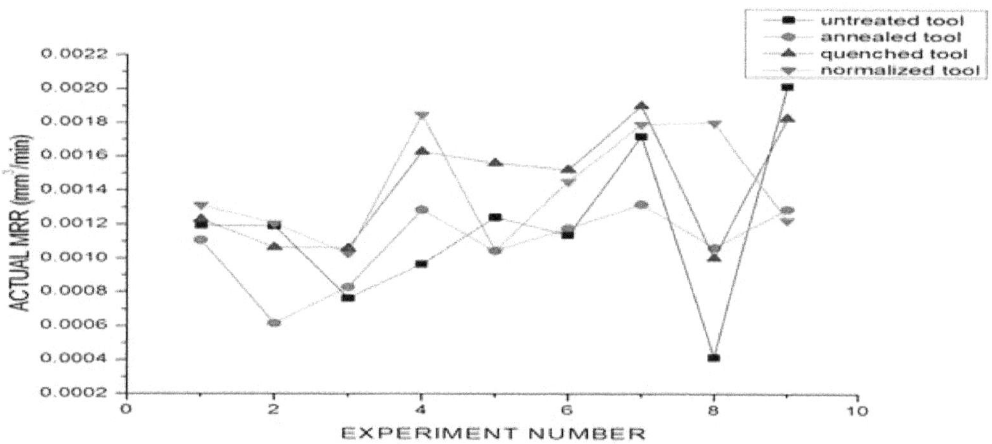

Influence of Process Parameters on Theoretical MRR

From the obtained mini tab plot figure 5, it is inferred that the frequency is the most influencing parameter for the untreated tool electrode where the rate of material removal increases as the frequency increases. At 50 Hz the MRR is less and at 70 Hz the MRR is high, this is because at high frequency level the pulse on time per periodic time is more compared to low frequency level. This results in less time taken for machining with increased MRR. In case of normalized tool electrode, the most influencing parameter is the electrolyte concentration where, the material removal rate increases with increase in the electrolyte concentration. This is because electrolyte concentration is the most influencing parameter. Higher concentration of electrolyte leads to more number of ions in the electrolytic medium which causes higher dissolution rate. As the no of ions in the medium increases there is an increase current density which leads to higher amount of material being removed.

In case of annealed tool electrode it is inferred that applied voltage is the influencing parameter for the annealed tool and electrode where the MRR increases with the increase in the applied voltage. The rate of erosion of the material from the work piece is influence by the applied voltage and increases linearly with current. And due to the higher current available at frontal gap between work piece and electrode thereby reducing the machining time and increasing the material removal rate. The most influencing parameter for the quenched tool is duty cycle where the material removal rate at 50% is high and at 66% it is observed to be less than 33%. This is because at 66% the load-on time is less and thus increasing the time taken for machining results in low MRR. And at 50% the load-on time and load-off time are equal so this gives the advantage of time for flushing away the by-products such as slurry, bubbles etc.

Figure 5. Main effect plot for theoretical MRR

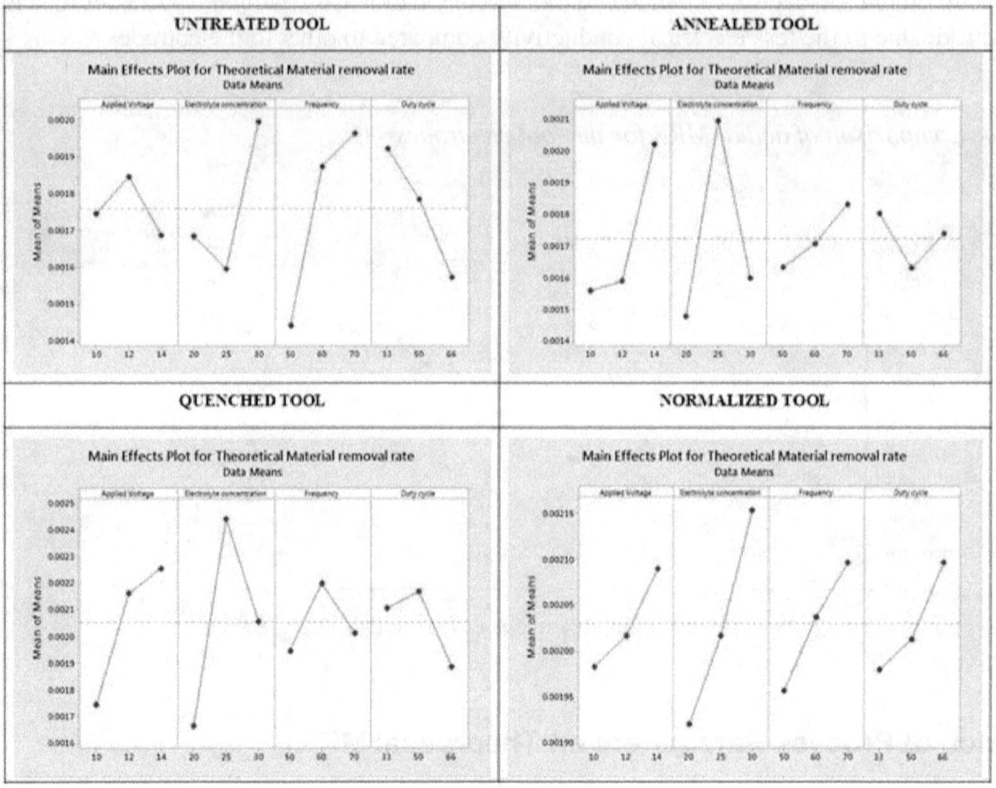

Effect of Heat Treated Tool on Theoretical MRR

From the obtained graph figure 6, we can infer that the quenched tool has high theoretical material removal rate. This is because the quenched tool electrode has high conductivity value compared to all other tool electrode. Also its fine grain structure facilitates the higher dissolution and removal of particles from the work piece thereby increasing the material removal rate. It was found that the MRR of quenched tool is

16.63% higher than untreated tool and normalized tool is 15.37% higher than untreated tool. Annealed tool has the least MRR having 1.99%lesser than untreated tool.

Figure 6. Comparison of theoretical MRR for the tool electrodes

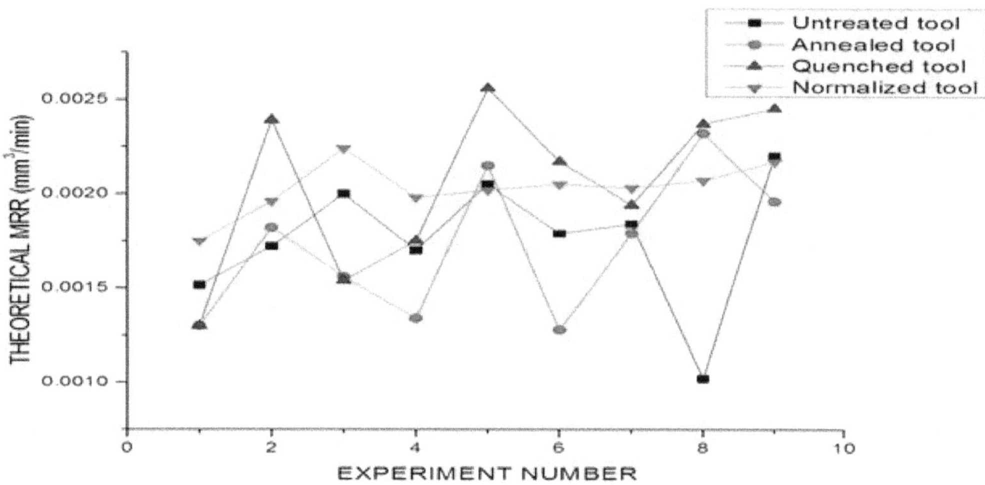

CONCLUSION

In this project, monel 400 alloy was machined using ECMM process parameters with Untreated and Heat treated tool electrode. The input characteristics such as applied voltage, electrolyte concentration, frequency and duty cycle were varied suitably to get the optimum values of response parameters like material removal rate, overcut, conicity and circularity. From the experiments that have been carried out using untreated and heat treated tool electrodes, the following conclusions have been drawn. High MRR has been obtained using quenched tool electrode than the other treated tools and untreated tool electrode because the quenched tool electrode has a higher electrical conductivity which means for the same applied voltage, the treated tool supplies more current through the electrode reducing the machining time, thus results in increasing material removal rate.

REFERENCES

Bhattacharyya & Mitra. (2002). Development of Electrochemical Micromachining (ECMM) Setup. *Robotics and Computer Integrated Manufacturing, 18*, 283–289.

Bhattacharyya, B., & Munda, J. (2003). Experimental investigation into electrochemical micromachining (EMM) process. *Journal of Materials Processing Technology, 140*(1-3), 287–291. doi:10.1016/S0924-0136(03)00722-2

Florence, T., Dharmalingam, S., Ragupathy, V., & Satishkumar, P. (2015). Machinablity study on Electrochemical Machining–A Review. *International Journal of Chemtech Research, 7*(6), 2596.

Kozak, J., Rajurkar, K. P., & Makkar, Y. (2004). Selected problems of micro-electrochemical machining. *Journal of Materials Processing Technology*, *149*(1-3), 426–431. doi:10.1016/j.jmatprotec.2004.02.031

Liu, W., Luo, Z., Yuan, T., Li, Y., Zhang, H., & Ao, S. (2018). The multi-physics analysis for a novel tool structure to improve the accuracy in electrochemical micro-machining. *International Journal of Advanced Manufacturing Technology*, *94*(5), 1991–2001. doi:10.100700170-017-0902-0

Millett, J. C. F., Whiteman, G., Park, N. T., Case, S., & Appleby-Thomas, G. (2019). The effects of heat treatment upon the shock response of a copper-beryllium alloy. *Acta Materialia*, *165*, 678–685. doi:10.1016/j.actamat.2013.01.021

Moon, Y. H., Lee, E. S., & Park, J. W. (2002). A study on electrochemical micromachining for fabrication of microgrooves in an air-lubricated hydrodynamic bearing. *International Journal of Advanced Manufacturing Technology*, *20*(10), 720–726. doi:10.1007001700200229

Pang, G., Xin, K., Bu, F., & Gao, T. (2019). Modeling and experiment study on inter-electrode gap in electrochemical machining with a floating cathode for rotary parts. *International Journal of Advanced Manufacturing Technology*, *101*(5), 2151–2164. doi:10.100700170-018-3051-1

Raja, K., Ravikumar, R., Thirunavukkarasu, V., & Kumaravel, S. (2016). Experimental Investigations on Machinability of Beryllium Copper C17200using Electrochemical Micro Machining. *Asian Journal of Research in Social Sciences and Humanities*, *6*(11), 367–373. doi:10.5958/2249-7315.2016.01198.9

Senthilkumar, C., Ganesan, G., & Karthikeyan, R. (2011). Parametric optimization of electrochemical machining of Al/15% SiCp composites using NSGA-II. *Transactions of Nonferrous Metals Society of China*, *21*(10), 2294–2300. doi:10.1016/S1003-6326(11)61010-8

Soundarrajan, M., & Thanigaivelan, R. (2020). Effect of coated and geometrically modified tools on performance of electrochemical micromachining. *Materials and Manufacturing Processes*, *35*(7), 775–782. doi:10.1080/10426914.2020.1740252

Thanigaivelan, R., & Arunachalam, R. M. (2010). Experimental study on the influence of tool electrode tip shape on electrochemical micromachining of 304 stainless steel. *Materials and Manufacturing Processes*, *25*(10), 1181–1185. doi:10.1080/10426914.2010.508806

Uttarwar, S. S., & Chopde, I. K. (2013). A Study of Influence of Electrochemical Process Parameters on the Material Removal Rate and Surface Roughness of SS AISI. *International Journal of Computational Engineering Research*, *3*(3).

Wang, M., Chen, X., Tong, W., Wang, J., & Wang, X. (2020). Influences of gap pressure on machining performance in radial ultrasonic rolling electrochemical micromachining. *International Journal of Advanced Manufacturing Technology*, *107*(1), 157–166. doi:10.100700170-020-05012-7

Zhu, D. C., Ke, T. A. N. G., Song, M. Z., & Tu, M. J. (2006). Effects of annealing process on electrical conductivity and mechanical property of Cu-Te alloys. *Transactions of Nonferrous Metals Society of China*, *16*(2), 459–462. doi:10.1016/S1003-6326(06)60078-2

Chapter 7
Elements of Industrial Automation and Robotics

Chandika Samynathan
Algonquin College, Canada

Kavitha G.
Muthayammal Engineering College, India

ABSTRACT

Automated manufacturing is an advanced manufacturing process which is enabled to correct its parameters autonomously during the machining process to achieve an assured objective. It can be developed through the institution of interactions with various systems, including machine tools, sensors and controller networks, actuators and simulation-based designs, as well as control algorithms. The automated manufacturing process can be realized in order to optimize process parameters automatically in real time, obtaining optimum processing performance and product quality.

AUTOMATIC CONTROL IN MANUFACTURING

There are many reasons automatic control is crucial to manufacturing. The following are the utmost significant reasons:

- To lower the cost of production
- To improve the quality
- To mitigate the drudgery of numerous routine, repetitive manual operations
- To provide a safe system

An automated control system consists of three basic elements: Input Sensors, PLC, and output actuators. The relationship between these elements is illustrated in figure 1.

DOI: 10.4018/978-1-7998-9574-9.ch007

- **Input Sensor** is a device that converts physical phenomena (example: force) to an electrical signal
- **Programmable Logic Controller** utilizes the sensors measurements and calculates control actions
- **Output actuator** is a device that converts the electrical signal to a physical action (example: solenoid valve or motor)

Figure 1. Elements of an automated control system

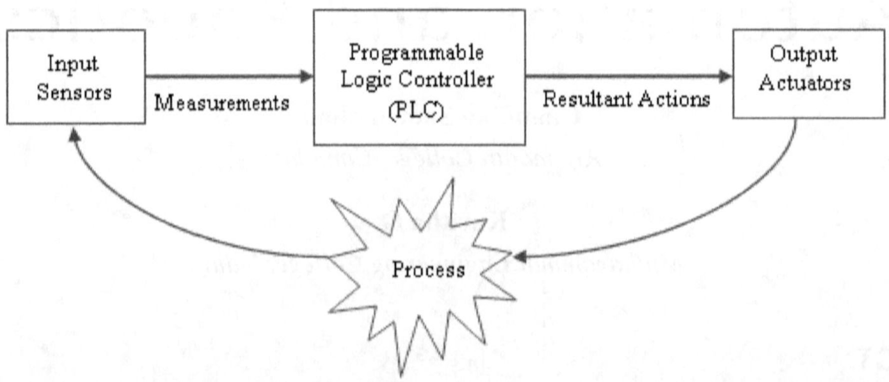

A process is a sequence of steps for the particular task. Industrial manufacturing processes are usually classified as continuous, batch, or discrete-parts manufacturing. Discrete manufacturing is a type of manufacturing process that involves different workstations, components, and sub-assemblies to produce the distinct items.

Example of a Discrete-Part Manufacturing Process

Today, numerous industries rely on robots and automation in manufacturing facilities. The figure 2 shows the layout of the automated line that machines the parts by the CNC machine. The CompactLogic controller PLC is used to control the sequence of operations. Ethernet communication enables connections with other industrial equipment such as Robots, Machine vision systems, 3D-Laser markers, etc. These connections can perform the timing and sequencing functions required to operate this equipment.

The raw materials (Part needs to be machined) are stacked at the Feed-In Dial Table station. It has four work stockers. Parts are stacked in the work stockers by an operator. For operator safety, the light curtains are installed at both ends of the Feed-In Dial Table station. The feed-in Dial Table and Main Dial Table is operated by the servo drives. 2D Robot transfers the Part from Feed-In Dial table to the Main Dial Table station 1. Station 2 has a Keyence machine vision camera for the part inspection. After the inspection, the part is transferred to station 3. The 6-axis robot unloads the part from station 3 and loads it into the CNC machine for machining. Further, the machined part is unloaded from the CNC machine and loaded into station 5. Stamping is done with the help of a 3D laser marker. After flipping, the part is transferred to the quality control and inspection station by a 2D axis Robot. The latter part is dipped into the oil station for rust protection and then to the packaging station.

Figure 2. Example of automated process centre

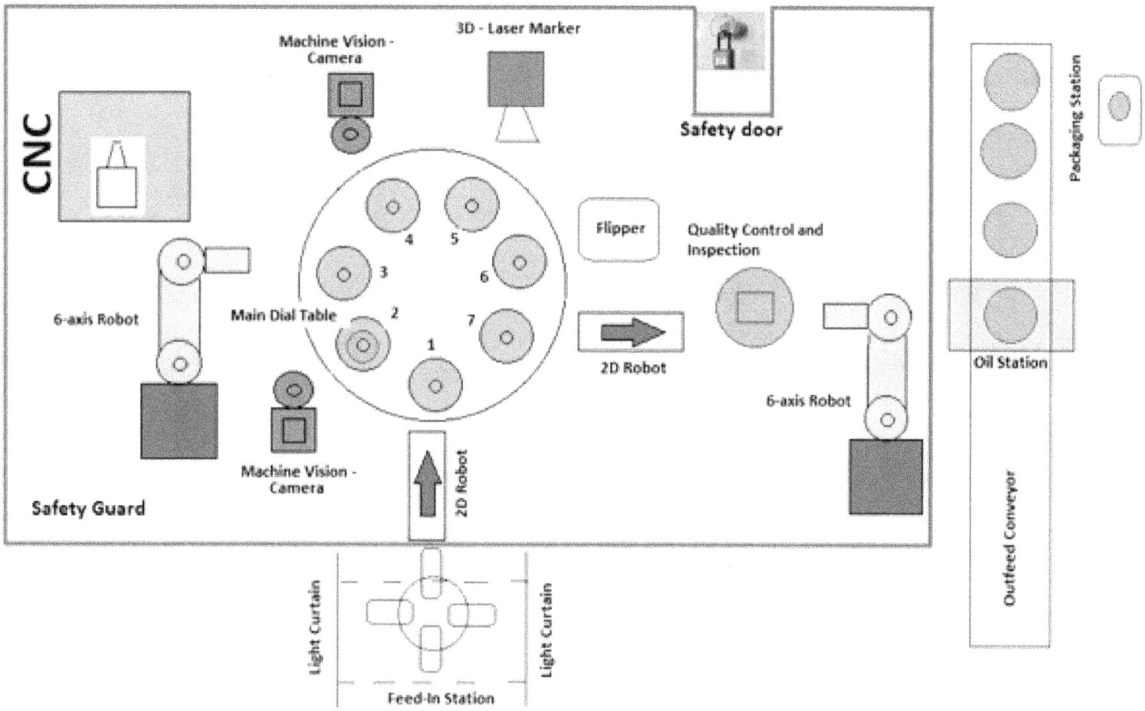

The Layout of the Automation Cell

Figure 3 shows the basic layout of an automated cell. Almost all automated systems are broken down into a combination of cells. One or more robots are integrated into a production system that includes a process machine and other support hardware, such as material handlers, conveyors, parts feeders, controllers, robots, and PLCs (Cell Controller). In most automated cells and machines, the PLC is the heart of the machine. It tends to make judgments about what is occurring in the cell. The main reason is, it has all signals communicated through it. That is the outputs of the HMI, CNC, and robot are input to the PLC. The flip is also accurate in that PLC outputs are often inputs to the other devices. These other devices normally have independent I/O of their own that relates only to their operation. They send and receive signals via the PLC only if the data is needed by another component of the cell.

Programmable Logic Controller (PLC)

PLC is a special-purpose industrial computer designed to monitor the state of input devices such as pushbuttons, switches, sensors, etc., and makes decisions based upon a ladder program to control the state of the output devices such as lights, solenoids, motors, etc. The basic architecture of a PLC is shown in figure 4. The main components of the PLC's are: the processor module, the power supply and the input/output modules. An input module receives signals (digital or analog) from various field devices (sensors, switches, pushbuttons, etc.) and converts these signals into an electrical signal. The CPU (or processor or Controller) section executes the program and makes decisions based on program instructions in memory (Allen-Bradley PLCs, 2013). Output modules convert control instructions from the processor into a digital or analog signal that can be used to control various field devices (motors, pumps, valves, etc.) (Gawali & Sharma, 2009). To input the desired instructions, a programming device is used. These instructions determine what the PLC will do for appropriate input.

Figure 3. Configuration of automation cell

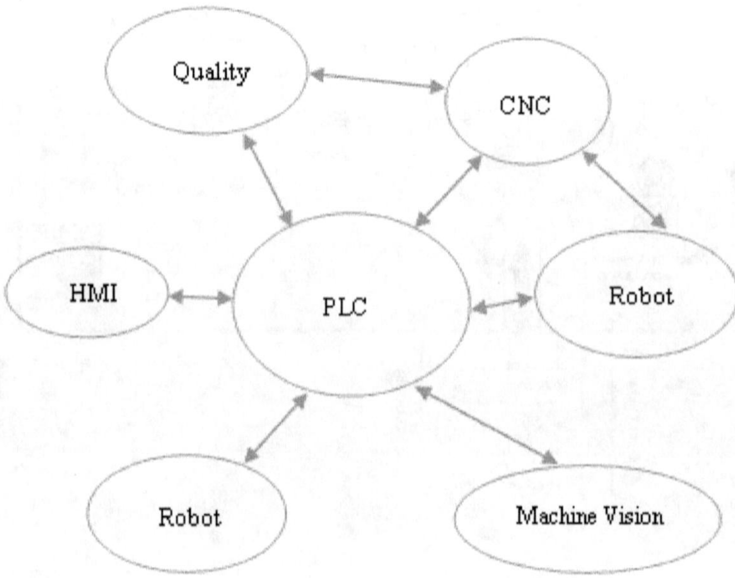

PLC ARCHITECTURE

How Does a PLC Work?

There are three fundamental steps in the operation of all PLCs: Input Scan, Program Scan, and Output Scan. During each operating cycle, the processor reads the states of all the input devices that are connected to the PLC, executes the user-created program, and excites or de-excites the output devices that are connected to the PLC. These tasks are strictly executed in that order in a boundless recurring manner. This process is called a program scan cycle. Figure 5 shows the scan cycle of the PLC.

A diversity of PLC from numerous manufactures is used in industry. For example:

- Allen Bradley
- Siemens
- Omron
- Schneider Electric
- Mitsubishi Electric
- Delta
- Pilz

Figure 4. Basic PLC architecture

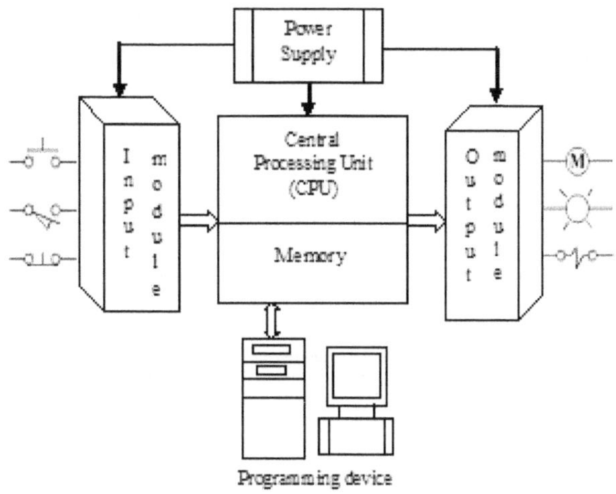

Figure 5. Scan cycle of a PLC
(Source: https://www.researchgate.net/figure/The-scan-cycle-of-a-PLC_fig6_338129116)

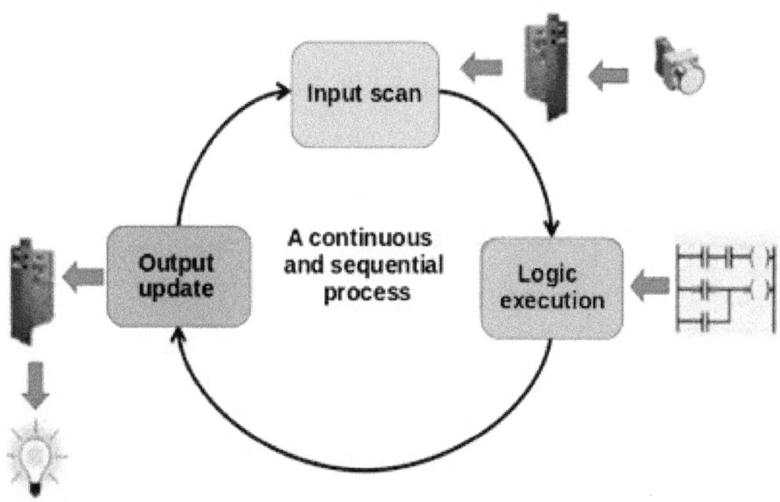

Each manufacturer will have a different configuration, programming language and communication networks.

PLC Programming Languages

The standard IEC 61131 was established to standardize the various languages associated with PLC programming. The following are the five most popular standard languages shown in figure 6:

- Ladder Diagram (LD): LD is the most popular PLC language because it was designed to replace the hardwired relay logic program. A ladder diagram is a symbolic representation of instructions arranged in rungs similar to ladder-formatted schematic diagrams.

Figure 6. Standard IEC 61131 languages associated with PLC programming

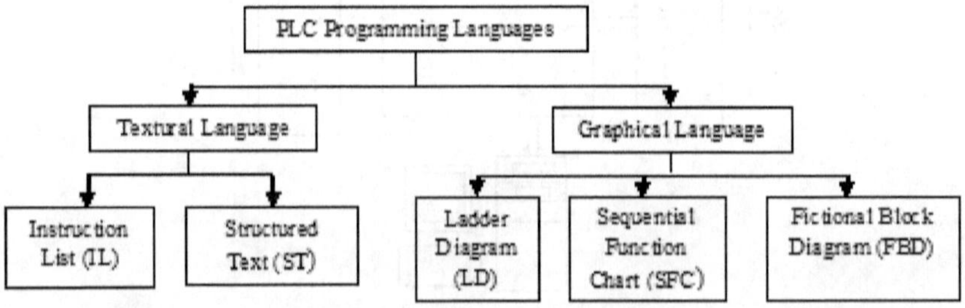

Example:

- Function Block Diagram (FBD): FBD is the second most popular PLC programming language. FBD is a graphical representation of process flow using simple and complex interconnecting blocks.

Figure 7. Example

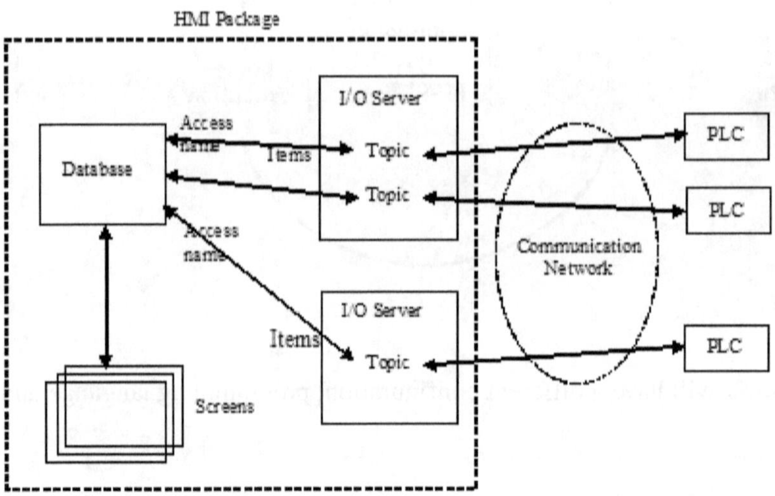

- Structured Text (ST):ST is the next on the list of PLC programming languages.ST is a high-level, text-based language and looks very similar to BASIC or C programming and specifically developed for industrial control applications.

Figure 8. Example

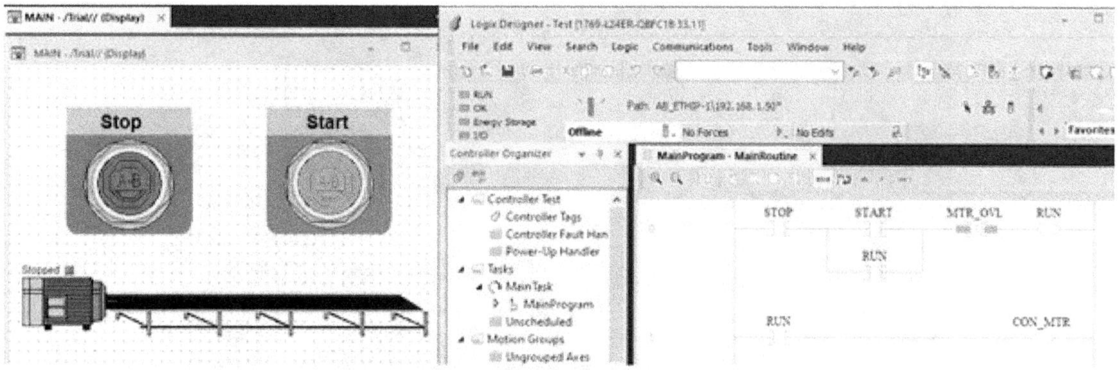

- Instruction List (IL): Next, on the list of PLC programming languages is Instruction List. IL is a low-level, text-based language that uses mnemonic instructions.

Figure 9. Example

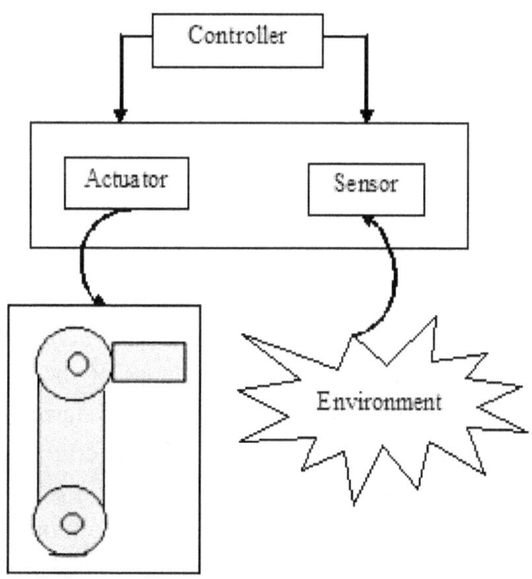

- Sequential Function Chart (SFC): SFC is the last option. It is a graphical representation of inter-connecting steps, actions, and transitions.

Figure 10. Example

Human Machine Interface (HMI)

A Human-Machine Interface is one of the component of an industrial control system. A Human-Machine Interface (HMI) is a user interface or dashboard that correlates a person to a machine, system, or device (Zhang, 2010). HMI technology is used by almost all industrial organizations, to interact with their machines and optimize their manufacturing processes. HMIs communicate with Programmable Logic Controllers (PLCs) and input/output sensors to acquire and display information for operators or users to view. HMI's are used to control and monitor the process and machines in the industry.

General Structure of HMI Software

The general structure of the HMI software is shown figure 7. There are three major components of the HMI package:database, HMI screen and I/O servers. The heart of the HMI software is the database. The graphic screen objects are programmed to interact with the data base variables, and the database variables interact with the I/O servers. For each type of PLC and communication network, a different I/O server is desirable. A "topic name" is fundamentally a communication channel established for a particular PLC. Ethernet communication is the most widely used communication protocol in industries. Communication among the database and the I/O server is usually modeled on the dynamic data exchange (DDE) or object linking and embedding (OLE) method of data exchange. The communication between the database and the I/O server is named as access name that defines the I/O server and the topic in the server. The present value of the database variables is obtained from the I/O server which communicates with the PLC. If the operator changes a variable in the database, its value is sent to the I/O server, which writes it to the PLC.

The following are the four major tasks to develop an HMI application.

1. Communication set up between HMI and PLCs that involves configuring the access names, I/O servers is shown in figure 8.

2. The heart of the HMI software is the tag database. The tag database is created from the PLC programming software and is imported into the HMI.
3. The graphical objects are created on the HMI screen and the corresponding tags are assigned. More complicated objects and symbols are commonly imported from a drawing package, or imported from a library of objects.
4. The final task is the animation. Animate the objects wherever is desirable.

Figure 11. General structure of HMI software

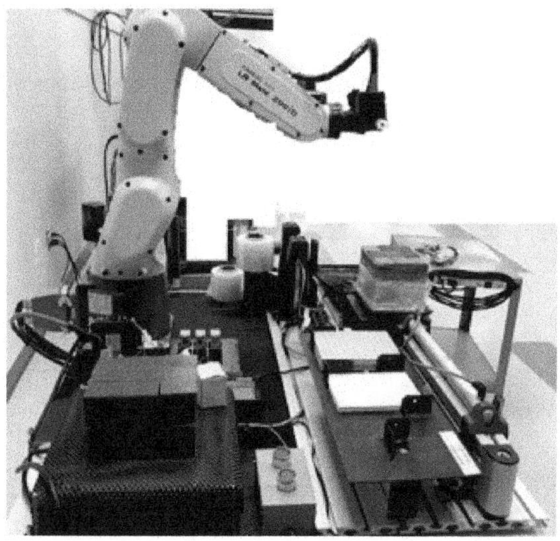

Figure 12. Communication setup between HMI and the PLC

Example Program

Industrial Robots

According to the International Federation of Robotics, industrial robots are automatically controlled, reprogrammable multipurpose manipulators programmable in three or more axes, where axes refer to the number of moveable joints. They are designed to handle specific automation applications within manufacturing companies, such as material handling, material processing, assembly and inspection, etc.

Based on the sensory signal received from the PLC controller, robots carry out programmed work or activity. Through network communication, Robot also communicate /interact with the other machines and the central computer.

A diversity of robots from numerous manufactures are used in industry. For example:

Fanuc
ABB
Panasonic
KUKA
CRS
Denso
Motoman

Each manufacturer will have a different configuration, programming language and communication networks.

How Does a Robot Work?

A robot has three main parts: Controller, Actuator, and Sensor as shown in figure 9. All of these parts work together to make a robot work.

1. Controller: The robot controller regulates the movements and functions of the robot. The robot controller is a feedback control system, run by a computer program that gives demands for the moving parts of the robot to follow. The controller stores data about the robot and its work environment and contains programs that operate the robot.
2. Actuator: The Actuators such as motors, grippers, etc. are used to create mechanical movements of the robot components to make the robot move, grab, turn or lift. The source of energy that the actuators need to produce motion is normally electricity, hydraulic fluid, or pneumatic pressure.
3. Sensors: It provides the robot the awareness of its environment or surroundings and other important information such as presence of the objects, gripper open and close positions, and details like sizes, shapes, space between objects, direction, etc. to the controller which processes them and then sends back the control signal making necessary changes. This is to activate the actuators to facilitate suitable movements of the robot.

Figure 13. Main parts of the robot

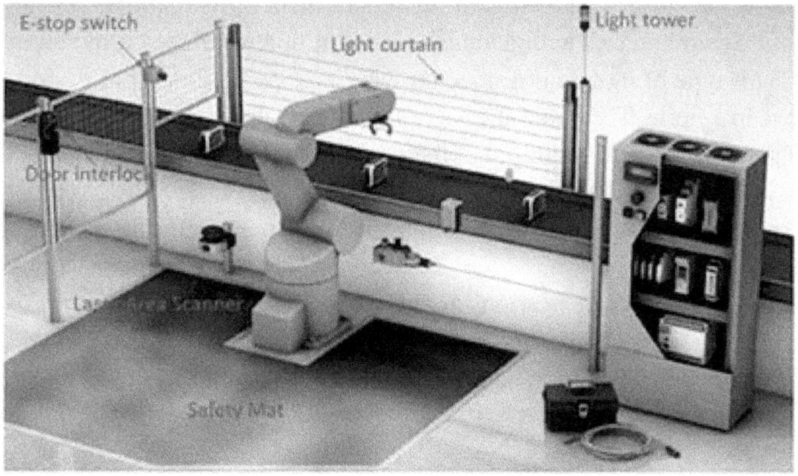

Types of Robot Control

There are two basic types of robot control: Point-to-point control and Continuous-path control (Abdel-Malek & Wolf, 1991).

Point-to-Point Control Robot

This type of robot control is attained by moving the robot from one point to another point. During the programming phase of the project, the point-to-point system records a series of locations and the robot moves sequentially from point to point when the robot runs the program. The positions are recorded in the control memory. When the robot reaches a point, it can energize or de-energize any output signals or send output signals that are used for interfacing to other equipment in the cell. Common applications include: spot welding, loading and unloading, assembly operations, etc.

Continuous-Path Control Robot

Continuous path control is used where the robot is required to follow a specified path such as in welding or spray painting. During the teach mode, each location in the path is recorded, and then the robot will follow each path during run mode. Since this type of robot must follow a precise path when it's spray painting, each location in the path the robot takes to move from point to point is recorded during the programming phase of the project and replayed when the robot is in the run phase. Typical applications include: spray painting, finishing, Arc welding operations.

Motion Types

It defines how the robot will move to the destination points. There are three motion types: Joint, Linear, and Circular.

Joint Motion: It is the fastest way to move the robot from the start point to the destination point. The robot axes just follow the closet's path without the requirement of maintaining a straight line.

Linear Motion: The robot axes will maintain a straight line when they move from the start point to the ending point. This type of motion is a slow, but very precise motion type.

Circular: Move in a circle. The circle always begins at the robot's present position and flows through the two positions given in the instruction. The first position in the circle instruction is the midpoint, and the second position is the endpoint. Limit the usage to 180 degrees.

Robots are commonly used in the following applications in manufacturing is shown in figure 10 &11:

1. **Material Handling:** it involves various activities such as:

- Picking and placing parts at desired locations
- Palletizing and de-palletizing: robots retrieve objects from one location, and deposit them on a pallet in a specific area of the pallet

Figure 14. Example of pick and place robot
[Source: https://images.app.goo.gl/Ae1K6uvrhZG6iW5N9]

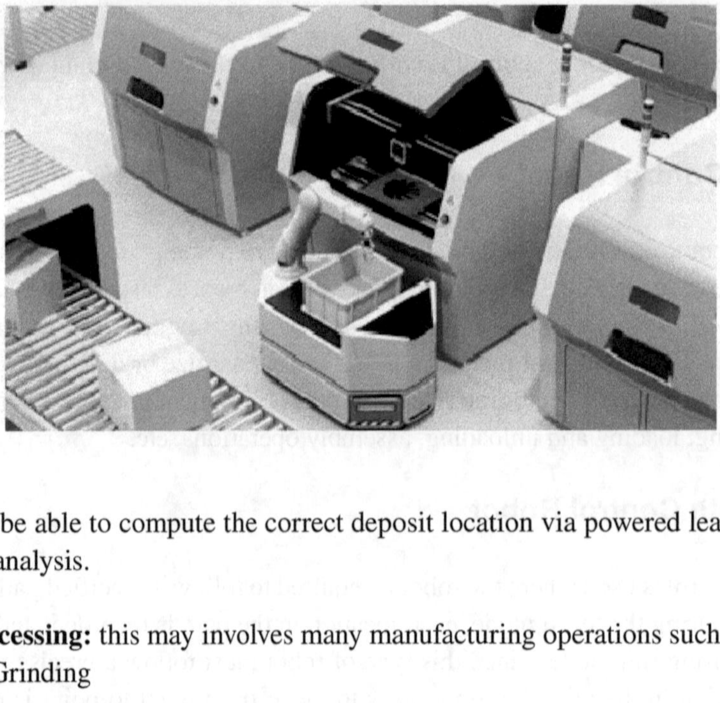

The robot must be able to compute the correct deposit location via powered lead- through method, or by dimensional analysis.

2. **Material processing:** this may involves many manufacturing operations such as:

- Drilling and Grinding
- Arc welding
- Deburring
- Spray painting
- Coating
- Polishing
- Heat treatment
- Loading and unloading of the parts on required machines

3. Assembly and Inspection

Automatic Quality Control and Inspection Systems (Machine Vision)

Vision systems have had a considerable impact on the use of robots. Camera systems with processing can allow a robot to accomplish quality control on products as shown in figure 12.

The basic aim of the manufacturing industry is to supply a good quality product or a system to the market. The product should meet the requirements of the customers and it must be reliable. To catch and possibly discard defective work units produced on the line (Bosman & Fernhaber, 2018). The quality can be achieved by building up the quality right from the product design stage, and maintaining the standards during the production stages.

Figure 15. Example of palletizing robot

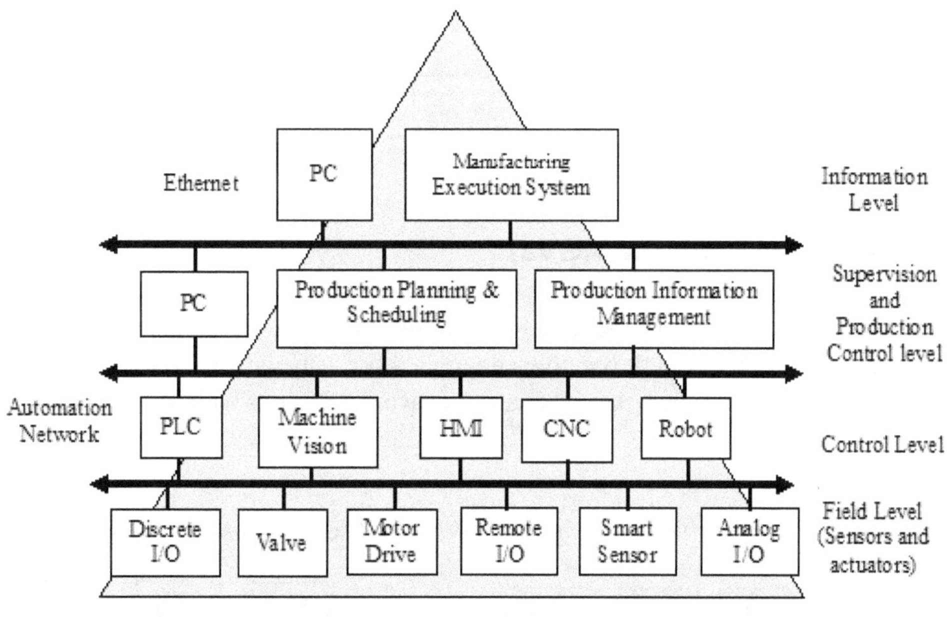

Several sensors and systems are available to monitor quality continuously with or without the assistance of the operator (Solowjow et al., 2020). Technologies available are data acquisition systems, machine vision systems, metrology instruments such as coordinate measuring machines (CMM), optical profilometers, digital calipers, and screw gauges, etc. Various physics-based simulation software are used to predict the performance of the product or the system to be developed.

Robot Safety

Safety systems are indispensable to robots designed to work with or operate within a safe distance from human workers. Machine safety systems are the systems that by-laws and codes must be designed and built into automated machinery and robotic cells. A great deal of the safety of an automated machine is designed based on the CSA (Canadian Standards Association) guidelines for automated machines and the CEC (Canadian Electrical Code).

The most common safety systems shown in figure 13 were used in the manufacturing industry. It consist of: Laser Scanner, safety mats, Door interlock, E-stop switches and light curtain.

Figure 16. Example of Quality control
[Source: https://www.fanuc.eu/pt/en/robots/accessories/robot-vision]

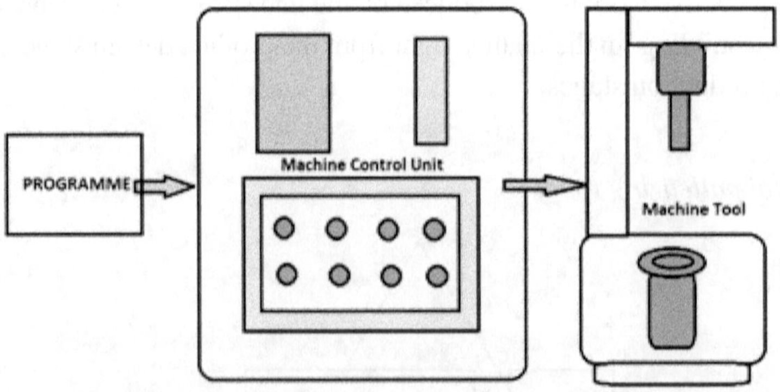

Autonomous Guided Vehicles (AGVs)

Material handling systems in factories are used for transportation of raw materials, work-in-process, and finished goods and storage/retrieval. It is equipped with a robot arm and a vision system and uses intelligence to find the shortest route, pick the right parts and avoid collisions with other vehicles or humans.

Figure 17. Safety devices
[Source: https://www.digikey.com/en/articles/reducing-robot-risk-how-to-design-a-safe-industrial-environment]

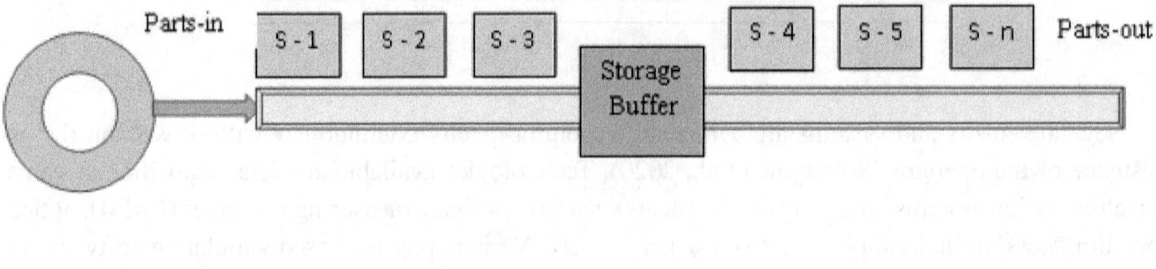

Factory Communications Hierarchy

An industrial communication network is a backbone for any automation system that provides data exchange, data controllability, and flexibility to connect various devices. Figure 15 illustrates an example of the hierarchy of an industrial automation system. Industrial networks can be classified in numerous different categories based on functionality: field-level networks at the bottommost level, control networks at the intermediate level, and enterprise networks at the topmost.

Figure 18. AGV example
[Source: http://www.ti.com/applications/industrial/industry-4-0.html#PLC]

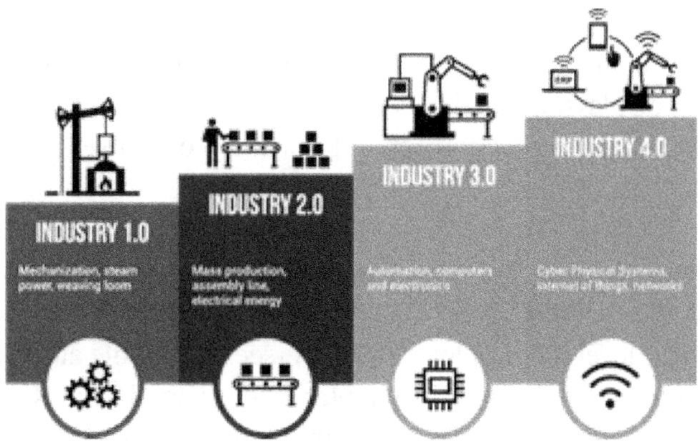

Figure 19. The industrial automation pyramid

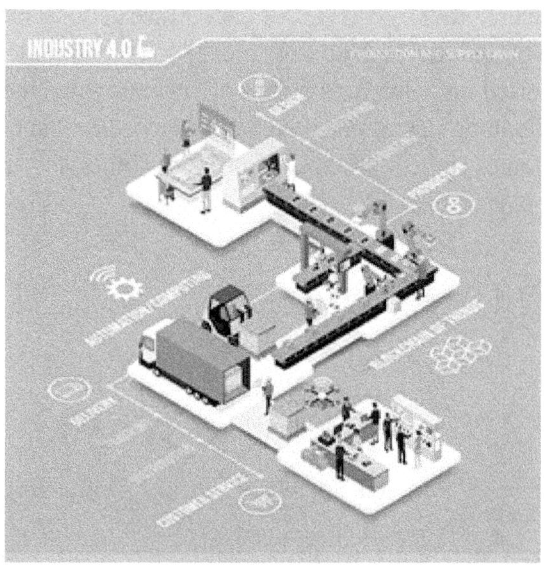

Field Level

This level is the bottommost level of the automation hierarchy, which comprises the devices such as actuators and sensors (Trsek, 2016). The main task of this level is to send the information between these devices and technical process elements such as PLCs, HMIs, etc (Daniyan & Mpofu, 2018). Devices operating at this level normally utilize discrete signal (ON/OFF), continuous signal (4-20 mA, 0-10V, etc.), or serial point-to-point communication options. The most common communication protocols used at this level is a device network such as DeviceNet, AS-i, Profibus-DP, HART, ControlNet, etc. This communication allows for distributed control among smart field devices and controllers.

Control Level

This level is an intermediate level that consists of industrial controllers such as PLCs, HMIs, Robots, etc. The tasks of this level include configuring automation devices, loading programs, parameters, and data, supervising control, displaying variables data on HMIs, etc. The most common communication protocols used at this level is an Automation network such as ControlNet, DH+, Modbus, etc.

Information Level

The information level is the topmost level of an automation hierarchy that collects the management information from the lowest level (control level) and oversees the whole automation system. Generally, this level handles large volumes of data that are not time-sensitive. The most common communication protocols used at this level is a Backbone network such as Ethernet, Ethernet/IP, Modbus/TCP, Profinet, etc.

Computer Numerical Control (CNC) Machines

CNC means operating a machine tool by a series of coded instructions such as numbers, letters of the alphabets, and symbols which the machine control unit (MCU) can understand (Trsek, 2016). These coded instructions are converted into electrical pulses of current which the machine's motors and controls follow to carry out machining operations on a work piece. Numbers, letters, and symbols are the coded instructions which denotes specific distances, positions, functions or motions which the machine tool can understand.

Figure 20. Elements of a CNC

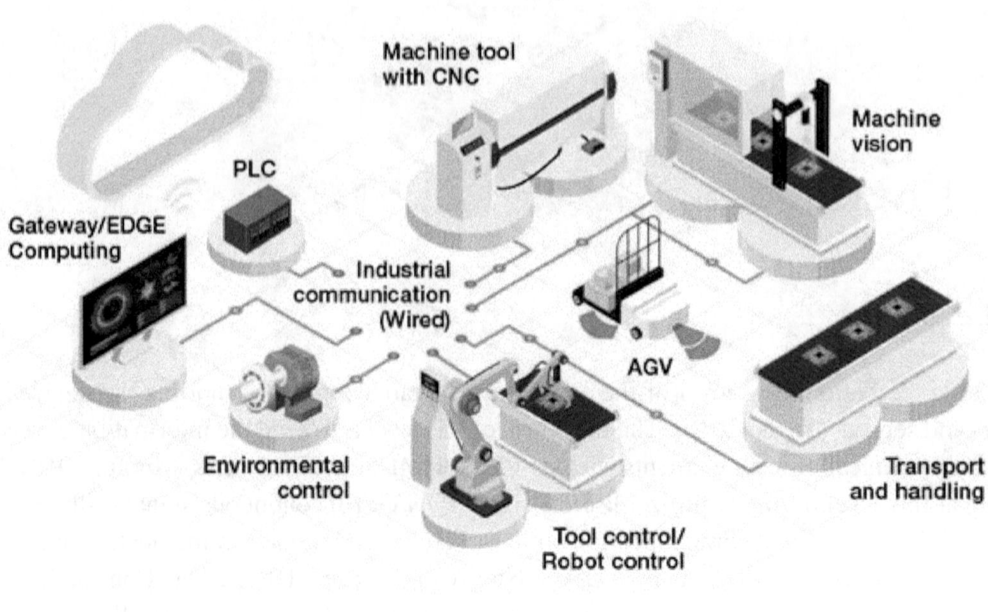

A CNC system shown in figure 16 consists of three basic components (Zhang & Chen, 1993):

1. Part program
2. Machine Control Unit (MCU)
3. Machine Tool (Lathe, drill press, milling machines etc)

CNC can control the motions of the workpiece or tool, the input parameters such as feed, depth of cut, speed, and functions such as turning spindle on/off, turning coolant on/off (Sahu et al., 2014). With the help of computers, the numerical controller automatically guides the axial movements of machine tools. The auxiliary operations such as coolant on-off, tool change, door open-close are automated with the help of controllers. Manual operation of table and spindle movements is automated by using numerical controllers and servomotors. The spindle speed and work feed can accurately be controlled and maintained at a programmed level by the controller.

Flexible Manufacturing System (FMS)

FMS is an automated manufacturing cell or system that consists of several automated workstations, interconnected with automated material handling and storage system.An example of automated Production line is shown in figure 17:

- Example 1:
    - Station 1: Forming Press (starting base parts)
    - Station 2: Machining (CNC Machine)
    - Station 3: Washing
    - Station 4: Assembly
    - Station 5: Welding

Storage buffer between two stages of a production line:

A location in the sequence of workstations where parts were collected and temporarily stored before preceding to the next processing workstation

- Station 6: Machining (CNC Machine)
- Station 7: Dipping (for rust free)
- Station 8: Quality check
- Station 9: Packing and shipping (completed parts)

Automated production line consists of number of automated workstations, which perform certain processing operations. This is an example of fixed automation.

Control functions in an automated production line (Bosman & Fernhaber, 2018):

1. Sequence control: to coordinate the sequence of actions of the transfer system and workstations
2. Safety monitoring: to avoid hazardous operation for workers and equipment
3. Quality control: to detect and possibly reject defective work units produced on the line

In each station, a specific operation is performed. The raw material is feed at one end of the line, continues through each station, and emerges at the other end as a completed product. These automated lines are controlled by PLC, which enable connections with industrial equipment (such as automated production lines) and can achieve the kinds of timing and sequencing functions needed to run such equipment.

Industry 4.0 in Manufacturing

Phases in the evolution of industrial manufacturing systems from manual work towards Industry 4.0 theory can be displayed as a path through the four industrial revolutions. The development is illustrated in figure 18. The first industrial revolution commenced with mechanization and mechanical power Generation. It brought the evolution from manual work to the first manufacturing processes; generally in the textile industry (Rojko, 2017). The Second Industrial Revolution was categorized by mass production and the substitution of steam by chemical and electrical energy. In order to encounter the emergent demand, numerous technologies such as the assembly line with automation and microelectronics enabling the increase of productivity in industry and mechanization have been evolved. In manufacturing, this facilitates flexible production, where a variety of products is manufactured on flexible production lines with PLC, Robot etc. This is the third industrial revolution. Today we are in the Fourth Industrial Revolution is also known as Industry 4.0 shown in figures 19 & 20, which refers to automation and data exchange in manufacturing technologies that focus profoundly on artificial intelligence (AI), robotics and automation, machine learning, Cyber-Physical Systems (CPS), the Internet of Things (IoT), and real-time data. (Pereira & Romero, 2017). The key purpose of Industry 4.0 is to accomplish improvements in terms of automation and operational efficiency as well as effectiveness. (Auer & May, 2020)

Implementing Industry 4.0 and exerting digital tools such as 3D modeling, virtual reality, and simulation in the manufacturing industries can pave the way for growing the production system and making different optimal decisions. These digital tools play an indispensable role in enhancing the production processes in manufacturing companies, which leads to achieving flexibility in the production process and reach customer demand (Monostori et al., 2016). In contemporary plant automation applications, industrial devices such as Programmable Logic Controllers (PLCs), Human Machine Interfaces (HMIs), robots, and Industrial PCs (IPCs) are changing from pre-programmed, stationary systems into machines capable of modifying their behaviour, based on interactions with the environment and other devices.

REFERENCES

Abdel-Malek, L., & Wolf, C. (1991). Evaluating flexibility of alternative FMS designs—A comparative measure. *International Journal of Production Economics*, *23*(1-3), 3–10. doi:10.1016/0925-5273(91)90042-R

Allen-Bradley PLCs. (2013). *An Emphasis on Design and Application, Kelvin T. Erickson*. Dogwood Valley Press, LLC.

Auer, M. E., & May, D. (Eds.). (2020). *Cross Reality and Data Science in Engineering: Proceedings of the 17th International Conference on Remote Engineering and Virtual Instrumentation* (*Vol. 1231*). Springer Nature.

Bosman, L., & Fernhaber, S. (2018). Example Engineering Curriculum: Online Discussion Prompts. In *Teaching the Entrepreneurial Mindset to Engineers* (pp. 65–90). Springer. doi:10.1007/978-3-319-61412-0_9

Daniyan, I. A., & Mpofu, K. (2018). Application of Design for Manufacturing and Assembly: Development of a Multifeedstock Biodiesel Processor. In *Applications of Design for Manufacturing and Assembly*. IntechOpen.

Gawali, D., & Sharma, V. K. (2009, December). FPGA based micro-PLC design approach. In *2009 International conference on advances in computing, control, and telecommunication technologies* (pp. 660-663). IEEE.

Monostori, L., Kádár, B., Bauernhansl, T., Kondoh, S., Kumara, S., Reinhart, G., Sauer, O., Schuh, G., Sihn, W., & Ueda, K. (2016). Cyber-physical systems in manufacturing. *CIRP Annals*, *65*(2), 621–641. doi:10.1016/j.cirp.2016.06.005

Pereira, A. C., & Romero, F. (2017). A review of the meanings and the implications of the Industry 4.0 concept. *Procedia Manufacturing*, *13*, 1206–1214. doi:10.1016/j.promfg.2017.09.032

Rojko, A. (2017). Industry 4.0 concept: Background and overview. *International Journal of Interactive Mobile Technologies*, *11*(5), 77. doi:10.3991/ijim.v11i5.7072

Sahu, A. K., Sahu, N. K., & Sahu, A. K. (2014). Appraisal of CNC machine tool by integrated MULTI-MOORA-IVGN circumferences: An empirical study. *Grey Systems: Theory and Application*.

Solowjow, E., Ugalde, I., Shahapurkar, Y., Aparicio, J., Mahler, J., Satish, V., . . . Claussen, H. (2020, August). Industrial robot grasping with deep learning using a programmable logic controller (plc). In *2020 IEEE 16th International Conference on Automation Science and Engineering (CASE)* (pp. 97-103). IEEE.

Trsek, H. (2016). *Isochronous wireless network for real-time communication in industrial automation*. Springer. doi:10.1007/978-3-662-49158-4

Zhang, D. T., & Chen, J. G. (1993). An NC lathe simulator for part programming and machine operation training. *Computers in Industry*, *21*(2), 139–148. doi:10.1016/0166-3615(93)90132-K

Zhang, P. (2010). *Advanced industrial control technology*. William Andrew.

Section 2
New Materials and Processes

Chapter 8
Fabrication and Application of Aluminum Metal Matrix Composites

Pradeep Kumar Krishnan
National University of Science and Technology, Muscat, Oman

ABSTRACT

A substantial effort has recently been made to fabricate numerous combinations of metal matrix composites. Aluminium metal matrix composite (AlMMCs) has been extremely popular among MMCs and is best suited for lighter materials in various industries and domestic applications. AlMMCs have significant faults in terms of production costs. More research is needed to develop AlMMCs that are both economic and suitable for a wide range of industrial applications. AlMMCs are manufactured using the casting technique, solid-state processing, infiltration method, accumulative roll bonding, spray deposition, friction stir processing, and other technologies. This chapter will summarize the current state of the art in aluminum-based metal matrix composites, as well as recent developments in processing and application, all in one location, so that relevant stakeholders can benefit the most and new advancements on such themes can be fostered.

INTRODUCTION

Metals are a naturally occurring resource found in the earth's crust, and their compositions differ depending on where they are located, resulting in spatial concentration differences. The qualities of the specific material, as well as environmental factors, influence metal accumulation and distribution in the environment (Jaishankar et al., 2014). Higher specific strengths and elastic modulus, increased tolerance to rising temperatures, decreased coefficients of thermal expansion, and, in some cases, improved wear resistance are all advantages of metal matrix composites over their base metal equivalents. On the negative side, they are more expensive and have a lower toughness than their base metal counterparts. Metal matrix composites also have several advantages over polymer matrix composites, including higher matrix dependent strength and elastic modulus, higher elevated temperature resistance (due to

DOI: 10.4018/978-1-7998-9574-9.ch008

their lack of moisture absorption), higher electrical and thermal conductivities, and the fact that they are nonflammable. Metal matrix composites (MMCs) are often more expensive than polymer matrix composites, and their manufacturing procedures are much more limited, especially for complex structural geometries. Metal matrix composites are used in a small number of commercial applications due to their high cost. There are some limited applications for discontinuously reinforced MMCs, but there are virtually no applications for continuously reinforced MMCs at this time. For today's engineering applications, stronger, lighter-weight, and lower-cost materials are necessary. In the field of engineering applications, AMCs are well-known for providing precisely the tailored property materials that are required. Advanced metal composites (AMCs) are rapidly replacing standard metal alloys in a variety of applications as their uses expand, notably in the automotive, aerospace, biomedical, and manufacturing industries (Maniraj et al., 2021). Metal matrix composites can be subdivided according to the type of reinforcement shown in Figure 1.

Figure 1. Metal matrix composite reinforcements
(F.C. Campbell, 2006)

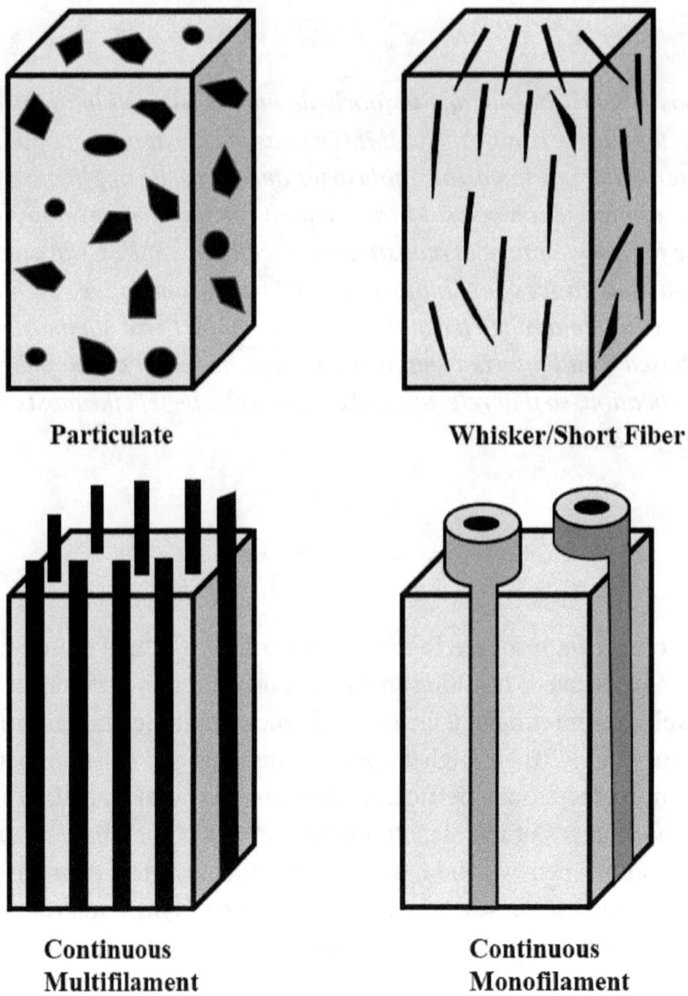

As reinforcing materials, it is possible to use particles, high-strength single crystal whiskers, short fibers that are typically random but may contain some degree of alignment, or long aligned multifilament or monofilament fibers.Aluminum and its alloys, which are lightweight and well-performing, are widely employed in automobile and aircraft applications because to their low weight and excellent performance. These materials have outstanding features such as high specific strength, excellent wear and corrosion resistance, low density, and high specific stiffness (Samal et al., 2020). The usage of aluminum alloys in the aerospace and automobile industries has increased dramatically as a result of its superior qualities, which include high specific strength, good low-temperature performance, exceptional corrosion resistance, and resistance to chemical attack, among others. The main disadvantages of aluminum alloys, on the other hand, are their poor high-temperature performance and low wear resistance. These issues have been addressed by the development of aluminum alloys reinforced with ceramic particles, which are referred to as Aluminum metal matrix composites (AMMCs). Composites made of aluminum matrix are lightweight and high-performing materials that have the potential to replace traditional materials in a wide range of sophisticated applications. (Panwar & Chauhan, 2018). Since the beginning of the last decade, there has been a significant increase in the amount of research that has taken place in the subject of aluminum MMCs. These amalgamations are used to replace metallic alloys in a variety of applications, including aircraft components such as wings and fuselages, as well as automobile components such as brake discs, drums, and pistons, at a lower cost of production (Samal et al., 2020). They feature high specific strength and modulus, moderate thermal expansion, and isotropic qualities, making them ideal for applications in the defense, automotive, and aerospace industries. A/C, thermal management, aerospace, industrial, recreational and infrastructure industries use MMC. Increasingly, aluminium-based composites are being used in the automotive, aerospace, maritime, and mineral processing industries due to their superior specific strength, wear resistance, thermal conductivity, and low coefficient of thermal expansion Composites incorporating hard ceramic particles, such as brake drums, cylinder liners, pistons, cylinder blocks and connecting rods, have shown promise as a wear-resistant and lightweight material. Aluminum based metal matrix composites (MMC) are gaining popularity due to their superior strength, stiffness, and light weight. Due to the scarcity of energy and fossil fuels, new lightweight materials for automotive and aviation uses. It has been used to construct everything from airplanes to bicycles, medical devices to electronic packages, and spacecraft to domestic products.Ceramic-reinforced aluminium matrix composite materials are replacing monolithic alloys in the automotive, marine, and aviation industries due to their superior tribological characteristics and corrosion resistance.(Panwar & Chauhan, 2018). Aluminum-based composites have mostly been produced through the application of two processing methods: solid-state processing and liquid-state processing. Using this procedure, the matrix and reinforcement are combined together for compacting, and the resulting compacted portion is sintered to increase its strength and durability (Samal et al., 2020). Adding reinforcement phase to the matrix creates MMCs. Included in this list are powder metallurgy, spray atomization and co-deposition, plasma spraying, stir casting, In engineering materials, MMCs can be produced through a unique technology called casting, which is affordable and offers many material and processing possibilities. They are injected into the solid or liquid matrix either by powder metallurgy or liquid metallurgy casting. Powder metallurgy or liquid metallurgy can be used to inject reinforcing particles into a solid/semi-solid or liquid matrix. The reinforcement phase is added to the matrix by powder metallurgy, spray atomization and co-deposition, plasma spraying, stir casting, or squeeze casting. (Panwar & Chauhan, 2018).

FABRICATION ASPECTS AND METHODS OF MMCS

Conversion of one form to the required form of the product with its proper dimensions occurs during the manufacturing process. Accordingly, several techniques are used to fabricate AMMC, as illustrated in Figure 2.

Figure 2. AMMC processing methods

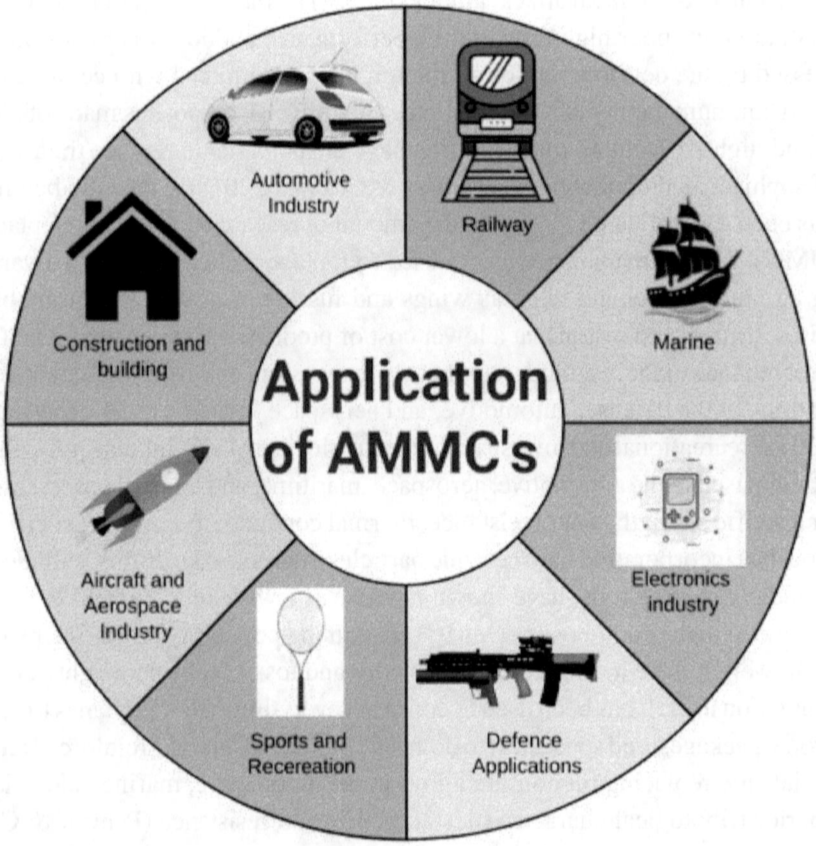

The cost of the process, its convenience of use, and the increase of the qualities of composites are all factors to consider while developing fabrication techniques. It is possible to choose the most appropriate fabrication process depending on the requirements since different fabrication procedures each have their own set of pros and disadvantages. There are several methods for producing MMCs, but the most prevalent are liquid state and solid-state manufacture, both of which are discussed in greater depth (Sharma et al., 2020). When compared to solid-state synthesis, the liquid-state approach to MMC manufacture is more straightforward and less expensive to implement. As a result of mutual diffusion between the matrix and reinforcing materials at high pressure and temperature during solid-state processing, MMCs are formed. Solid-state-produced MMCs can also be deformed by cold or hot forging, extrusion, or other processes in addition to the solid-state approach. The following section highlights a number of the solid-state processing technologies that are now available for the production of MMC's.

1. **Powder Metallurgy Method:** This technology enables the correct ratio of metal powder and reinforcement alloys to be blended by machine. In order to achieve the desired shape and size depicted in Figure 3, the material is poured and forged in standard equipment utilizing compact dies. This forging technique can be carried out using either cold or hot pressing. In the following step, the green compacts are handled in an enclosed space in order to undergo the sintering process, which converts them into high-density products. This method enables a decrease in the amount of metal removed. (Senthil et al., 2020).

Figure 3. Powder metallurgy process (High-energy ball milling and sintering)

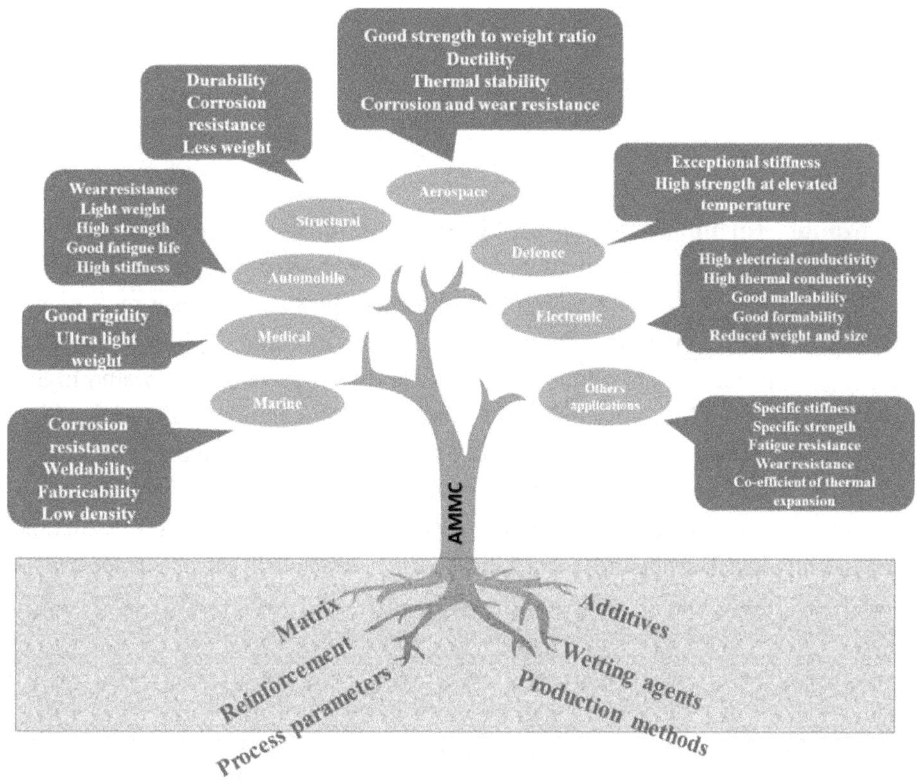

2. **Friction stir processing (FSP):** It is a relatively new technique for producing surface composites with a refined grain structure (Ma et al., 2013). In contrast to friction stir welding, it does not require the connecting of two metal plates to be done. It just influences the microstructure, yet it improves the qualities of the material. When performing FSP, a non-consumable tool with a pin and shoulder is stirred and thrust into the workpiece, causing significant friction to be generated. As a result of this, the frictional heat softens the metals, causing plastic flow and grain refinement to occur. Figure 4 depicts the process of friction stir processing (Samal et al., 2020).

Figure 4. Friction stir processing

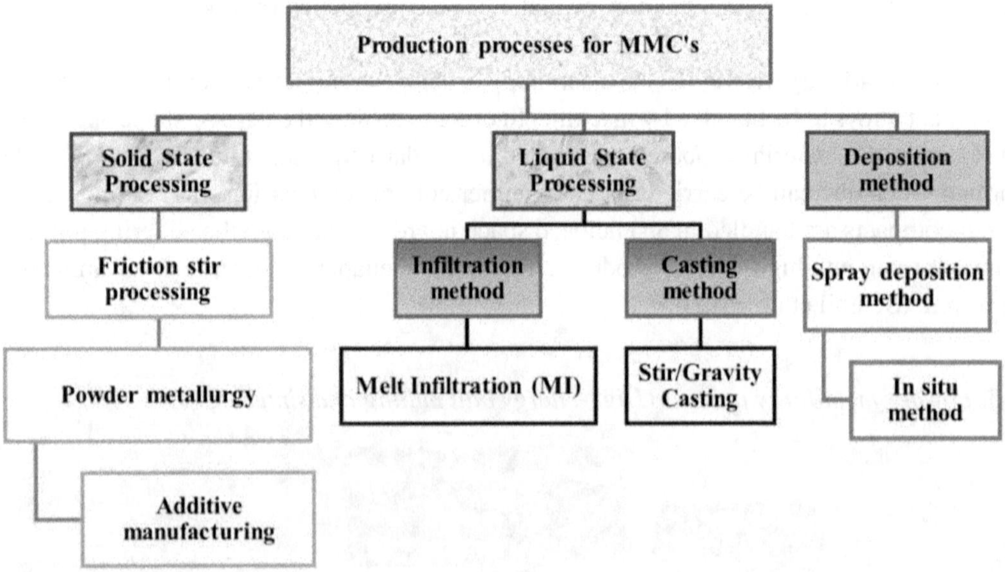

3. **Additive Manufacturing:** Through point-by-point material consolidation, additive manufacturing techniques like as selective laser melting can solve this shortcoming (Lathabai, 2018). However, process complexity could be a substantial bottleneck, as the selective laser melting process is quite complex due to the multiple process parameters indicated in Figure 5. Aluminum alloy powders, in particular, are intrinsically light, with limited flowability, high reflectivity, and high thermal conductivity when compared to other Selective Laser Melting candidate materials. Clearly, a higher laser power is necessary to melt, but also to overcome the rapid heat dissipation that occurs during heating, but rapid heat dissipation occurs more frequently with solid substrates than with Al powders. Additionally, Al alloys are very oxidative and thus porous, posing significant obstacles during selective laser melting. Despite these restrictions, advancements in the use of selective laser melting to the processing of aluminum metal matrix composites have been made. (Behera et al., 2020).

Powder

4. **Stir Casting Method:** Figure 6 depicts the manufacturing process of AMMC and its alloy using the stir casting method, which is widely used for large-scale production at a low cost and with a flexible production strategy. AMMC and its alloy are manufactured using the stir casting method. At the melting temperature, the appropriate reinforcing is incorporated into the molten metal. Stirring is then performed with the assistance of an external device in order to properly distribute the reinforcement throughout the molten metal in order to avoid heterogeneous distribution and to reduce porosity.
5. **Squeeze Melt Infiltration Method (MI):** By applying pressure to the squeeze melt infiltration technique under vacuum conditions, the molten aluminum not only penetrates the bundle effectively, but also avoids oxidation of the aluminum matrix and reinforcement. Figure 7 illustrates the squeeze melt infiltration process schematically showing the direction of the applied load. Aluminum

is lightly treated with acidic solutions at low concentrations to eliminate the oxide layer that forms on the surface of aluminum. Typically, the oxide layer obstructs the flow of aluminum melt. After applying the treated aluminum on the reinforcement, it is contained in a graphite die. Finally, the graphite die got transferred to the crucible furnace. A suitable squeeze pressure is given to the aluminum and reinforcement during solidification. After that, the load is maintained until the furnace reaches room temperature. (SHA et al., 2021).

Figure 5. Process parameters in selective laser melting for additive manufacturing

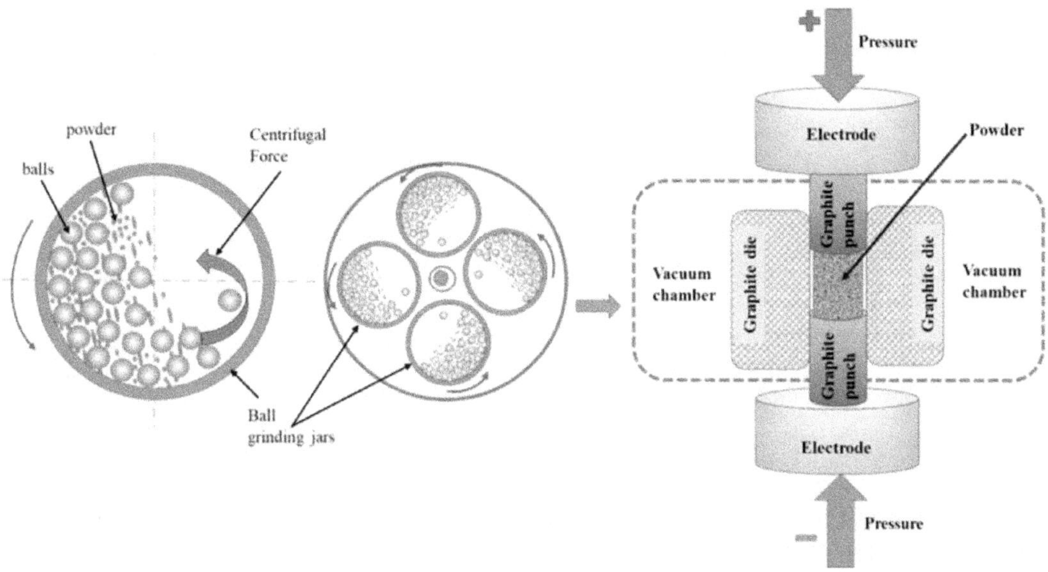

Figure 6. Stir casting method

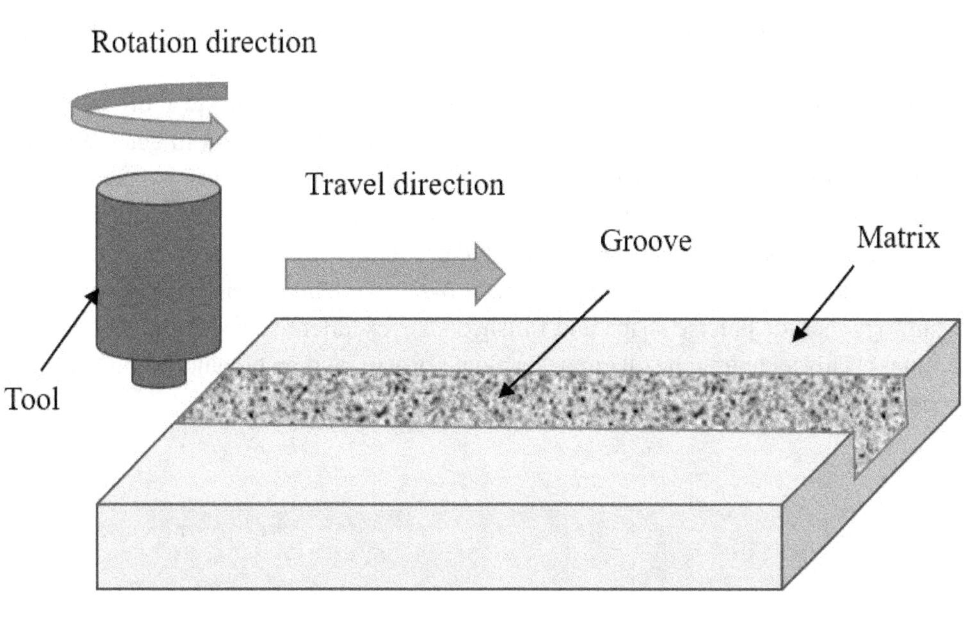

Figure 7. Schematic illustration of squeeze melt infiltration technique

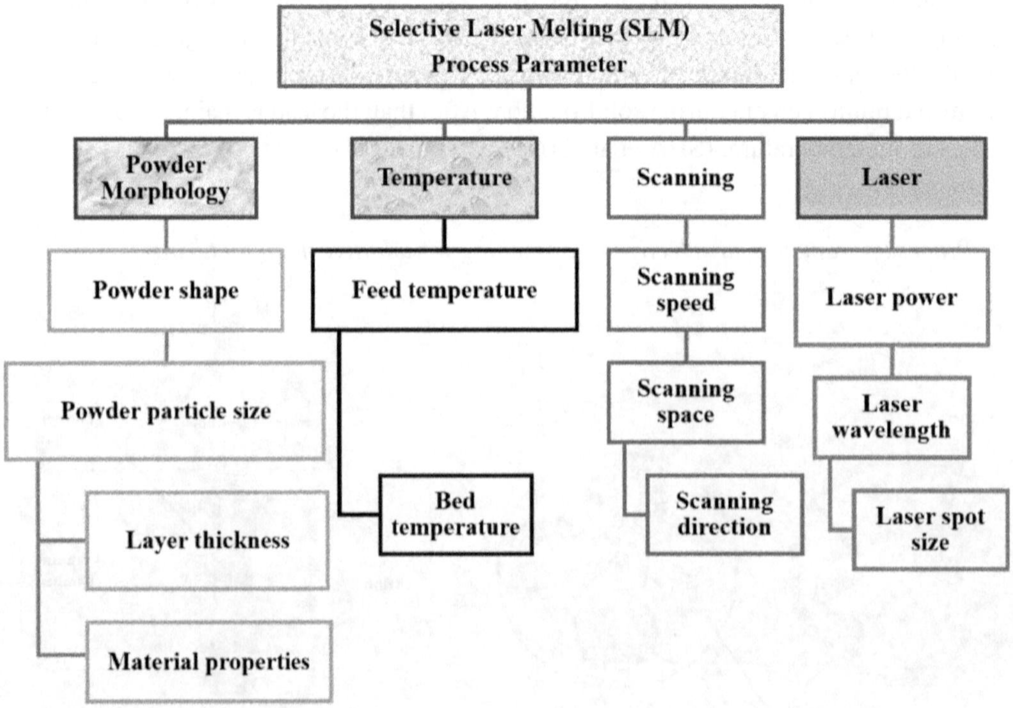

6. **Spray Deposition Method:**This technique is also known as the Osprey Process. As illustrated in Figure 8, atomized thin molten metal droplets are sprayed over a stationary metallic substrate with reinforcement using a high-speed inert gas jet. After then, the metal is allowed to solidify. To obtain a high-density deposit of MMC or the desired MMC, numerous parameters such as the deposition rate, the distance between the spray nozzle and the substrate, the inert gas used, the spraying pressure, and the angle of the spray must be considered. This procedure is really inexpensive (Peat et al., 2016).
7. **In-Situ method:**In-Situ technique As illustrated in Figure 9, during the fabrication process, a regulated chemical reaction occurs between the metal and reinforcement to generate composites. Through the use of fine powder or gaseous phase, the molten metal alloy is reacted between reinforcement. Later in the process, the reinforcement and metal matrix are incorporated into the isolated MMC. Good bonding and finely structured products with the proper interface are the result of the homogeneous distribution of reinforcement over the metal matrix. The primary benefits of this technology include precise shape and finished size, excellent surface quality, and metallurgical soundness. This technique results in a very low porosity and an enhanced contact between the reinforcement and the aluminum matrix (Govindarao et al., 2015).

Figure 8. Spray deposition method

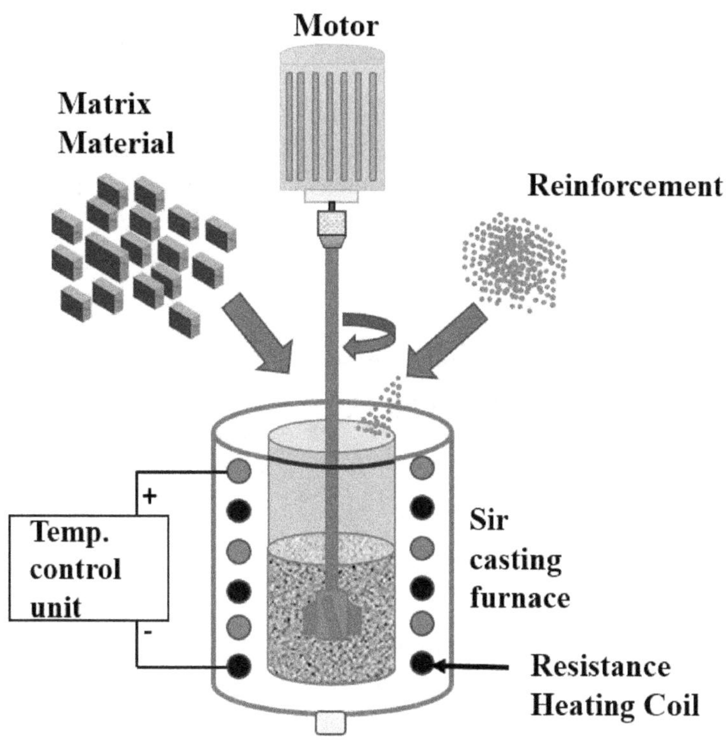

Figure 9. In situ method

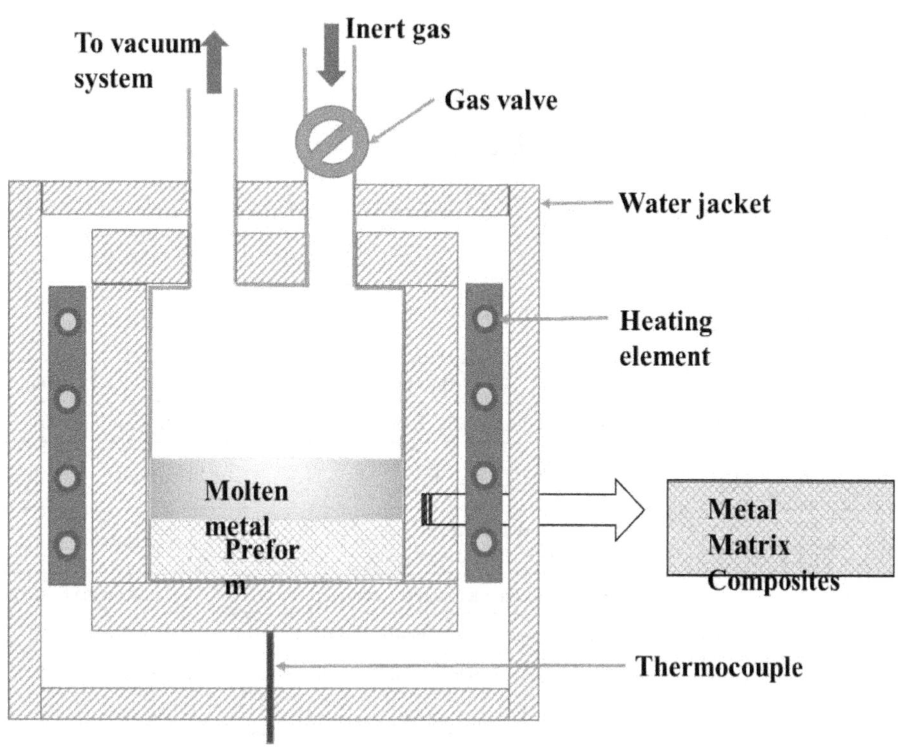

APPLICATION

AlMMCs have the potential to be widely used in a variety of industrial applications in the automotive industry due to their high strength, low cost, and light weight (Mehta & Vadher, 2020) AlMMCs have a number of advantages over conventional materials, including superior mechanical qualities and a wide range of applications in the aviation, mechanical, electrical, automotive, electronics, and transportation industries (Sharma et al., 2020). The various applications of AMMCs are depicted in Figure 10.

Figure 10. Different applications of aluminium MMC's

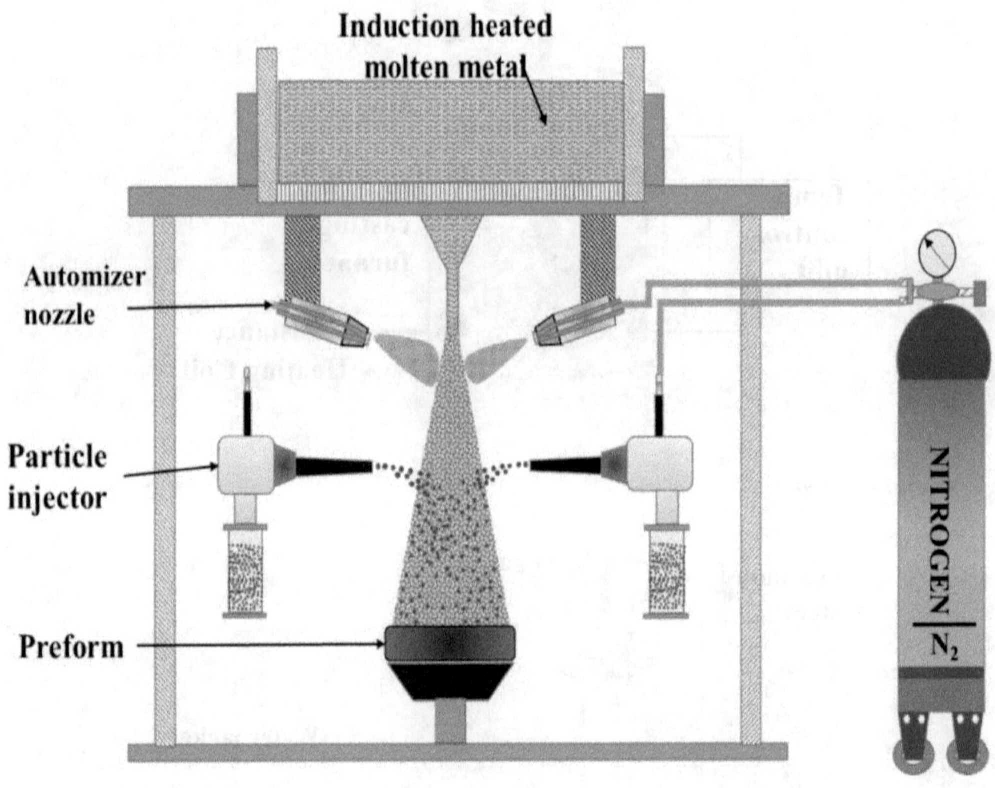

Table 1. Current applications of AMMCs

S.No	Processing methods	Applications	References
1	• Powder metallurgy process	•Motor housing, wheel, piston Space	(Maclin John Vasanth et al., 2020)
2	• Friction stir processing	•Railways and aerospace application Nozzle,	(Altinkok et al., 2013)
3	•Additive Manufacturing	•Structural applications, especially in construction and marine, etc.	(Singh et al., 2019)
4	• Stir Casting Method	• Pistons,Cylinder heads, pump housing	(Shanmughasundaram et al., 2011)
5	• Squeeze Melt Infiltration Method	• Space and industrial application	(Samal et al., 2020)
6	• Spray Deposition Method	• pistons, base plates, connecting rods	(Peat et al., 2016)
7	• In-Situ method	• Automobile application	(Govindarao et al., 2015)

Table 1 summarizes the present applications of several AMMCs. From this table, it is clear that AMMCs are commercially viable and should be explored by researchers in order to further their uses.

Figure 11. Applications of AMCs in various fields

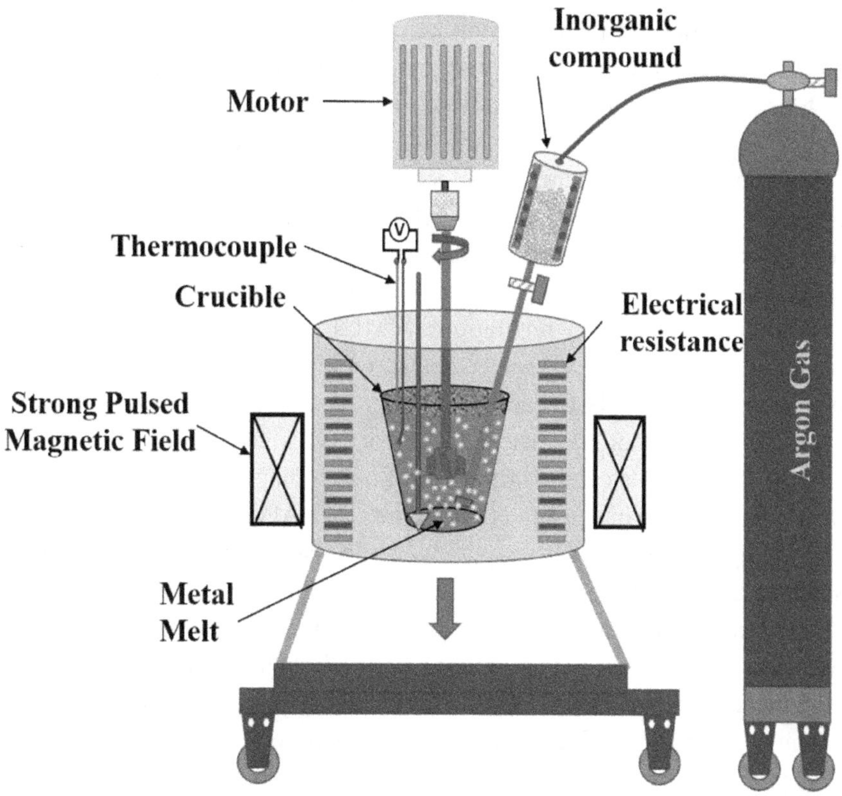

Hybrid AMMCs, as well as those reinforced with nanoparticles, are being investigated, and several studies are now being published by researchers worldwide. As a result, the next generation of AMMCs will be hybrid composites with superior properties. The tree diagram in Figure 11 illustrates the potential applications of AMMCs in various sectors. The roots denote the critical elements affecting various applications.

1. **Automotive Industry:** The majority of metal matrix composites used today are made of aluminum and alloys for automotive applications, primarily engine components, pistons, connecting rods, and piston pins. MMCs are primarily used to reduce the mass of reciprocating components in automotive engines, resulting in reduced noise and vibration. Aluminum's widespread use in automotive alloys, in compared to other lightweight materials such as titanium and magnesium, is mostly due to its lightweight and low cost. Leading car manufacturers began commercializing AlMMCs due to their superior wear resistance, strength, thermal characteristics, light weight, and low cost. MMCs are employed in the production of valves and light-weight brake calipers. The primary application of AlMMCs has been in cars. At the earliest, research on AlMMCs in the automobile sector

began. Toyota, a Japanese company, successfully manufactured reinforced AlMMC piston rings for automotive engines, connecting rods for automobiles, and other parts in the 1980s. When SiC particles are added to AlMMCs used in the manufacture of automotive brake discs, wear resistance is increased while noise is significantly reduced. Aluminum-based composite materials can be used to make brake rotors, brake pistons, brake pads and calipers, and other brake system components, as well as vehicle drive shafts, rocker arms, and other automotive components. Aluminum and aluminum-based composite materials can be used in the automobile sector to improve product performance, weight reduction, and energy economy. AlMMCs are typically utilized to build engine pistons, connecting rods, and components for vehicle brake systems. Indirectly, the use of aluminum has resulted in an increase in the life span of components such as the engine, car frame, gearbox function, braking system, and other automotive structures. Automotive manufacturers are attempting to meet customer needs, which include greater inside conveniences and advanced electronic systems for safety, navigation, and entertainment, which all add unneeded weight in the absence of lightweight materials. Additionally, a significant reason for the increased use of MMCs is to meet rigorous fuel economy regulations. MMCs enable car manufacturers to overcome some difficulties by meeting unique and severe design requirements. (Bulei et al., 2018).

2. **Piston and Connecting rods:**Because the piston operates at extremely high temperatures and pressures inside the cylinder, the material used for both must be highly thermally conductive and resistant to wear. AlMMCs can be utilized for this, as they provide great quality at a reasonable cost, and weight reduction of a piston is also achievable. AlMMCs in the engine block help engines to reach operating temperatures more quickly, while also providing improved wear resistance and weight savings. By comparison, the MMC casting method is far simpler than the standard piston manufacturing technique. MMCs have a high rate of success when used as pistons. Silicon carbide is frequently utilized as a reinforcing material, particularly in motorsports vehicles. When connecting rods are made using AlMMCs, a significant weight decrease is noticed. Vibrations throughout the operation are minimized. This also lessens the load on the shaft, which results in decreased fuel consumption and increased engine output (Senthil et al., 2020).
3. **Brake and Chassis:**Automotive disk brakes and brake calipers are an area where substantial weight reduction can be achieved. In order to lower their costs and maximize machinability generally AlMMCs used, ceramic reinforced AlMMCs used in braking disks of high-speed trains and in the racing car braking systems, For the manufacturing of brake disks and pipes Specific casting methods are used. A growing number of car manufactures are now using AlMMCs in production of braking systems. The specific mechanical properties of the material used to construct the chassis can have an impact on the performance of the vehicle, including its strength and toughness, as well as the properties that determine the safety of the occupants in serious crashes. As a result, materials that improve both torsional rigidity and energy absorption for better vehicle dynamics are commonly used in chassis construction. (Badini et al., 2000).
4. **Rail and Marine Transportation:**AlMMCs are used to reduce the weight of rail cars. Aluminum remains a superior material for railroad cars. Due to its small weight, this product has a good resistance to corrosion. Marine transportation has also been altered by the use of aluminum alloys and their composites. Utilization of these materials has resulted in an increase in the speed, size, and fuel efficiency of ships and boats. Aluminum-based materials frequently provide increased flexibility and access to low-draught ports. Aluminum-based composite materials meet stringent requirements for high-speed ship construction, protection, and fire hazard reduction. Aluminum

is capable of effectively transporting high-speed waterborne cargoes. As the aerospace sector has evolved significantly, the requirements for material qualities have become increasingly stringent. Aluminum MMCs are used in a wide variety of aeronautical applications. Due to the harsh climate conditions in space, new materials with desirable properties such as low thermal expansion coefficient (CTE), high, dimensionally stable structures, high specific stiffness, and low weight are required. Obviously, normal phenomena such as vacuum and radiation such as thermal and ionizing are discovered and met by spacecraft in near-earth orbit. Aluminum matrix composites enabled the manufacture of high-performance aircraft and satellites. Aluminum-based composites have a number of unique properties that enable them to be widely used in the aerospace, aviation, and military industries. Lockheed-martin, an American firm, was the pioneer in the use of MMC in military aircraft. Aluminum matrix composites have been widely employed in the fabrication of aircraft, helicopters, and other big aircraft wings, rudders, flaps, fuselages, and other components. AlMMCs are also utilized to build fan outlet guide vanes for aircraft engines. These composites can also be used to construct aircraft airframe structures, camera lens direction frames, hydraulic pipes, helicopter rotor systems, landing gear, and valve bodies. AlMMCs are suitable for satellite applications due to their high conductivity and low thermal expansion coefficient. Al matrix composites are also employed in the fabrication of optical and electrical components, as well as ultra-light spacecraft. These types of applications demonstrate the importance of these composites, which are predicted to have a long life. Railway carriages are manufactured due to the adaptability of the metal matrix choices. It has a high fuel efficiency rating and a greater load carrying capacity (Nwanji& Babalola, 2018).

5. **Construction and Building Industry:** AlMMCs are extensively used in the construction and building industries. The two key advantages of aluminum are its strength and light weight. Aluminum has established itself as a prominent material over time due to the fact that metal and its alloys are 100% recyclable; in fact, a considerable amount of aluminum utilized in modern structures is recycled. Advanced AlMMCs can be used to boost the structure's capacity. Due to the superior rigidity and strength of Aluminium Metal Matrix Composites, they are utilized in the construction of bridge decks, fall sealing, and as a shield against the sun for buildings, window frames, door panels, and roof structures (Senthil et al., 2020).
6. **Electronics and Electrical Industry:**The most recent generation of modern electrical equipment creates significantly more heat than previous generations of such equipment. Therefore, heat dissipation in electrical applications has emerged as a crucial issue to be addressed. Heat is a primary cause of electrical device failure, and it is therefore necessary to eliminate or dissipate as much heat as possible from the system to the greatest extent possible. Increasingly common in electronic devices, heat sinks are becoming nearly vital in computer equipment, particularly when it comes to processor units. Thermal fatigue is caused by a slight difference in the thermal expansion coefficient between the material and the thermal sink. Excellent thermomechanical and thermophysical properties are required in the ideal material for electrical and electronic applications in order to allow for maximal heat transport with the least amount of thermal distortion. Modern new materials, such as metal matrix composites with high thermal conductivity and low thermal expansion coefficient, have emerged in recent years and have proven to be crucial in the resolution of heat transmission and dissipation issues associated with electronic devices. Aluminum is generated when a variety of reinforcing elements, such as nitrides, oxides, and carbides, are combined to form a material that has minimal thermal expansion while also having a high thermal conductivity. The low coef-

ficient of thermal expansion, low density, and high thermal conductivity of AlMMCs, particularly those reinforced with SiC, have made them popular in electronic applications such as electronic packaging and heat sinks. Material used in electronic packaging must be able to structurally support electronic components while also protecting them from the destructive effects of the environment and absorbing excess heat generated by electronic components. One of the most important qualities of such materials is their great stiffness, which is combined with their extremely low density. AlMMCs have become increasingly popular in electronic packaging, notably in aviation, where the high cost of the material can be justified by the weight reductions achieved by using it. Power transmission from producing stations to businesses and households is made possible by the excellent electrical conductivity, light weight, and corrosion resistance of aluminum-based composites used in the transmission of electricity. We are able to get good results in the electrical field by utilizing the upgraded AMMC, which has increased thermal conductivity, reduced thermal expansion, is lightweight, has increased strength and stiffness, and has low thermal stress (Nayak et al., 2020).

7. **Sports and Recreation Applications:**To manufacture sporting equipment, a variety of materials, including plastics, ceramics, and polymers, are used in conjunction with a variety of fabrication procedures, depending on the desired shape and design of the equipment. Aluminum metal matrix composites are very versatile materials that can be used to create goods for sporting activities. In many different types of sporting goods, MMCs have been employed; in particular, the use of AlMMCs in sporting goods has been steadily expanding due to their low cost and lightweight nature. Recreational products, such as those used in golf, baseball, skiing, and several other activities, as well as extremely competitive sporting tasks, always require high performance materials at a low cost, and AlMMCs are a valuable asset for these types of applications (Samal et al., 2020).
8. **Defense Applications:** Using composite material in huge quantities of conventional weapons is a possibility. Because of their excellent overall performance, fiber reinforced aluminum composites are one of the most extensively used composite materials in the manufacture of firearms. AlMMCs are being used in important components of missiles that were formerly built of beryllium. MMCs provide several advantages, including decreased prices and the avoidance of toxicity issues associated with beryllium. As a result of their high rigidity, MMCs are essential in the production of directed gun fins, which are extremely critical components. This contributes to improving the accuracy of a weapon. To endure high temperatures, aluminium–silicon carbide (whiskers) is employed in tank construction. Wings and allied components for commercial, military, and helicopter aircrafts are also manufactured (Pol N Verma G Pandey R Shanmugasundaram T, 2018).
9. **Marine Applications:** Due to the reduced weight of Aluminium Metal Matrix composites, they can be used to construct more fuel-efficient and quicker boats and ships. Additionally, corrosion resistance is exceptional. (Nwanji& Babalola, 2018).
10. **Packaging and Containers Making:**AMMC is a prominent player in this industry, specializing in the packaging of beverages, food container foils, and cold beverage cans(Moungomo et al., 2016).

CONCLUSION

We have provided a full overview of AMMCs in this book chapter, beginning with the fabrication characteristics and procedures of metal matrix composites and concluding with their application. The liquid state fabrication approach is one of the simplest and most recognized methods for producing AMMCs.

Metal matrix nanocomposites with superior mechanical properties have been developed for automotive and aerospace applications. Thus, this paper is designed to provide a broad understanding of the many features of AMMCs, including the fabrication process and use. Understanding the various fabrication techniques will considerably aid in the production of high-quality AMMCs for future industrial applications requiring high performance. Modern technologies provide a variety of solutions and guidelines for material selection to manufacturers of products for a number of uses. With current technological advancements, different materials are necessary to make products with a variety of features. Gradual advancements in material manufacturing technology enable the development of lightweight MMCs for a variety of applications at a lower cost with a greater quality. Aluminum and its alloys are highly sought materials since they meet the majority of contemporary needs and frequently exhibit reasonable mechanical properties. Aluminum-based composites find widespread use in transportation vehicles, aerospace, defense weapons, and electronics and optical devices. In the coming years, this field will see increased development, and new materials with superior mechanical, thermal, and electrical qualities will become available. AlMMCs are still under investigation but are predicted to grow at a healthy rate in the future. These composite classes have an excellent future, as industry has focussed on building low-cost, high-performance MMCs. Several of the most significant findings are presented below.

1. Among the available production processes for MMCs, stir/squeeze casting appears to be the most promising for this application. The squeeze pressure is the most important parameter in the stir/squeeze casting process since it has the greatest influence on the mechanical properties.
2. Aluminum is a more preferred matrix material due to the ease with which it can be handled during the manufacturing process.
3. For the manufacturing of AMMCs reinforced with any reinforcing material, a bottom tapping stir casting furnace with ultrasonic stirring in conjunction with a squeezing attachment would be optimal.

ACKNOWLEDGMENT

The authors thank the National University of Science and Technology (NUST) Muscat, Sultanate of Oman, for providing research support through National University Faculty Research Grant (NUFRG-2020), Project ID (NUFRG20.11).

REFERENCES

Altinkok, N., Özsert, I., & Findik, F. (2013). Dry sliding wear behavior of Al2O3/SiC particle reinforced aluminium based mmcs fabricated by stir casting method. *Acta Physica Polonica A*, *124*(1), 11–19. doi:10.12693/APhysPolA.124.11

Badini, C., La Vecchia, M., Fino, P., & Cobelli, P. (2000). Gravity casting processing of brake discs: Use of partially recycled (Al-Si-Mg)/SiCp composite. *Materials Science and Technology*, *16*(6), 681–686. doi:10.1179/026708300101508270

Behera, M. P., Dougherty, T., Singamneni, S., & de Silva, K. (2020). Selective laser melting of aluminium metal-matrix composites and the challenges. *Materials Today: Proceedings*, *33*, 5729–5733. doi:10.1016/j.matpr.2020.05.086

Bulei, C., Todor, M. P., & Kiss, I. (2018). Metal matrix composites processing techniques using recycled aluminium alloy. *IOP Conference Series. Materials Science and Engineering*, *393*(1), 012089. Advance online publication. doi:10.1088/1757-899X/393/1/012089

Campbell, F. C. (2006). Metal Matrix Composites. In *Manufacturing Technology for Aerospace Structural Materials* (1st ed., pp. 419–458). Elsevier Science. doi:10.1016/B978-185617495-4/50009-3

Govindarao, R., Ganguly, R. I., Dash, R. R., Rao, P. S. P., Reddy, G. S., & Singh, S. K. (2015). Development of a Novel Aluminium Based Metal Matrix Composite using Insitu Ternary Mixture (Al 2 O 3 -SiC-C) Prepared by Thermal Treatment of Fly-ash. *Transactions of the Indian Institute of Metals*, *68*(5), 951–958. doi:10.100712666-015-0532-x

Jaishankar, M., Tseten, T., Anbalagan, N., Mathew, B. B., & Beeregowda, K. N. (2014). Toxicity, mechanism and health effects of some heavy metals. *Interdisciplinary Toxicology*, *7*(2), 60–72. doi:10.2478/intox-2014-0009 PMID:26109881

Lathabai, S. (2018). Additive Manufacturing of Aluminium-Based Alloys and Composites. In *Fundamentals of Aluminium Metallurgy*. Elsevier Ltd., doi:10.1016/B978-0-08-102063-0.00002-3

Ma, Z. Y., Liu, Z. Y., Zhang, Q., Ni, D. R., & Xiao, B. L. (2013). Fabrication of metal matrix composites via friction stir processing. *Proceedings of the 1st International Joint Symposium on Joining and Welding*, 395–399. 10.1533/978-1-78242-164-1.395

Maclin John Vasanth, K., Lokendar Ram, P. S., Pon Anand, V., Prabu, M., & Rahul, S. (2020). Experimental investigation of mechanical and tribological properties of Aluminium metal matrix composites fabricated by powder metallurgy route - A review. *Materials Today: Proceedings*, *33*, 1058–1072. doi:10.1016/j.matpr.2020.07.057

Maniraj, S., Thanigaivelan, R., Viswanathan, R., & Elumalai, P. (2021). Experimental investigation of MRR and ROC in aluminium metal matrix composites. *Materials Today: Proceedings*, *45*, 1102–1106. doi:10.1016/j.matpr.2020.03.190

Mehta, M. G., & Vadher, J. A. (2020). Mechanical characterization of aluminum LM25 based metal matrix composite made with conventional process. *Materials Today: Proceedings*, *38*, 3091–3096. doi:10.1016/j.matpr.2020.09.490

Moungomo, J. B. M., Kouya, D. N., & Songmene, V. (2016). Aluminium machining chips formation, treatment & recycling: A review. *Key Engineering Materials*, *710*(1977), 71–76. doi:10.4028/www.scientific.net/KEM.710.71

Nayak, K. C., Pandey, A. K., & Date, P. P. (2020). Mechanical and physical characterization of powder metallurgy based aluminium metal matrix hybrid composite. *Materials Today: Proceedings*, *33*, 5408–5413. doi:10.1016/j.matpr.2020.03.134

Nwanji, E. C., & Babalola, P. O. (2018). Synthesizing a Metal Matrix Composite (Al-Al2O3) from Aluminium (AA1237) Reinforced with Alumina (Al2O3) Particulates. *IOP Conference Series. Materials Science and Engineering*, *413*(1), 012043. Advance online publication. doi:10.1088/1757-899X/413/1/012043

Panwar, N., & Chauhan, A. (2018). ScienceDirect Fabrication methods of particulate reinforced Aluminium metal matrix composite-A review. *Materials Today: Proceedings*, *5*(2), 5933–5939. doi:10.1016/j.matpr.2017.12.194

Peat, T., Galloway, A., Toumpis, A., Mcnutt, P., & Iqbal, N. (2016). The erosion performance of cold spray deposited metal matrix composite coatings with subsequent friction stir processing. *Applied Surface Science*. Advance online publication. doi:10.1016/j.apsusc.2016.10.156

Pol, N., Verma, G., Pandey, R., & Shanmugasundaram, T. (2018). *Fabrication of AA7005/TiB2-B4C surface composite by friction stir processing: Evaluation of ballistic behaviour*. Defence Technology. doi:10.1016/j.dt.2018.08.002

Samal, P., Vundavilli, P. R., Meher, A., & Mahapatra, M. M. (2020). Recent progress in aluminum metal matrix composites: A review on processing, mechanical and wear properties. *Journal of Manufacturing Processes, 59*(November), 131–152. doi:10.1016/j.jmapro.2020.09.010

Senthil, S., Raguraman, M., & Manalan, D. T. (2020). Manufacturing processes & recent applications of aluminium metal matrix composite materials: A review. *Materials Today: Proceedings*, *45*, 5934–5938. doi:10.1016/j.matpr.2020.08.792

Sha, J., Lü, Z., Sha, R., Zu, Y., Dai, J., Xian, Y., Zhang, W., Cui, D., & Yan, C. (2021). Improved wettability and mechanical properties of metal coated carbon fiber-reinforced aluminum matrix composites by squeeze melt infiltration technique. *Transactions of Nonferrous Metals Society of China*, *31*(2), 317–330. doi:10.1016/S1003-6326(21)65498-5

Shanmughasundaram, P., Subramanian, R., & Prabhu, G. (2011). Some studies on aluminium - Fly ash composites fabricated by two step stir casting method. *European Journal of Scientific Research*, *63*(2), 204–218.

Sharma, A. K., Bhandari, R., Aherwar, A., Rimašauskiene, R., & Pinca-Bretotean, C. (2020). A study of advancement in application opportunities of aluminum metal matrix composites. *Materials Today: Proceedings*, *26*, 2419–2424. doi:10.1016/j.matpr.2020.02.516

Singh, N., Singh, R., Ahuja, I. P. S., Farina, I., & Fraternali, F. (2019). Metal matrix composite from recycled materials by using additive manufacturing assisted investment casting. *Composite Structures*, *207*, 129–135. doi:10.1016/j.compstruct.2018.09.072

ADDITIONAL READING

Arunachalam, R., & Krishnan, P. K. (2021). Compressive Response of Aluminum Metal Matrix Composites. In Encyclopedia of Materials: Composites (pp. 1–21). Elsevier Ltd. doi:10.1016/B978-0-12-803581-8.11818-1

Christy, J. V., Arunachalam, R., Mourad, A.-H. I., Krishnan, P. K., Piya, S., & Al-Maharbi, M. (2020). Processing, Properties, and Microstructure of Recycled Aluminum Alloy Composites Produced Through an Optimized Stir and Squeeze Casting Processes. *Journal of Manufacturing Processes*, *59*, 287–301. Advance online publication. doi:10.1016/j.jmapro.2020.09.067

Fuhaid, M. S., Murali, V., Maleque, M. A., Krishnan, P. K., & Rahaman, M. A. (2021). Synthesis and Characterization of Aluminium Composite with Graphene Oxide Reinforcement Synthesis and Characterization of Aluminium Composite with Graphene Oxide Reinforcement. *IOP Conference Series. Materials Science and Engineering*, *1057*(1), 1–12. doi:10.1088/1757-899X/1057/1/012005

Krishnan, P., Arunachalam, R., Husain, A., & Al-Maharbi, M. (2020). Studies on the Influence of Stirrer Blade Design on the Microstructure and Mechanical Properties of a Novel Aluminium Metal Matrix Composite. *Journal of Manufacturing Science and Engineering*, 1–25. doi:10.1115/1.4048266

Krishnan, P. K., Christy, J. V., Arunachalam, R., Mourad, A.-H. I., Muraliraja, R., Al-Maharbi, M., Murali, V., & Chandra, M. M. (2019). Production of aluminum alloy-based metal matrix composites using scrap aluminum alloy and waste materials: Influence on microstructure and mechanical properties. *Journal of Alloys and Compounds*, *784*, 1047–1061. Advance online publication. doi:10.1016/j.jallcom.2019.01.115

Mathan Kumar, N., Senthil Kumaran, S., & Kumaraswamidhas, L. A. (2016). Aerospace application on Al 2618 with reinforced – Si 3 N 4, AlN and ZrB 2 in-situ composites. *Journal of Alloys and Compounds*, *672*, 238–250. doi:10.1016/j.jallcom.2016.02.155

Rajan, T. P. D., Pillai, R. M., Pai, B. C., Satyanarayana, K. G., & Rohatgi, P. K. (2007). Fabrication and characterisation of Al – 7Si – 0. 35Mg / fly ash metal matrix composites processed by different stir casting routes. *Composites Science and Technology*, *67*(15-16), 3369–3377. doi:10.1016/j.compscitech.2007.03.028

Rao, V. R., Ramanaiah, N., & Sarcar, M. M. M. (2014). Fabrication and investigation on Properties of TiC reinforced Al7075 metal matrix composites. *Applied Mechanics and Materials*, *594*, 349–353. doi:10.4028/www.scientific.net/AMM.592-594.349

Rayapandi Thevar, T. (2020, April 1). V, R., Krishnan, P. K., R, T., & Arunachalam, R. (2020). A novel approach for the production and characterisation of aluminium–alumina hybrid metal matrix composites. *Materials Research Express*, *7*(4), 046512. Advance online publication. doi:10.1088/2053-1591/ab8657

KEY TERMS AND DEFINITIONS

Additive: Adds material layer by layer or point by point to create an object

Alloy: A metallic material made up of two or more chemical compounds, at least one of which is a metal.

Casting: A technique in which hot liquid metal flows into a mold cavity and solidifies.

Composite: A mixture of two or more insoluble materials that exhibits qualities superior to those of either of the component materials.

Crucible: A ceramic pot consisting of materials with a relatively high thermal conductivity, such as graphite, and joined with clay or carbon. It is used to melt metals.

In-Situ: During processing, the reinforcement is generated within the matrix as a result of reaction.
Infiltration: Filtration allows a liquid to permeate into something.
Metal Matrix: Acts as a continuous phase and accommodates the reinforcement material.
Metallurgy: Properties of metals and their production and purification
Reinforcement: Material that enhances the strength or other mechanical properties of the composite.

Chapter 9
Emergence of Advanced Manufacturing Techniques for Engineered Polymeric Systems in Cancer Treatment

Mohamad Taleuzzaman
https://orcid.org/0000-0002-5201-3503
Faculty of Pharmacy, Maulana Azad University, India

Ali Sartaz
Independent Researcher, India

Md. Jahangir Alam
Radiant Institute of Pharmaceutical Sciences, India

Md. Noushad Javed
https://orcid.org/0000-0002-5775-8256
K.R. Mangalam University, India

ABSTRACT

Clinical performances of chemotherapeutic drugs which are used to manage different stages of cancers are usually facing numerous pharmacological challenges such as tumor microenvironment, high dose requirements, poor selectivity towards cancer cells, life-threatening cytotoxicity, and frequent drug resistance incidences, in addition to pharmacotechnical issues such as poor aqueous solubility, uncontrolled drug-release, low stability, non-specific bio-distribution, and erratic bioavailability profiles. The chapter aims to provide a brief account of advancements made in nanotechnology-enabled manufacturing engineering tools for manipulating polymeric materials as efficient carriers so that loaded anti-cancer drugs would exhibit better therapeutic applications and optimized clinical significance in cancers.

DOI: 10.4018/978-1-7998-9574-9.ch009

INTRODUCTION

Cancers are uncontrolled cell division and the spreading of abnormal cells; which leads to globally incidences as a second most common reason of deaths (Pottoo et al 2021a). The report suggested at least 1.8 million new cases and 0.6 million deaths by the end of 2020 year with cancers, only in the United States alone (Siegel et al 2020). Clinical choices for the selection of treatment modalities and therapies such as radiation, surgery, immunotherapy and chemotherapy in management of cancers, are mainly determined by the types and stages of cancers, presence of different biomarkers levels, as well as physiological status of patients (Christofi et al 2019). However, among all, roles of anti-cancer drugs based interventions in therapeutical management are slightly dominating because of in comparison to normal cells, these drugs exhibit relatively higher specificity and cytotoxicity towards cancerous cells (Bharti et al 2021). However, the convention chemotherapies have lots of drawbacks for their therapeutic uses like severe side effects, low bioavailability, drug resistance, non-specific biodistribution (Nurgali et al 2018). While, although drugs used in conventional treatments offers significant extent of protection and efficacy against cancers but owing to their associated challenges which minimizing therapeutic values and efficacy, are always a critical concern for clinicians so they prefer to explore novel techniques, tools and alternative therapeutics for the same (Pottoo et al 2021b).

Subsequently, in order to efficiently manage any disease clinically, researchers always emphasis upon emerging evidences and efforts made towards understanding both pathology and mechanisms as a very important aspect, including in cancer as well (Waziri et al 2021). CNS acting drugs anti-inflammatory and anti-cancer drugs are facing additional barriers because of either organ specific characteristics such as Blood Brain Barrier (BBB) (Pottoo et al 2019b). or disease based change in the pathological changes such BBTB (Blood Brain Tumor Barrier), however immunotherapy based targeted delivery approaches showing significant impacts and relevancy to address such challenges (Pottoo et al 2021c; Abu-Izneid et al 2020). Here, selection of different materials, which are being exploited in drug delivery systems, are based on their pharmaceutical and pharmacological properties; hence these may bear significant advantages and attracted researchers with their promising roles in delivering loaded drugs (Wong et al 2020). Furthermore, approaches are towards exploiting various advantages and benefits of nanotechnology approached manufacturing of novel drug delivery systems (NDDS) to improve the final performances with high level of safety indexes (Javed et al 2020a). Such emerging roles for high potential of nanotechnology approached DDS (drug delivery systems) in various diseases, including cancers, are being attributed due to unique potential of such carriers to deliver such drugs at targeted site of actions, high concentration in targeted cells, low deposition of drug at non-pathological sites, high drug clearance, modulation of physiological properties in microenvironment, control drug release with reduction in dose requirements (Pottoo et al 2020a). Briefly, similar to other inflammatory disorders such as IBD (inflammatory bowel diseases), epilepsy, and Parkinson's; in cancer too different inflammatory signaling and proteins are involved as a major cause of degeneration of cells as well as progression of disease states (Pandey et al 2020; Pottoo et al 2019a). Additionally, novel carriers include favoring binding of drugs at specific site in protein as well as other interfering in existing pathologically significant biochemical signals, which offers promising in targeted treatments and tissue engineering steps (Aslam et al 2021). These nanoparticles are being made in a way which favor easily passing of carriers within targeted cells as well as enhanced retention of accumulated loads; such phenomena is also called enhanced permeability and retention (EPR) effects, and one of the most exploited approaches in DDS (Javed et al 2018). These characteristics features are bear some special significance in all metabolic and inflammatory disorders, however very critical especially in case of cancers (Mishra et al 2019; Sharma et al 2019).

Among different nanotechnology integrated advanced manufacturing techniques to develop chemotherapy loaded carriers; approaches for fabrication of materials into nanocarriers are bearing unique abilities such as delivering loaded drugs at targeted sites, reduced undesirable bio-distribution in non-pathological tissues, minimizing systemic side effects and lowering incidences of drug resistance cases; so these are proven great promising to clinicians to integrate such systems in clinical usages. Figure 1 briefly illustrated roles, types and significances of these nanotechnology processed polymeric materials in the cancer management. The characteristic features of polymeric carriers are based on composition, types, surface charge, morphology, stability, encapsulation efficiency and biopharmaceutical profiles of both cores and peripheries, in independent manners (Alam et al 2019). Briefly, depending upon similarity in nature and type of both drugs and excipients (polymers); loaded drugs may either link at the surface of the NPs or load within polymeric systems of Polymeric NPs may be solid polymeric NPs, polymeric micelles, polymeric conjugates, dendrimers, polymersomes, polyplexes, and polymer-lipid hybrid systems all are identified by their physicochemical structures (Prabhu et al 2015). These nanotechnology fabricated delivery systems of polymers are gaining immense attentions in cancer, due to excellent control over sustain release kinetics and enhanced half-life of loaded drugs; while, targeted delivery at cancerous cells but with reduced retention in non-pathological cells (Javed et al 2021). Similarly, many polymers are being investigated as a building block material, in association of lipids, to fabricate novel polymer-lipid hybrid nanoparticles (PLHNPs) so that integration of polymers would potentiate performances of lipids based nanoparticles in developed hybrid carriers. Due to composition of both lipid and polymeric materials in PLHNPs, these have characteristic properties of both systems such as relatively small particle size, co-encapsulation, high drug loading capacity, modified drug release profile and high stability so such versatility of hybrid carriers offer efficient and targeted delivery of chemotherapeutic drugs to their sites of action (Rizwanullah et al 2020).

Additionally, polymers are also being used to functionalize NPs also help of loaded drugs with further enhancing their target-specificity, reducing non-specific bio-distribution as well as increased systemic residence time. Usually, these specific targets in cancer cells are surface antigens while polymers such as peptides, nucleic acid sequences, antibody fragments, and small molecules, bear specific functional groups that severed as a purpose of ligand by attaching at specific sites on cell surfaces (Senapati et al 2018).

This chapter would further revealed how different nanotechnology based manufacturing techniques offer approaches to fabricated native polymers to developed into promising carriers for cancer targeting, especially in solid tumors which have additional challenges of low penetrability of chemotherapeutics and high level of drug resistances; followed by their clinical success (Makino et al 2014).

IN-VIVO TARGETING OF NPS IN SOLID TUMORS

The various approaches of drug-loaded NPs which act on target site by a various mechanism of action by using different strategies in cancer treatment mentioned below.

Passive and Active Targeting

It involves the release of the drug to the tumor site, which is characterized by the vascularized the tumor tissues and allows to pass the drug molecules utilizing a simple passive diffusion mechanism in the interstitial area and inside the cells, as shown in figure 2A. Whereas in active targeting, the NPs are bind with some specific antibodies, proteins, or peptides, which are chosen based on the receptors overexpress in tumor cells, as shown in figure 2B (Yang et al 2015).

Enhanced Permeation and Retention (EPR) Effect

When tumor volume increases, the demand of the tumor cells also increased, so atypical tortuosity, basement membrane abnormalities, and the absence of pericytes lining endothelial cells are the main features of EPR which cause holed vessels with increased gap sizes. Due to high interstitial pressure in the core compared to the peripheral makes them poor drainage to by the lymphatic systems. These characteristics feature the EPR effect; in this, NPs can target the tumor cells and entrap inside the tumor due to their high retention ability (Browning et al 2017).

Tumor Microenvironment

The microenvironment surrounding the tumor cells is well defined from the normal cells because high metabolic rates result in an acidic environment. So, the NPs systems are designed accordingly and stable at physiological pH 7.4, but when exposed to such an environment, the where pH is less than physiological and NPs degraded to release the drug in that particular site Another parameter like temperature-sensitive polymer, which exhibits low critical solution temperature (LCST) and releases the drug from the NPs system when the temperature is found higher to LCST. Sometimes to precipitated the payload; the polymers are triggered by an external method using ultrasound or photothermal (Karimi et al 2016).

Surface Charge

Most of the tumor cells bear a negative charge on their surface compared to the normal cells due to which the positive charge containing NPs favors the targeting the tumor cells in which NPs bind the electrostatically negative charged phospholipids headgroup. For cationic NPs which deliver siRNA to specific gene present in cancer cells, and it also purifies cancer cell towards the treatment of Paclitaxel and improved the anti-cancerous activity (Serrano-Sevillaet al 2019).

Ligand-Based Targeting

In this system, ligand (for example, an antibody, nucleic acid, peptides, vitamins, and carbohydrates) were bound with the surface of NPs and specifically targeted to the receptor present on tumor cells. Targeting ligand plays a critical role in the uptake of NPs through the endocytosis process (Yu et al 2016).

Tumor Endothelium Targeting

The inhibition of the making of new blood vessels so that angiogenesis for enough blood supply can result in solid tumor growth inhibition. Thus, designing to target the blood vessels production using NPs approaches prove to be very useful for the prevention of further growth of solid tumor. The signaling pathway of$\alpha v\beta 3$ integrin, which activates the vascular endothelial growth factors (VEGFs), induces the tumor angiogenesis, and these receptors seem to be promising targeting of NPs system (de la Torre et al 2020).

NPS SYSTEMS FOR SOLID TUMORS

Due to the non-specific biodistribution of chemotherapeutics of an anti-cancerous drug, the NPs strategy is used for the bioaccumulation to a target site or tumor cells due to the EPR effect. The nanoparticles also have the ability to reduce the exposure to the normal cells hence reduces the side effects (Canavese et al 2018). Hapca et al investigated that polymeric nanoparticles of PLA (Polylactic acid) attached with monoclonal antibodies which have great utility in lymphomas and ovarian cancer. Even a small quantity of bilirubin is used for preventing the oxidation of cells if formulated into a nanoparticle. Hence drug-loaded nanoparticles are utilized for solid tumor prevention (Binjawadagi et al 2014). Some of the advantages over conventional treatment shown in table 1.

Natural Polymers

The source of natural polymers is plants, animals, fungi, and bacteria. Based upon nature, it is of two types; one is polysaccharides, and the second is protein-based polymers. Both can be used for drug delivery and act as a scaffold of the extracellular matrix (ECM). The drug delivery with high drug loading can be possible with minimal invasive behavior (Dragan et al 2019). The different functional groups present in the polymers used for the functionalization after slight modification for better target drug delivery and shown promising results. The excellent biocompatibility because easily broken down by enzymatic degradation and this can be tuned for rate drug release profile. In this section, the most common naturally found polymers used for drug delivery for the solid tumor are being discussed (Song et al 2019; Kumar et al 2017).

Polysaccharides

Polysaccharides are composed of long chains of simple monosaccharide (sugar). The monosaccharides are multiple replicate units linked through a glycosidic bond, as shown in table 2. Polysaccharides obtained from animal-like chitosan, chondroitin, plants like pectin, guar gum, mannan, algae-like alginate, and from microbes like dextran, xanthan gum. Among them, chitosan, dextran, alginates, and hyaluronic acid are commonly used in the drug delivery system due to its non-toxicity and biocompatibility. Polysaccharides are containing a large number of reactive sites in their structure, large molecular weight variation, and varying composition makes them diversified property (Alam et al 2017). Polysaccharide nanoparticle was generally prepared by crosslinking polyelectrolyte interaction and self-assembly. Because of the EPR effect, the polysaccharide nanoparticles accumulated the drug into the tumor cells and decreased their associated side effect and toxicity (Zhu et al 2019). Each category of polysaccharides is mentioned with their formulation to the targeted delivery of solid tumor cells shown in tables 3, 4, 5, & 6.

Protein-Based Polymers

Proteins are composed of amino acids, and they are attached by peptide bonds. The tertiarystructure of proteins which stabilized by the various interaction and bonds like hydrophobic interaction, hydrogen bonding, disulfide bonding, and salt bridges. The properties of a protein are good compatibility and degradability. Because of natural degradation, the byproduct of drug delivery systems is least among all protein albumin, collagen, and gelatin most commonly used, as shown in Table-7 (Huo et al 2020).

Protein-based nanoparticles are relatively safe and also easy preparation methods. During their preparation, they easily modifying their functional group and targeting capabilities so that drug delivery would be efficient (Hong et al 2020). Some of the protein-based nanoparticlulated formulations with their suitability in drug loading is shown in Table-8.

FATE OF NPS FORMULATION

The novel formulation includes various ingredients like polymers, lipids, and protein used as a carrier for the therapeutic molecules to deliver at target sites. The carriers are classified as naturals or synthetic materials; the natural material is found naturally, inexpensive, biocompatible, biodegradable, non-toxic, and non-immunogenic (Hasnain et al 2019). These benefits make them promising approaches for drug delivery. Although some of the drawbacks like a batch-to-batch variation and their structural complexity (Ulbrich et al 2016; Oladeji et al 2012). The formulation and method of preparation of NPs depend upon the chosen polymers. Generally, the polymers dissolves in the organic phase and mixed with surfactant. Now organic phase is added to the aqueous phase, and then sonicated to form the nanodroplets. At last, stir continuously to evaporate the organic phase to leave behind the solid polymeric nanoparticles. The solid polymeric NPs coated with favorable surface properties (Zielińska et al 2020).

LIMITATIONS AND EMERGING TRENDS

Although, natural polymer-based delivery systems have shown favorable results in contrast to the conventional therapies. Furthermore, due to natural origin of most of such polymers, they have lot of advantages like good degree of biocompatibility, biodegradability, non-immunogeneticity, and non-toxic in nature etc. (Kumar et al 2020). Their distinctive properties enabled them to enhance the therapeutic effects of loaded active anti-cancer molecules. Apart from above, chemical structures of these polymers also offer excellent opportunities of functionalization purposes (Ulbrich et al 2016). However, despite all the successes of polymeric DDS, still there is a gap between preclinical and clinical outputs for assessment of therapeutic efficacy. Hence, it is quite challenging and difficult to translate the experimental approaches to good clinical outcomes. Subsequently, applications and scopes of polymeric DDS are still very limited owing to variations, reproducibility and validation related regulatory challenges. However, with the furtherance of nanotechnology these issues need to be address in judicious manners, so that such carriers would be integrated with anti-cancer therapeutics. For example, although Abraxane® is still one of the natural polymer (albumin) based nanoparticlulated delivery, approved by the FDA in 2005 for cancer therapy. While; many natural polymers like Chitosan, which is however also being classified with a remark of GRAS by the FDA but still not being approved in cancer management (Boboet al 2016).

The natural polymers originated from a natural source, so achieving a high degree of purity due to batch to batch variation is still challenging (Oladeji et al 2012). Another challenging situation when crosslinkers used in DDS are not completely digested by the natural enzymes. Although many attempts are made towards discovery of different types of digestible polymeric complexes, however these problems still bear limited successes in many cases. Apart from the above-mentioned issues, there are some technical issues like scale-up from laboratory research to large quantity during industrial manufacturing. Particular equipment in their production can be costly. There may be chances of stability issues during production. Here QbD based approaches reported significant improvement in process, reproducibility, shelf-life and clinical performances, in cost effective manners.

CONCLUSION

Cancers are uncontrolled cell division and the spreading of abnormal cells; and second most leading reason for global deaths. Although different clinical choices such as radiation, surgery, immunotherapy and chemotherapy in management of cancers, are available but their successes are very limited because difference in types and stages of cancers, presence of different biomarkers levels, as well as physiological status of patients. Still, anti-cancer therapeutics are a preferred choice for interventions but their associated challenges such as severe side effects, low bioavailability, drug resistance, non-specific biodistribution of convention delivery systems put them with limited successes and minimized therapeutic performance in a very significant manner. While, among the cancers, some cancer such as solid tumors bear challenges of low penetrability and high level of drug resistances while, drugs used in the brain cancer are facing additional challenges of Blood brain tumor barrier (BBTB) and Blood brain barrier (BBB).

Natural polymers are abundantly present and with bear significant evidences of their biocompatibility, low cytotoxicity, different surface properties and cheaper alternatives. Hence, among the nanocarriers; polymeric nanoparticles are preferred owing to their characteristics such as wide range of possible compositions, types (core, conjugation, complexes), surface charge, morphology, stability, encapsulation efficiency, as well as biopharmaceutical profiles. Hence, nanotechnology integrated advanced manufacturing techniques are nowadays emerged as a promising approach to develop nanocarriers with attributes of high pay-off of loads (drugs) at pre-specified targeted pathological sites, desired bio-distribution profiles, minimized systemic side-effects and lowering drug resistance incidences. these polymeric systems such as polymeric NPs, polymeric micelles, polymeric conjugates, dendrimers, polymersomes, polyplexes, and polymer-lipid hybrid systems are gaining immense attentions in especially cancer, where conventional delivery systems however facing additional challenges; owing to their excellent control over sustain release kinetics; temperature mediate system degradation (LCST and UCST); enhanced half-life and site-targeted delivery abilities; ligand bindings, offering of both active and passive targeting in tumor microenvironment and at tumor endothelium sites; EPR effects and surface charge mediated localization.

There are different polymeric systems which are already approved for different diseases, including cancers such as anti-cancer drug Abraxane®, approved by the FDA in 2005. However, there is still a gap between preclinical and clinical outputs for assessment of therapeutic efficacy of many of such polymeric systems, which limit their clinical translations. For example, many natural polymers based systems such as Chitosan, are although classified as a GRAS by the FDA but were not being approved in cancer management; owing to different regulatory issues such as validation of raw materials (degree of purity), high degree of batch-to-batch final product variations, reproducibility issues and other related regulatory challenges. However, trends are emerging towards integration of QbD based approaches to control quality and performance of process as well as final products also; in judicious and cost effective manners. Another important issue with polymers is related to digestibility of cross-linkers by the natural enzymes. However, many attempts are made towards discovery of different types of digestible polymeric complexes, which lead to some significant and promising successes, in many of the cases as well.

Overall, current chapter was focused on technological advancements made and engineered tools developed, to fabricate different polymers using different nanotechnology based manufacturing techniques into promising carriers for cancer targeting, especially in solid tumors

Conflicts of Interest: The authors declare no conflict of interest.

REFERENCES

Abu-Izneid, T., AlHajri, N., Ibrahim, A. M., Javed, M. N., Salem, K. M., Pottoo, F. H., & Kamal, M. A. (2020). Micro-RNAs in the regulation of immune response against SARS COV-2 and other viral infections. *Journal of Advanced Research*. PMID:33282419

Alam, M. S., Garg, A., Pottoo, F. H., Saifullah, M. K., Tareq, A. I., Manzoor, O., Mohsin, M., & Javed, M. N. (2017). Gum ghatti mediated, one pot green synthesis of optimized gold nanoparticles: Investigation of process-variables impact using Box-Behnken based statistical design. *International Journal of Biological Macromolecules*, *104*, 758–767. doi:10.1016/j.ijbiomac.2017.05.129 PMID:28601649

Alam, M. S., Javed, M. N., Pottoo, F. H., Waziri, A., Almalki, F. A., Hasnain, M. S., Garg, A., & Saifullah, M. K. (2019). QbD approached comparison of reaction mechanism in microwave synthesized gold nanoparticles and their superior catalytic role against hazardous nirto-dye. *Applied Organometallic Chemistry*, *33*(9), e5071. doi:10.1002/aoc.5071

Aslam, M., Javed, M. N., Deeb, H. H., Nicola, M. K., Mirza, A., Akhtar, M. H., & Waziri, A. (2021). Lipid Nanocarriers for Neurotherapeutics: Introduction, Challenges, Blood-brain Barrier, and Promises of Delivery Approaches. *CNS & Neurological Disorders - Drug Targets*, *20*. Advance online publication. doi:10.2174/1871527320666210706104240 PMID:34967302

Bajaj, G., Kim, M. R., Mohammed, S. I., & Yeo, Y. (2012). Hyaluronic acid-based hydrogel for regional delivery of paclitaxel to intraperitoneal tumors. *Journal of Controlled Release*, *158*(3), 386–392. doi:10.1016/j.jconrel.2011.12.001 PMID:22178261

Bharti, C, Alam, MS, Javed, MN, Khalid, M, Saifullah, FA, & Manchanda, R. Silica Based Nanomaterial for Drug Delivery. (2021). Nanomaterials: Evolution and Advancement towards Therapeutic. *Drug Delivery*, *2*(2), 57.

Bi, Y. G., Lin, Z. T., & Deng, S. T. (2019). Fabrication and characterization of hydroxyapatite/sodium alginate/chitosan composite microspheres for drug delivery and bone tissue engineering. *Materials Science and Engineering C, 100*, 576–583. doi:10.1016/j.msec.2019.03.040 PMID:30948094

Binjawadagi, B., Dwivedi, V., Manickam, C., Ouyang, K., Wu, Y., & Lee, L. J. (2014). Adjuvanted poly (lactic-co-glycolic) acid nanoparticle-entrapped inactivated porcine reproductive and respiratory syndrome virus vaccine elicits cross-protective immune response in pigs. *International Journal of Nanomedicine, 24*(9), 679–694. PMID:24493925

Bobo, D., Robinson, K. J., Islam, J., Thurecht, K. J., & Corrie, S. R. (2016). Nanoparticle-Based Medicines: A Review of FDA-Approved Materials and Clinical Trials to Date. *Pharmaceutical Research, 33*(10), 2373–2387. doi:10.100711095-016-1958-5 PMID:27299311

Browning, R. J., Reardon, P. J. T., Parhizkar, M., Pedley, R. B., Edirisinghe, M., Jonathan, C., Knowles, J. C., & Stride, E. (2017). Drug Delivery Strategies for Platinum-Based Chemotherapy. *ACS Nano, 11*(9), 8560–8578. doi:10.1021/acsnano.7b04092 PMID:28829568

Cai, S., Thati, S., Bagby, T. R., Diab, H. M., Davies, N. M., Cohen, M. S., & Forrest, M. L. (2010). Localized doxorubicin chemotherapy with a biopolymeric nanocarrier improves survival and reduces toxicity in xenografts of human breast cancer. *Journal of Controlled Release, 146*(2), 212–218. doi:10.1016/j.jconrel.2010.04.006 PMID:20403395

Canavese, G., Ancona, A., Racca, L., Canta, M., Dumontel, B., Barbaresco, F., Limongi, T., & Cauda, V. (2018). Nanoparticle-assisted ultrasound: A special focus on sonodynamic therapy against cancer. *Chemical Engineering Journal, 340*, 155–172. doi:10.1016/j.cej.2018.01.060 PMID:30881202

Chaubaroux, C., Vrana, E., Debry, C., Schaaf, P., Senger, B., Voegel, J. C., Haikel, Y., Ringwald, C., Hemmerlé, J., Lavalle, P., & Boulmedais, F. (2012). Collagen-based fibrillar multilayer films cross-linked by a natural agent. *Biomacromolecules, 13*(7), 2128–2135. doi:10.1021/bm300529a PMID:22662909

Chen, G., Svirskis, D., Lu, W., Ying, M., Huang, Y., & Wen, J. (2018). N-trimethyl chitosan nanoparticles and CSKSSDYQC peptide: N-trimethyl chitosan conjugates enhance the oral bioavailability of gemcitabine to treat breast cancer. *Journal of Controlled Release, 277*, 142–153. doi:10.1016/j.jconrel.2018.03.013 PMID:29548985

Choi, K. Y., Jeon, E. J., Yoon, H. Y., Lee, B. S., Na, J. H., Min, K. H., Kim, S. Y., Myung, S.-J., Lee, S., Chen, X., Kwon, I. C., Choi, K., Jeong, S. Y., Kim, K., & Park, J. H. (2012). Theranostic nanoparticles based on PEGylated hyaluronic acid for the diagnosis, therapy and monitoring of colon cancer. *Biomaterials, 33*(26), 6186–6193. doi:10.1016/j.biomaterials.2012.05.029 PMID:22687759

Christofi, T., Baritaki, S., Falzone, L., Libra, M., & Zaravinos, A. (2019). Current Perspectives in Cancer Immunotherapy. *Cancers (Basel), 11*(10), 1472. doi:10.3390/cancers11101472 PMID:31575023

de la Torre, P., Pérez-Lorenzo, M. J., Alcázar-Garrido, Á., & Flores, A. I. (2020). Cell-Based Nanoparticles Delivery Systems for Targeted Cancer Therapy: Lessons from Anti-Angiogenesis Treatments. *Molecules (Basel, Switzerland), 25*(3), 715. doi:10.3390/molecules25030715 PMID:32046010

Din, F. U., Aman, W., Ullah, I., Qureshi, O. S., Mustapha, O., Shafique, S., & Zeb, A. (2017). Effective use of nanocarriers as drug delivery systems for the treatment of selected tumors. *International Journal of Nanomedicine*, *12*, 7291–7309. doi:10.2147/IJN.S146315 PMID:29042776

Dragan, E.S., & Dinu, M.V. (2019). Polysaccharides constructed hydrogels as vehicles for proteins and peptides. A review. *CarbohydrPolym.*, *1*(225).

Elzoghby, A.O., Saad, N.I., Helmy, M.W., Samy, W.M., & Elgindy, N.A. (2013). Ionically-crosslinked milk protein nanoparticles as flutamide carriers for effective anticancer activity in prostate cancer-bearing rats. *Eur J Pharm Biopharm.*, *85*(3 Pt A), 444-51.

Feldman, D. (2019). Polymers and Polymer Nanocomposites for Cancer Therapy. *Applied Sciences (Basel, Switzerland)*, *9*(18), 3899. doi:10.3390/app9183899

Fu, C., Yang, R.M., Wang, L., Li, N.N., Qi, M., & Xu, X.D. (2017). Surface functionalization of superparamagnetic nanoparticles by an acid-liable polysaccharide-based prodrug for combinatorial monitoring and chemotherapy of hepatocellular carcinoma. *RSC Adv.*, *66*.

Golla, K., Bhaskar, C., Ahmed, F., & Kondapi, A. K. (2013). A target-specific oral formulation of Doxorubicin-protein nanoparticles: Efficacy and safety in hepatocellular cancer. *Journal of Cancer*, *4*(8), 644–652. doi:10.7150/jca.7093 PMID:24155776

Gulfam, M., Kim, J. E., Lee, J. M., Ku, B., Chung, B. H., & Chung, B. G. (2012). Anticancer drug-loaded gliadin nanoparticles induce apoptosis in breast cancer cells. *Langmuir*, *28*(21), 8216–8223. doi:10.1021/la300691n PMID:22568862

Han, J., Wang, Q., Zhang, Z., Gong, T., & Sun, X. (2014). Cationic bovine serum albumin based self-assembled nanoparticles as siRNA delivery vector for treating lung metastatic cancer. *Small*, *10*(3), 524–535. doi:10.1002mll.201301992 PMID:24106138

Hasnain, M. S., Javed, M. N., Alam, M. S., Rishishwar, P., Rishishwar, S., Ali, S., Nayak, A. K., & Beg, S. (2019). Purple heart plant leaves extract-mediated silver nanoparticle synthesis: Optimization by Box-Behnken design. *Materials Science and Engineering C*, *99*, 1105–1114. doi:10.1016/j.msec.2019.02.061 PMID:30889643

Hong, S., Choi, D. W., Kim, H. N., Park, C. G., Lee, W., & Park, H. H. (2020). Protein-Based Nanoparticles as Drug Delivery Systems. *Pharmaceutics*, *12*(7), 604. doi:10.3390/pharmaceutics12070604 PMID:32610448

Hong, S. H., Li, Y., Eom, J. B., & Choi, Y. (2018). Responsive alginate-cisplatin nanogels for selective imaging and combined chemo/radio therapy of proliferating macrophages. *Quantitative Imaging in Medicine and Surgery*, *8*(8), 733–742. doi:10.21037/qims.2018.09.01 PMID:30306054

Hosseinifar, T., Sheybani, S., Abdouss, M., Hassani Najafabadi, S. A., & Shafiee Ardestani, M. (2018). Pressure responsive nanogel base on Alginate-Cyclodextrin with enhanced apoptosis mechanism for colon cancer delivery. *Journal of Biomedical Materials Research. Part A*, *06*(2), 349–359. doi:10.1002/jbm.a.36242 PMID:28940736

Huang, S., & Huang, G. (2019). Preparation and drug delivery of dextran-drug complex. *Drug Delivery, 26*(1), 252–261. doi:10.1080/10717544.2019.1580322 PMID:30857442

Huo, M., Wang, H., Zhang, Y., Cai, H., Zhang, P., Li, L., Zhou, J., & Yin, T. (2020). Co-delivery of silybin and paclitaxel by dextran-based nanoparticles for effective anti-tumor treatment through chemotherapy sensitization and microenvironment modulation. *Journal of Controlled Release, 321*, 198–210. doi:10.1016/j.jconrel.2020.02.017 PMID:32044390

Ige, Umoru, & Aribo. (2012). Natural Products: A Minefield of Biomaterials. *International Scholarly Research Notices.*

Javed, M. N., Ashraf, G. M., Barreto, G. E., & Naim, M. (2020b). Current Trends for Rationalizing Brain Targeting Nanoparticles in Neurological Disorders (Part-I). *Current Drug Metabolism, 21*(9), 648. doi :10.2174/138920022108201113094659 PMID:33327901

Javed, M.N., Dahiya, E.S., Ibrahim, A.M., Alam, M.S., Khan, F.A., & Pottoo, F.H. (2020a). Recent Advancement in Clinical Application of Nanotechnological Approached Targeted Delivery of Herbal Drugs. *Nanophytomedicine*, 151-172.

Javed, M. N., Kohli, K., & Amin, S. (2018). Risk assessment integrated QbD approach for development of optimized bicontinuous mucoadhesive limicubes for oral delivery of rosuvastatin. *AAPS PharmSciTech, 19*(3), 1377–1391. doi:10.120812249-018-0951-1 PMID:29388027

Javed, M.N., Pottoo, F.H., Shamim, A., Hasnain, M.S., & Alam, M.S. (2021). Design of Experiments for the Development of Nanoparticles, Nanomaterials, and Nanocomposites. *Design of Experiments for Pharmaceutical Product Development*, 151-169.

Jiang, H., Liang, G., Dai, M., Dong, Y., Wu, Y., Zhang, L., Xi, Q., & Qi, L. (2020). Preparation of doxorubicin-loaded collagen-PAPBA nanoparticles and their anticancer efficacy in ovarian cancer. *Annals of Translational Medicine, 8*(14), 880. doi:10.21037/atm-20-5028 PMID:32793724

Karimi, M., Eslami, M., Sahandi-Zangabad, P., Mirab, F., Farajisafiloo, N., Shafaei, Z., Ghosh, D., Bozorgomid, M., Dashkhaneh, F., & Hamblin, M. R. (2016). pH-Sensitive stimulus-responsive nanocarriers for targeted delivery of therapeutic agents. *Wiley Interdisciplinary Reviews. Nanomedicine and Nanobiotechnology, 8*(5), 696–716. doi:10.1002/wnan.1389 PMID:26762467

Kumar, A., Walia, H., Pottoo, F.H., & Javed, M.N. (2020). Insights of Nanophytomedicines as a Combinatorial Therapy in Disease Diagnosis and Treatment. *Nanophytomedicine*, 113-132.

Kumar, D., Pandey, J., Raj, V., & Kumar, P. (2017). A Review on the Modification of Polysaccharide Through Graft Copolymerization for Various Potential Applications. *The Open Medicinal Chemistry Journal, 11*(1), 109–126. doi:10.2174/1874104501711010109 PMID:29151987

Lachowicz, D., Karabasz, A., Bzowska, M., Szuwarzyński, M., Karewicz, A., & Nowakowska, M. (2019). Blood-compatible, stable micelles of sodium alginate – Curcumin bioconjugate for anti-cancer applications. *European Polymer Journal, 113*, 208–219. doi:10.1016/j.eurpolymj.2019.01.058

Lee, K. Y., & Mooney, D. J. (2012). Alginate: Properties and biomedical applications. *Progress in Polymer Science, 37*(1), 106–126. doi:10.1016/j.progpolymsci.2011.06.003 PMID:22125349

Li, J., Cai, C., Li, J., Li, J., Li, J., Sun, T., Wang, L., Wu, H., & Yu, G. (2018). Chitosan-Based Nanomaterials for Drug Delivery. *Molecules (Basel, Switzerland)*, *23*(10), 2661. doi:10.3390/molecules23102661 PMID:30332830

Li, J., Huo, M., Wang, J., Zhou, J., Mohammad, J. M., Zhang, Y., Zhu, Q., Waddad, A. Y., & Zhang, Q. (2012). Redox-sensitive micelles self-assembled from amphiphilic hyaluronic acid-deoxycholic acid conjugates for targeted intracellular delivery of paclitaxel. *Biomaterials*, *33*(7), 2310–2320. doi:10.1016/j.biomaterials.2011.11.022 PMID:22166223

Li, J., Yang, X., Yang, P., & Gao, F. (2017). Hyaluronic acid–conjugated silica nanoparticles for breast cancer therapy. *Inorganic and Nano-Metal Chemistry*, *47*(5), 777–782. doi:10.1080/15533174.2016.1218509

Lin, T., Zhao, P., Jiang, Y., Tang, Y., Jin, H., Pan, Z., He, H., Yang, V. C., & Huang, Y. (2016). Blood-Brain-Barrier-Penetrating Albumin Nanoparticles for Biomimetic Drug Delivery via Albumin-Binding Protein Pathways for Antiglioma Therapy. *ACS Nano*, *10*(11), 9999–10012. doi:10.1021/acsnano.6b04268 PMID:27934069

Lohcharoenkal, W., Wang, L., Chen, Y. C., & Rojanasakul, Y. (2014). Protein nanoparticles as drug delivery carriers for cancer therapy. *BioMed Research International*, *2014*, 180549. doi:10.1155/2014/180549 PMID:24772414

Makino, A., & Kimura, S. (2014). "Solid Tumor-Targeting Theranostic Polymer Nanoparticle in Nuclear Medicinal Fields", The Scientific World Journal, vol. 2014. *Article ID*, *424513*, 12.

McAtee, C. O., Barycki, J. J., & Simpson, M. A. (2014). Emerging roles for hyaluronidase in cancer metastasis and therapy. *Advances in Cancer Research*, *123*, 1–34. doi:10.1016/B978-0-12-800092-2.00001-0 PMID:25081524

Mishra, S., Sharma, S., Javed, M. N., Pottoo, F. H., Barkat, M. A., Alam, M. S., Amir, M., & Sarafroz, M. (2019). Bioinspired nanocomposites: Applications in disease diagnosis and treatment. *Pharmaceutical Nanotechnology*, *7*(3), 206–219. doi:10.2174/2211738507666190425121509 PMID:31030662

Mohanta, B.C., Javed, M.N., Hasnain, M.S., & Nayak, A.K. (2020). Polyelectrolyte complexes of alginate for controlling drug release. *Alginates in Drug Delivery*, 297-321.

Motiei, M., Kashanian, S., Lucia, L. A., & Khazaei, M. (2017). Intrinsic parameters for the synthesis and tuned properties of amphiphilic chitosan drug delivery nanocarriers. *Journal of Controlled Release*, *260*, 213–225. doi:10.1016/j.jconrel.2017.06.010 PMID:28625671

Murugan, C., Rayappan, K., Thangam, R., Bhanumathi, R., Shanthi, K., Vivek, R., Thirumurugan, R., Bhattacharyya, A., Sivasubramanian, S., Gunasekaran, P., & Kannan, S. (2016). Combinatorial nanocarrier based drug delivery approach for amalgamation of anti-tumor agents in breast cancer cells: An improved nanomedicine strategy. *Scientific Reports*, *6*(1), 34053. doi:10.1038rep34053 PMID:27725731

Nurgali, K., Jagoe, R. T., & Abalo, R. (2018). Editorial: Adverse Effects of Cancer Chemotherapy: Anything New to Improve Tolerance and Reduce Sequelae? *Frontiers in Pharmacology*, *9*, 245. doi:10.3389/fphar.2018.00245 PMID:29623040

Okamoto, Y., Taguchi, K., Sakuragi, M., Imoto, S., Yamasaki, K., & Otagiri, M. (2019). Preparation, Characterization, and in Vitro/in Vivo Evaluation of Paclitaxel-Bound Albumin-Encapsulated Liposomes for the Treatment of PancreaticCancer. *ACS Omega*, *4*(5), 8693–8700. doi:10.1021/acsomega.9b00537 PMID:31459959

P, R, S,, N. R., A. K., & Raj. (2016). Preparation, characterization and biological evaluation of curcumin loaded alginate aldehyde-gelatin nanogels. *Materials Science and Engineering C*, *68*, 251–257. doi:10.1016/j.msec.2016.05.046 PMID:27524019

Paliwal, S. R., Paliwal, R., Agrawal, G. P., & Vyas, S. P. (2016). Hyaluronic acid modified pH-sensitive liposomes for targeted intracellular delivery of doxorubicin. *Journal of Liposome Research*, *26*(4), 276–287. doi:10.3109/08982104.2015.1117489 PMID:26784587

Pandey, M., Saleem, S., Nautiyal, H., Pottoo, F.H., & Javed, M.N. (2020). PINK1/Parkin in neurodegenerative disorders: crosstalk between mitochondrial stress and neurodegeneration. *Quality Control of Cellular Protein in Neurodegenerative Disorders*, 282-301.

Pei, M., Jia, X., Zhao, X., Li, J., & Liu, P. (2018). Alginate-based cancer-associated, stimuli-driven and turn-on theranostic prodrug nanogel for cancer detection and treatment. *Carbohydrate Polymers*, *183*, 131–139. doi:10.1016/j.carbpol.2017.12.013 PMID:29352868

Pottoo, F.H., Abu-Izneid, T., Ibrahim, A.M., Javed, M.N., AlHajri, N., & Hamrouni, A.M. (2021c). Immune system response during viral Infections: Immunomodulators, cytokine storm (CS) and Immunotherapeutics in COVID-19. *Saudi Pharmaceutical Journal*.

Pottoo, F.H., Barkat, M.A., Ansari, M.A., Javed, M.N., Jamal, Q.M., & Kamal, M.A. (2021b). Nanotechnological based miRNA intervention in the therapeutic management of neuroblastoma. *Seminars in Cancer Biology, 69*, 100-108.

Pottoo, F. H., Javed, M., Barkat, M., Alam, M., Nowshehri, J. A., Alshayban, D. M., & Ansari, M. A. (2019a). Estrogen and serotonin: Complexity of interactions and implications for epileptic seizures and epileptogenesis. *Current Neuropharmacology*, *17*(3), 214–231. doi:10.2174/1570159X16666180628164432 PMID:29956631

Pottoo, F.H., Javed, M.N., Rahman, J.U., Abu-Izneid, T., & Khan, F.A. (2021a). Targeted delivery of miRNA based therapeuticals in the clinical management of Glioblastoma Multiforme. *Seminars in Cancer Biology, 69*, 391-398.

Pottoo, F. H., Sharma, S., Javed, M. N., Barkat, M. A., Harshita, Alam, M. S., Naim, M. J., Alam, O., Ansari, M. A., Barreto, G. E., & Ashraf, G. M. (2020). Lipid-based nanoformulations in the treatment of neurological disorders. *Drug Metabolism Reviews*, *52*(1), 185–204. doi:10.1080/03602532.2020.1726942 PMID:32116044

Pottoo, F. H., Tabassum, N., Javed, M. N., Nigar, S., Rasheed, R., Khan, A., Barkat, M. A., Alam, M. S., Maqbool, A., Ansari, M. A., Barreto, G. E., & Ashraf, G. M. (2019b). The synergistic effect of raloxifene, fluoxetine, and bromocriptine protects against pilocarpine-induced status epilepticus and temporal lobe epilepsy. *Molecular Neurobiology*, *56*(2), 1233–1247. doi:10.100712035-018-1121-x PMID:29881945

Prabhu, R. H., Patravale, V. B., & Joshi, M. D. (2015). Polymeric nanoparticles for targeted treatment in oncology: Current insights. *International Journal of Nanomedicine*, *2*(10), 1001–1018. PMID:25678788

Prabhu, R. H., Patravale, V. B., & Joshi, M. D. (2015). Polymeric nanoparticles for targeted treatment in oncology: Current insights. *International Journal of Nanomedicine*, *10*, 1001–1018. PMID:25678788

Raman, S., Mahmood, S., Hilles, A. R., Javed, M. N., Azmana, M., & Al-Japairai, K. A. (2020). Polymeric nanoparticles for brain drug delivery-a review. *Current Drug Metabolism*, *21*(9), 649–660. doi:10.2174/1389200221666200508074348 PMID:32384025

Ramya,, A.N., Joseph,, M.M., Maniganda,, S., & Karunakaran,, V., T., T. S., & Maiti, K. K. (2017). Emergence of Gold-Mesoporous Silica Hybrid Nanotheranostics: Dox-Encoded, Folate Targeted Chemotherapy with Modulation of SERS Fingerprinting for Apoptosis Toward Tumor Eradication. *Small*, *13*(31). Advance online publication. doi:10.1002mll.201700819 PMID:28671767

Rippe, M., Cosenza, V., & Auzély-Velty, R. (2019). Design of Soft Nanocarriers Combining Hyaluronic Acid with Another Functional Polymer for Cancer Therapy and Other Biomedical Applications. *Pharmaceutics*, *11*(7), 338. doi:10.3390/pharmaceutics11070338 PMID:31311150

Rizwanullah, M., Alam, M., Harshita, Mir, S. R., Rizvi, M. M. A., & Amin, S. (2020). Polymer-Lipid Hybrid Nanoparticles: A Next-Generation Nanocarrier for Targeted Treatment of Solid Tumors. *Current Pharmaceutical Design*, *26*(11), 1206–1215. doi:10.2174/1381612826666200116150426 PMID:31951163

Safdar, M. H., Hussain, Z., Abourehab, M. A. S., Hasan, H., Afzal, S., & Ei Thu, H. (2018). New developments and clinical transition of hyaluronic acid-based nanotherapeutics for treatment of cancer: Reversing multidrug resistance, tumour-specific targetability and improved anticancer efficacy. *Artificial Cells, Nanomedicine, and Biotechnology*, *46*, 8. PMID:29082766

Sagnella, S.M., Duong, H., MacMillan, A., Boyer, C., Whan, R., & McCarroll, J.A. (2014). Dextran-Based Doxorubicin Nanocarriers with Improved Tumor Penetration. *Biomacromolecules*, *15*(1), 262-275.

Senapati, S., Mahanta, A. K., Kumar, S., & Maiti, P. (2018). Controlled drug delivery vehicles for cancer treatment and their performance. *Signal Transduction and Targeted Therapy*, *3*(1), 7. doi:10.103841392-017-0004-3 PMID:29560283

Serrano-Sevilla, I., Artiga, Á., Mitchell, S. G., De Matteis, L., & de la Fuente, J. M. (2019). Natural Polysaccharides for siRNA Delivery: Nanocarriers Based on Chitosan, Hyaluronic Acid, and Their Derivatives. *Molecules (Basel, Switzerland)*, *24*(14), 2570. doi:10.3390/molecules24142570 PMID:31311176

Sharma, S., Javed, M. N., Pottoo, F. H., Rabbani, S. A., Barkat, M., Sarafroz, M., & Amir, M. (2019). Bioresponse inspired nanomaterials for targeted drug and gene delivery. *Pharmaceutical Nanotechnology*, *7*(3), 220–233. doi:10.2174/2211738507666190429103814 PMID:31486751

Siegel, R. L., Miller, K. D., Goding Sauer, A., Fedewa, S. A., Butterly, L. F., Anderson, J. C., Cercek, A., Smith, R. A., & Jemal, A. (2020). Colorectal cancer statistics, 2020. *CA: a Cancer Journal for Clinicians*, *70*(3), 145–164. doi:10.3322/caac.21601 PMID:32133645

Sinha, A., Chakraborty, A., & Jana, N. R. (2014). Dextran-gated, multifunctional mesoporous nanoparticle for glucose-responsive and targeted drug delivery. *ACS Applied Materials & Interfaces*, *6*(24), 22183–22191. doi:10.1021/am505848p PMID:25458145

Song, R., Murphy, M., Li, C., Ting, K., Soo, C., & Zheng, Z. (2018). Current development of biodegradable polymeric materials for biomedical applications. *Drug Design, Development and Therapy*, *12*, 3117–3145. doi:10.2147/DDDT.S165440 PMID:30288019

Sun, M., Li, J., Zhang, C., Xie, Y., Qiao, H., Su, Z., Oupický, D., & Ping, Q. (2017). Arginine-Modified Nanostructured Lipid Carriers with Charge-Reversal and pH-Sensitive Membranolytic Properties for Anticancer Drug Delivery. *Advanced Healthcare Materials*, *6*(8), 1600693. doi:10.1002/adhm.201600693 PMID:28187243

Taghavi, S., Ramezani, M., Alibolandi, M., Abnous, K., & Taghdisi, S. M. (2017). Chitosan-modified PLGA nanoparticles tagged with 5TR1 aptamer for in vivo tumor-targeted drug delivery. *Cancer Letters*, *400*, 1–8. doi:10.1016/j.canlet.2017.04.008 PMID:28412238

Thambi, T., You, D. G., Han, H. S., Deepagan, V. G., Jeon, S. M., Suh, Y. D., Choi, K. Y., Kim, K., Kwon, I. C., Yi, G.-R., Lee, J. Y., Lee, D. S., & Park, J. H. (2014). Bioreduciblecarboxymethyl dextran nanoparticles for tumor-targeted drug delivery. *Advanced Healthcare Materials*, *3*(11), 1829–1838. doi:10.1002/adhm.201300691 PMID:24753360

Thomas, R. G., Moon, M., Lee, S., & Jeong, Y. Y. (2015). Paclitaxel loaded hyaluronic acid nanoparticles for targeted cancer therapy: In vitro and in vivo analysis. *International Journal of Biological Macromolecules*, *72*, 510–518. doi:10.1016/j.ijbiomac.2014.08.054 PMID:25224289

Ulbrich, K., Holá, K., Šubr, V., Bakandritsos, A., Tuček, J., & Zbořil, R. (2016). Targeted Drug Delivery with Polymers and Magnetic Nanoparticles: Covalent and Noncovalent Approaches, Release Control, and Clinical Studies. *Chemical Reviews*, *116*(9), 5338–5431. doi:10.1021/acs.chemrev.5b00589 PMID:27109701

Wang, G., Wang, J. J., Tang, X. J., Du, L., & Li, F. (2016). In vitro and in vivo evaluation of functionalized chitosan-Pluronic micelles loaded with myricetin on glioblastoma cancer. *Nanomedicine (London)*, *12*(5), 1263–1278. doi:10.1016/j.nano.2016.02.004 PMID:26970027

Wang, Y., Zhou, J., Qiu, L., Wang, X., Chen, L., Liu, T., & Di, W. (2014). Cisplatin-alginate conjugate liposomes for targeted delivery to EGFR-positive ovarian cancer cells. *Biomaterials*, *35*(14), 4297–4309. doi:10.1016/j.biomaterials.2014.01.035 PMID:24565522

Waziri, A., Bharti, C., Aslam, M., Jamil, P., Mirza, A., Javed, M. N., Pottoo, U., Ahmadi, A., & Alam, M. S. (2021). Probiotics for the Chemoprotective Role Against the Toxic Effect of Cancer Chemotherapy. *Anti-cancer Agents in Medicinal Chemistry*, *21*. Advance online publication. doi:10.2174/1871520621666210514000615 PMID:33992067

Wong, K. H., Lu, A., Chen, X., & Yang, Z. (2020). Natural Ingredient-Based Polymeric Nanoparticles for Cancer Treatment. *Molecules (Basel, Switzerland)*, *25*(16), 3620. doi:10.3390/molecules25163620 PMID:32784890

Wu, J., Tang, C., & Yin, C. (2017). Co-delivery of doxorubicin and interleukin-2 via chitosan based nanoparticles for enhanced antitumor efficacy. *Acta Biomaterialia*, *47*, 81–90. doi:10.1016/j.actbio.2016.10.012 PMID:27729232

Yang, N. J., & Hinner, M. J. (2015). Getting across the cell membrane: An overview for small molecules, peptides, and proteins. *Methods in Molecular Biology (Clifton, N.J.)*, *1266*, 29–53. doi:10.1007/978-1-4939-2272-7_3 PMID:25560066

Yu, N., Li, G., Gao, Y., Jiang, H., & Tao, Q. (2016). Thermo-sensitive complex micelles from sodium alginate-graft-poly(N-isopropylacrylamide) for drug release. *International Journal of Biological Macromolecules*, *86*, 296–301. doi:10.1016/j.ijbiomac.2016.01.066 PMID:26806647

Yu, X., Trase, I., Ren, M., Duval, K., Guo, X., & Chen, Z. (2016). Design of Nanoparticle-Based Carriers for Targeted Drug Delivery. *Journal of Nanomaterials*, *2016*, 1087250. doi:10.1155/2016/1087250 PMID:27398083

Zhang, X., Zhang, T., Ma, X., Wang, Y., Lu, Y., Jia, D., Huang, X., Chen, J., Xu, Z., & Wen, F. (2020). The design and synthesis of dextran-doxorubicin prodrug-based pH-sensitive drug delivery system for improving chemotherapy efficacy. *AJPS*, *15*(5), 605–616. doi:10.1016/j.ajps.2019.10.001 PMID:33193863

Zhen, X., Wang, X., Xie, C., Wu, W., & Jiang, X. (2013). Cellular uptake, antitumor response and tumor penetration of cisplatin-loaded milk protein nanoparticles. *Biomaterials*, *34*(4), 1372–1382. doi:10.1016/j.biomaterials.2012.10.061 PMID:23158934

Zhu, Y., Liu, Y., Jin, K., & Pang, Z. (2019). Polysaccharide nanoparticles for cancer drug targeting. In *Polysacch. Carriers Drug Deliv* (pp. 365–396). Elsevier. doi:10.1016/B978-0-08-102553-6.00013-1

Zielińska, A., Carreiró, F., Oliveira, A. M., Neves, A., Pires, B., Venkatesh, D. N., Durazzo, A., Lucarini, M., Eder, P., Silva, A. M., Santini, A., & Souto, E. B. (2020). Polymeric Nanoparticles: Production, Characterization, Toxicology and Ecotoxicology. *Molecules (Basel, Switzerland)*, *25*(16), 3731. doi:10.3390/molecules25163731 PMID:32824172

APPENDIX

Table 1. Comparison between the conventional and nanoparticles based delivery approaches in cancer

S. No	Parameters	Nanoparticles approaches	Conventional treatment
1	Drug delivery	Targeted	Non-specific
2	Drug protection	Encapsulated	Easy to degrade
3	Effectiveness	Good retention at active site	Low retention
4	Clearance	Long time for the clearance	Fast clearance
5	Bioavailability	High	Low
6	Dose	proportionality is less	High
7	Toxicity	Low	High

(Rippe et al 2019)

Table 2. Natural polymers based delivery systems in cancer

Type	Description	Monomers	Linkage	Properties	Drug delivery system (DDS)	References
Chitosan	Linear cationic amino polysaccharide, derived from chitin a mucopolysacchride.	β-d-glucosamine and N-acetyl-d-glucosamine units	1, 4 glycosidic	Biocompatibility, mucoadhesive, antibacterial and antifungal properties, and non-toxicity. Aminopolysaccharide have strong positive charge attract negative charge.	Nanoparticles, microspheres, hydrogels and micelles to target cancerous cells	(Motiei et al 2017)
Hyaluronic Acid	Linear anionic glycos-aminoglycan. Non-sulfated glycosaminoglycan found throughout the connective, epithelial, and neural tissues.	glucuronic acid and N-acetylglucosamine units	β-1,4 and β-1,3 glycosidic linkages	Tumor cell division and production of new blood vessels. carboxylic acid, hydroxyl, and N-acetyl groups, and can easily be combined with other chemical.	It used as a carrier for cancer treatment	(McAtee et al 2014)
Alginates	Linear anionic polymers. Biocompatibil, Low toxicity, mild gelation by addition of divalent cations such as Ca^{+2}.	β-d-mannuronic acid and α-L-guluronic	1,4 glycosidic linkages	Formation of gel without using a toxic solvent. Delivery of bioactive agents such as small chemical drugs and proteins, and cell transplantation	Hydrogel used for the tumor cell drug delivery..	(Lee et al 2012; Mohanta et al 2020)
Dextran	Branched glucan. Biosynthesized by the nonpathogenic organism Leuconostoc mesenteroides NRRL B-512.	D-glucose units	α-1,6 linkage	The branching of dextran may affect the biological activities. Soluble in water and organic solvents, can be used in the preparation of different structures through blending dextran with bioactive agents or hydrophobic polymers	Micelles and hydrogels for cancer treatment	(Wong et al 2020)

Table 3. Chitosan-based nanoparticles for targeted delivery in cancer

S. No	Chitosan/ derivatives	DDS	Drug	Application	References
1.	Cationic trimethyl chitosan	Hydrophilic NPs	DOX and interleukin-2	Enhanced antitumor efficacy	(Wu et al 2017)
2.	Oleoyl-chitosan	Nanosphere	DOX and Roussin's black salt (RBS)	Synergistic therapeutic effect	(Li et al 2018)
3.	Chitosan and poly-acrylic acid	Mesoporous silica nanoparticles	Topotecan and quercetin	Effective in breast cancer treatment.	(Murugan et al 2017)
4.	Chitosan-pluronic	Micelles	Myricetin	Drug delivery for glioblastoma cancer	(Wang et al 2016)
5.	Arginine-chitosan	NLC	Hydroxycamptothecin	Improved antitumor therapy	(Sun et al 2017)
6.	N-trimethyl chitosan	Nanoparticles	Gemcitabine	Enhance the oral bioavailability in breast cancer	(Chen et al 2018)
7.	Chitosan-folic acid conjugate	Gold-Mesoporous	Doxorubicin	Superior antitumor response	(Ramya et al 2017)
8.	PLGA-Chitosan	Nanoparticles	Epirubicin	Showed significantly high therapeutic efficiency in MCF7 cells (breast cancer cell)	(Taghavi et al 2017)

Table 4. Hyaluronic acid-based nanoparticles for targeted delivery in solid cancer

S. No	Type	DDS	Drug	Application	References
1.	Hyaluronic Acid (HA)	Micelles	Paclitaxel	Exhibited more pronounced cytotoxic effect. A redox-sensitive nano-sized drug delivery system.	(Safdar et al 2018)
2.	Hyaluronic Acid	Nanocarrier	Doxorubicin	Improves survival and reduces toxicity in xenografts of human breast cancer. Reduces dose-limiting cardiac toxicity. Minimal toxicity in normal tissues.	(Cai et al. 2010)
3.	5B-cholanic acid-HA	Micelles	Paclitaxel	Cancer cells expressing CD44 targeted with this approach which is specific and efficient chemotherapeutic treatment for cancer.	(Thomas et al 2015)
4.	Deoxycholic acid-HA	Micelles	Paclitaxel	Hold great potential as targeted intracellular delivery carriers of lipophilic anticancer drugs.	(Li et al 2012)
5.	5B-cholanic acid-HA-PEG	Micelles	Irinotecan	Early detection, targeted therapy, and therapeutic monitoring of colon cancer.	(Choi et al 2012)
6.	DSPE-PEG-HA	Liposomes	Doxorubicin	Intracellular drug delivery carrier developed with reduced toxicity.	(Paliwal et al 2016)
7.	Hyaluronic Acid	Hydrogel	Paclitaxel	Temporal availability of the drug-enhanced antitumor effect.	(Bajaj et al 2012)
8.	Hyaluronic Acid	Novel silica nanoparticles	Paclitaxel	Minimal toxic side effects and strong tumor-suppression potential	(Jinmao et al 2017)
9.	SPION-HA	Nanoparticles	Doxorubicin	Effective chemotherapy of hepatocellular carcinoma	(Fu et al, 2017)

Table 5. Alginate based nanoparticles for targeted delivery in solid tumor

S. No	Composition	DDS	Drug	Application	References
1.	Alginate aldehyde-gelatin	Nanogel	Curcumin	Stable and gives sustained release profile	(Sarika et al 2016)
2.	Alginate	Nanogel	Cisplatin	Showed highly pH-responsive drug release behavior	(Hong et al 2018)
3.	Alginate with modified beta-cyclodextrin	Nanogel	5-Fluorouracil	Higher 5-Fu intracellular accumulation	(Hosseinifar et al 2018)
4.	Hydroxyapatite/sodium alginate/ chitosan (HA/SA/CS)	Microsphere	Doxorubicin	Better cell adhesion and proliferation capacity	(Bi et al 2019)
5.	Alginate	Micelles	Curcumin	Decreased the viability of cancer cells	(Lachowicz et al 2019)
6.	Alginate	Nanogels	Doxorubicin	The cytotoxicity and cellular uptake were enhanced	(Pei et al 2018)
7.	Alginate	Liposome	Cisplatin	Enhanced efficacy and fewer adverse effects.	(Wang et al 2014)
8.	Alginate-graft-poly (N-isopropylacrylamide	Micelles	5-Fluorouracil	Metal ions may induce the polymer of micelles and expected to be utilized in controlled delivery	(Yu et al 2016)

Table 6. Dextran based nanoparticles for targeted delivery in solid tumor

S. No	Dextran/ derivatives	DDS	Drug	Application	References
1.	Dextran	Micelles	Hydroxycamptothecin	Reduced toxicity in mice and showed significant, sustained release	(Huang et al 2019)
2.	PEG-grafted dextran	Nanoparticles	Hydroxycamptothecin	Overcome the bladder toxicity	(Feldman D 2019)
3.	Carboxymethyl dextran	Nanoparticles	Doxorubicin	Higher antitumor efficacy	(Thambi et al 2014)
4.	Aldehyde dextran	Nanoparticles	Doxorubicin	Enhanced penetration ability combined with their improved efficacy	(Sagnella et al 2014)
5.	Dextran	Magnetic mesoporous silica	Camptothecin	Used for advanced drug delivery applications for cancers with more efficient therapy	(Sinha et al 2014)
6.	Dextran	Micelles	Doxorubicin	Better tissue permeability and tumor-suppressive effects in vivo	(Zhang et al 2020)
7.	Dextran	Nanoparticles	Silybin and Paclitaxel	Inhibit tumor growth through an enhanced intratumoralpenetration	(Huo et al 2020)

Table 7. Peptide-based natural polymers and their properties against the solid tumor

Peptide-based polymer	Source	Properties	Features for DDS	References
Collagen	Abundant protein invertebrates	High biocompatibility and biodegradability	Stability and the release profile can be controlled.	(Chaubaroux et al 2012)
Gelatin	Hydrolysis of animal collagen	Biocompatible and biodegradable biopolymer	Altering the release profile of gelatin.	(Wong et al 2020)
Albumin	Human, bovine, and rat serum albumin	Highly soluble proteins with no toxicity and low immunogenicity	Albumin-based nanoparticles were able to penetrate the blood-brain barrier (BBB).	(Lin et al 2020; Javed et al 2020b)

Table 8. Protein-based nanoparticle drug delivery system for solid tumor targeting

S. No	Protein derivatives	Drug	Application	References
1.	Collagen-poly (3-acrylamidophenylboronic acid)	Doxorubicin	Drug-loaded nanoparticles showed a superior inhibitory effect to free DOX on the growth of human ovarian cancer.	(Jiang et al 2020)
2.	Human serum albumin	Paclitaxel	It can be given with higher dosing compared to standard dose of paclitaxel (Taxol) marketed formulation.	(Okamoto et al 2019)
3.	Bovine serum albumin (Cationic)	siRNA	Inducing apoptosis of cancer cells and inhibiting tumor growth by efficient gene-silencing effect.	(Han et al 2014)
4.	PEGylated gelatin	Doxorubicin	Highly efficient in inhibiting tumor growth compared to plain doxorubicin or doxorubicin-loaded into non-PEGylated nanoparticles.	(Din et al 2017)
5.	Gliadin	Cyclophosphamide	Drug release for long period of time and induces apoptosis of breast cancer cells.	(Gulfam et al 2012)
6.	Zein	5-Fluorouracil and Quantum dot (QD) fluorophores	Enhanced drug delivery and imaging of breast cancer.	(Lohcharoenkal et al 2014)
7.	Casein	Cisplatin	Easily penetrate the cell membranes, target the tumor cells, and hence inhibit the tumor growth.	(Zhen et al 2013)
8.	Casein	Flutamide	Reduce tumor growth efficiently and it also reduced the Prostate-Specific Antigen (PSA) serum level.	(Elzoghby et al 2013)
9.	Lactoferrin	Doxorubicin	Improved efficacy, bioavailability, and safety.	(Golla et al 2013)

Figure 1. An overview of different active and passive targeting approaches of polymeric systems in cancer

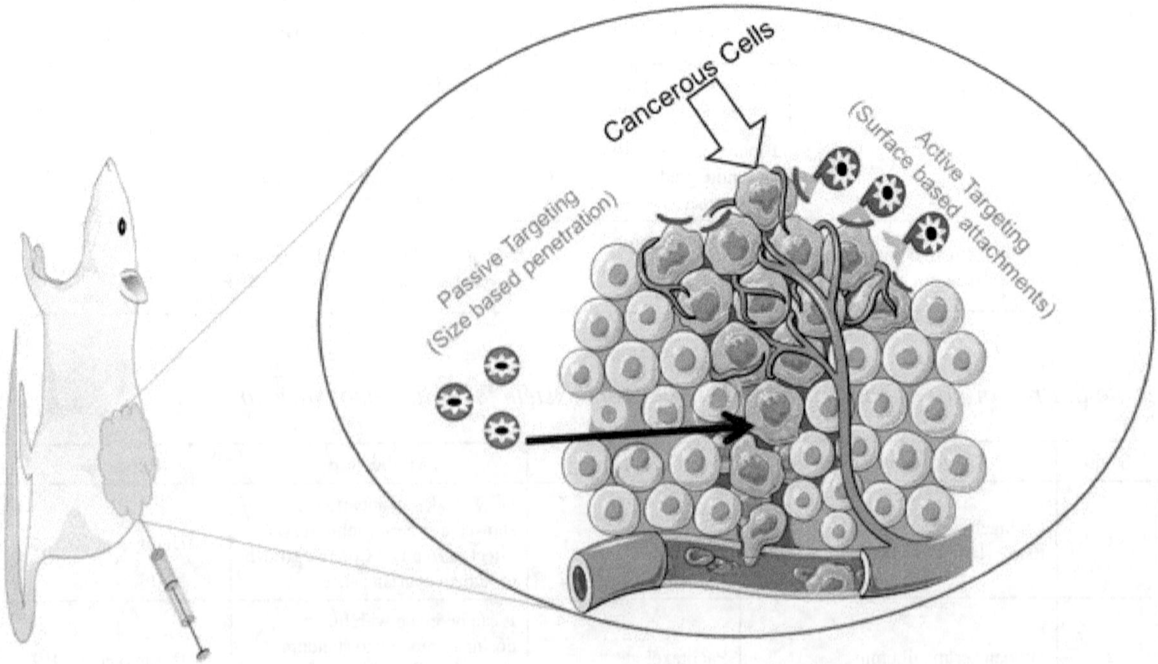

Figure 2. A schematic representation of drug targeting. Figure 2A (Passive targeting): Transport of drugs loaded nanoparticles to cancerous cells by the permeation and retention effect (EPR), in non-specific manner. Figure 2B (Active targeting): Transport of functionalized nanoparticles to cancerous cells in site-specific manners

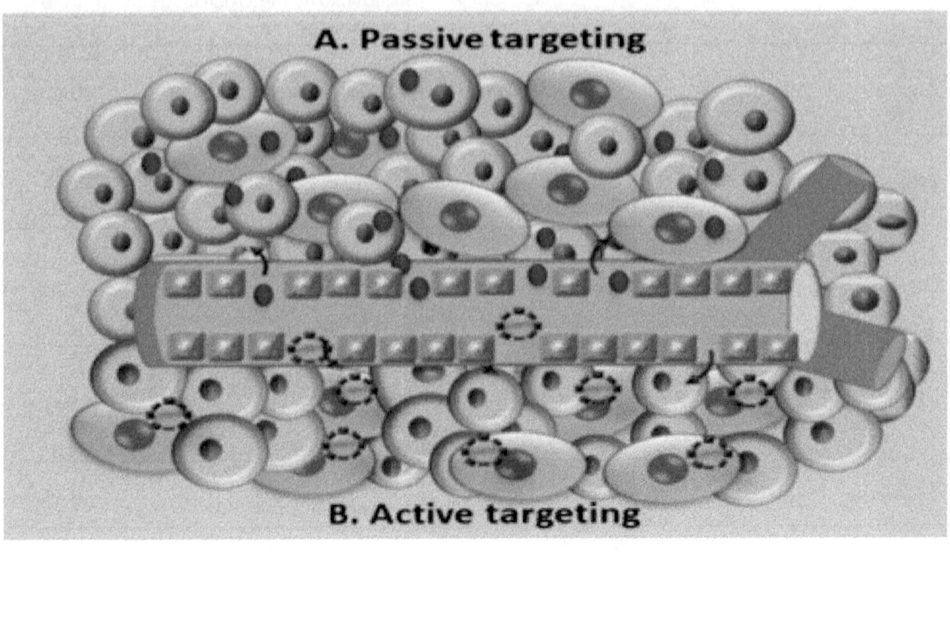

Chapter 10
Monitoring of the Friction Stir Welding Process:
Upgrading Towards Industry 4.0

Sudhagar Sukkasamy
Muthayammal Engineering College, India

Gopal P. M.
Karpagam Academy of Higher Education, India

ABSTRACT

Most of the manufacturers in the world are already being adopted to industrial digitization in the perspective of Industry 4.0. Every conventional manufacturing method is being converted into cyber-physical systems in order to govern the process digitally. Friction stir welding is an advanced solid state welding method that has been utilized in many industries, and the process is no exception to digital transformation. This chapter aims to discuss the various aspects of digitizing the friction stir welding process by the application of different sensors. Implementation of various sensors such as current, force, sound, and vision in friction stir welding machine are able to collect valuable data that can be used for adopting Industry 4.0.

INTRODUCTION

The term Industry 4.0 is defined in number of ways by different organizations and industry experts which is generally used as a notation for fourth industrial revolution. One such definition for Industry 4.0 is utilizing information and communication technology (ICT) for smart networking of machines and processes for industry. By this view it can be stated that the process and machines are networked and can communicate together that leads to novel routes of manufacturing, value addition and concurrent process optimization. The advancements in ICT tool technologies makes possibilities for smart industries. These tools are almost similar to the tools used in Industrial Internet of Things (IIOT) for tracking and remote

DOI: 10.4018/978-1-7998-9574-9.ch010

monitoring like purposes. Industry 4.0 can also be stated as the name used for exposing the present development in data communication and automation in manufacturing that consists of various ICT tools such as cyber-physical systems, the Internet of things, cloud computing and cognitive computing in road to developing the smart factory (Pires, Cachada, Barbosa, Moreira, & Leitão, 2019).

Industry 4.0 proposes more industrial automation when compared to third revolution and also tries to bridge the gap between digital and physical world all the way through cyber-physical system, facilitated by IIoT. It also shifts to the system where smart products or devices make the decision on production process instead of traditional central industrial control system. Further it proposes closed-loop data forms and control systems and enables more customization of products (Cao, Giustozzi, Zanni-Merk, de Bertrand de Beuvron, & Reich, 2019). Overall, it can be stated as the system/model that aims to facilitate the automatic decision making, synchronized asset and process monitoring and also facilitates all the stakeholders to involve in value addition at the earlier part of the process through networking.

At present, the Industry 4.0 is really revolutionizing the entire industrial sector and it transformed the approach industries used to produce, enhance and dispense their products. Many industries integrated the Industry 4.0 facilitating technologies such as IoT, cloud computing, AI and machine learning into their various sections of operations and some industries implemented in all sections of their facility. These industries utilize different kind of sensors, software and robotic tools for data collection & analysis, decision making and concurrent rectification (Chen, Han, Cao, Zheng, & Xu, 2020). It is used for factory automation, predictive maintenance, optimization and process enhancement and all are in single focus of customer satisfaction.

This advanced concept also reduces machine down time by initiating predictive maintenance based on the continuous data collected from the shop floor and also helps in real time visibility of machine performance (Dinardo, Fabbiano, & Vacca, 2018). Utilization of Artificial Intelligence based quality assessments instead of manual checkup results in greater product quality. remote monitoring of process flow and machine performance also an added advantages of this concept and it is possible to detect errors in any part of production or machine in quick time with the aid of machine learning concepts which will protect the machine break down and saves lot of wealth (Farahani et al., 2019).

Friction Stir Welding (FSW) is new and advanced solid state welding technique developed by The Welding Institute in 1991. Since its inception, the technique has been widely employed in many type of industries, majorly in automobile and aerospace industries. In Industry 4.0 context, the basic building block is data and without enough data the process cannot be adoptable to smart factory. In order to make the FSW Industry 4.0 ready process, different sensors are implemented in existing FSW machine (Mishra et al., 2020). The valuable information collected from sensors can be utilized for precise control of the process.

FSW PROCESS

Friction Stir Welding (FSW) is a solid state joining process combining deformed material by the action of mechanical force to attain high quality weld joint. In FSW, a non-consumable specially designed welding tool encompassed with a shoulder and pin plays a major role. The rotating welding tool is plunged into interface of plates to be joined and moved forward along interface to seam the plates. The various stages of FSW process is schematically represented in the Figure 1. The shoulder in welding tool is responsible for generating heat due to friction between tool and work piece. The frictional heat plasticize base material and the plasticized material is combine by stirring action of pin in welding tool. During stirring action, material present in front of tool is extruded around the pin and deposited behind the pin and forged into solid state. It is most important to maintain sufficient contact between shoulder and work material, since shoulder is the critical part that generate heat for mixing the material. The movement of material in this process is majorly in longitudinal direction only and a small amount of vertical movement takes place when hotter welding condition and threaded tool pin is used (Sudhagar, Sakthivel, Ashok, & Ajith Arul Daniel, 2017).

Figure 1. Schematic of FSW process

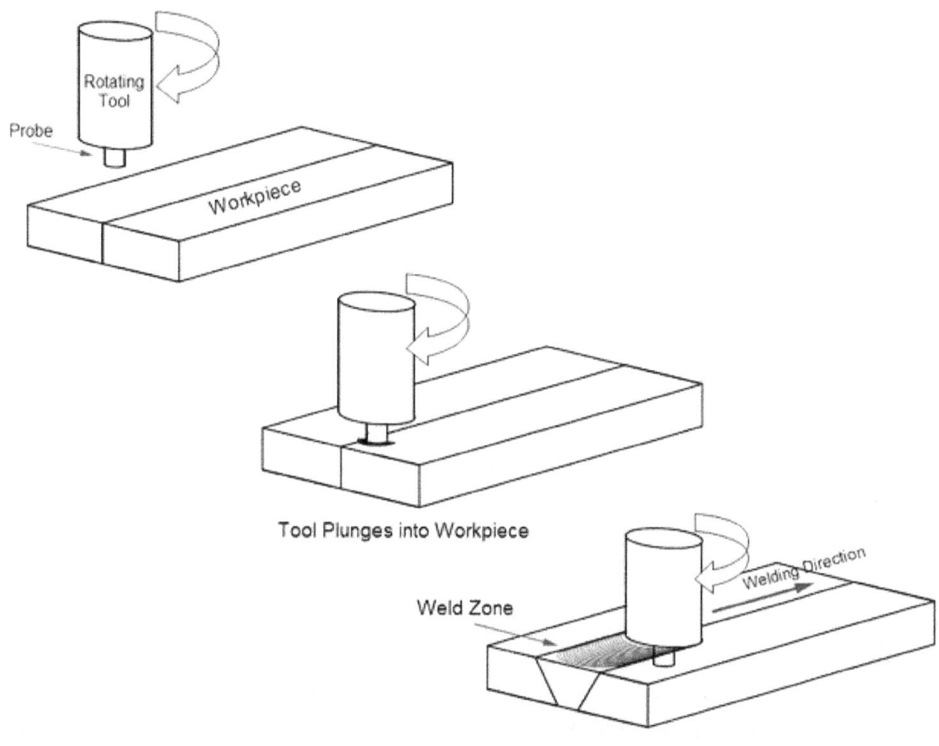

Parameters Involved in FSW

There are many numbers of variables involved in FSW, which are illustrated in Figure 2. These variables can be broadly classified into three categories such as machine variables, tool variables and material variables. Among these variables, machine parameters only be online controllable during welding process and tool parameters and material parameters are offline controllable. The tool rotational speed, measures in rpm determine the amount of heat generated during welding process. Meanwhile, the welding speed measures in mm/min controls the amount of heat input to workpiece. The other parameters such as tool plunge depth, axial force, tool tilt angle and tool offset also affect the quality of weld joint. Tool parameters include shoulder diameter, pin diameter, pin profile, shoulder profile and tool material. Material properties such as composition, melting point and thickness are highly influence the weld quality (Sudhagar, Sakthivel, Mathew, & Daniel, 2017).

Figure 2. FSW process variables

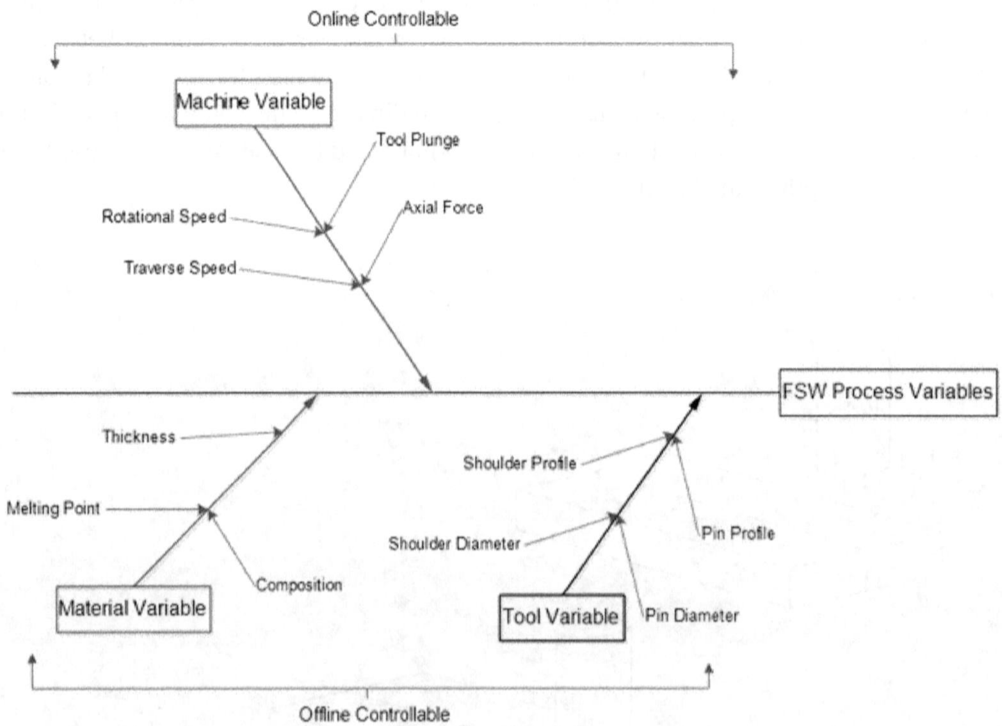

MONITORING OF FSW

Monitoring of FSW process is the mining of required information from welding machine in order to identify the excellence of joining process. The information from welding machine is acquired by various means such as different sensors, image processing, physics based techniques, etc. architecture for developing monitoring technique for FSW process has been illustrated in block diagram Figure 3.

Development of monitoring process for any manufacturing process is initiated by setting objectives i.e. selection of characteristics and properties of process that are to be observed. Next step is identification of signals to be acquired, from which information about the selected parameter can be extracted. After identifying the signals, suitable sensors are to be deployed in machines to gather signals. The signals measured should be processed using suitable techniques to extract useful information or features and any fluctuation in these features indicates the abnormalities in process which help to control the process. Monitoring of a process can be broadly classified into two types such as direct and indirect monitoring. The direct method of monitoring involves measuring the characteristics of process directly using laser, vision camera, X-ray, optical, etc. Optical and camera vison method can be utilized to measure the tool wear in a process. The image of the tool which give information about shape and size is captured before starting the process and compared with the images captured during the process. The change in shape and size of tool indicate tool wear during process and this method of monitoring technique directly give information about the parameter to be monitored. X-ray radiographic is best method to monitor the arising of internal defects in weld joint but severe radiation makes it unsuitable for using in continuous monitoring. Likewise, in high luminance working environment, utilization of camera vision is difficult.

Figure 3. Block diagram of FSW monitoring technique

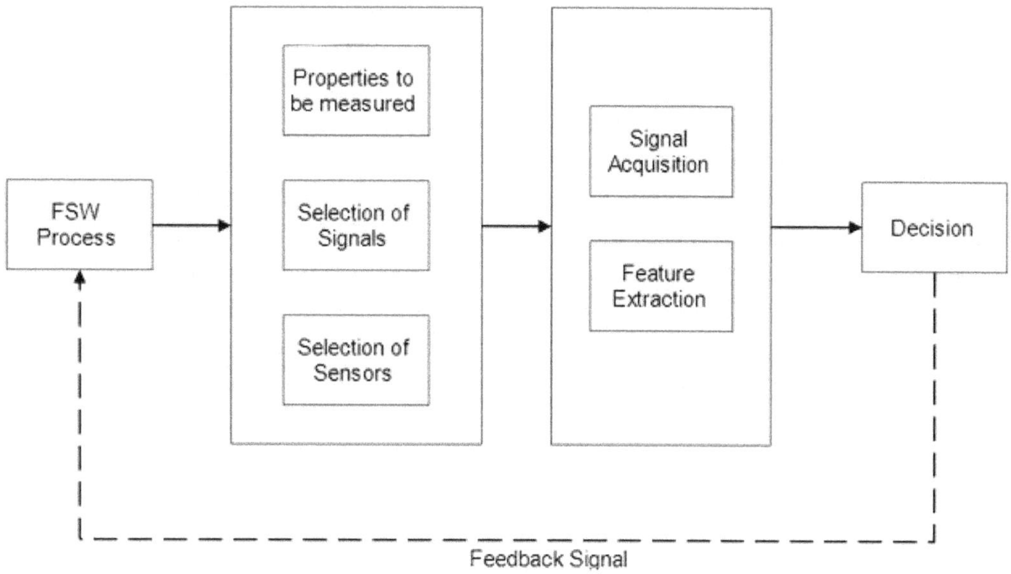

The difficulties in direct monitoring method is overcome by measuring the current signals, torque, force, sound emitted and vibration generated and these signals are correlated with the characteristics to be studied. This method of monitoring a process is called indirect monitoring. Compared to direct monitoring, indirect monitoring are less precise but more economical and can be deployed in any type of working environment. For example, many manufacturing process involves forces for achieving final shape of product and any sharp change in force result in tool failure which can be eliminated by continuously monitoring force signals. The following section discuss various direct and indirect monitoring method developed for friction stir welding process.

Sensor Based Monitoring

Sensor is a device that measures the physical quantity of a system and convert it into analog or digital signals. Using these signals, characteristics of the system can be studied and controlled easily. Various types of sensors that can be deployed in FSW machine for monitoring is shown in Figure 4.

In FSW process, a rotating tool is made to rub against the material that are to be joined and the rubbing action generate frictional heat to plasticize material. The linear traverse movement of rotating tool mix and forges the plasticized material thus a sound joint is formed. In order to accomplish the above process, huge amount of force and torque is required. The continuous monitoring of force and torque during FSW process gives better insight and precise control over welding process. Since FSW process is very much resemblance to conventional milling process, the same force measurement system can be employed. A cutting force dynamometer can be attached to FSW machine to measure the various force and torque developed during welding process. The signal from dynamometer is analyzed to identify occurrence of any defects during welding process. Using force signals, tool position, gap between plates, internal voids can be easily detected in FSW process. (Bipul Das, Pal, & Bag, 2016) developed a monitoring technique for identification of internal defects during FSW by measuring and analysing real time force signals. The formation of tunnel defect is due to improper plasticization of material which offers higher resistance to rotating tool. The higher resistance to rotating tool result in increased welding force; hence the formation of defect can be studied by using force signals. (Kumar et al., 2015) discussed the detection of defects during FSW using force and torque signals. The welds are produced at various welding speed and tool rotational speed and real time data is acquired using load cell and data acquisition system. Information from the measured signal is extracted by applying discrete wavelet transform and formation of defect during welding leads to sudden change in features of force signals. This change in features of signal is used to monitor and control FSW process. (Bipul Das, Pal, & Bag, 2017a) developed a low cost force measurement for FSW process using stain gauges and measured real time welding force signals for quality monitoring of weld joint. The force signals are analysed in time-frequency domain using wavelet packet transform and features are correlated with mechanical strength of weld joint for controlling the process.

Figure 4. Sensor based monitoring of FSW process

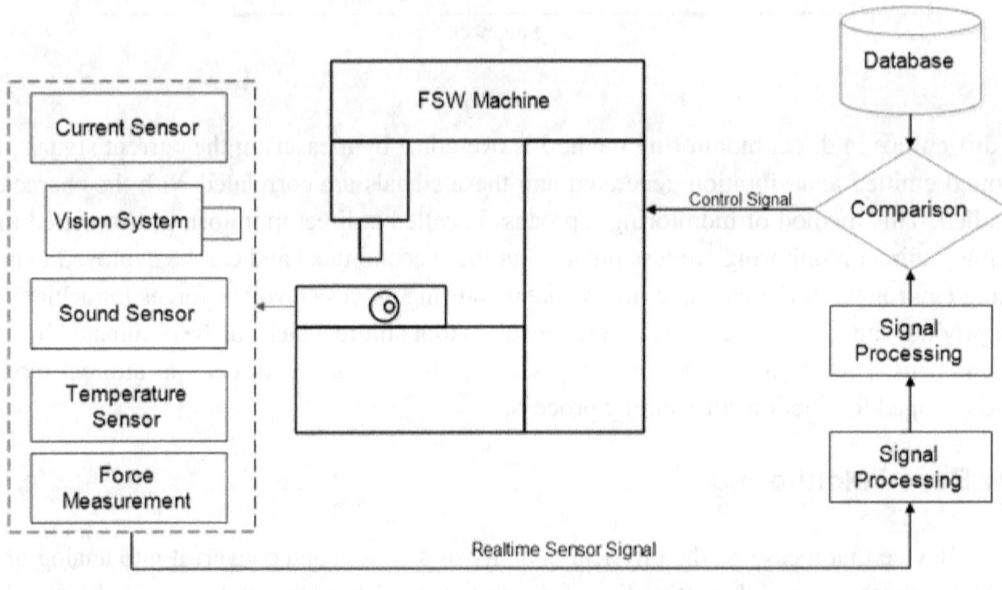

During FSW process, all the forces and torques are generated by spindle and feed motor of welding machine. The change in current and power consumption pattern of machine indicates the variation in welding process and assessing of these signals can be effectively utilized to control the process. Many researchers were attempted to develop monitoring technique for FSW process using current sensors. (Longhurst, Wilbur, Osborne, & Gaither, 2018) developed indirect monitoring system for FSW process by measuring and assessing frequency of spindle motor current. The current signals are measured by a clamp on current meter and Fourier transform is used to decompose current signal into various frequencies. The experimental result shows that there is no change in frequency spectrum when the weld does not possess voids. But when the weld produces voids, there is consistent change in current signal and frequency spectrum of signal increase upto 4 Hz. (Mehta, Chatterjee, & De, 2013) proposes a methodology for online measurement of torque and transverse force indirectly using current and power transient of drive motors in FSW machine. The welding is done at different combination of tool rotational speed and tool shoulder diameter and respective current signals were recorded and analyzed. The features from current signals were correlated with force and torque developed during process. (Bipul Das, Bag, & Pal, 2017) presented a monitoring technique for FSW of aluminium AA1100 using signals measured during welding. The various signal measured that contained information about welding process are current and tool rotational signals. The information from signal is extracted in terms of fractal dimension by using fractal theory. The fractal dimension calculated from signals is compared with corresponding tensile strength of weld joint to study the relation between them. The result reveals that fractal dimension is in direct proportion with tensile strength of weld joint. Hence in order to obtain high strength joint, fractal dimension of current signals measured during welding should be high. (Bipul Das, Pal, & Bag, 2017b) predicted weld quality using spindle motor and feed motor current signals measured during FSW and applied wavelet analysis for feature extraction. 65 experiments are conducted by varying tool rotational speed, welding speed and shoulder diameter. The features extracted from signals by discrete wavelet transform are used as input for training Artificial Neural Network to predict tensile and yield strength of joint.

Temperature Based Monitoring

Friction heat generated between rotating tool and work piece in FSW process plays a major role in production of sound weld joint. Too much of heat generation leads to more plasticized material that create turbulence in material and too low heat generation leads to lesser plasticization result in improper mixing. Both welding condition produce low strength weld joint, hence proper frictional heat generation is important in FSW. Heat generated can be evaluated by measuring the temperature of the process and using which welding process can be precisely controlled. The temperature during welding process can be measured using different means such as thermocouple, thermometer, pyrnometer and thermal image analyzer.

(Serio, Palumbo, Galietti, De Filippis, & Ludovico, 2016) monitored FSW process by means of thermography during actual welding process. The experiments are conducted at various weld conditions by varying tool rotational speed and welding speed. The temperature distribution during welding process is recorder by using two infrared thermal image cameras. The temperature developed inside workpiece is measured by using twelve N type thermocouples. The thermal data acquired during welding is related with process parameters and found this method is effective for monitoring of FSW process. Imam et al. 2013 used the temperature developed during FSW process for monitoring quality by directly correlating the measured temperature value with tensile strength of weld joint. The study suggests that strength of weld joint produced is function of welding temperature; therefore it can be successfully used for monitoring of welding process. (Kruger, Miller, Shih, & van Niekerk, 2015) discussed the development of Bluetooth based Wireless Rotating Process Monitoring (WRPM) system for real time thermal feedback control FSW process. Temperature developed in welding tool during FSW process is measured via WRPM and the measured data is transmitted to computer wirelessly through Bluetooth. The measured data is used to develop feedback signals for controlling spindle speed which responsible for heat generation and plastic deformation.

Vision Based Monitoring

In machine vision system actual image of the object is captured and digital image processing has been carried out to extract the features that give information about the object. Image acquisition system is the primary part of the machine vision system that comprise of camera and illumination system. Type of image sensor used in camera determines the quality of image captured for processing. There are two type of image sensors used in camera namely Charge Coupled Diode (CCD) and Complementary Metal Oxide Semiconductor (CMOS). Machine vision based monitoring technique developed for FSW has been discussed in this section.

(Bhat, Kumari, Dutta, Pal, & Pal, 2015) made an attempt to monitor FSW process to detect defective and non-defective weld by processing surface image of weld. The useful information from weld image that describe defective and non-defective weld has been extracted through discrete wavelet transform technique. Three features such as energy, variance and entropy of images are extracted through wavelet analysis. The extracted features are given as input for Support Vector Machines (SVM) for classification of weld and the SVM model performs well in weld classification. (B. Das, Pal, & Bag, 2016) proposed that change in welding condition shows different pattern of semi-circular ring that is choose to assess the quality of joint. The information from weld image is extracted using fractal theory and extracted information is presented as fractal dimension. The fractal dimension and tensile strength of weld joint possess negative trend with each other and minimizing the fractal dimension during welding process improves the tensile strength of weld joint. (Ranjan et al., 2016) attempted to identify and classify surface defect in FSW through an image processing approach. The surface defects in FSW process such as voids, cracks, grooves, key hole and flash are identified using image pyramid and image reconstruction algorithm. Further these defects are classified based on their unique features. (Sudhagar, Sakthivel, & Ganeshkumar, 2019) developed a monitoring technique for FSW using vision system. The experiments were conducted at different welding condition which shows significant difference in surface appearance of weld joint. Surface image of weld joint is related with tensile strength of joint and classification model was developed in SVM to predict the defective and non-defective weld.

CONCLUSION

The current era of Industry 4.0, industrial digitization and data is the fundamental theme. In this context, most of the manufacturer in the world are involved in digitization of their factory. Therefore, it is essential to develop a methodology for digitizing the existing manufacturing process. Friction Stir Welding is an advanced solid state joining technique which finds applications in automobile, aerospace and many other industries. Automation and digitization of FSW process is the current need of manufacturing industries and this article discuss the various aspects of monitoring and control of FSW process. The detailed study revealed that most of monitoring technique developed for FSW process is open loop and sensor based monitoring technique is most economic and easily deployable. Development of hardware system for process monitoring of FSW has been well explored whereas software for analyzing the real time data and controlling of machine autonomously need to be explored more.

REFERENCES

Bhat, N. N., Kumari, K., Dutta, S., Pal, S. K., & Pal, S. (2015). Friction stir weld classification by applying wavelet analysis and support vector machine on weld surface images. *Journal of Manufacturing Processes*, *20*, 274–281. doi:10.1016/j.jmapro.2015.07.002

Cao, Q., Giustozzi, F., Zanni-Merk, C., de Bertrand de Beuvron, F., & Reich, C. (2019). Smart condition monitoring for industry 4.0 manufacturing processes: An ontology-based approach. *Cybernetics and Systems*, *50*(2), 82–96. doi:10.1080/01969722.2019.1565118

Chen, Y., Han, Z., Cao, K., Zheng, X., & Xu, X. (2020). Manufacturing upgrading in industry 4.0 era. *Systems Research and Behavioral Science*, *37*(4), 766–771. doi:10.1002res.2717

Das, B., Bag, S., & Pal, S. (2017). Probing weld quality monitoring in friction stir welding through characterization of signals by fractal theory. *Journal of Mechanical Science and Technology*, *31*(5), 2459–2465. doi:10.100712206-017-0444-2

Das, B., Pal, S., & Bag, S. (2016). Monitoring of friction stir welding process using weld image information. *Science and Technology of Welding and Joining*, *21*(4), 317–324. doi:10.1080/13621718.2015.1109805

Das, B., Pal, S., & Bag, S. (2016). A combined wavelet packet and Hilbert-Huang transform for defect detection and modelling of weld strength in friction stir welding process. *Journal of Manufacturing Processes*, *22*, 260–268. doi:10.1016/j.jmapro.2016.04.002

Das, B., Pal, S., & Bag, S. (2017a). Design and development of force and torque measurement setup for real time monitoring of friction stir welding process. *Measurement: Journal of the International Measurement Confederation*, *103*, 186–198. doi:10.1016/j.measurement.2017.02.034

Das, B., Pal, S., & Bag, S. (2017b). Weld quality prediction in friction stir welding using wavelet analysis. *International Journal of Advanced Manufacturing Technology*, *89*(1–4), 711–725. doi:10.100700170-016-9140-0

Dinardo, G., Fabbiano, L., & Vacca, G. (2018). A smart and intuitive machine condition monitoring in the Industry 4.0 scenario. *Measurement*, *126*, 1–12. doi:10.1016/j.measurement.2018.05.041

Farahani, S., Brown, N., Loftis, J., Krick, C., Pichl, F., Vaculik, R., & Pilla, S. (2019). Evaluation of in-mold sensors and machine data towards enhancing product quality and process monitoring via Industry 4.0. *International Journal of Advanced Manufacturing Technology*, *105*(1), 1371–1389. doi:10.100700170-019-04323-8

Kruger, G. H., Miller, S. F., Shih, A. J., & van Niekerk, T. I. (2015). Robustness of 802.15 Wireless Interface for Real-Time Thermal Feedback Control of the Friction Stir Welding Process. *ASME 2015 International Manufacturing Science and Engineering Conference*, V001T02A085-V001T02A085. American Society of Mechanical Engineers.

Kumar, U., Yadav, I., Kumari, S., Kumari, K., Ranjan, N., Kesharwani, R. K., ... Chakravarty, D. (2015). Defect identification in friction stir welding using discrete wavelet analysis. *Advances in Engineering Software*, *85*, 43–50. doi:10.1016/j.advengsoft.2015.02.001

Longhurst, W. R., Wilbur, I. C., Osborne, B. E., & Gaither, B. W. (2018). Process monitoring of friction stir welding via the frequency of the spindle motor current. *Proceedings of the Institution of Mechanical Engineers. Part B, Journal of Engineering Manufacture*, *232*(4), 720–730. doi:10.1177/0954405416654089

Mehta, M., Chatterjee, K., & De, A. (2013). Monitoring torque and traverse force in friction stir welding from input electrical signatures of driving motors. *Science and Technology of Welding and Joining*, *18*(3), 191–197. doi:10.1179/1362171812Y.0000000084

Mishra, D., Gupta, A., Raj, P., Kumar, A., Anwer, S., Pal, S. K., ... Pal, A. (2020). Real time monitoring and control of friction stir welding process using multiple sensors. *CIRP Journal of Manufacturing Science and Technology*, *30*, 1–11. doi:10.1016/j.cirpj.2020.03.004

Pires, F., Cachada, A., Barbosa, J., Moreira, A. P., & Leitão, P. (2019). Digital twin in industry 4.0: Technologies, applications and challenges. *2019 IEEE 17th International Conference on Industrial Informatics (INDIN)*, *1*, 721–726.

Ranjan, R., Khan, A. R., Parikh, C., Jain, R., Mahto, R. P., Pal, S., Pal, S. K., & Chakravarty, D. (2016). Classification and identification of surface defects in friction stir welding: An image processing approach. *Journal of Manufacturing Processes*, *22*, 237–253. doi:10.1016/j.jmapro.2016.03.009

Serio, L. M., Palumbo, D., Galietti, U., De Filippis, L. A. C., & Ludovico, A. D. (2016). Monitoring of the friction stir welding process by means of thermography. *Nondestructive Testing and Evaluation*, *31*(4), 371–383. doi:10.1080/10589759.2015.1121266

Sudhagar, S., Sakthivel, M., Ashok, A., & Ajith Arul Daniel, S. (2017). Investigation of discontinuities in friction stir welding of 2024 aluminum alloy using radiography. *Materials Evaluation*, *75*(8).

Sudhagar, S., Sakthivel, M., & Ganeshkumar, P. (2019). Monitoring of friction stir welding based on vision system coupled with Machine learning algorithm. *Measurement*, *144*, 135–143. doi:10.1016/j.measurement.2019.05.018

Sudhagar, S., Sakthivel, M., Mathew, P. J., & Daniel, S. A. A. (2017). A multi criteria decision making approach for process improvement in friction stir welding of aluminium alloy. *Measurement*, *108*, 1–8. doi:10.1016/j.measurement.2017.05.023

Chapter 11
Advances in Welding Techniques for Similar and Dissimilar Materials

Soundarrajan Madesh
https://orcid.org/0000-0002-7100-7192
Muthayammal Engineering College, India

Arul T. G.
St. Mary's Engineering College, India

Varatharaju Perumal
Ethiopian Technical University, Ethiopia

ABSTRACT

The need of welding in the current scenario of manufacturing is increasing due to the application of various components. New innovations in welding are created in order to meet benefits such as high production rate, quality surface finish, high mechanical strength of joint, difficult geometry, new materials for new components. This chapter provides insight on the current advancements of welding to explore the aforementioned benefits. Welding process parameters, novel implements, research backgrounds, applications, and ability of welding process along with its merits and demerits are covered in this chapter. In addition, advanced welding techniques in the areas of similar and dissimilar welding are presented. Moreover, this chapter would unfold the latest trends in advanced welding techniques for current requirements in industries.

INTRODUCTION

Welding is one of essential part in manufacturing sector for assembling and make complete product. There are various conventional welding techniques exist and implemented successfully in manufacturing. Simultaneously, their modifications and advancements are increasing tremendously with accelerated

DOI: 10.4018/978-1-7998-9574-9.ch011

commercial necessities. The modifications with conventional welding provide an excellent improvement in various factors of welding process such as improved welding strength, considerable reduction in cost, increasing productivity and ability to deal with various materials in manufacturing. Inventive ideas implementation with welding technology offers a new open up by various materials such as tough metals, composites, ceramics, non-metallic materials, similar and dissimilar materials. The main objective of this book chapter is given a preliminary knowledge of current innovations in the area of welding process along with its general concepts and distinct features of various techniques, assuming the common thoughtfulness of the researchers with respect to welding process. In last four decades, many welding techniques are evolved and practiced for the joining the metals by the researchers. Figure 1 represents the various welding techniques and its advancements practiced in industry. Advanced welding process can be classified into the three major categories such as fusion welding, non fusion welding and hybrid welding process.

Figure 1. Types of advanced welding processes

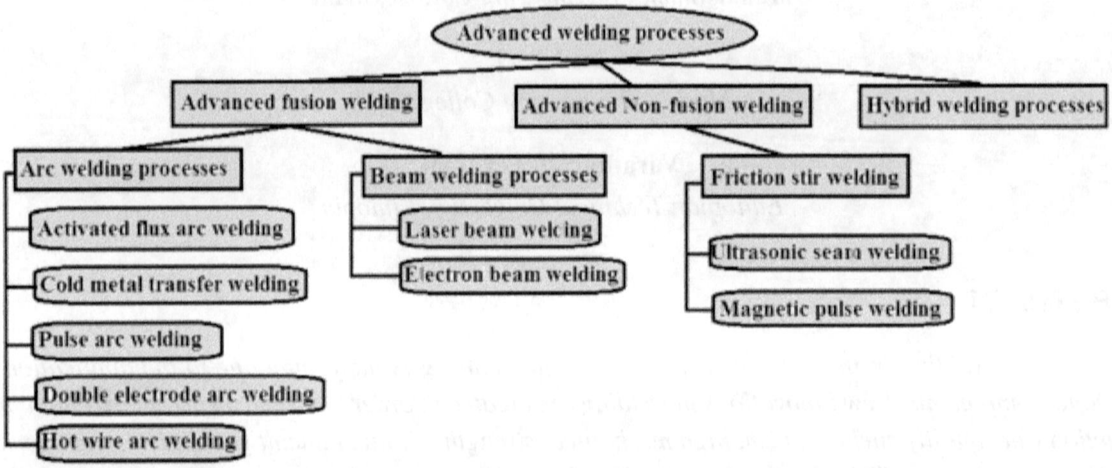

Advanced Arc Welding Processes

Arc welding process comes under the type of fusion welding method which is employed to join a various material by means of melt and solid of parent materials using an arc with the application of filler material. Advancements of arc welding processes are discussed in the forthcoming sections (Chai & Zhu, 2010; Huang, 2010; Kuo et al., 2013; Vidyarthy & Dwivedi, 2016).

Flux Coated Arc Welding

In flux coated arc welding process, work piece has been coated with activated flux which is prepared by mixing of inorganic compounds such as metallic oxides, chlorides and fluorides and applied on the work surface as present in figure 2. The flux mixing is prepared as a liquid with help of solvents such as acetone or ethanol. This liquid sprayed on the work surface and welding torch moved along in it. The flux coated welding increases the welding penetration and diminishes weldment width. Moreover, heat affected zone of welding reduced much better than the use of non flux coated welding. The principal of activated flux coating techniques employed in tungsten inert gas welding (TIG) process, gas metal arc welding, plasma welding and laser welding processes (Tümer et al., 2021; Xu et al., 2018).

Figure 2. Process involved in flux coated arc welding
(Sivakumar et al., 2021)

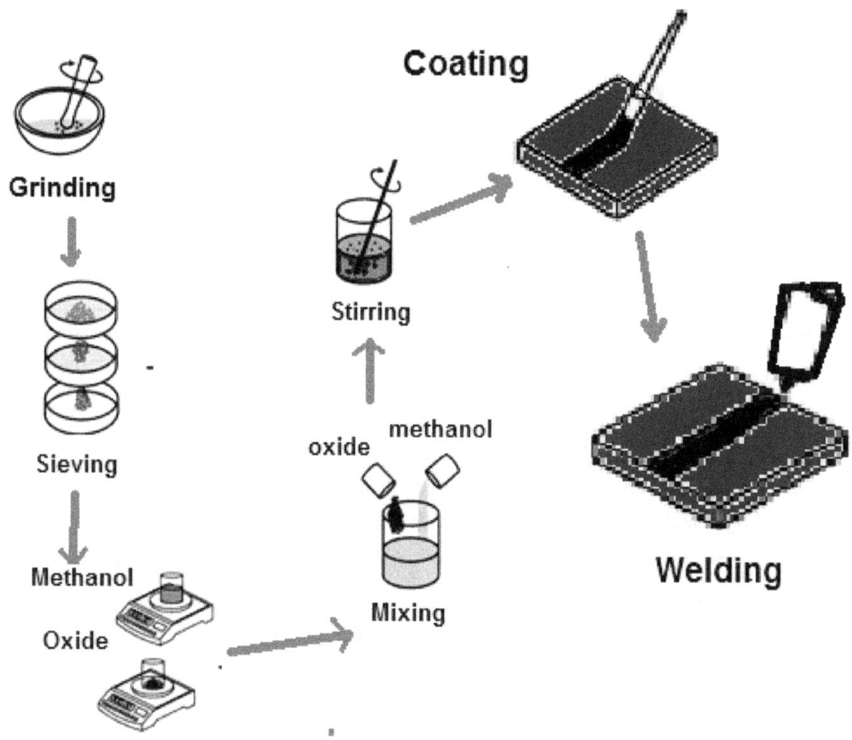

Flux assisted or coated TIG welding process is significantly improving the welding penetrations around up to 12 mm which is not conceivable to attain in traditional TIG method (Zhang & Zhang, 2009). Enhancement of welding diffusion is obtained using the influence of reverse Marangoni convection (also known as surface tension driven convection). The surface tension of molten weld metal plays a major role on weldment diffusion when flux coated on the work piece. Since, direction of weldment flow would deviate from the welding center to apart. This factor extremely increases the diffusion range and diminishes the weldment width on the work surface which is represented in figure 3. Also, arc constriction mechanism delivers focused arc using high power and resulting the higher diffusion (Berthier et al., 2012). The flux coated TIG welding technique is also employed for the repair work due to its advantages such as non- use of filler wire, not necessary to make V grooves, reduced weldment shrinkage, maximum welding length up to 10-12 mm could be obtained by autonomous mode.

Figure 3. Welding penetration

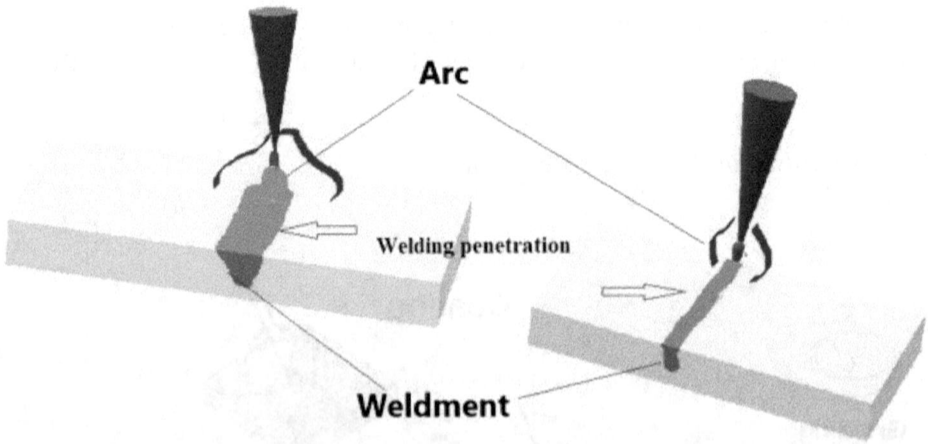

The performance of TIG welding is affected majorly with the factors such as type of flux, welding speed, input parameters and type work material to be welding. The flux material and its composition selection are very important which should not react with parent work material by chemically. In addition, properties of molten weldment such as surface tension of weldment, magnetic and electromagnetic nature, buoyancy force due to the pressure of molten weld, aerodynamic drag force due to self-weight of the molten weld are plays significant role on the weldment characteristics (Oreper & Szekely, 1984). Also, welding speed and applied current influence the heat production during arc length, electrode shape, gas composition controls welding performance are significantly affects the shape of the welding joint (Li et al., 2012). However, due the various advantages of TIG welding it has been employed in various tough metallic metals and composites such as stainless steels, nickel based super alloys, chromium, magnesium etc.

Cold Metal Transfer Arc Welding

Cold metal transfer (CMT) arc welding technique is an automatic process and created for the metal inert gas welding (MIG). The units such as wire feed, arc production is fully controlled automatically and transferring of metal take place in cold condition. This process provides benefits such as low heat supply, less spatter and high stable arc (Selvi et al., 2018). The controlled short circuit is created between the work and electrode. In first stage of this process, filler wire is diffused and transferred to the worksurface using short circuit controller. The filler wire moves towards down direction and get back its original position when short circuit occurs which transfer the molten metal in cold condition. Moreover, the metal transferring takes place such as drop by drop and this process repeat again and again for certain the time interval. In this process, electrode movements that forward and reverse play's major role on the performance of welding. Along with that electrical parameter such as low voltage and currents are considered as a major parameter to obtain better welding by CMT process (Kumar et al., 2016). The process of CMT presented in figure 4. In advanced CMT, filler movements created using pulse technique and this pulsed cycle improves the heat delivers. Also, combining pulse cycle with CMT process improves the welding quality significantly. Moreover, further developments in CMT process taken up such as polarity reversal with short circuit which enhances the heat supplies and controls the welding spatters over the work surface.

Figure 4. Processes of Cold metal transfer welding technique

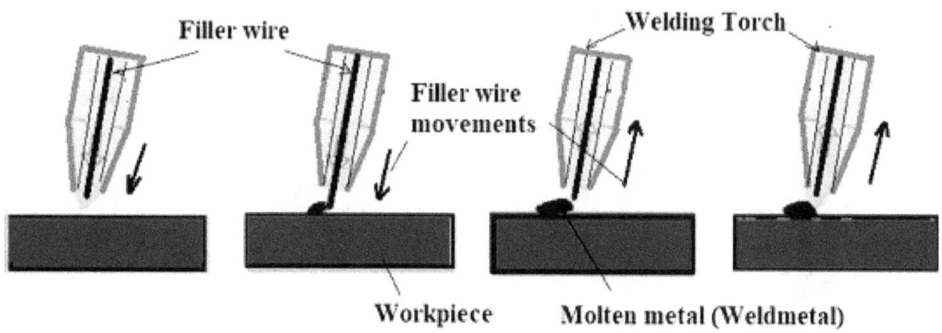

Pulse Arc Welding

Gas tungsten arc welding (GTAW) process is integrated with pulse technique for welding of various similar and dissimilar metallic materials. Pulse GTAW process operates based on the pulsed power supply to enhance the welding characteristics such as welding diffusion, rate of welding deposition and mechanical strength (Qi et al., 2013). The solidification of weld metal over the work surface has been enhanced through two steps which are during pulse off time and gap created at time of next pulse. The most influencing factors of pulse GTAW such as duration of pulse, peak, mean and base current which are subsequently influence the velocity of molten weld metal droplet. In addition, process parameters like diameter of filler, feed rate of filler, shielding gas and speed of welding controls the quality of pulse GTAW significantly. Improper selection of these process parameters leads to attained defects such as spatters, high rough weld surface, more under cut and stub (Kutelu et al., 2018). Also, Pulse GTAW sprayed out the molten metal spray with diminished heat input. This factor is hindered the spatter on the weld surface and heat affected zone reduced significantly using this technique.

Double Electrode Arc Welding

Double electrode arc welding is one of the prominent and novel methods for welding of similar and dissimilar metallic materials. In this process, both kinds of electrodes such as consumable and non consumable are involved to attain quality weld with less power consumption. The secondary power supply is taken from the same source whereas passed two separate parts. The power supply from one point is connected to the two portals such as work and bypass torch. As well as, another one terminal connected to the primary welding torch that gas metal arc welding (GMAW) (Lu et al., 2014). In addition to monitor the power supply of the bypass and GMAW torch control system is connected between them which can give the signals to the welder about the use of power. The bypass current supply to another electrode diminishes the production of heat at the welding spot. The working principle of double electrode setup is presented with the figure 5. Also, this double electrode techniques employed in welding methods such as plasma, submerged arc welding processes (Li & Zhang, 2007).

Figure 5. Double arc welding mechanism

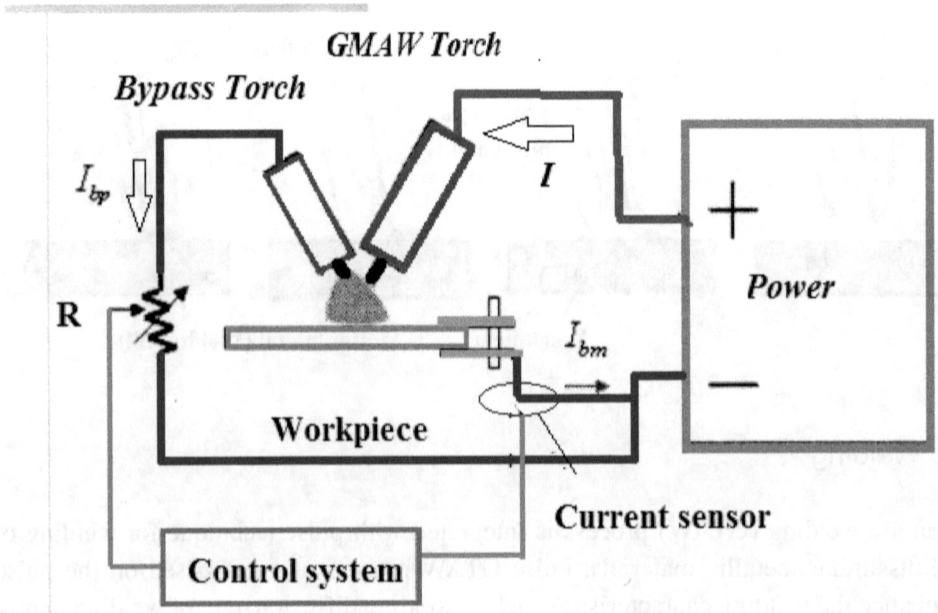

Gas Tungsten Arc Welding

Gas tungsten arc welding (GTAW or TIG) process is also known as tungsten inert gas welding which is presented in figure 6. In this welding, molten metal produced through heating with arc among the tungsten electrode and work metal. The main purpose of tungsten electrode is to maintain the arc intensity in the welding spot. The current passes among the parent material during bypass current passes via double GTAW bypass path. Double torches are fitted with GTAW system to give bypass current for the melting the metal. Combinations of helium and argon gases are act like shielding gas on the welding zone (Traidia et al., 2011). This welding technique provides better welding strength compare to GMAW and SMAW due to the less fume production. Also, this process comes under the category of double electrode welding technique. Moreover, this process diminishes the heat supply without influencing the deposition of filler which brings to the reduced heat affected zone (Chen et al., 2009). However, due to the need of double bypass current flow it is difficult to fabricate the setup.

Plasma Arc Welding

Plasma arc welding method created from the mechanism of tungsten inert gas welding (TIG) technique through fixing a narrow sharp edge and passes a minimal amount of argon or helium (shielding gas) gas using the narrow gap which is presented in figure 7. The confined arc detaches the shielding gas into the anode and cathode charged electrons to induces the plasma. Also, if plasma gas passes apart from the arc portion it creates neutral atoms and provides the energy as heat. Initially less current is used to create the arc and supplied between electrode and narrow sharp edge. The shielding gas passing via narrow sharp edge is ionized and creates the primary arc among the electrode and base metal during higher range of current. The welding zone has been covered and protected from the atmospheric reactions using shielding gases. Generally, the GTAW reaches the temperature nearly 11000°C where as the plasma assistance in it can touch the 20 000°C (Wu et al., 2014). This high temperature produces the excellent weld diffusion for high range of depth to width with lesser distortion of base metal. This plasma welding can be applied the materials such as high carbon steel, stainless steels, nickel-based super alloys for the thickness range of 2.5 mm to 20 mm (Yin et al., 2020). Moreover, the welding speed is 150% higher than the normal TIG welding and very stable welding than TIG arc welding.

Figure 6. Gas tungsten arc welding process

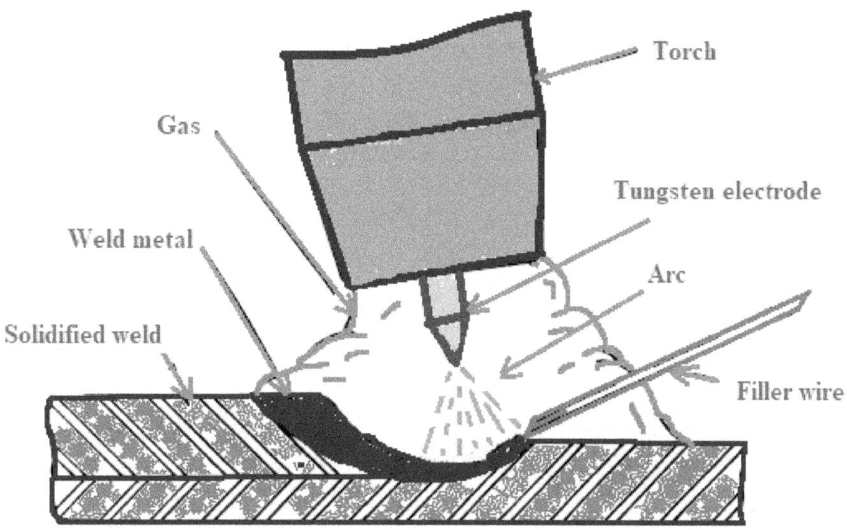

Figure 7. Plasma arc welding

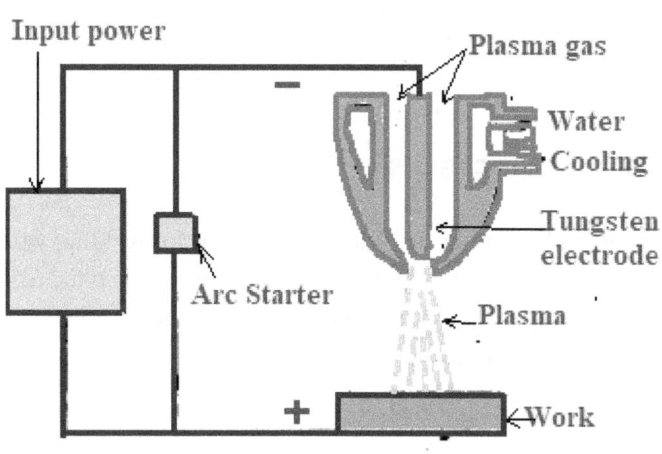

Submerged Arc Welding

Submerged arc welding (SAW) process widely employed in various sectors such as bridge welding, inline butt weld, pipe line, flange welding etc. Continuous filler wire supplied during the welding to make a weld pool. The welding is submerged using flux feeding through a hopper. During the weld, flux material is used to cover the molten metal immediately and helps to form a slag which protecting the welding from the atmosphere. Also, the excess flux material collected and recycled for another welding. SAW is generally automated or robot controlled welding which is widely employed in industries for fabrications process. the another name of this process is carbon dioxide welding where carbon dioxide gas used as a shielding gas in a welding zone like TIG welding process. According to the base metal of welding, flux material has been considered. Since, flux should be the non reactive at higher temperature and slag formation over the metal will diminish the quality of weld especially for the materials like chromium, nickel, stainless steels.There more flux materials used in SAW where very less cost flux such as silica is employed to weld the chromium in order to recover the base metal and deposit ferrous molecules is almost same like predicted from the filler wire composition. Another category of flux is consisting of chromium metallic compounds which increase the weld deposition rate (Li et al., 2017; Wada & Cox, 1983). In case of duplex stainless steel the fluxes such as silica is considered. The function of SAW is illustrated in figure 8.

Figure 8. Submerged arc welding

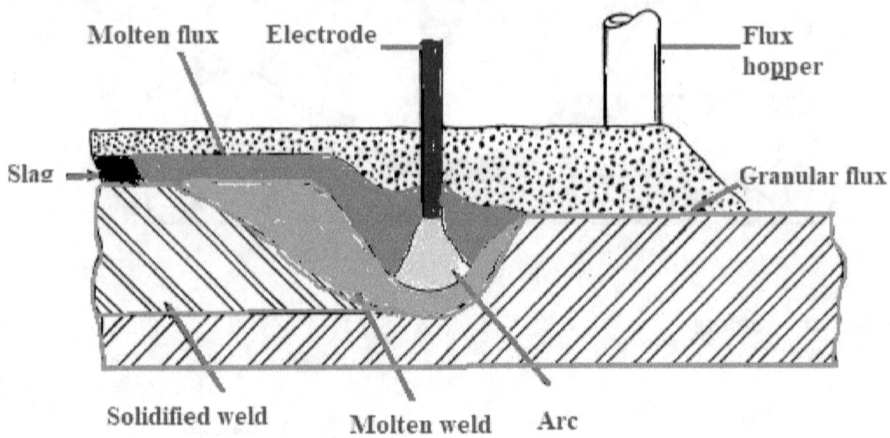

Advanced Beam Welding Processes

The high intensity heat energy also known as beam is employed for welding various metals. The beam welding technique is one of the fusion welding method in this metals is melted using intense, heat energy through bulge of photons or electrons or plasma. The beam welding processes are categorized such as laser, plasma and electron beam welding.

Laser Beam Welding

Laser beam welding (LBW) is a better technique and widely used in several field of manufacturing engineering and it is presented in figure 9. LBW method considered like versatile welding process in automobile, manufacturing areas. LBW is created for its various merits such as high accuracy, good deep weld ratio, less heat supply, less heat affected zone, quick cooling and can be weld difficult geometries. Also, due to an excellent welding nature laser beam has been employed in various fields such as narrow gas welding, non metals, dissimilar welding, plastic welding and various repair applications. The ultra narrow gap welding is one of the advanced welding techniques to weld thick parts in lesser cost. In this welding method, welding base metals prepared with one half of width has been joined with slight angles. Less than 2 mm narrow width among the work spices are considered for ultra narrow welding. Also, less amount of weld metal is employed to join the cavity which can reduces the cost of welding and time consumptions. In case of plastic welding, diode lasers are majorly used for the plastic welding which can provide more stable and prominent welding for thermoplastics. The thermoplastics are welded through the lasers without the application of absorber and lasers wavelengths of up to 2000 nm are considered for the welding. However, based on the literature this process much expensive than the other welding process (Gillner et al., 2021).

Figure 9. Laser welding method
(Gillner et al., 2021)

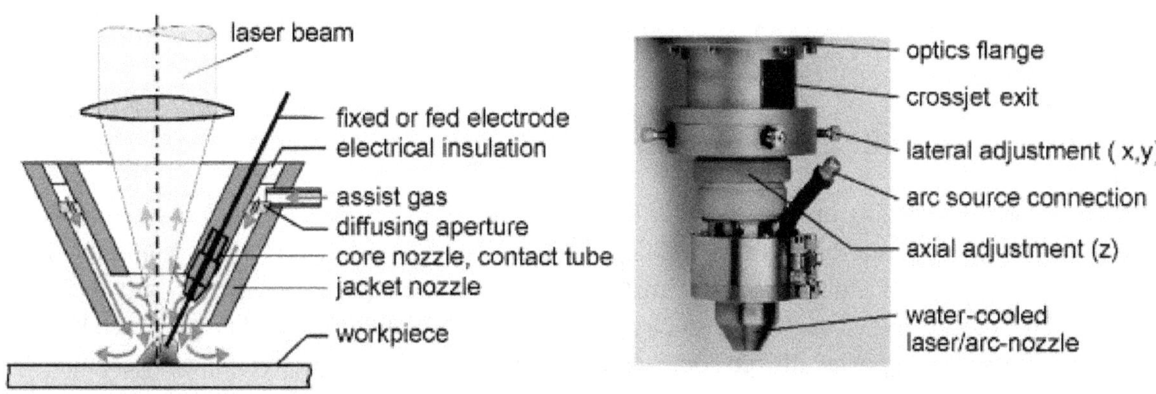

Electron Beam Welding

Electron beam welding (EBW) process is employed the stream of electrons on the work material which exhibits the heat energy to melting the base metal. The high energy EBW process are find applications in space, automobile, medical, bio medical and oil refinery industries. The welding of metal matrix composite becomes very significant role in all kind of welding process. This welding process needs a vacuum chamber to control the electron beam (Schultz, 1993). EBW is having ability to operate on very thin thick materials to higher thick materials for the various metals and non metals such as plastics, magnesium, tungsten and composites etc. The stream of electron beam is focused on welding work prices and heat energy production is lesser than the laser welding technique. Also, this process employed in various functions such as repair applications, surface modification etc. Furthermore, welding thickness and fusions are obtained based on the work material employed and maintainability of welding process. Welding of bigger size parts like turbines, compressor blades and engine case could be possible through EBW process (Azar et al., 2001). During the welding external powders and filler metals are employed for welding like the laser powder welding process. These powders and filler metal addition increases the rate of diffusion and welding strength. The processes of EBW are represented with the figure 10.

Figure 10. Electron beam welding

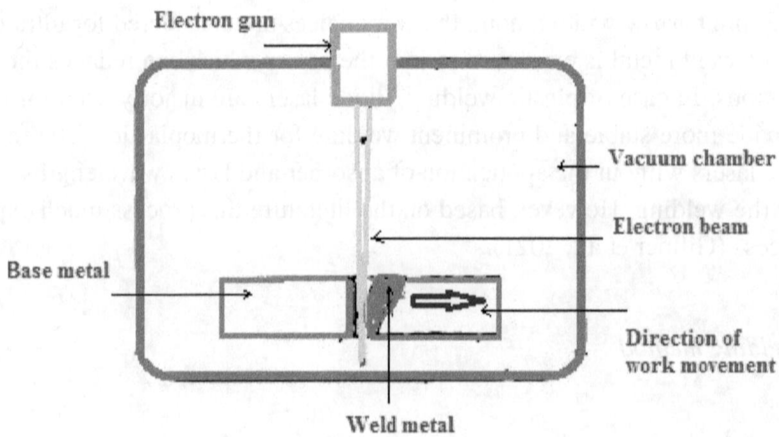

Friction Stir Welding

It is one of the solid welding methods initially invented for welding the materials such as aluminum and aluminum-based composites. Due to its welding nature friction stir welding (FSW) method modified and employed with various materials tough materials, nickel based super alloys, similar and dissimilar welding. This process has characteristics such as fumes free, no shielding gas and no melt metal. In addition, around 5% lesser power utilized compare to laser welding, less weight welding and consumables free method for better welding (Thomas et al., 1999). FSW method and its functions are illustrated in figure 11.

The electrode is made up of hardened steel material with rotating mechanism and pin along with shoulder head which can be mounted with moving end. The electrode pin fully implanted into the work material and shoulder head rotates by self which induces the heat effect due to the high friction. This friction causes for the high heat production and metal molecular collision. Therefore, the consequences of high friction create the slight metal melting and its transverse movement joins the metal efficiently. The specially prepared tool electrodes for the FSW are presented in figure 12 which is used in current industries for welding (Cui et al., 2007). Also, because of technological growth and market needs FSW process modified based on the welding method such as stable shoulder FSW, under water FSW, bit joining etc (Suresh et al., 2020a; Suresh et al., 2020b; Suresh et al., 2019).

Figure 11. Friction stir welding process
(Mishra & Ma, 2005)

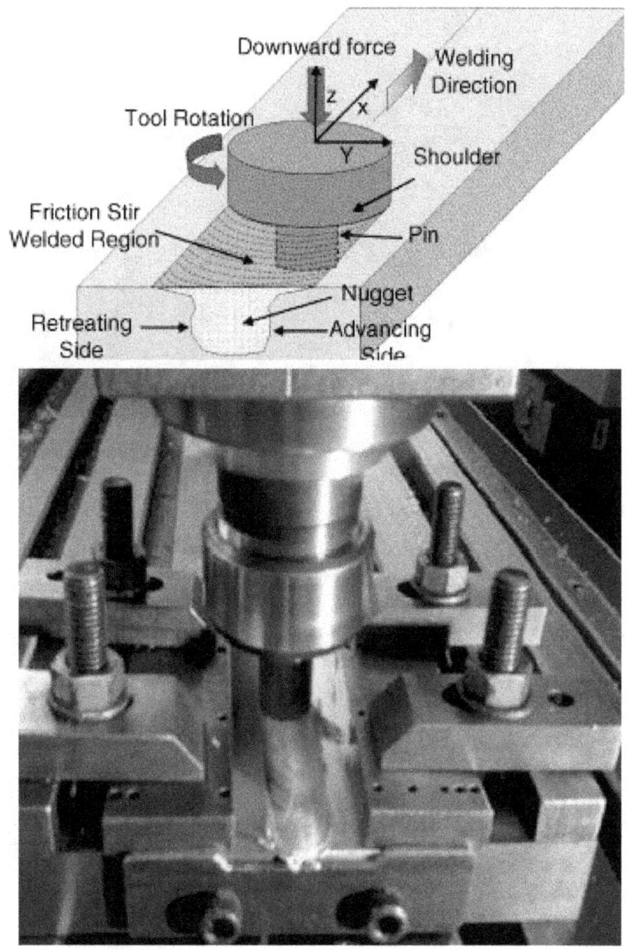

Ultrasonic Welding

Ultrasonic welding (USW) is one of the solid welding methods in thisheat energy provided by the frictional force using ultrasonic vibrations over work materials. Welding is obtained due to plastic deformation without the help of molten metal which is caused by the ultrasonic vibrations. USW is considered for the complex welding process due its various merits such as less power consumption, quick and automatic method, filler and gas free, non slag process and also posses' ability to welding dissimilar and similar metals (Yeh, 2013). The principal of USW is illustrated in figure 13. The main components of USW are piezoelectric transducer, horn, sonotrode are connected as presented in figure and produced the vibrations and then supplied into the work material.

Figure 12. Tool used for friction stir welding process
(Besharati-Givi & Asadi, 2014)

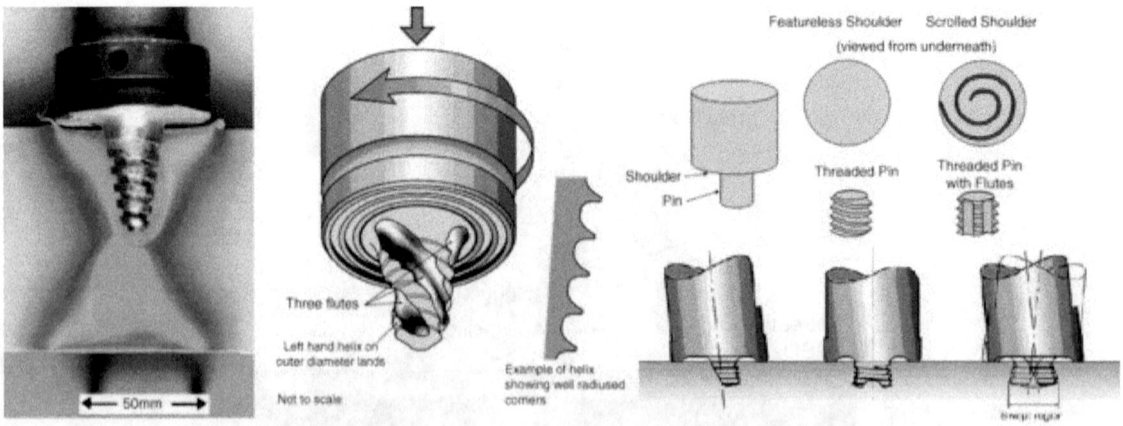

Figure 13. Ultrasonic welding process

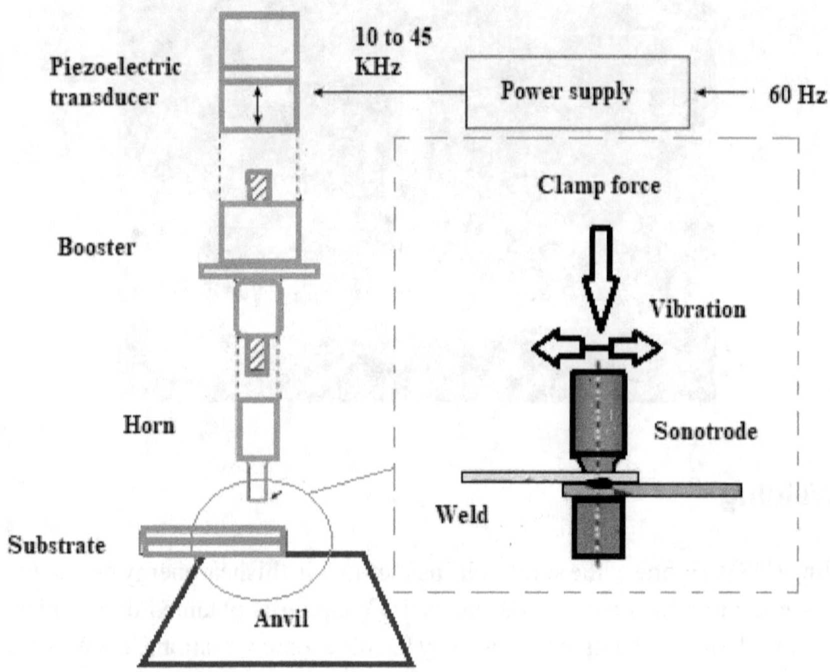

The major process parameters like frequency, amplitude, pressure and weld area are considered for the experiment and analyzed by the researchers. Also, piezoelectric transducer or electromechanical converter is used to convert high energy current to the mechanical vibration. The booster and horn provide the movement through the sonotrode with help of vibrations which is delivered from the transducer. The clamping force is employed over the work materials which can enhance the weld by each layer of work sheet. Moreover, the ultrasonic vibration range 10 to 40 kHz with the range between 5 to 50 µm.In addition, USW process is employed for the welding nonmetallic materials such as thermoplastics due to the solid-state welding method (Yeh, 2013). Ultrasonic vibrations are inducing the materials softening with higher temperature. Due to the nature of its welding, high thermal and high electric conductive materials such as copper and aluminum are being considered for welding in various research applications by the researchers. Also, nonmetallic materials such as ceramic, glass, plastics are welded with metals for some certain applications using USW technique (Zhang et al., 2016). Thickness of work material maintained between 0.025 to 0.250 mm which increase the welding strength.

CONCLUDING REMARKS

The abovementioned chapter that advanced welding techniques are shows the excellent welding ability with new developed techniques, specifically for similar and dissimilar materials. The traditional welding method provides complexity for the welding of similar or dissimilar metals when it is being as super alloys. Therefore, academicians and researchers need to know the recent developments of the welding process for the further developments. Hence, in order to provide the quality weld for the industry needs various modifications are carried out such as flux addition to the filler, double electrode, pulse power supply, heat addition, vibration assistance etc,. Although, due the enormous applications of welding still many developments and modifications can be done with welding techniques. In line up that, current advancements of welding methodologies are discussed in aforenoted chapter. The special structures of development circumstantial, parameters, competences, innovative aspects and alternatives are presented for the recently progressing welding methods in the industries. Advancements in the field of arc, beam, laser, friction and ultrasonic welding process are covered. The developments in welding methods provides the concrete merits such as excellent welding properties, good quality, capable to weld dissimilar materials weld, repairing applications, increasing the production rate, enhances the geometrical shapes and increase the ability to face fast growing industry need with the variations in materials to produce new components.

REFERENCES

Azar, P., Li, P., Patnaik, P. C., Thamburaj, R., & Immarigeon, J. P. (2001). Electron beam weld repair and qualification of titanium fan blades for military gas turbine engines. *RTO AVT Specialsits' Meeting on Cost Effective Application of Titanium Alloys in Military Platforms.*

Berthier, A., Paillard, P., Carin, M., Valensi, F., & Pellerin, S. (2012). TIG and A-TIG welding experimental investigations and comparison to simulation: Part 1: Identification of Marangoni effect. *Science and Technology of Welding and Joining*, *17*(8), 609–615. doi:10.1179/1362171812Y.0000000024

Besharati-Givi, M. K., & Asadi, P. (2014). *Advances in friction-stir welding and processing*. Elsevier.

Chai, G., & Zhu, Y. (2010). Spectra and thermal analysis of the arc in activating flux plasma arc welding. *Guangpuxueyuguangpu fen xi = Guangpu, 30*(4), 1141–1145.

Chen, J. H., Chen, P. N., Hua, P. H., Chen, M. C., Chang, Y. Y., & Wu, W. (2009). Deposition of multicomponent alloys on low-carbon steel using gas tungsten arc welding (GTAW) cladding process. *Materials Transactions*, *50*(3), 689–694. doi:10.2320/matertrans.MRP2008276

Cui, L., Fujii, H., Tsuji, N., & Nogi, K. (2007). Friction stir welding of a high carbon steel. *Scripta Materialia*, *56*(7), 637–640. doi:10.1016/j.scriptamat.2006.12.004

Gillner, A., Petring, D., & Olowinsky, A. (2021). Laser joining. In *Advanced Joining Processes* (pp. 3–34). Elsevier. doi:10.1016/B978-0-12-820787-1.00001-2

Huang, H.-Y. (2010). Effects of activating flux on the welded joint characteristics in gas metal arc welding. *Materials & Design*, *31*(5), 2488–2495. doi:10.1016/j.matdes.2009.11.043

Kumar, N. P., Vendan, S. A., & Shanmugam, N. S. (2016). Investigations on the parametric effects of cold metal transfer process on the microstructural aspects in AA6061. *Journal of Alloys and Compounds*, *658*, 255–264. doi:10.1016/j.jallcom.2015.10.166

Kuo, M., Sun, Z., & Pan, D. (2013). Laser welding with activating flux. *Sci Technol Weld Joining*, *6*(1), 17–22.

Kutelu, B. J., Seidu, S. O., Eghabor, G. I., & Ibitoye, A. I. (2018). Review of GTAW welding parameters. *Journal of Minerals & Materials Characterization & Engineering*, *6*(05), 541–554. doi:10.4236/jmmce.2018.65039

Li, D., Lu, S., Dong, W., Li, D., & Li, Y. (2012). Study of the law between the weld pool shape variations with the welding parameters under two TIG processes. *Journal of Materials Processing Technology*, *212*(1), 128–136. doi:10.1016/j.jmatprotec.2011.08.015

Li, K., Wu, Z., Zhu, Y., & Liu, C. (2017). Metal transfer in submerged arc welding. *Journal of Materials Processing Technology*, *244*, 314–319. doi:10.1016/j.jmatprotec.2017.02.004

Li, K., & Zhang, Y. (2007). *Metal transfer in double-electrode gas metal arc welding*. Academic Press.

Lu, Y., Chen, S., Shi, Y., Li, X., Chen, J., Kvidahl, L., & Zhang, Y. M. (2014). Double-electrode arc welding process: Principle, variants, control and developments. *Journal of Manufacturing Processes*, *16*(1), 93–108. doi:10.1016/j.jmapro.2013.08.003

Mishra, R. S., & Ma, Z. Y. (2005). Friction stir welding and processing. *Materials Science and Engineering R Reports*, *50*(1-2), 1–78. doi:10.1016/j.mser.2005.07.001

Oreper, G. M., & Szekely, J. (1984). Heat-and fluid-flow phenomena in weld pools. *Journal of Fluid Mechanics*, *147*(-1), 53–79. doi:10.1017/S0022112084001981

Qi, B. J., Yang, M. X., Cong, B. Q., & Liu, F. J. (2013). The effect of arc behavior on weld geometry by high-frequency pulse GTAW process with 0Cr18Ni9Ti stainless steel. *International Journal of Advanced Manufacturing Technology*, *66*(9-12), 1545–1553. doi:10.100700170-012-4438-z

Schultz, H. (1993). *Electron beam welding* (Vol. 93). Woodhead Publishing.

Selvi, S., Vishvaksenan, A., & Rajasekar, E. (2018). Cold metal transfer (CMT) technology-An overview. *Defence Technology*, *14*(1), 28-44.

Sivakumar, J., Vasudevan, M., & Korra, N. N. (2021). Effect of activated flux tungsten inert gas (A-TIG) welding on the mechanical properties and the metallurgical and corrosion assessment of Inconel 625. *Welding in the World*, *65*(6), 1061–1077. doi:10.100740194-020-01061-4

Suresh, S., Venkatesan, K., Natarajan, E., & Rajesh, S. (2020a). Performance Analysis of Nano Silicon Carbide Reinforced Swept Friction Stir Spot Weld Joint in AA6061-T6 Alloy. *Silicon*, 1-14.

Suresh, S., Venkatesan, K., Natarajan, E., & Rajesh, S. (2020b). Influence of tool rotational speed on the properties of friction stir spot welded AA7075-T6/Al2O3 composite joint. *Materials Today: Proceedings*, *27*, 62–67. doi:10.1016/j.matpr.2019.08.220

Suresh, S., Venkatesan, K., Natarajan, E., Rajesh, S., & Lim, W. H. (2019). Evaluating weld properties of conventional and swept friction stir spot welded 6061-T6 aluminium alloy. *Materials Express*, *9*(8), 851–860. doi:10.1166/mex.2019.1584

Thomas, W. M., Threadgill, P. L., & Nicholas, E. D. (1999). Feasibility of friction stir welding steel. *Science and Technology of Welding and Joining*, *4*(6), 365–372. doi:10.1179/136217199101538012

Traidia, A., Roger, F., Chidley, A., Schroeder, J., & Marlaud, T. (2011). Effect of helium-argon mixtures on the heat transfer and fluid flow in gas tungsten arc welding. *Journal of Chemistry and Chemical Engineering*, *5*(9), 854–861.

Tümer, M., Mert, T., & Karahan, T. (2021). Investigation of microstructure, mechanical, and corrosion behavior of nickel-based alloy 625/duplex stainless steel UNS S32205 dissimilar weldments using ERNiCrMo-3 filler metal. *Welding in the World*, *65*(2), 171–182. doi:10.100740194-020-01011-0

Vidyarthy, R., & Dwivedi, D. (2016). Activating flux tungsten inert gas welding for enhanced weld penetration. *J Manuf Process, 22*, 211–228.

Wada, T., & Cox, T. B. (1983). *Preparation and evaluation of submerged-arc weld in 4 inch thick 3Cr-1. 5Mo-0. 1V steel plate*. Final report (No. DOE/NBM-4005684). AMAX Metals Group.

Wu, C. S., Wang, L., Ren, W. J., & Zhang, X. Y. (2014). Plasma arc welding: Process, sensing, control and modeling. *Journal of Manufacturing Processes*, *16*(1), 74–85. doi:10.1016/j.jmapro.2013.06.004

Xu, L. Y., Jing, H. Y., & Han, Y. D. (2018). Effect of welding on the corrosion behavior of X65/Inconel 625 in simulated solution. *Welding in the World*, *62*(2), 363–375. doi:10.100740194-018-0549-y

Yeh, H. J. (2013). Ultrasonic welding of medical plastics. In *Joining and assembly of medical materials and devices* (pp. 296–323e). Woodhead Publishing. doi:10.1533/9780857096425.3.296

Yin, X., He, G., Meng, W., Xu, Z., Hu, L., & Ma, Q. (2020). Comparison study of low-heat-input wire arc-fabricated nickel-based alloy by cold metal transfer and plasma arc. *Journal of Materials Engineering and Performance*, *29*(7), 4222–4232. doi:10.100711665-020-04942-3

Zhang, Y., Li, Y., Luo, Z., Yuan, T., Bi, J., Wang, Z. M., Wang, Z. P., & Chao, Y. J. (2016). Feasibility study of dissimilar joining of aluminum alloy 5052 to pure copper via thermo-compensated resistance spot welding. *Materials & Design*, *106*, 235–246. doi:10.1016/j.matdes.2016.05.117

Zhang, Z., & Zhang, F. (2009). Spectral analysis of welding plasma of magnesium alloy using flux coated wire. *Materials Transactions*, *50*(8), 1909–1914. doi:10.2320/matertrans.M2009023

Chapter 12
An Overview of Welding Methods for Advanced Materials: Testing and Evaluation

Thiraviam Rayapandi
PT Saipem Indonesia, India

Suresh P.
Muthayammal Engineering College, India

ABSTRACT

Recent developments in the engineering industry require joining of like and unlike materials with different properties such as melting point, coefficient of thermal conductivity, solubility, difference in electrochemistry, etc. as part of machines, tools, and more specific applications. Materials including those similar and dissimilar in nature are successfully joined by fusion and solid-state welding processes. In accordance with ASME Sec IX and AWS D1.1 codes and API 1104 standard, welding procedures specifications (WPS) through procedure qualifications (PQR) are required prior to commencing any fabrication work pertaining to pressure vessels, piping and pipeline, storage tanks, offshore platform structural parts, and so on. A specific welding process must be chosen based on the design of the component, the material, thickness, production, availability of equipment, people, and other factors. Weldment testing, including destructive and non-destructive examinations, are crucial during procedure qualification, welder qualification, and the production welding process.

INTRODUCTION

Welding is a joining process which involves heating materials to the welding temperature, either with or without the application of pressure, and with or without the use of filler metal. In industry, there are various welding methods available; however, arc welding is particularly prevalent and widely employed

DOI: 10.4018/978-1-7998-9574-9.ch012

in the oil, gas, chemical, and power industries. A master chart of welding and related processes utilized in current metal fabrication and maintenance services is shown in Figure 1 (Jeffus, 2020).

Figure 1. Master chart of welding and allied processes

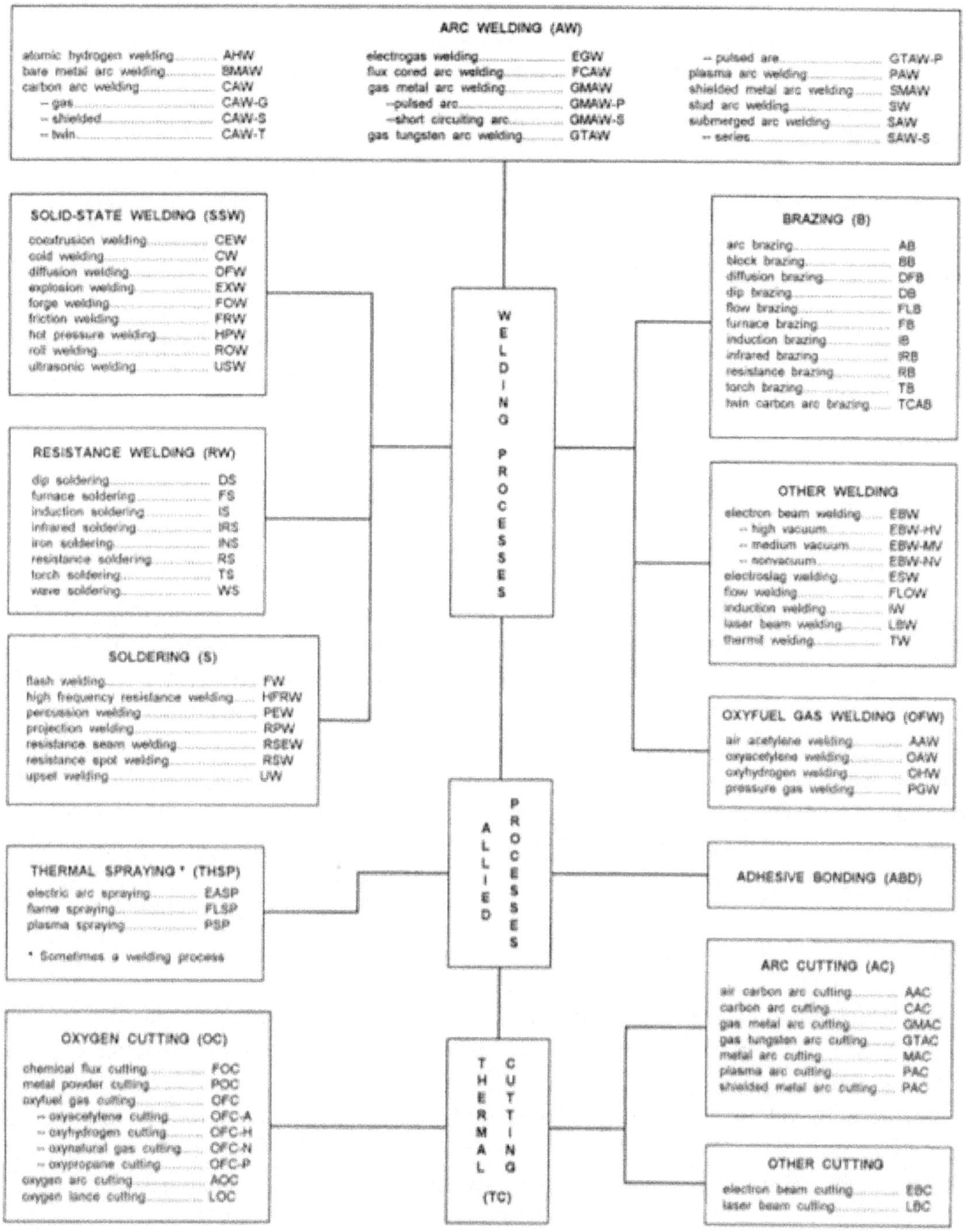

OVERVIEW OF WELDING PROCESS

Gas Welding

A gas flame is used as a source of heat in one of the most prevalent welding procedures. Heat is generated by burning a combustible gas, such as MAPP (methylacetylene-propadiene) or acetylene, combined with oxygen in the oxyfuel gas welding process (Abdulateef et al 2010). Because oxygen and fuel cylinders are easy to transport, gas welding is commonly employed in maintenance and repair operations.

Arc Welding

This is a method of joining metals using an electric arc. The concentration of heat is a notable benefit of arc welding over gas welding. The flame in gas welding spreads out across a vast area, which can cause thermal distortion. Arc welding's heat concentration is advantageous because reduced heat diffusion lowers buckling and warping (Kumar et al. 2009). This heat concentration also increases penetration depth and speeds up the welding process; as a result, arc welding is typically more practical and cost-effective than gas welding. There are three things that all arc welding procedures have in common: a heat source, filler metal, and shielding. Arc welding generates heat by arcing an electrical current between two conductors.

Shielded Metal Arc Welding (SMAW)

By striking an arc between a coated-metal electrode and the base metal, shielded metal arc welding is done. Once the arc is produced, the molten metal from the electrode's tip joins the molten metal from the base metal's edges to form a sound junction. Stick electrode welding, coated electrode welding, and manual metal arc welding are all terms used to describe this procedure. Because the electrode coating produces a protective covering over the weld deposit, sheltering it from contamination, the method is known as shielded metal arc welding. The main benefits of shielded metal arc welding are that it produces high-quality welds quickly and at a reasonable cost. The method is very versatile, and it can be used for any position and any metal, including low carbon steel, low alloy steels, quenched and tempered steels, high alloy steels, corrosion resistant steels, and so on. It has a limited application in aluminium welding (Matusiak & Pfeifer (2013)).

Gas Shielded Arc Welding

The type of shielding employed distinguishes shielded metal arc welding from gas shielded arc welding. Both the arc and the molten puddle are shielded by an inert gas in gas shielded arc welding. The inert gas barrier protects the weld from contamination by the atmosphere, resulting in a superior weld. Helium, argon, or carbon dioxide are the main gases employed in this process (Kah & Martikainen, (2013)). A blend of these gases is sometimes utilised. Gas tungsten arc welding (GTAW), gas metal arc welding and plasma welding method are three types of gas shielded arc welding technologies.

Tungsten Inert Gas Welding, or GTAW

It was first used in the industry to join aluminium and magnesium alloys. Its application spread to practically all metals after then. A non-consumable tungsten electrode is employed in this method, which is surrounded by an inert shielding gas envelope. Both ac and dc power sources are employed. Arc initiation in GTAW used to be done by touching the electrode to the graphite. Modern machines now include a high frequency unit that automatically triggers the arc (Norrish, (2006)). This method results in a clean, high-quality weld. It is used to weld in all positions. This procedure is used to weld the majority of the materials.

Gas Metal Arc Welding – GMAW

After GTAW, gas metal arc welding was devised to boost weld productivity. In this method, a 0.8-2 mm diameter consumable wire wound on a spool is fed into a welding torch with an electrical connection and shielding gas. The utilisation of an arc voltage power source and tiny welding wire makes the system more sensitive. The power supply is always dc, and electrode positive is the preferred polarity. GMAW is a semi-automated, all-position welding method with automatic versions on the market (Kolodziejczak, (1987)).

Plasma Welding

An inner plasma gas and an outside shielding gas are used in plasma welding. The plasma gas circulates around a tungsten electrode with a retracting centre electrode. The shielding gas passes through the outer jet in the same way that TIG welding does. Because a plasma arc is significantly straighter and more focused than a TIG arc, the approach is less susceptible to arc length fluctuations. The process can be used to weld the same materials as can be welded by TIG welding, with the exception of magnesium (Weman, (2011)).

Submerged Arc Welding

It is a high-productivity welding technology that is usually done with mechanical welding methods and can be used with 1-3 continuous wire electrodes. SAW welding is similar to MIG welding in that an arc is formed between a constantly fed bare wire electrode and the workpiece. A flux is used to produce protective gases and slag, as well as to introduce alloying components to the weld pool. It is not necessary to use a shielding gas. A small layer of flux powder is applied to the workpiece surface prior to welding. Excess flux is recycled through a hopper as the arc advances along the joint line. After welding, the remaining fused slag layers can be simply removed. Heat loss is minimal since the arc is totally encased in the flux layer. This can result in a 60 percent thermal efficiency (compared with 25 percent for manual metal arc). There is no apparent arc light, there are no spatters, and no fume extraction is required (Choudhary, (2006)).

Resistance Welding

It is one of the oldest methods of welding, and it's usually quick, efficient, and low-polluting. There is no need for filler. High initial costs and a limited range of applications are some of the disadvantages. Heat is produced when an electric current pass through the resistance established by two metal surfaces in contact. The current density is so great that it creates a local pool of molten metal that connects the two halves. The current is usually between 1000 and 1,000,000 A, and the voltage is between 1 and 30 V.

Solid State Welding Process

Solid state welding is a welding technique in which two work pieces are bonded at a temperature essentially below the melting point of the base metal under a pressure that provides close contact between them. There is no metal utilised as filler. Diffusion of the materialscontactatoms causes bonding. Cold welding, diffusion welding, explosion welding, forge welding, friction welding, hot pressure welding, roll welding, and ultrasonic welding are under this category of welding. Explosion welding and friction welding are the most popular welding methods.

Welding of Similar Materials

The heat input to which the metal is exposed in any kind of welding causes specific changes in the material, some of which are permanent. Structure changes during heating and cooling, as well as changes in shape or size due to thermal stresses, are examples of these modifications. Good weldability refers to the ability of a material to be welded without harmful repercussions.If, on the other hand, the changes caused by a normal welding process pose a significant risk of failure in a welded component, or if actual defects such as cracking occur during welding, and can be welded without risk in most cases if certain precautions are taken or certain pre- or post-welding treatments are performed. Welding can be done on any material if the proper metallurgical conditions are used. However, in some cases, these criteria may be impossible to achieve in real-world production.

Weldability of Steel

Weldability is a function of many inter-related factors and may be summarised as:

- Process and technique.
- Composition of parent plate.
- Joint design and size.

Influence of Process

Each welding method has distinct power intensity. Higher-power-intensity processes are advantageous in fusion welding because essential melting can be achieved without excessive heat inputs and the resulting thermal expansion of the parent metal. Filler wire containing a deoxidant is frequently fed into the weld pool for successful welding, hence techniques that do not employ filler wire are limited in use.

Heat Input

Heat input is a combination of voltage, amperage and travel speed.

$$\text{Heat Input} = \frac{\text{Voltage x Amerpage}}{\text{Travel speed}}$$

As heat input increases: grain size increases, toughness decreases, brittleness increases, hardness decreases, incidence of cracking decreases (Gharibshahiyan et al (2011)).

Preheating

There are four general factors that must be assessed to determine the level of preheat.

a. Material type.
b. Combined thickness.
c. Heat input potential (welding process).
d. Joint type.

Influence of Composition

The composition of the steel and its effect on weldability may be divided into two parts (Dev et al. 2018):

a. Segregation effects, particularly that of sulphur

When a steel solidifies, the iron tends to solidify quickly, accumulating the alloying elements in the ingot's centre. This arrangement is maintained even after extended and severe rolling, resulting in high sulphur concentrations in the plate's middle layers. These layers are weak and will crack if they are pressured.

b. The tendency of the steel to harden.

The hardening characteristics of steel are mentioned in the heat treatment section. Low ductility and hardened microstructures are more likely to form when the carbon or alloy content of the steel increases, and if stresses exist, cracking will result.

Welding of Carbon Steel

Carbon-manganese and simple carbon steels are other names for these materials. They can typically be welded without difficulties with mild steel consumables, but weldability reduces when carbon levels or C.E. (carbon equivalent) values increase. When welded at high speeds, very low carbon steels (C 0.15%) can cause porosity. By increasing the heat input, this problem can be resolved. Welding steels with carbon concentrations of 0.15 percent to 0.25 percent are recommended because, with low impurity levels, they rarely cause weldability issues (Wu et al. (2018)). If other hardening elements like Mn and Si are present in relatively high percentages, steels with carbon levels more than 0.25 percent may be susceptible to cracking. Low hydrogen procedures or electrodes are optimal for welding these steels, and large sections may require some pre-heating to slow down the cooling process. Wear resistance and hardness are two advantages of high carbon steels (for example tool steels). Pre-heating, interpass heating, and post-weld stress reduction are all required to keep them from cracking. Depending on the carbon content and joint thickness, most medium carbon steels require one or more of the above treatments. Low hydrogen techniques are used to weld them.They are always welded with low hydrogen methods or consumables, and a weldability test is usually performed prior to production welding.

Welding of Low Alloy Steels

Nickel steels, carbon-molybdenum alloys, and chromium-molybdenum alloys are among the most prevalent low alloy steels. At low temperatures, nickel in the range of 2% to 5% in a 0.15 percent to 0.25 percent carbon steel gives a combination of high strength and toughness. Welding without pre-heating is possible if the carbon content is less than 0.18 percent. Similar measures should be taken above this level as with medium carbon steels. Because of their great creep resistance and strength, carbon-molybdenum and chromium-molybdenum steels are commonly utilised in high-temperature applications.Pre-heating is not necessary below 0.18 percent carbon, however thicker parts with greater carbon levels harden in air and are hence break prone. An austenitic stainless-steel consumable can occasionally be utilised for very high carbon levels (0.55 percent) where welding is not recommended. The weld will be more ductile and less strong than the parent plate, allowing some internal tension to be relieved. However, because the HAZ is still brittle, pre-heating is required. In case of chromium-molybdenum steels, preheating and post heating treatments are required to avoid any cracking in the weldments.

Welding of High Yield Quench and Tempered Steels

Heat treatment during manufacturing produces martensitic features in these steels. HY80, which is utilised on submarine hulls, is one of the most commonly used steels (Yayla, (2007)). The formation of strong martensitic structures in steels is dependent on rather high cooling rates. If the cooling rate is too low, strength can be lost. Weld cracking can occur if the cooling rate is too fast. To obtain the correct structure without cracking, both the minimum pre-heat and maximum inter-pass temperatures must be carefully managed. It's common to utilise consumables that match. To keep the heat input low, weaving should be confined to no more than 15mm.

Welding of High Strength Low Alloy Steels

The strength of the majority of these steels is not determined by heat treatment. They have ferritic structures with low carbon content (0.2%) and modest amounts of alloying elements in solution (Mn, Cr, Ti, V, Nb, N). There are numerous varieties, each with varied levels of strength, toughness, and corrosion resistance. Although they are normally weldable, their great strength makes them more susceptible to cracking than low carbon steel. Pre-heating is occasionally necessary and high heat inputs are normally advised (Mohandas (1999)).

Welding of Austenitic Stainless Steels

Austenitic stainless steels are iron-based alloys with low carbon, chromium between 15% and 32%, and nickel between 8% and 37%. They are employed because of their resilience to corrosion and high temperature deterioration. Austenitic stainless steels are known for their weldability and can be welded using any standard welding process or technique. Solidification cracking, hot cracking, distortion, and preserving corrosion resistance are the most essential factors when welding austenitic stainless steels. Weld metal chemistry and the metallurgical phases that occur in the weld metal are intimately related to solidification cracking and hot cracking (also known as hot shortness). Both solidification cracking and hot cracking have the same cracking mechanism. Solidification cracking occurs in the fusion zone, while hot cracking occurs in the partially melted zone (Shankar et al.2003). The ferrite number of the weld metal is the most typical indicator of weldability and susceptibility to heat cracking. To withstand cracking, austenitic welds require a minimal level of delta ferrite. The amount of ferrite in the weld metal is principally determined by the chemistry of both the base metal and the weld metal. The base metal chemistry should be acceptable for welds done without filler metal in order to produce the minimal amounts of ferrite required to prevent cracking. Filler metal is recommended to develop appropriate ferrite in the weld metal if the base metal chemistry does not allow for ferrite production. Welding settings and methods can also influence the development of ferrite. Small amounts of nitrogen, for example, absorbed into the weld metal, can inhibit the production of ferrite. To avoid cracking, a minimum of 5% - 20% ferrite is recommended by a number of literature and codes & standards. Sulfur, phosphorus, selenium, silicon, and columbium, all of which increase hot cracking susceptibility, can impair the weldability of austenitic stainless steels. Welding of austenitic stainless steels is more prone to distortion than welding of carbon or low-alloy steels. Austenitic stainless steels distort more than carbon steels due to less thermal conductivity.In 304/316 austenitic stainless steels, HAZ corrosion resistance can be reduced by welding. Temperatures between 800°F and 1650°F (427°C and 900°C) can precipitate chromium carbides at grain boundaries if exposed to these temperatures for long enough (Rezaee et al.2011). As a result of the chromium depletion, corrosion resistance suffers. This issue can be avoided by using stainless steels with a low carbon content, such as Type 304L or 316L, or stable grades, such as Type 321 and 347. To avoid a loss of corrosion resistance, it's also crucial to choose the right filler metal. It is preferable to utilise low carbon electrodes or stabilised grades of bare filler metal.The corrosion resistance of austenitic stainless steels can be harmed by oxidation of the backside of welds done without sufficient shielding. The root of the weld should be shielded with an inert backing gas to avoid a reduction in corrosion resistance (Lippold, & Kotecki, (2005)).

Ferritic-Austenitic Steels (Duplex)

Duplex steels are ferritic-austenitic steels that have grown in popularity since the 1970s. They are notable for combining great corrosion resistance with strong mechanical strength. These steels are mostly used in the offshore industry, where their characteristics make them ideal for usage in high-chloride environments. In comparison to austenitic steels, their high strength results in lighter structures. They are also becoming more commonly employed in the cellulose, chemical, and petrochemical sectors.This series of steels has chromium concentrations of up to 29%, nickel of 5-8%, and molybdenum contents of around 14%. Carbon content is often less than 0.03 percent. Nitrogen, which can be present in amounts up to 0.4 percent, has been proven to be essential in determining the weldability of ferritic-austenitic steels. When welding, the weld pool initially solidifies as a completely ferritic structure, with austenite developing later as the temperature drops.The presence of nitrogen as an effective austenite forming element speeds up this process, allowing the weld to achieve the desired metallurgical structure without the use of hot welding or additional heat treatment. With little nitrogen and a large amount of ferrite, older ferritic-austenitic steels, become practically completely ferritic in a zone around the fusion line, compromising both mechanical and corrosion resistance. To rectify this, heat treatment between 100^{0}C and 200^{0}Cis required. Neither too low nor too high heat input should be used when welding ferritic-austenitic steels (Suutala et al., (1980)). A low heat input can lead to an excessively quick cooling rate, which can cause chromium nitrides to develop, which can compromise corrosion resistance. A high heat input combined with a high temperature between passes, on the other hand, causes the sigma phase, which affects corrosion resistance and mechanical qualities, among other things. Welding should be done using comparable filler, but with a nickel concentration of about 2-3% higher, to aid austenite production and generate a ferrite concentration in the weld metal that is similar to the base metal. This guarantees that the weld metal's corrosion resistance and mechanical properties are at their best.

Welding of Titanium

Titanium is widely used in the chemical, aerospace, marine, and medical industries due to its light weight, superior corrosion resistance, and high strength-to-weight ratio (Dey et al. (2013)). Recent advancements include the use of titanium alloys in the petrochemical industry and the production of sports equipment. The use of various processes to weld titanium is common, and the service performance of welds has been verified by a long and growing list of accomplishments. Welding research, technology, and economics are all progressing thanks to newer titanium-compatible procedures.Fusion welding is possible with most titanium alloys, and solid-state operations are possible with all alloys. Unfortunately, some engineers still believe titanium is difficult to weld, either because of its specific needs for argon or helium gas shielding, or because it was previously exclusively handled by specialized fabricators.

Welding and joining methods appropriate to titanium include:Arc Welding Processes, Tungsten Inert Gas (TIG), Metal Inert Gas (MIG), Plasma Arc (PAW), Power Beam Processes, Laser and Electron Beam (EB)welding, Resistance Welding, Spot, Seam and the Resistance cladding process, Friction Welding, Rotary, Radial, Linear, Orbital, Stir, Stud and other specific jointing techniques, Diffusion Bonding, Forge Welding, Explosive Bonding, Brazing, Soldering& Adhesive Bonding (Kou, (2003)).

Weld Joint Preparation

The edge preparation for titanium weld joints is often done by machining or grinding, as it does for other metals. The weld joint surfaces must be clean before welding and remain so during the welding process. Welding wire used as filler metal must meet the same standards. Detergent cleaners or non-chlorinated solvents should be used to remove contaminants like oil, grease, and fingerprints. Acid pickling can remove light surface oxides, however grit blasting and pickling may be required for larger oxides.

INERT GAS SHIELDING

Proper gas shielding is a need for successfully arc welding titanium. Until the weld metal temperature drops below 426°C (800°F), care must be made to maintain inert atmosphere protection. During welding, three independent gas streams are maintained to do this. The initial shield gas stream, also known as the primary shield gas stream, emerges from the torch and shelters the molten pool and surrounding surfaces. During cooling, the secondary or trailing gas shield protects the solidified weld metal and the heat-affected zone. During welding and cooling, the third or backup shield protects the weld underneath. These trailing and backup shields are made using a variety of approaches, with a torch trailing shield being one example. Many different types of backup shields exist. Copper backing bars with gas ports serve as a heat sink and shielding gas source, and are often used for straight seam welds. Complex workpiece configurations, as well as specific shop and field conditions, necessitate some ingenuity in devising backup shielding mechanisms. Plastic or aluminium foil enclosures or "tents" taped to the backside of the weld and flooded with inert gas are common examples.

Welding of Nickel Alloys

Nickel alloys such as Alloy C276 and Alloy 625 have similar issues as austenitic stainless steels. Most nickel alloys are thought to be less weldable than austenitic stainless steels. Some nickel alloys, like Alloy 825, 600, and 625, exhibit welding properties that are similar to austenitic stainless steels. In comparison to austenitic stainless steels, Alloy 200, Alloy 400, and Alloy B2 will have substantially different welding characteristics (Avery & Tuthill,(1994)). The tendency for nickel alloy and carbon steels to be sluggish during welding is one of the key contrasts between them and austenitic stainless steels.This means that the molten weld pool in nickel alloys will not flow as freely as it does in other metals. Because of this sluggish propensity, the welder should weave or oscillate the weld pool to achieve excellent sidewall fusion. If no oscillation is employed, the weld contour will be high convex, resulting in sidewall lack of fusion, weld undercut, or slag inclusions. Centerline cracking will be less likely if a slightly concave weld bead profile is formed. It is also crucial for nickel alloys to have a wide enough bevel angle to allow for the welding torch essential oscillation.In terms of weld penetration, the broader weld bevel will also help. In comparison to carbon steels and austenitic stainless steels, nickel alloys also have shorter weld penetration. The weld joint is adjusted with a broader bevel and a thinner root face to overcome this. Nickel alloys, like austenitic stainless steels, are vulnerable to heat cracking. As the weld pool cools and hardens, the hot tearing will occur. The weld connection should be designed to reduce constraint and the weld should be allowed to cool as rapidly as possible to prevent hot cracking.The less time a nickel alloy weld is exposed to the temperature range where it can tear, the faster it solidifies. As a result, pre-heating, which reduces the pace at which the weld cools, is actually hazardous since it increases the chances of hot tearing. The presence of large levels of low melting point components like sulphur, phosphorus, zinc, copper, and lead can impair the weldability of nickel alloys, just as it can with austenitic stainless steels. All of these pollutants can cause weld and base metal cracking.

Welding of Dissimilar Metals

When design requirements ask for a variety of qualities, manufacturability, or aesthetics that arenot possible with a single material, dissimilar materials permit the achievement of function. Welding stainless steel to plain carbon or low-alloy steels as dissimilar welds is frequently required for producing stainless steel structures and/or incorporating them into other machinery. Particular attention must be made to the metallurgy of the weld metal in order to achieve the greatest performance from such junctions.The transition zone between the metals and the intermetallic compounds generated in this transition zone are the source of the challenge of producing welds between dissimilar metals. It is critical to look into the phase diagrams of the two metals involved in fusion welding techniques. Dissimilar couplings can be created successfully if the two metals have mutual solubility. The weld joint will fail if the two metals to be connected have little or no solubility. The crack sensitivity, ductility, and corrosion susceptibility of the intermetallic compounds generated between different metals must all be examined.

This intermetallic compound's microstructure is particularly important. In some circumstances, using a third metal that is soluble with both metals is required to create a good junction. The coefficient of thermal expansion of both materials is another aspect to consider when calculating the service life of a dissimilar metal junction. Internal stresses will form in the intermetallic zone at any temperature change in the weldment if they are significantly different. Service failure may develop quickly if the intermetallic zone is highly brittle. The melting temperatures of the two metals to be connected must also be taken into account.When two metals are exposed to the same heat source, one will melt much sooner than the other. When joining metals with various melting temperatures and thermal expansion rates, a welding method that uses a high heat input and produces a fast weld has an advantage. A composite insert is used between the two metals at the weld joint as another technique of connecting dissimilar metals. The composite insert is made out of a non-heated welding method that creates a transition junction between two different metals. Metals can be welded to resemble composite inserts by selecting the appropriate materials for the composite insert.

Welding of Carbon Steel – Stainless Steel

Although stainless steel is a popular choice for many applications, the expense of fabricating large parts entirely out of stainless steel might be prohibitive when it comes to heavy fabrication. Using a lower-cost carbon steel for non-essential elements and structure can assist minimise overall production costs. This allows engineers to employ stainless steel exclusively in select locations, such as high-heat or corrosive areas, while mild steel is used in the rest of the time. Although joining carbon steel with stainless steel is not particularly difficult, the two metals require special care in terms of heat, filler materials, and joint design.

In comparison to carbon steel, stainless steel has a high coefficient of thermal expansion, which is a measurement of how quickly a material expands as temperature changes. Stainless steel also has a thermal conductivity of about half that of carbon steel. A piece of heated stainless steel will stay hot for considerably longer due to the lack of thermal conductivity, as it does not conduct heat away from the source as quickly. Heat conducts along that piece rather quickly because carbon steel has a higher thermal conductivity, pushing heat away from the weld zone. It takes an experienced team of welders and fabricators to prepare and plan the joining of two different metals together.

Important Factors to be Considered During Welding of Carbon Steel – Stainless Steel

To avoid problems while welding stainless steel to carbon steel, pay attention to the chemistry, mechanical qualities, and corrosion resistance. The most usually recommended filler metal is 309L. During the welding process, the weld is diluted with stainless steel from one side of the joint and mild steel from the other, resulting in material from both sides of the weld being mixed in. The goal is to make a final weld deposit that is chemically compatible on both sides of the weld joint.The mechanical properties of each type of material in the dissimilar combination must also be matched. When welding any type of stainless steel to carbon steel, the filler metal should match or slightly surpass the weaker of the two materials' mechanical qualities.Finally, while welding stainless steel and mild steel, it's critical to keep the weld joint and adjoining stainless steel base metal corrosion-free.

Important Welding Parameter: Heat Input

It's critical to choose a welding process that restricts the heat input to the weld and stainless base material to address chemistry, mechanical qualities, and corrosion resistance. The weld deposit is less diluted with the mild steel component of the weld joint when the heat input is limited. This, in turn, contributes to the weld deposit's alloy content and desirable corrosion resistance.Moderate heat input also protects corrosion resistance in some stainless steels by preventing undesired phases from forming on the stainless-steel side of the joint. If stored too long in a crucial temperature range of 800 degrees to 1400 degrees Fahrenheit, 300 series austenitic stainless steels, for example, can develop carbide precipitation. This problem can be avoided by reducing the amount of time spent in this range and using a low-carbon base and filler metal. The use of stable grades of filler metals (such as ER321 or ER347) may also be indicated, and can provide additional assurance against carbide precipitation.If maintained at a high temperature for too long, some stainless-steel grades may produce undesired phases that result in brittleness or poor corrosion resistance. At high temperatures, sigma phase (a brittle, intermetallic phase with a high hardness) can occur in some stainless steels, compromising mechanical and corrosion resistant qualities (Kobayashi & Wolynec (1999)). The heat input, for example, is responsible for the balance between ferrite and austenite in the final weld and heat-affected zone in duplex stainless steels (HAZ). The right amount of heat input can assist keep the amount of each phase in the completed weld and base metal HAZ where it belongs.

Welding of Aluminum to Various Metals

Aluminum melts at 650°C, while steel melts at around 1538°C. Before the steel has melted, the aluminium will melt and flow away. A multitude of complicated brittle intermetallics are generated. Iron - aluminium alloys with more than 12% iron have little to no ductility, according to research. Furthermore, aluminium and steel have vastly different coefficients of linear expansion, thermal conductivity, and specific heats. Thermal stresses of a significant magnitude will result as a result of this. The most effective way involves using an aluminum-steel transition insert with one metal soldered to the other.The metal used to coat the steel, its thickness, and the bond between the coating and the steel surface all have a role in the success of this sort of joint. The aluminium can be welded to the steel using a zinc coating and a gas tungsten arc welding method. The filler wire should be made of high-silicon aluminium. The welder will benefit from pulsing if the arc is directed toward the aluminium. Transition inserts are provided for welding aluminium to stainless steel. Coating can also be used. Pure aluminium coating, which is applied by dipping clean stainless steel into molten aluminium, is a coating for stainless steel.The aluminium surface can then be welded with gas tungsten arc welding. The arc should be aimed at the metal, and the welder will benefit from pulsing. A copper-aluminum transition insert piece is used to fuse aluminium to copper.

Welding of Copper to Various Metals

Welding copper and copper-base alloys to mild and low-alloy steels, as well as stainless steels, is possible. The gas tungsten arc welding procedure can be utilised with a high-copper-alloy filler rod for thinner parts in the gauge metal thickness. It is simpler to get a good weld when you use the pulsed mode. To avoid iron pickup, the arc should be directed to the copper part. In the thicker part, the steel is first overlaid or buttered with the same filler metal, and then the copper is welded to the overlaid surface. Because iron pickup in copper alloys causes brittleness, it is crucial to minimise extensive penetration into the steel portion of the joint. Preheating the copper is required.A nickel-base electrode can also be used to cover the copper. For thicker materials, a second overlay or layer is suggested. The copper should be warmed to 540°C (1000°F) while making the overlay welds on thick copper. To achieve a homogeneous joint preparation, smooth the copper part's overlay or buttered surface. Copper dilution caused by the nickel electrode should be avoided as much as possible. It is possible to employ the shielded metal arc, gas tungsten arc, or gas metal arc welding techniques. The option chosen will be determined by the available equipment as well as the thickness of the material to be bonded. Brass and copper can be bonded together, as can copper and stainless steel.

Welding of Nickel-Base Alloys to Steels

By employing the Monel analysis of filler material while using any of the arc welding procedures, nickel-base alloys such as Monel and Inconel can be successfully welded to low-alloy steel. The Inconel base electrode would be employed in Inconel to mild or low-alloy steel applications. Welding Inconel or Monel to stainless steels is the same issue. This electrode is made of Inconel or Monel.

Welding of Plastics (Polymers and Composites)

Even simple devices like mobile phones have over 100 joints in them. Welding, adhesive bonding, and mechanical fastening are all methods that can be used to join polymers and composites. The material to be joined, the joint configuration, the needed joint strength, the required level of seal, process cost and speed, and manufacturing quantity all influence the process selection. Welding relies on heat at the joint to melt the nearby polymer, with a weld developing as the material cools. As a result, welding is only a possibility for thermoplastic polymers and composites, whereas thermoset systems must rely on adhesives or mechanics.

Methods for welding plastics can be classified into three categories (Stokes (1989))

1. Employing an external heat source
2. Using heat is generated by mechanical movement
3. Deploying electromagnetism

Polymer composites can benefit from a number of these strategies as well. However, because composites' better qualities are derived from the reinforcement, the weld is invariably the weak point in the system because the fibre is discontinuous across the joint contact. The parameters of time, temperature, and pressure are used to control all welding processes. The welding settings must be carefully optimised for each application in order to generate high-quality welds.

Selection of Welding Consumable

When choosing an electrode, the first criterion is to choose one that generates weld metal quality that is equivalent to or better than the base material and is approved for the material in issue. Other considerations that influence electrode selection include the welding position and kind of joint, as different electrodes have varied qualities in different welding positions and types of joints. The electrodes are divided into groups based on the type of alloy that has been deposited. There are various examples where multiple distinct electrodes are intended for welding the same type of steel within each group of mild, low-alloy, and stainless steel electrodes.As a result, there are often a vast number of electrode types to choose from for each steel grade, many of which generate identical weld metal compositions but have various coatings, welding qualities, welding speeds, and weld metal quality. This wide range of options allows you to select the electrode that produces the best weld metal quality for the least amount of money. Electrode manufacturers all around the world, including ESAB, BOHLER, LINCOLIN, METRODE, KOBE, and others, offer a wide range of electrodes on the market.

The coatings are made up of several finely powdered chemicals and minerals that are bound together by a binder. The coating protects the metal droplets and weld pool from interactions with the air, which is supplied by the molten slag and the gases produced by the coating. Arc stability is being improved. The arc would be difficult to control and produce a lot of spatters if the coating didn't include arc-stabilizing ingredients in it. It would also be easy to extinguish, especially if welding with AC. Titanium, zirconium, and magnesium are all arc-stabilizing or ionising chemicals.Coated electrodes leave a coating of slag on the joint, which must be removed after welding. This needs the use of a chipping hammer or wire brush, and depending on the type of electrode coating, it might be easy or difficult. adding alloying and/or anti-oxidizing materials to the weld pool. To increase yield, the coating may additionally contain iron powder. ensuring enough penetration of the base material while welding Materials that may emit a considerable amount of heated gas, such as carbonates or cellulose compounds, decide penetration.

The chemical composition of the slag divides the electrodes into three categories: acid, basic, and rutile.

WELDING DESIGN

General

The main purpose of this part is to concentrate on all of the factors that go into making an educated, sound, and cost-effective welding design selection. Most of the aspects that must be considered when making an engineering decision will be identified. Two basic welding design factors necessitate unique and competent engineering attention: First, the welds must perform the specific design role assigned to them in the structure or product, as well as maintain their integrity reliably under predicted handling, shipping, and eventually service loads. Second, the welded joints must fully meet the requirements of maximum economy in their execution and maximum inspection access. Table 6.1 illustrate the basic weld symbol given by AWS.

Design Applications

Steels and their alloys, stainless steels of various types, aluminium, copper, cobalt, titanium, and other metals and alloys are all used in welding design. Non-metallic materials, such as plastics, ceramics, and composite materials, are also used in welding design. The following list, while not exhaustive, provides some insight into the welding industry's scope:Medical and bioengineering equipment, Electronics, Household utensils, Automobiles and all transportation equipment, Pressure vessels and the energy industry, oil and gas industries, Plastics, Machinery, Buildings & Bridges, Ship building and Aerospace industry.

PRECAUTION TO BE TAKEN BEFORE, DURING AND AFTER WELDING

General Welding Requirements

The requirements for structural (nonpressure boundary) welding should be in accordance with AWS D1.1 or AWS D1.6, or applicable regulations and standards. Welding techniques for pressure boundary equipment's pressure vessel, pipelines, storage tanks, and other components may be qualified in accordance with ASME BPVC Section IX. Prior to the start of fabrication or construction, all welding process specifications (WPSs) and procedure qualification records (PQRs) must be reviewed and approved by the purchaser. The submittal must contain weld maps, comparable guides, or fabrication drawings that clearly indicate the application of each WPS, stating where and how these WPSs will be employed. During fabrication, weld maps, related guidelines, drawings, or other documentation must be updated to clearly identify the welder(s) or welding operator(s) who created each weld (O'brien (2017)). Record the location of any NDE performed in the same way, and update the information as needed during fabrication.

Before Welding

Cleaning and Surface Preparation

The WPS must include information on how to clean base metal surfaces before welding. Any additional weld surface preparation requirements for equipment that will be coated after manufacture should be specified by the purchaser. Figure 2 shows the welding symbols. When the base metal surface is wet or moist, welding should be avoided. Cleaning/gouging the backside of double-welded joints to sound metal is required. Unless the consumer specifies otherwise, all slag must be removed from the backside of each finished austenitic stainless steel or nickel-base alloy weld.For weld joint surface protection, aluminium flake weld-through primers can be utilised. The weld-through primer kind and brand should be included in the WPS. Other types of weld-through primers or coatings are not allowed unless the buyer approves them, and the buyer may require additional method qualifications or weldability tests. Carbon steel wire brushes and other tools, as well as brushes and tools that have previously been used on carbon or low-alloy steel, must not be used on stainless steel or non-ferrous materials (O'brien (2017)). The surface of stainless steels must be honed to a bright finish if carbon-arc cutting or gouging are utilised.

Figure 2. AWS standard welding symbols

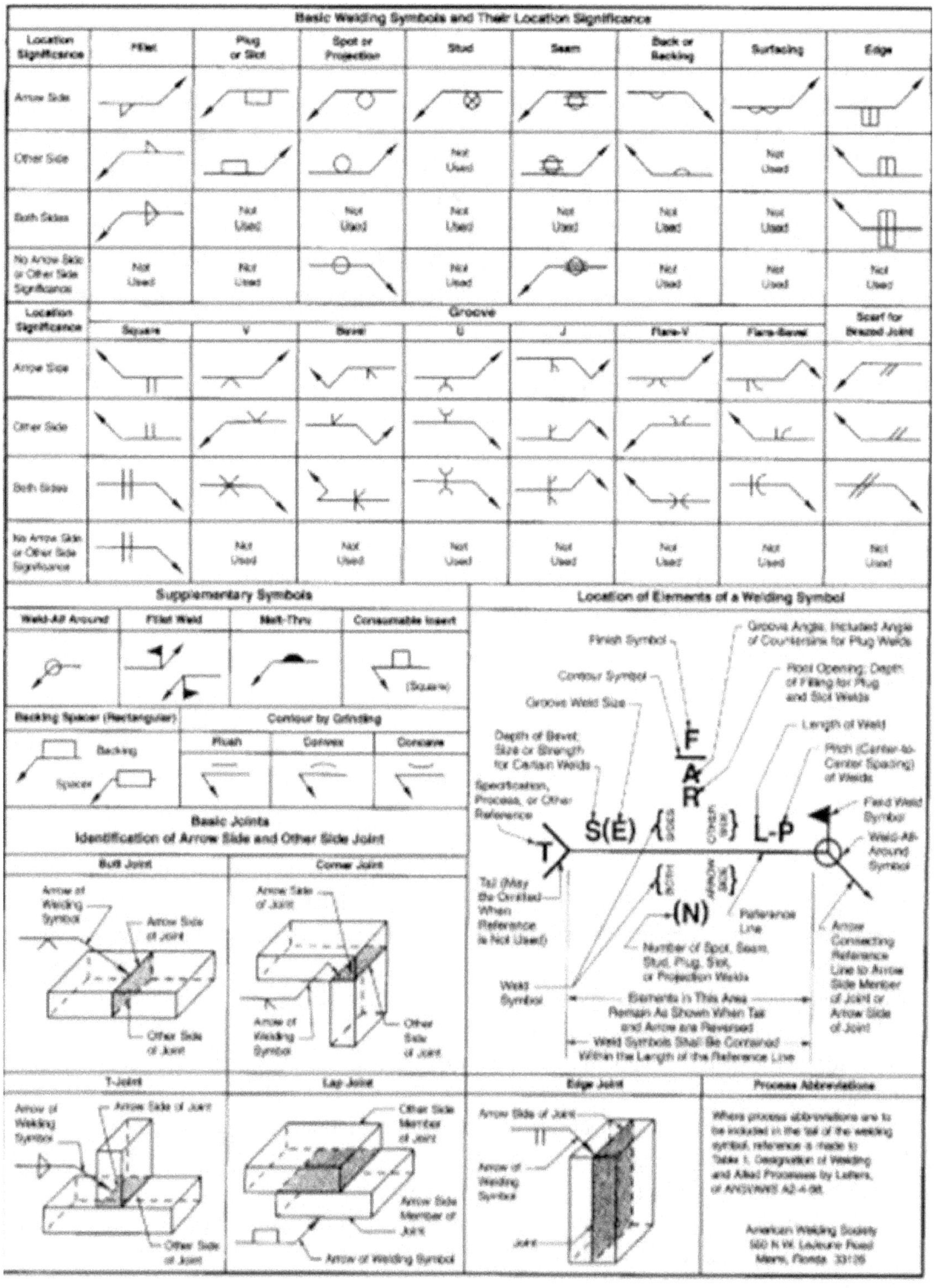

Shielding and Purging Gases

When shielding gases are employed, the WPS must specify the shielding gas (or gas mixture), the percent composition of the gas(es), and the flow rate. The purity standards of ASME/AWS SFA/A5.32/5.32M must be met by shielding gases. When only one gas is used, the purity should be reported on the PQR and WPS. If the joint is not ground or back gouged to sound metal, back purging is necessary for the GTAW and GMAW processes when welding materials with a nominal chromium concentration greater than 2-1/4 percent. When using a back purge, the WPS must specify the gas used and the flow rate. When a back purging gas is used to prevent oxidation or scale formation on the weld's bottom, it must be kept on until at least 1/4 in. (6.5 mm) of weld metal has been deposited. The back purging must be maintained throughout the welding operation for socket, seal, and any other attachment welds on base materials less than 1/4 in. thick (Funderburk, (1999)).

Handling and Storage of Consumables

The manufacturer's recommendations must be followed for storing and handling welding consumables. Welding consumables must be stored and baked separately in different ovens. The ovens must be electrically heated and feature an automatic temperature control system. A prominent temperature indication is required in welding consumable storage and bake ovens. All welding consumables must be stored, segregated, distributed, and returned according to a documented method by the fabricator. The identification of the filler metal must be preserved.

Inspection Before Welding

Before Assembly:

- All applicable documents.
- Quality plan is authorised and endorsed with signature, date and company stamp.
- Application standard is up to date with the latest edition, revision or amendment.
- The drawings are clear, the issue number is marked and the latest revision is used.
- Welding procedure sheets (specifications) are available, have been approved and are employed in production.
- Welder qualifications with identification and range of approval are verified and that only approved welders as required are employed in production.
- Calibration certificates, material certificates (mill sheets) and consumer certificates areavailable and valid.
- Parent material identification is verified against documentation and markings.
- Material composition, type and condition.
- Correct methods are applied for cutting and machining.
- Identification of welding consumables such as electrodes, filler wire, fluxes, shielding and backing gases and any special requirements (e.g. drying) are met.
- Plant and equipment are in a safe condition and adequate for the job.
- Safety permits e.g. hot work permit, gas free permit, enclosed space certificate are available and valid.

During Welding

Preheating and Interpass Temperature

All welding, tack welding, and thermal cutting require preheating when it is necessary. Appendix R of ASME BPVC Section VIII Division 1, Table 330.1.1 of ASME B31.3, API 934-A, API 934-C, and Annex XI of AWS D1.1 are examples of relevant codes and suggested practises for minimum preheat requirements (Funderburk, (1999)). Any heating guidelines or requirements in the applicable code must be followed. For low-alloy steels, the preheat temperature must be applied and maintained until PWHT throughout the weld thickness and at least 3 in. (75 mm) on each side.Lowering the preheat temperature below Mf before PWHT should be taken into consideration. The temperature of the preheat and interpass shall be monitored using thermocouples, temperature indicating crayons, pyrometers, or other appropriate means. For austenitic stainless steels, duplex stainless steels, non-ferrous alloys, and carbon and low-alloy steels, the maximum interpass temperature must be indicated in the WPS and PQR when impact testing is necessary. Interpass temperatures are suggested in Table 7.1.

Post Weld Heat Treatment (PWHT)

The PWHT utilised for a PQR must follow a method based on the applicable code and purchase order requirements. Prior to PWHT, the customer must examine and authorise the heat treatment technique for production use. The following information should be included in all WPSs that specify PWHT: a) maximum heating rate, b) holding temperature range, c) holding period, and d) maximum cooling rate. Alternatively, the WPS could relate to a project-specific PWHT technique. Prior to production, the purchaser must authorise the PWHT technique for unique heat treatment methods such as induction and internally fired.When testing is undertaken for reasons other than service-related issues, the purchaser must specify the testing requirements. The purchaser may need production hardness testing to ensure that heat treatments are adequate. The buyer can provide testing specifications or company-specific criteria. The purchaser must approve PWHT of austenitic stainless steel, duplex stainless steel, or non-ferrous alloys.When the stainless steel is heated over 1300°F (705°C), welding procedure qualification tests for austenitic stainless steel to ferritic steel welds must use the maximum PWHT temperature limit indicated in the welding procedure, except for weld overlays (Funderburk, (1999)). If cladding or overlay welds on low-alloy steels are repaired without PWHT, a minimum remaining clad or overlay thickness of 3/16 in. (5 mm) is needed unless no additional HAZ can be established.

Inspection During Welding

- The welding process must be monitored.
- Preheat and interpass temperatures must be monitored.
- Interpass cleaning - chipping, grinding, gouging, must be monitored.
- Root and subsequent run sequence.
- Essential variables such as current, voltage, travel speed to be monitored.
- Filler metals, fluxes and shielding gases are correct.
- Welding is in compliance with weld procedure sheet and application standard.

After Welding

Once the welding activity is completed, the inspection shall be carried as followed and required in line with Inspection testing plan (ITP). Recommended Maximum Interpass Temperatures is provided in the table 1.

- Visual inspection to be carried out to ascertain acceptability of appearance of welds.
- Dimensional accuracy to be ascertained.
- Conformity with drawings and standards requirements.
- Post weld heat treatment, if any, monitored and recorded.
- NDT carried out and reports assessed.
- Assess defects as to either repairing, or application for concession.
- Carry out any necessary repairs.
- Control of distortion

After Assembly

- Dimensions, tolerances, preparation, fit-up and alignment are in accordance with the
- Approved drawings and standards.
- Tack welds, bridging pieces, clamping and type of backing - if any used are correct.
- Cleanliness of work area is maintained.
- Preheat in accordance with procedure.
- NOTE Good inspection prior to welding can eliminate conditions that lead to the formation of defects.

Table 1. Recommended maximum interpass temperatures

Material Group	Maximum Interpass Temperature
P-1 (carbon steels)	600 °F (315 °C)
P-3, P-4, P-5A, P-5B, and P-5C (low-alloy steels)	600 °F (315 °C)
P-6 (Type 410)	600 °F (315 °C)
P-6 (CA6NM)	650 °F (345 °C)
P-7 (Type 405/410S)	500 °F (260 °C)
P-8 (austenitic stainless steel)	350 °F (175 °C)
P-10H (duplex stainless)	300 °F (150 °C)*
P-41, P-42	300 °F (150 °C)
P-43, P-44, and P-45	350 °F (175 °C)
NOTE Interpass temperature may vary depending on material grades.	

(Funderburk, (1999))

Welding Repairs

- Repair procedure and welding code should be authorised.
- Defect area should be marked positively and clearly.
- Check when partially removed and fully removed (visual and NDT).
- Re-welding should be monitored.
- Re-inspect completed repair.

Finally, collate all documents and reports and pass the document package on to a higher authority for final inspection, approval and storage.

QUALIFICATION OF WELDING PROCEDURES SPECIFICATION

Before manufacturing begins, all welding activities must be thoroughly planned. This entails creating welding procedure specifications for all welded connections in compliance with ASME sec IX regulations, as well as providing as much detail as the certification method requires. All critical aspects that could influence the welded joint's qualities must be considered. Variations that are allowed must be indicated. The welding process specification is categorised as preliminary, pWPS, until it has been authorised in compliance with relevant regulations and standards.

A Welding Procedure Specification (WPS) is a written document that instructs a welder or welding operator on how to make production welds that meet Code criteria. Any WPSs used by a manufacturer or contractor who will be responsible for the operational control of production welding must be qualified by that manufacturer or contractor in accordance with Article II of ASME sec IX, or an AWS Standard Welding Procedure Specification (SWPS) listed in Appendix E and adopted by that manufacturer or contractor in accordance with Article X of ASME sec X. Both WPSs and SWPSs state the welding conditions (including any ranges, if any) that must be met. The permissible base metals, the filler metals that must be used (if any), the preheat and post weld heat treatment requirements, and so on are all part of these conditions. Welding "variables" are the term used in this section to describe these conditions. When a manufacturer or contractor prepares a WPS, it must address the specific variables, both necessary and non-essential, as specified in Article II of ASME Sec IX for each process to be used in production welding.In addition, supplemental critical variables must be addressed in the WPS when other sections of the Code demand notch toughness qualification. The goal of a WPS qualification is to see if the proposed weldment can provide the needed qualities for its intended application.

The Procedure Qualification Record (PQR) records what happened during the welding of the test coupon as well as the results of the testing. The attributes of the weldment are determined by welding technique qualification, not the expertise of the welder or welding operator. The PQR must at a minimum document the important factors and other specific information defined for each welding method utilised on the test coupon, as well as the results of the needed testing. When notch toughness testing is required for method qualification, the supplemental critical factors for each process must also be reported.P-Numbers are assigned to base metals based on characteristics such as composition, weldability, and mechanical properties, where this is logical; and Group Numbers are assigned to P-Numbers for steel and steel alloys (table QW/QB-422 of ASME sec IX) to reduce the number of welding procedure qualifications required. These Group Numbers are used to categorise metals within P-Numbers for method qualification where notch-toughness requirements are specified (Wang & Yang (2016)). The thickness constraints and test specimen for method qualification in compliance with ASME Sec IX code are summarised in Table 2.

Table 2. Procedure qualification thickness limits and test specimen

GROOVE-WELD TENSION TESTS AND TRANSVERSE-BEND TESTS

Thickness T of Test Coupon, Welded, in. (mm)	Range of Thickness T of Base Metal, Qualified, in. (mm) [Notes (1) and (2)]		Maximum Thickness t of Deposited Weld Metal, Qualified, in. (mm) [Notes (1) and (2)]	Type and Number of Tests Required (Tension and Guided-Bend Tests) [Note (2)]			
	Min.	Max.		Tension, QW-150	Side Bend, QW-160	Face Bend, QW-160	Root Bend, QW-160
Less than $^{1}/_{16}$ (1.5)	T	$2T$	$2t$	2	...	2	2
$^{1}/_{16}$ to $^{3}/_{8}$ (1.5 to 10), incl.	$^{1}/_{16}$ (1.5)	$2T$	$2t$	2	Note (5)	2	2
Over $^{3}/_{8}$ (10), but less than $^{3}/_{4}$ (19)	$^{3}/_{16}$ (5)	$2T$	$2t$	2	Note (5)	2	2
$^{3}/_{4}$ (19) to less than $1^{1}/_{2}$ (38)	$^{3}/_{16}$ (5)	$2T$	$2t$ when $t < ^{3}/_{4}$ (19)	2 [Note (4)]	4	...	...
$^{3}/_{4}$ (19) to less than $1^{1}/_{2}$ (38)	$^{3}/_{16}$ (5)	$2T$	$2T$ when $t \geq ^{3}/_{4}$ (19)	2 [Note (4)]	4	...	...
$1^{1}/_{2}$ (38) to 6 (150), incl.	$^{3}/_{16}$ (5)	8 (200) [Note (3)]	$2t$ when $t < ^{3}/_{4}$ (19)	2 [Note (4)]	4	...	...
$1^{1}/_{2}$ (38) to 6 (150), incl.	$^{3}/_{16}$ (5)	8 (200) [Note (3)]	8 (200) [Note (3)] when $t \geq ^{3}/_{4}$ (19)	2 [Note (4)]	4	...	...
Over 6 (150)	$^{3}/_{16}$ (5)	$1.33T$	$2t$ when $t < ^{3}/_{4}$ (19)	2 [Note (4)]	4	...	...
Over 6 (150)	$^{3}/_{16}$ (5)	$1.33T$	$1.33T$ when $t \geq ^{3}/_{4}$ (19)	2 [Note (4)]	4	...	...

NOTES:

(1) The following variables further restrict the limits shown in this table when they are referenced in QW-250 for the process under consideration: QW-403.9, QW-403.10, QW-404.32, and QW-407.4. Also, QW-202.2, QW-202.3, and QW-202.4 provide exemptions that supersede the limits of this table.

(2) For combination of welding procedures, see QW-200.4.

(3) For the SMAW, SAW, GMAW, PAW, and GTAW welding processes only; otherwise per Note (1) or $2T$, or $2t$, whichever is applicable.

(4) See QW-151.1, QW-151.2, and QW-151.3 for details on multiple specimens when coupon thicknesses are over 1 in. (25 mm).

(5) Four side-bend tests may be substituted for the required face- and root-bend tests, when thickness T is $^{3}/_{8}$ in. (10 mm) and over.

(Wang & Yang (2016))

Welder Performance Qualification

The main criterion for welder certification is to verify the welder's ability to deposit sound weld metal. The goal of the welding operator's performance qualification test is to determine the welding operator's mechanical ability to operate welding equipment. Each manufacturer or contractor is responsible for the welding performed by his organisation and must conduct the testing necessary to qualify the welders and welding operators he employs in the weldments' construction. Each manufacturer or contractor must keep a record of the welding procedure results as well as the qualifications of welders and welding operators. When choosing a product type (plate/sheet or pipe), the enterprise must take into account the requirements in the standard for welder testing renewal: welders must have completed welding work covered by their approval with reasonable consistency. Acceptance of a welder's ability, in general, comprises approval of any welds done by the welder that are deemed to be easier than the test weld. Welder testing must be performed according to the welding procedure standard or welding data sheet. Welder testing must be overseen by an examiner or an examining test body that has been approved by the contractual authority.When examination is done by an approved certification organisation for people certification, the highest level of approval is awarded.

The essential characteristics listed in QW-360 of ASME Sec IX code for each type of weld limit welding operator qualification. A welder or welding operator may be qualified through volumetric NDE of a test coupon, initial production welding within the limits of QW-304 and QW-305, or bend tests obtained from a test coupon accordance with ASME Sec IX code. When one of the following events occurs, a welder's or welding operator's performance qualification is affected: (a) His credentials for a procedure will expire if he has not welded with it for six months or longer; (b) the welding operator has welded with that process using machine or automatic welding, under the supervision and control of the qualifying manufacturer or contractor or participating organization(s) as identified in QW-300.3 of ASME Sec IX ; this will extend his qualification for an additional 6 months (c) the welder has welded with that process using manual or semiautomatic welding, under the supervision and control of the qualifying manufacturer or contractor or participating organization(s) as identified in QW-300.3; this will extend his qualification for an additional 6 months.

Testing of Weldments

The standard specifies the shape and minimum dimensions of standardised test pieces to be used in connection with the welding procedure. The test pieces must be sufficiently large to ensure that there is sufficient material to conduct away the heat. When impact testing of the heat-affected zone is required, the test pieces must be marked with the rolling direction. All welding of test pieces must be carried out in accordance with the preliminary WPS, and under the same conditions as can be expected in production. Working positions and angles of slope and rotation must be as specified in relevant codes and standards. Tack welding must be included in the test welds if it is to be used in production. Welding and testing must be supervised by an examiner or examining body.

Non-destructive and destructive testing are used as needed in the examination and testing process, as follows: visual examination Ultrasonic or radiographic testing detection of cracks tensile test in the transverse plane impact testing transverse bend test of toughness examination at the macro and micro levels. The position of the test pieces is specified in the standard. Retesting the results cannot be authorised if the welding method test parts do not meet all of the test conditions. A procedure test can be performed again if necessary. If a single test piece fails to fulfil the specifications due to geometrical flaws, two additional test pieces may be chosen for retesting.The thicknesses of the base material, the welding procedure, the welding position, and the type of joint consumables are all important variables. current heat input pre-heat temperature intermediate pass temperature post heat treatment type of welding Documents relating to welding technique tests that have been authorised (WPAR) All information required for approval must be included in the welding and testing records. The examiner must sign the Welding Procedure Approval Records (WPAR). The WPAR model forms are included in the standard. Older welding procedures were put to the test.

Destructive Testing

Destructive tests on welded joints are usually made as part of the approval of a welding procedure or a welderbased on the code requirements.

Commonly used destructive testing methods are: Bend test, Tensile test, Charpy test, Fracture tests, Micro&macro examination and Corrosion test (Potter (1990)).

Non-Destructive Examination (NDE)

In addition to destructive tests, non-destructive examination has also been used for procedure, welder and production welding requirements. Non-destructive Examination is a prominent testing procedure that allows you to evaluate welds and neighbouring areas without compromising their functionality (Matthews (2005)). The basic NDE procedures utilised in industrial practises are listed below and will be explored.

1. Visual Examination (VT)
2. Penetrant testing (PT)
3. Magnetic particle testing (MT)
4. Radiographic testing (RT)
5. Ultrasonic testing (UT)
6. Eddy Current testing (ET)

Visual Examination

It is the most common non-destructive examination. It is used to verify the integrity of most welds. Visual examination is the most widely used non-destructive assessment approach since it is simple to use, quick, and frequently requires no additional equipment other than excellent eyesight and a few inexpensive tools. AWS B1.11, Guide for the Visual Inspection of Welds, contains a comprehensive treatment of visual examination. For several elements of visual weld testing, specific tools are required. The dimensions of the welds are checked using various measurement scales and gauges. Fillet weld gauges come in a variety of shapes and sizes, and they are used all around the world to determine the size of fillet welds. Root holes, material clearance measurements, backing materials, and alignment and fit-up of the work pieces are all checked using measuring gauges/devices. Preheat and interpass temperatures are checked using temperature indicators.In areas with limited access, borescopes, video scopes, flashlights, and mirrors are used. Flexible fiber optic inspection systems have been developed, allowing inspectors to visually evaluate places that are inaccessible to other instruments. If performed before, during, and after welding, visual welding examination can help prevent discontinuities that would otherwise show up in non-destructive testing or cause service failure. The inspector should keep track of all inspections and their outcomes.

Liquid Penetrant Testing

It is a surface testing method for detecting discontinuities in non-porous materials' surfaces. If the discontinuities are clear and open to the surface, penetrant examination provides a sensitive means of detecting and finding them. The approach uses a penetrating liquid dye that is applied on a thoroughly cleaned surface to be inspected and enters the discontinuity by capillary action. The surplus penetrant is removed from the surface and the part is dried after a proper dwell time. The penetrant is then drawn out of the discontinuity using a developer, which functions as a blotter.The presence and position of a discontinuity are shown by a penetrant drawn from an aperture on the surface. The penetrant method is divided into two categories, each of which operates on the same basis. The first utilises a visible dye, whereas the second uses a fluorescent dye that only shows up when exposed to ultraviolet light. To contrast against the white developer background, visible penetrant is usually red in hue. Seeing the discontinuities in normal white light is typically enough. It is relatively affordable to do a penetrant examination. The procedure is straightforward, and operators will have little trouble mastering it. The success of penetrant examination, like most other examination methods, depends on the visual acuity of the inspector.

Magnetic Particle Testing

It is a surface testing method for ferromagnetic materials that is used to reveal discontinuities in or immediately below the surface. Surface or near-surface discontinuities in ferromagnetic materials are located using this NDE method. Magnetic particle examination is based on the idea that a change in material continuity, such as a discontinuity that creates a magnetic field or flux leakage, will distort magnetic lines of force. A weldment can be magnetised by running an electric current across the weld region (direct magnetization) or by putting it in a magnetic field (indirect magnetization) (indirect magnetization). As a method of in-process evaluation, magnetic particle examination can be highly useful. Obtaining assurance of a sound weld before the weld is done may save money on final product repairs. Because of the portability of new lightweight equipment, in-process magnetic particle examination has become more popular. This benefit contributes in the reduction of production time. The MT technique is only applicable to ferromagnetic materials. Even if the welds themselves are sound, welded connections between metals with different magnetic properties can produce unnecessary magnetic particle indicators. After removing slag, spatter, and other contaminants, most weld surfaces are suitable for magnetic particle analysis.

Eddy Current Testing

It also known as inductive testing, is a technique for detecting discontinuities on or near the surface of electrically conductive materials. Testing must also follow established test procedures. Electromagnetic (Eddy current) examination (ET) is a non-destructive electromagnetic testing method in which tiny electrical currents are created in a material and any changes in the flow of these currents are detected by a nearby coil for further electronic processing and presentation.Eddy current techniques have also been used to determine the thickness of coatings and plating on various materials, as well as to measure conductivity, grain size, hardness, and thickness; to identify materials with different compositions, microstructures, magnetic permeability, and heat treatment conditions; and to determine the thickness of coatings and plating on various materials. An alternating current is passed through a coil located near the weld during eddy current investigations. The material is subjected to an alternating magnetic field created by the coil's alternating current. Electrical currents (eddy currents) are created in the material by the alternating magnetic field in the weld.These eddy currents, which change in response to the magnetic field, generate their own magnetic field that interacts with the initial field. To identify any changes in this field interaction, the test coil, or in certain situations a second "pickup" coil, is electrically monitored. Weld discontinuities will change the amount and direction of the eddy currents, allowing the test signal to identify them. Depending on the equipment and application, the signal is shown on an analogue metre, digital metre, cathode ray tube (CRT), X-Y plotter, or strip chart recorder.

Radiography

Internal volumetric discontinuities are primarily revealed through radiography. The thickness of the weld, which should not exceed 50 mm, limits the application of this technology. RT is a non-destructive evaluation technique that uses radiation to penetrate a weld and disclose details about its internal state. When penetrating radiation strikes a weld, part of it is absorbed, some diffused, and some is delivered through the weld to a recording device. Exposures that record a permanent image on photographic film are utilised in the majority of traditional RT procedures today, however other image recording technologies are also used. The films are compared to the required degree of quality.X-ray machines and radioactive isotopes are two types of radiation sources used frequently in weld inspection. Portable, low-energy systems capable of radiographing relatively thin items to gigantic linear accelerators and betatrons capable of radiographing thick steel welds up to 500 mm (20 in.) of steel are all capable of producing X-radiation. Radioisotopes such as Cobalt 60 and Iridium 192 emit gamma radiation. Cobalt 60 can penetrate steel up to a thickness of around 125 mm (5 in.), whereas Iridium 192 can only penetrate steel up to about 75 mm (3 in.). Various amounts of radiation are absorbed by different parts of the weld throughout the radiography process. The quantity of mass represented by the metal and the penetrating power (measured by the energy) of the radiation source are two essential elements that cause differential absorption rates. The mass of a metal is proportional to its thickness and density or composition. The X-ray machine's instrument settings or the characteristic energy level and intensity of the particular isotope utilised for gamma radiography determine the radiation source's penetrating power.The dark and light areas on the radiograph are caused by changes in absorption that occur throughout the exposure procedure.

Ultrasonic Examination (UT)

Internal plane discontinuities are best detected using ultrasonic testing. Ultrasonic examination (UT) is quickly becoming one of the most popular non-destructive testing techniques. Its main use is to discover and characterise internal discontinuities. It's also utilised to spot surface flaws, characterise bond properties, and measure thickness. Welds are often examined using the pulse-echo method with A-scan data display. The examination information is shown on a cathode ray tube (CRT) or digital screen with this system.Probes, amplifiers, and other basic components of the pulse-echo method Surface and subsurface discontinuities are detected using high-frequency sound waves injected into the material. Sound waves are reflected at interfaces after losing some energy (attenuation) as they travel through the material. To identify the presence and location of discontinuities, the reflected sound beam is detected and evaluated. Longitudinal waves (straight beam) or shear waves are used to perform ultrasonic examinations (angle beam). Sound beams with angles of 0, 45, 60, and 70 degrees measured from a line perpendicular to the material surface are the most widely used frequencies, which range between 1 and 5 MHz. A straight beam search unit introduces sound in the form of ultrasonic vibrations into the part perpendicular (normal) to the entry surface in longitudinal beam testing (often used to analyse plate material). A back reflection appears on the display screen when the entry surface and the back surface are parallel. On the display screen, a discontinuity between the front and back surfaces will also be seen. A reference level can be established by measuring the height of the reflection on the display screen from a real or artificial discontinuity of a known size, allowing reflections from discontinuities of unknown sizes to be evaluated.

REFERENCES

Jeffus, L. (2020). *Welding: Principles and applications*. Cengage Learning.

Abdulateef, O. F., Mustafa, F. F., & Salman, S. A. (2010). Prediction the Effect of Flame Cutting Parameters on the Quality of Metal Surface In CNC Flame Cutting Machine Using Artificial Neural Network. *The Iraqi Journal For Mechanical And Material Engineering, 10*(3).

Kumar, R., Dilthey, U., Dwivedi, D. K., & Ghosh, P. K. (2009). Thin sheet welding of Al 6082 alloy by AC pulse-GMA and AC wave pulse-GMA welding. *Materials & Design, 30*(2), 306–313. doi:10.1016/j.matdes.2008.04.073

Matusiak, J., & Pfeifer, T. (2013). The research of technological and environmental conditions during low-energetic gas-shielded metal arc welding of aluminium alloys. *Welding International, 27*(5), 338–344. doi:10.1080/09507116.2011.600040

Kah, P., & Martikainen, J. (2013). Influence of shielding gases in the welding of metals. *International Journal of Advanced Manufacturing Technology, 64*(9-12), 1411–1421. doi:10.100700170-012-4111-6

Norrish, J. (2006). *Advanced welding processes*. Elsevier. doi:10.1533/9781845691707

Kolodziejczak, G. C. (1987). *Pulsed GMAW parameter variation to minimize interference form tack welds* (Doctoral dissertation). Massachusetts Institute of Technology.

Weman, K. (2011). *Welding processes handbook*. Elsevier.

Choudhary, S. K. (2006). *Study of the Effects of Heat Input on Microstructure and Grain Size of Submerged Arc Weilding (Saw)*. Welds by Digital Image Analysis Technique.

Gharibshahiyan, E., Raouf, A. H., Parvin, N., & Rahimian, M. (2011). The effect of microstructure on hardness and toughness of low carbon welded steel using inert gas welding. *Materials & Design, 32*(4), 2042–2048. doi:10.1016/j.matdes.2010.11.056

Dev, S., Ramkumar, K. D., Arivazhagan, N., & Rajendran, R. (2018). Investigations on the microstructure and mechanical properties of dissimilar welds of inconel 718 and sulphur rich martensitic stainless steel, AISI 416. *Journal of Manufacturing Processes, 32*, 685–698. doi:10.1016/j.jmapro.2018.03.035

Wu, D., Hua, X., Huang, L., Li, F., & Cai, Y. (2018). Elucidation of keyhole induced bubble formation mechanism in fiber laser welding of low carbon steel. *International Journal of Heat and Mass Transfer, 127*, 1077–1086. doi:10.1016/j.ijheatmasstransfer.2018.07.107

Yayla, P. A. Ş. A., Kaluc, E., & Ural, K. (2007). Effects of welding processes on the mechanical properties of HY 80 steel weldments. *Materials & Design, 28*(6), 1898–1906. doi:10.1016/j.matdes.2006.03.028

Mohandas, T., Reddy, G. M., & Kumar, B. S. (1999). Heat-affected zone softening in high-strength low-alloy steels. *Journal of Materials Processing Technology, 88*(1-3), 284–294. doi:10.1016/S0924-0136(98)00404-X

Shankar, V., Gill, T. P. S., Mannan, S. L., & Sundaresan, S. (2003). Solidification cracking in austenitic stainless steel welds. *Sadhana, 28*(3-4), 359–382. doi:10.1007/BF02706438

Rezaee, A., Najafizadeh, A., Kermanpur, A., & Moallemi, M. (2011). The influence of reversion annealing behavior on the formation of nanograined structure in AISI 201L austenitic stainless steel through martensite treatment. *Materials & Design*, *32*(8-9), 4437–4442. doi:10.1016/j.matdes.2011.03.065

Lippold, J. C., & Kotecki, D. J. (2005). Welding metallurgy and weldability of stainless steels. Academic Press.

Suutala, N., Takalo, T., & Moisio, T. (1980). Ferritic-austenitic solidification mode in austenitic stainless steel welds. *Metallurgical Transactions. A, Physical Metallurgy and Materials Science*, *11*(5), 717–725.

Dey, H. C., Albert, S. K., Bhaduri, A. K., & Mudali, U. K. (2013). Activated flux TIG welding of titanium. *Welding in the World*, *57*(6), 903–912.

Kou, S. (2003). *Welding metallurgy*. Academic Press.

Avery, R. E., & Tuthill, A. H. (1994). *Guidelines for the Welded Fabrication of Nickel Alloys for Corrosion-resistant Service*. Nickel Development Institute.

Kobayashi, D. Y., & Wolynec, S. (1999). Evaluation of the low corrosion resistant phase formed during the sigma phase precipitation in duplex stainless steels. *Materials Research*, *2*(4), 239–247.

Stokes, V. K. (1989). Joining methods for plastics and plastic composites: An overview. *Polymer Engineering and Science*, *29*(19), 1310–1324.

O'Brien, R. L. (2017). Jefferson's Welding Encyclopedia (18th ed.). American Welding Society.

Funderburk, R. S. (1999). Key concepts in welding engineering. *Welding Innovation*, *16*(1), 1–4.

Wang, Z., & Yang, Y. (2016, July). Comparison of welder performance qualification rules between Chinese regulation and ASME BPVS Sec. IX-2015. In *Pressure Vessels and Piping Conference* (*Vol. 50367*, p. V01BT01A025). American Society of Mechanical Engineers.

Potter, J. M. (1990). *Fatigue and fracture testing of weldments* (Vol. 1058). ASTM International.

Matthews, C. (2005). *ASME engineer's data book*. Amer Society of Mechanical.

KEY TERMS AND DEFINITIONS

Coalescence: Joining of two pieces of metal together by heating process.
Code: A set of rules and specifications for the correct methods and materials.
Defects: A size of discontinuity which is not accepted by code/standard.
Destructive Testing: Testing with destroying the component specimen.
Discontinuity: Deviation in the geometry/size.
Heat Affected Zone: Portion of metal structure affected by heating.
Magnetization: Induce magnetic field in longitudinal & transverse direction.
Non-Destructive Examination: Testing without destroying the component.

Piping: System of pipes used to carry fluids from one place to another including flange, valve, instrument, supports, etc.

Solidification: The process of transformation of a liquid to a solid.

Standard: A set of procedure or system to perform the work.

Weldments: A component fabricated and assembled after welding operation.

Chapter 13
Influence of Additive Manufacturing in Reentry Launch Vehicle

Prasad G.
Dayananda Sagar University, India

Deepika G. N.
Dayananda Sagar University, India

Sahana Jawahar
Delft University of Technology, The Netherlands

Mirzada Mahaz Ahmed
Dayananda Sagar University, India

Sripad Kulkarni S.
Dayananda Sagar University, India

Likitha S.
Dayananda Sagar University, India

ABSTRACT

The two most significant and recently employed techniques in the reusable launch vehicle are computational fluid dynamics and additive fabrication. The design and flow simulation of the reentry capsule were done in this study utilizing computational fluid dynamics in order to optimize the appropriate reentry capsule design. Computational fluid dynamics may be utilized to model real-time supersonic flow during reentry, which aids in forecasting critical shock wave properties. The task was done in Mach 4.3. Furthermore, as compared to prior efforts, the production of shock waves has been delayed in the current experiments. It should be mentioned that the current aerospace sector takes advantage of additive printing to create prototypes for testing reentry launch vehicles. Additive manufacturing is a recently used technique in the reentry launch vehicle. The space industry has been progressively utilizing additive manufacturing in their production in order to optimize their processes and manufacture components that are not achievable with traditional manufacturing methods.

DOI: 10.4018/978-1-7998-9574-9.ch013

INTRODUCTION

Computational fluid dynamics has long been used in the aerospace industry. Additive Manufacturing (AM) has also emerged as a critical technology in aircraft applications. Many businesses are utilising this technology to create components for launch vehicles and satellites that are lighter and less costly when compared to traditional manufacturing processes. Rocket makers have embraced 3D printing because it is a faster and more cost-effective method owing to the reduction in human labour. It also allows for the rapid modification of vehicle designs. In the realm of aeronautical engineering, additive manufacturing has an exponential growth prospect (Schiller G. J. 2015). Topology optimization, or determining which features make a component optimum, is the most difficult element of this procedure. Topology optimization is used to minimise the bulk of a spacecraft component. The Federal Aviation Administration (FAA) is assisting in the development of additive manufacturing by implementing regulations for certification, production, and maintenance. A FAA team has created a draught Additive Manufacturing Strategic Roadmap that outlines the expansion of additive manufacturing through education and research and development. FAA has immediate requirement for an AM certification path. The fuel nozzle was handled as a point solution, and applicants are starting to seek approval for AM parts. The FAA usually uses American Society for Materials (ASM) guidelines to determine whether a plane is safe to fly. Those standards are being developed, but they are currently unavailable. Advocating the creation of a National Additive Manufacturing Roadmap in collaboration (Sun et al 2018). Methodologies for determining durability and damage tolerance are emphasised. Wright-Patterson main concerns were reliability and safety. In high-volume production scenarios, AM parts must be repeatable. Extensive in collaboration and characterise the variability of Ti-6Al-4V. The electron beam powder bed procedure is a technique that uses an electron beam to create a powder bed.

A launch vehicle's stages and payload bay, which houses the payload, payload adapter, avionics, and fairing, make up the vehicle. The propellant tanks, inter tank, inter stage, thrust frame, methods to separate itself from the upper stage or payload, engine(s), thrust-vector control (TVC) system, and propellants are all part of each stage's structure. The reusable launch vehicle optimization tool is a framework for creating a new virtual launch vehicle based on mission-level factors including payload mass and delta-v needs. Based on these requirements, the framework assembles a full rocket and calculates the characteristics of all internal subsystems (e.g., fuel tanks) based on the projected loads. The tool can select from a variety of propellant combinations, a variable number of stages, and various engines per stage, in general. The launch vehicle's first stage could be reusable and land through retro propulsive landing. The general programme flow is described in the first part of the next section, with an emphasis on a two-stage partially reusable launch vehicle. The second section uses a genetic algorithm to discover the best launch vehicle configuration for given mission criteria and restrictions. The same method had previously been used to investigate various engine cycles.

Additive Manufacturing (AM) technology will be used selectively by established companies with large launch vehicles, and a greater percentage of the vehicle's components will be made using this technology. With small and medium satellite launchers, New Space companies will be more daring and innovative, to the point of almost making the entire vehicle out of the technology. Costs and lead times will be reduced to new lows, and the vehicles will be more reliable as a result. In order to remain competitive, companies that do not make use of this technology risk falling behind their more technologically advanced competitors.

General aviation, commercial and military aircraft, missiles, satellites and space launch and in-space systems are part of the aerospace industry. Aerospace sector revenue will fall from $342.2 billion in 2019 to $298 billion in 2020 as a result of the recent COVID-19 pandemic. Restrictions on commercial aviation during the ongoing COVID-19 pandemic were a primary factor in this decline. As a result of this, the aerospace industry is forecasted to grow to $430.87 billion in revenue in 2025. Increased global military spending, high market activity in the space sector and extensive research and development have all contributed to this growth, which is expected to continue for many years to come.

Functional performance, lead time reduction, lightweighting, complexity, cost management and sustainment are all intertwined in aerospace manufacturing. In order to come up with the best possible design solution, each of these objectives must be taken into account in its entirety. In terms of relative importance, these goals will vary depending on the specific aerospace application, but in general, they can be summarised as:

1. Safety-critical components must be delivered in small production volumes with a relatively tight delivery schedule for the aerospace sector. Consequently, lead time reduction in the aerospace sector is extremely important because it facilitates product certification and allows for the design of high value components to remain flexible.
2. The performance of aerospace structures can be improved through the use of lighter materials. As a result, reducing the mass of an aerospace system has a direct correlation to improving economic and technical performance, including lower fuel costs, lower emissions, increased payload capacity, and longer range.
3. Economic constraints limit the scope of lightweighting, since only a limited amount of money can be allocated to specific design goals. These three objectives are intertwined and the allowable system cost generally rises for solutions that achieve lightweighting or lead time reduction objectives. There are numerous ways in which a company can save money by reducing the risk associated with certification and maintenance.
4. Regulatory bodies must be convinced that AM systems are well understood and can be designed and inspected in a repeatable manner so that reliability and safety expectations can be met. This is a major challenge for metal additive manufacturing (AM) applications in aerospace Depending on the level of criticality of the proposed AM system, these certification requirements vary. Conventional manufacturing standards must be integrated with newer, emerging standards for additive manufacturing (AM).
5. Maintaining aerospace structures can be a significant challenge due to their high complexity and low volume. Part availability for ageing aircraft, remanufacturing and repair, and the re-certification of existing aircraft for alternate uses are some of the challenges faced by the aviation industry.
6. Systems and subsystems are often composed of hundreds, or even thousands of parts, all of which work together to achieve a desired functional performance in a given environment. It's not uncommon for a plane's entire flight system to have millions of components. These components have a high level of complexity in order to provide functional performance (structural, flow, thermal, reliability, durability, compatibility, etc.) as well as a weight reduction.

As a result of the constant effort to reduce costs, lead times, and the mass of flight components, high-performance materials and increasingly complex designs are being used in increasingly complex designs. To meet commercial or mission requirements, this must be done at a reasonable cost and in a timely manner. However, AM has had a profound impact on design and manufacturing for many decades, and will continue to do so for many more years to come. AM's digital transformation, dubbed "Industry 4.0," is expected to boost the aerospace sector's market to $3.187 billion by 2025, growing at a CAGR of 20.24% on average.

Complex parts previously unmanufacturable can be improved by incorporating internal features like conformal cooling channels on combustion chambers or turbine blades, for example, using optimization techniques for mechanical, thermal and other design approaches. Aerospace applications are increasingly turning to additive manufacturing (AM) due to its ability to produce parts in a fraction of the time required by traditional methods.

AEROSPACE INDUSTRY ADDITIVE MANUFACTURING

Functional performance, lead time reduction, light weighting, complexity, cost management, and sustainment are all interacting technical and economic objectives in the aircraft industry. Each of these goals has a strong relationship with the others, thus concerns from each component must be carefully addressed when choosing the best design solution. Although the relative relevance of these objectives varies depending on the individual aerospace application, they can be summarised as follows:

- The aerospace industry necessitates the supply of safety-critical components in small production volumes with a reasonably rigid delivery schedule in order for them to function in their intended environment (functional performance). As a result, lead time reduction is extremely important in the aerospace industry, as it allows for speedy product certification while maintaining design flexibility for high-value components.
- The technical and economic performance of aircraft structures is influenced by their weight. Aerospace structures' technical performance and permitted mission-defined payload are physically constrained, which means that reducing system mass leads to improved economic and technical performance, such as lower fuel costs, lower emissions, greater payloads, and increased range.
- The economic limits of cost management, in which a defined financial resource is available for a specific design purpose, temper the light weighting goal. Cost management, light weighting, and lead time reduction goals are all intertwined, with the permissible system cost increasing for solutions that meet light weighting or lead time reduction goals. Cost management encompasses all elements of component use, including certification and maintenance, as well as cost-saving options provided by lower certification risk.
- The hurdle of certification, which requires regulatory bodies to be confident that metal additive manufacturing (AM) systems are fundamentally well understood and can be repeatable designed and inspected in order to meet reliability and safety expectations, is a critical challenge for metal additive manufacturing (AM) applications in aerospace. These certification requirements differ depending on whether the proposed AM system is safety-critical, mission-critical, or otherwise. Practical certification necessitates alignment with both existing and emerging standards for traditional manufacturing and additive manufacturing processes.
- Aerospace structures are often high-complexity, low-volume systems, and the challenges of long-term maintenance can be significant. Part availability for ageing aircraft, remanufacture and repair, and re-certification of existing aircraft for alternate applications are all issues in maintenance.

- Designing systems and subsystems with hundreds or thousands of elements to achieve the desired functional performance in the intended environment is inherently complex. These systems are integrated into the overall flight system, which can have millions of parts. The complexity of these components is high to allow for both functional performance (structural, flow, thermal, reliability, durability, compatibility, and so on) and weight reduction.

High-performance materials are utilised with increasingly complicated designs in the quest to increase efficiency continually through cost, lead time reduction, and attempts to reduce the bulk of flight components (Fujimoto et al 2018). To meet commercial orders or mission criteria, this must be done at a reasonable cost and on time. Traditional manufacturing technologies and tactics have been established over decades to meet these aerospace design objectives for a variety of application types; however, additive manufacturing is having and will continue to have a significant impact on design and manufacture. With a compound annual growth rate (CAGR) of 20.24 percent, this AM digital transformation, also known as Industry 5.0, will raise the aerospace sector's market size to $3.187 billion by 2025. Over the last decade, research into additive manufacturing in aerospace has grown at an exponential rate. 1 Aside from scholarly publications, there is a wealth of relevant material in the form of technical papers, popular literature, and publicity articles from commercial aerospace suppliers, with technical specifics occasionally withheld for business reasons.

The aerospace sector has benefited greatly from the use of additive manufacturing, particularly in the production of launch vehicles and satellites. Because of the intricacy of geometry and size of some components, manufacturing aircraft components has always been a difficult process (Chen et al 2015). Manufacturing and machining such complicated components using traditional methods like casting and forging has increased not only the cost and time of production but also the raw material usage. Because of its various advantages, the use of additive manufacturing for complicated components such as those of launch vehicles has proved advantageous. The laser sintering method reduces weight by 40 to 60 percent. Every gram of weight saved in the production of aircraft components saves aerospace firms hundreds of dollars. Vehicle component weight reduction has also resulted in decreased fuel usage and, as a result, fewer carbon dioxide emissions. The usage of additive manufacturing in the fabrication of aeroplane parts has been used by prominent aerospace manufacturers. Norsk Titanium 3D printed components have been placed on the Boeing 787 Dreamliner. Airbus Defence and Space creates 3D printed satellite components for the International Space Station.

Figure 1. Examples of large scale directed energy deposition integral channel nozzle1.52 m diameter and 1.78 m height with NASA HR-1 alloy nozzle
(Courtesy: NASA)

Space Industry is aiming for future additive expansion and increased production rates. The new Additive Manufacturing (AM) in particular has been anticipated to assist all AM teams in focusing on design simplification and subassembly consolidation. The goal is to improve scheduling, output, quality defect elimination, cost reduction, and technician-led safety efforts. Figure 1. Examples of Large Scale Directed Energy Deposition integral channel nozzle. Guarantee in terms of printing optimization, notably in terms of zero machine downtime, as well as effective and speedy post-processing, highlights the fact that Space Industry is looking at significant production through AM (Hollis Borrelli 2012). The Additive Manufacturing job has been keep in progress to a minimum while ensuring that all internal AM teams have sufficient consumables on hand (Natarajan et al 2013).

Table 1. Detailed literature survey on additive manufacturing and its potential applications

Authors	Inference	Application Areas	Year
Godec M.,et al.	For future space applications, hybrid additive manufacturing of Inconel 718 is needed.	Spacecraft	2021
D'Emilia G.,et al.	Selective laser melting: a case study of an aerospace part's uncertainty assessment.	Aerospace, Case Study	2020
Fereiduni E., Ghasemi A., Elbestawi M.	Aerospace industry applications for selective laser melting of aluminium and titanium matrix composites.	Aerospace	2020
Gao L.,et al.	Manufacturing Process and Properties of 5356 Aluminum Alloy Rocket Booster Module Transition End Frame are examined in this study.	Rocket	2020
Gu D.,et al.	Metal Aerospace Components Produced by Laser Additive Manufacturing.	Aerospace	2020
Hehr, A.,et al.	Ultrasonic Additive Manufacturing Liquid Cold Plate Heat Exchangers Hot Isostatic Pressed.	Case Study, Heat Transfer	2020
Kuntanapreeda S., Hess D.	Increasing the use of 3D printing in the design and production of launch vehicles.	Spacecraft, Design	2020
Oyesola M.O.,et al.	Maintenance repair overhaul in the aerospace industry can now be supported by a hybrid-additive manufacturing cost model.	Aerospace, Repair	2020
Shi G.,et al.,	Thermo-elastic topology optimization was used to design an aerospace bracket that was manufactured using additive manufacturing.	Aerospace, Design, Topology Optimization	2020
Stolt R., Elgh F.	A new design for selective laser melting is being introduced in the aerospace industry.	Aerospace, Design	2020

REUSABLE LAUNCH VEHICLE

The creation of reusable launch vehicles is one of the accomplishments in the area of Aerospace Technology as shown in Figure 2. Aerodynamic heating and high structural stresses are the two primary problems that a reusable launch vehicle faces during re-entry (Kaushikh et al 2018). During the reentry phase, the vehicle is protected by an effective heat shield. The vehicle's form and aerodynamics are critical in decreasing aerodynamic heating. Good mono-stability qualities are another key feature that the reentry vehicle. The mono stability properties of the reentry capsule have been improved utilising flap and strake. (Rose et al., 2014) have investigated the aerodynamic characterisation of reentry vehicles in hypersonic flow conditions using both computational fluid dynamics and numerical techniques. Aerodynamic heating occurs during the reentry phase as a result of air friction on the vehicle's surface. The reentry vehicles have ideally blunt bodies, resulting in bow shock at supersonic speeds.

Figure 2. Orion CEV crew module

These bow shocks aid in the dissipation of heat generated by friction into the surrounding environment. The Apollo Command Module's reentry aerodynamics and heat transfer rates must be thoroughly examined. Used a computer approach to forecast supersonic heating of reentry capsules. The propagation of sound pressure is another important parameter that has been measured during vehicle reentry. The prediction of the reentry vehicle's aero acoustics aids in mitigating structural vibrations caused by sound wave propagation (Murugan et al 2021).

An increasing number of manufacturers are turning to additive manufacturing, or 3D printing, as it's more commonly known. Companies in the space industry are taking advantage of this technology in novel ways to gain a competitive edge. When it comes to making highly customised parts for the space industry, additive manufacturing (AM) is ideal.

Building components by depositing material layer by layer using digital 3D design data is known as additive manufacturing. Subtractive manufacturing, on the other hand, utilises CNC machines to remove material from a solid block before reassembling it. When it comes to printing, subtractive techniques like milling and additive deposition can be combined in a hybrid printer.

There is less labour, more automation, and a shorter time to manufacture components with 3D printing, which is why it is becoming increasingly popular. There are some cases where the reduction in manufacturing lead times can be more than 50 percent. For space launch vehicles, AM reduces the weight of components, thereby increasing payload capacity. AM reduces the number of parts in a system, which increases the system's overall reliability. There are fewer issues to deal with because there are fewer parts to put together. Since printing is more automated, labor-related issues have decreased significantly. Supplier management is also simplified because there are fewer suppliers. Manufacturing costs and lead times are reduced as a result of these changes.

Directed energy deposition (DED) and powder bed fusion (PBF) are two of the seven process categories included in the international standard ISO/ASTM 52900 for additive manufacturing, both of which allow for the production of metal components with near-full density. Due to the fact that many of these processes are still in the early stages of development, they are not discussed in detail (TRL). An overview of the processes, resulting structure/microstructure and properties of manufactured metals is presented in. Because of its fine feature sizes and large build volumes, L-PBF (laser powder bed fusion) is becoming more popular in the industry. When the manufacturing process is optimised, the resulting components can have mechanical properties that are superior to those of conventionally produced goods.

Design for additive manufacturing: four essential principles Rocket nozzles can be made using additive manufacturing. These are:

1. Design for the correct additive manufacturing process
2. Design for minimal material usage
3. Design for improved functionality
4. Design for part consolidation

ANALYSIS AND DESIGN

Through the use of high-speed, computer-based computational fluid dynamics analytical tools, precise determination of flow-induced pressure and temperature loads can be achieved well before real hardware testing. The flow of fluids and gases may be properly modelled using CFD techniques, and designs can be created in an electronic three-dimensional computer-aided design format. These kinds of computer-based simulations may be completed so quickly that designs can be altered in real time, even before the hardware is built. CFD techniques are being successfully used as diagnostic tools to provide insight into existing rocket engine components and to develop optimum designs of liquid rocket engine pump components such as impellers, diffuser vanes, and shrouds; turbine components such as turbine blades, turbine staging, volutes, and turbine wheels; and launch vehicle base thermal protection configurations.

The geometry of the Crew Exploration Vehicle (CEV) to forecast the performance of the reentry capsule in this work. The CFD study was done because experimental prediction using a wind tunnel is not only expensive but also takes a long time, whereas supersonic flow can be reproduced in the test section in a matter of seconds. (Anderson, 2006) documented the problems associated with hypersonic vehicles. Figure 3 shows the geometry generated using ANSYS and carefully meshed with triangular and quad components. The reentry capsule's boundary conditions were anticipated to be at 15 km altitude above mean sea level, with a gauge pressure of 12043 Pa and a temperature of 205.65. The analysis is performed at Mach 4.3 supersonic speed. For the specified flow parameters, the shock wave development over the reentry capsule is also modelled. Under the gradient criterion, the Least Squares cell-based technique is employed in the solution methods to get a linearly changing solution. The courant number, which controls the time step throughout the flow, is set to 5.

Figure 3. CEV geometry and meshing for simulation

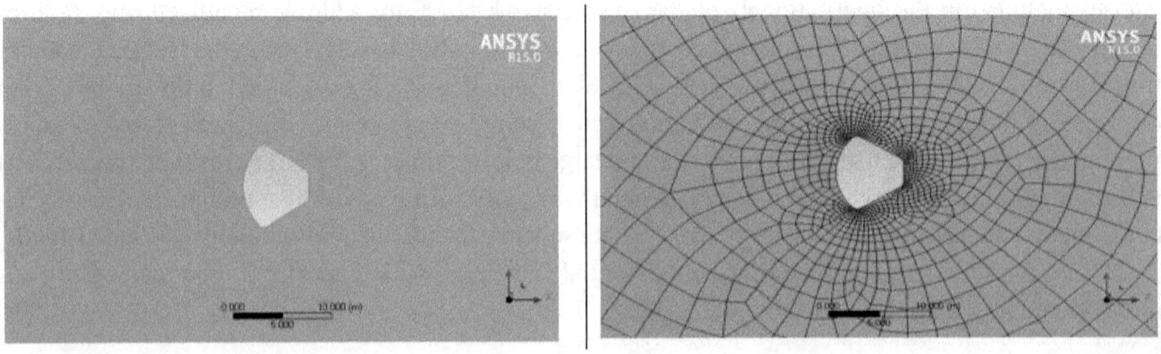

OUTCOMES AND DISCUSSION

The CFD simulation of the CEV at Mach 4.3 produced a variety of findings, including the contours of static temperature, coefficient of pressure, and velocity shown in figures 4 and 5. The creation of a bow shock at the front of the vehicle can be seen due to the CEV's blunt design. NASA is collaborating with Aerojet Rocketdyne to enhance 3D printing and metal additive manufacturing technology for liquid rocket engines in landers and on-orbit stages/spacecraft. The RDT team at NASA's Marshall Space Flight Centre in Huntsville, Alabama, is designing and manufacturing innovative and lightweight combustion chambers, nozzles, and injectors that will incorporate automated robotic deposition 3D-printing technologies such as cold spray deposition, laser wire direct closeout, laser powder bed fusion, and laser powder directed encapsulation (Liu et al 2015). Through hot fire testing, the goal is to enhance these processes utilising weight-optimized materials to demonstrate operability, performance, and reusability. The flow characteristics behind the shock wave change as a result of the bow shock. Figure 4 shows that the static temperature and pressure coefficient behind the shock wave have increased dramatically, but velocity has decreased substantially. The static temperature close to the capsule's walls has risen from 205K to 1220K as a result of shock wave generation. The front and back surfaces of the capsule reached a maximum static temperature of 1220K. A tiny fraction of the flow in front of the capsule has a pressure coefficient of $6.35e^{+06}$, but the majority of the flow in the frontal area has pressure coefficients ranging from $300e^{+06}$ to $500e^{+06}$.

Figure 4. Static temperature and pressure coefficient contours at Mach 4.3

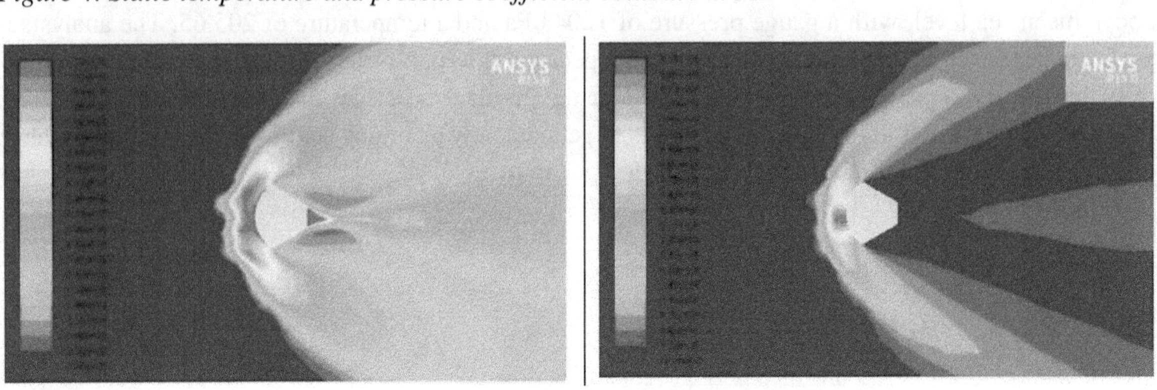

Figure 5 depicts the velocity contours at stagnation velocity situation near to the CEV's wall. Furthermore, the flow towards the front of the capsule has slowed to velocities ranging from 215m/s to 574m/s. The flow begins to increase after a certain distance behind the capsule, which might be due to the tapering shape of the CEV.

Figure 5. Velocity contours at Mach 4.3

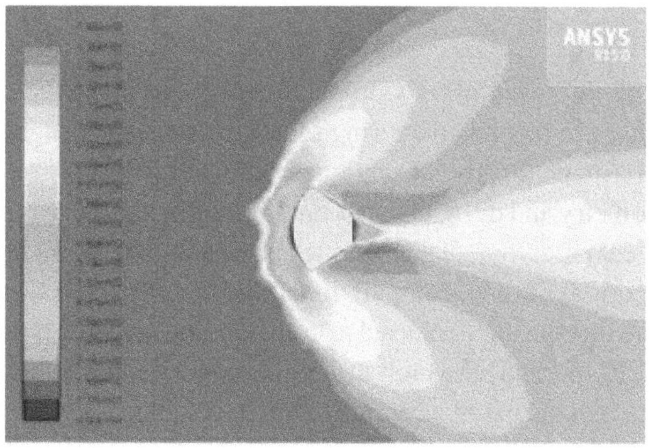

The graphs in Figure 6 illustrate how the Mach number and static temperature change with the location of the CEV. As previously stated, the Mach number is near to zero at the frontal position due to stagnation circumstances and is observed to slowly grow over the surface of the capsule. The highest static temperature has been displayed at the capsule's upstream and downstream positions. The abrupt drop in static temperature at the -1 location is caused by flow separation at that point of the capsule. The book Hypersonic and high-temperature gas dynamics by (Anderson, 2006) has been used in this research for data references for high speed flow.

Figure 6. Plots of the CEV's position versus Mach number and static temperature

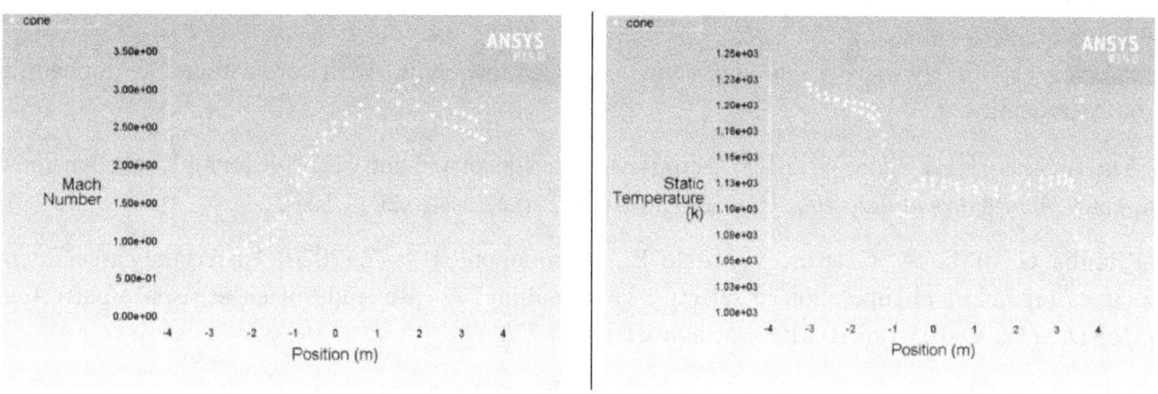

The use of computer-based computational fluid dynamics methods will help construct dependable, high-performance designs of space launch vehicles and their components, as well as eliminate preliminary development testing. CFD can be used to mimic anomalies that occur in actual space vehicle testing or flights to better understand the anomalies and how to remedy them, in addition to design verification and optimization. As a result, the space ship and propulsion system are more reliable and trouble-free.

CONCLUSION

The launch vehicle may be improved using Computational Fluid Dynamics, and the Additive Manufacturing method can be used in the manufacturing sector of Aerospace engineering due to its amazing benefits. The use of additive manufacturing techniques in the fabrication of reusable launch vehicles can be used only when the geometry and design have been thoroughly optimised using Computational Fluid Dynamics. Because physical modelling of supersonic and hypersonic flows in wind tunnels can only be done for extremely brief periods of time, computational flow simulation is the best option. Based on the research, we can determine that the material used to construct the reusable launch vehicle should be able to resist temperatures over 1220K and have high structural stability. Furthermore, additive manufacturing may be used to create the components of a reusable launch vehicle.

ACKNOWLEDGMENT

We acknowledge all the staff members of Aerospace Engineering, Dayananda Sagar University, Bangalore, India for the constant support and encouragement to carry out the research. I also thank Dr. K.N.B. Murthy (Vice Chancellor), Dr. A. Srinivas (Dean), Dr. BVN Ramakumar (Head of the Department), Dayananda Sagar University, Bangalore, India. I also thank my family members, especially wife Mrs. J. Kavya for the understanding and encouragement to preserve the research work. This research received no specific grant from any funding agency in the public, commercial, or not-for-profit sectors.

REFERENCES

Anderson. (2006). *Hypersonic and high-temperature gas dynamics*. American Institute of Aeronautics and Astronautics.

Chen, B., Zhan, H., & Zhou, W. (2015). Aerodynamic design of a re-entry capsule for high-speed manned re-entry. *Acta Astronautica*, *106*, 160–169. doi:10.1016/j.actaastro.2014.10.036

D'Emilia, G., di Ilio, A., Gaspari, A., Natale, E., & Stamopoulos, A. G. (2020). Uncertainty assessment for measurement and simulation in selective laser melting: A case study of an aerospace part. *Acta IMEKO*, *9*(4), 96–105. doi:10.21014/acta_imeko.v9i4.720

Fereiduni, E., Ghasemi, A., & Elbestawi, M. (2020). Selective laser melting of aluminum and titanium matrix composites: Recent progress and potential applications in the aerospace industry. *Aerospace (Basel, Switzerland)*, *7*(6), 77. doi:10.3390/aerospace7060077

Fujimoto, K., Negishi, H., Nakamura, R., & Nakakita, K. (2018). *Aero-Acoustics CFD Prediction for Re-entry Capsule at Subsonic to Supersonic Regine*. AIAA. doi:10.2514/6.2018-0520

Gao, L., Yu, S., Yu, R., He, T., & Shi, Y. (2020). Study on Arc additive manufacturing process and properties of 5356 aluminum alloy rocket booster module transition end frame. *Jixie Gongcheng Xuebao*, *56*(8), 28–36. doi:10.3901/JME.2020.08.028

Godec, M., Malej, S., Feizpour, D., Donik, Č., Balažic, M., Klobčar, D., Pambaguian, L., Conradi, M., & Kocijan, A. (2021). Hybrid additive manufacturing of Inconel 718 for future space applications. *Materials Characterization*, *172*, 172. doi:10.1016/j.matchar.2020.110842

Gu, D., Zhang, H., Chen, H., Zhang, H., & Xi, L. (2020). Laser additive manufacturing of high-performance metallic aerospace components. *Chinese Journal of Lasers*, *47*(5).

Hehr, A., Norfolk, M., Wenning, J., Sheridan, J., Ahlfors, M., Mastropietro, A. J., Maghsoudi, E., & Roberts, S. N. (2020). Hot isostatic pressing of ultrasonic additive manufacturing liquid cold plate heat exchangers. *Journal of Spacecraft and Rockets*, 1–5.

Hollis, B. R., & Borrelli, S. (2012). Aerothermodynamics of blunt body entry vehicles. *Progress in Aerospace Sciences*, *48–49*, 42–56. doi:10.1016/j.paerosci.2011.09.005

Kaushikh, K., Arunvinthan, S., & Pillai, S. N. (2018). Aerodynamics and Aerothermodynamics of undulated re-entry vehicles. *Acta Astronautica*, *142*, 95–102. doi:10.1016/j.actaastro.2017.10.024

Kuntanapreeda, S., & Hess, D. (2020). Opening access to space by maximizing utilization of 3D printing in launch vehicle design and production. *Appl. Sci. Eng. Prog.*, *14*(2), 1–3. doi:10.14416/j.asep.2020.12.002

Liu, G., Li, D., Xiang, J., & Luo, W. (2015). Design, Modeling and Analysis of a Sharp-edge Hypersonic Stealthy Re-entry Vehicle. *Procedia Engineering*, *99*, 163–167. doi:10.1016/j.proeng.2014.12.521

Murugan, K., & Guruprasad, B. (1964). Varatharaju Perumal, and Arul T G,(2021). Renewable Energy Based Grid System Using UPQC for PQ Improvement. *Journal of Physics: Conference Series*, 062094.

Natarajan, N., Arunachalam, R. M., & Thanigaivelan, R. (2013). Experimental study and analysis of micro holes machining in EDM of SS 304. *International Journal of Machining and Machinability of Materials*, *13*(1), 1–16. doi:10.1504/IJMMM.2013.051905

Oyesola, M. O., Mpofu, K., Mathe, N. R., & Daniyan, I. A. (2020). Hybrid-additive manufacturing cost model: A sustainable through-life engineering support for maintenance repair overhaul in the aerospace. *Procedia Manufacturing*, *49*, 199–205. doi:10.1016/j.promfg.2020.07.019

Rose, J. B. R., Saranya, P., & Benifa, J. V. B. (2018). Investigation of computational flow fields and aero-acoustic characteristics over a re-entry command module. *Proceedings of the Institution of Mechanical Engineers. Part G, Journal of Aerospace Engineering*, *232*(3), 532–544. doi:10.1177/0954410016682272

Schiller, G. J. (2015).Additive manufacturing for Aerospace. *IEEE Aerosp. Conf. Proc.*, 1-8.

Shi, G., Guan, C., Quan, D., Wu, D., Tang, L., & Gao, T. (2020). An aerospace bracket designed by thermo-elastic topology optimization and manufactured by additive manufacturing. *Chinese Journal of Aeronautics*, *33*(4), 1252–1259. doi:10.1016/j.cja.2019.09.006

Stolt, R., & Elgh, F. (2020). Introducing design for selective laser melting in aerospace industry. *Journal of Computational Design and Engineering*, *7*(4), 489–497. doi:10.1093/jcde/qwaa042

Sun, D., Qu, F., & Yan, C. (2018). An effective flux scheme for hypersonic heating prediction of re-entry vehicles. *Computers & Fluids*, *176*, 109–116. doi:10.1016/j.compfluid.2018.05.028

Chapter 14
Study of Different Additive Manufacturing Processes and Emergent Applications in Modern Healthcare

Ranjit Barua
https://orcid.org/0000-0003-2236-3876
Center for Healthcare Science and Technology, Indian Institute of Engineering Science and Technology, Shibpur, India

Amit Roychowdhury
Indian Institute of Engineering Science and Technology, Shibpur, India

Pallab Datta
National Institute of Pharmaceutical Education and Research, Kolkata, India

ABSTRACT

The additive manufacturing process denotes modern manufacturing technologies that create a practical model from digital data. These days, the 3D (three-dimensional) printing technology signifies a great prospect to support medicinal and healthcare firms to produce new definite medicines, allowing quick manufacture of medicinal transplants, and moving the approach that specializes surgeon and physician strategy measures. For example, currently, in the practice of modern medical treatment, patient-specific anatomical models (3D-printed) are used. Soon, functional implantable organs by 3D (three-dimensional) printed process will possibly be offered, decreasing the queue time and growing the total of lives protected. This modern manufacturing technology for healthcare and medical is still required to a great extent of work in development; however, it is applied in numerous dissimilar habits in a medicinal and therapeutic area that previously reeled below a huge burden concerning optimum presentation.

DOI: 10.4018/978-1-7998-9574-9.ch014

INTRODUCTION

The additive manufacturing process is basically a rapid prototyping procedure, which is also denoted as a three-dimensional (3D) fabrication or printing process (Phan et al., 2016) (Mondschein t al., 2017) (Salmi et al., 2013). Additive manufacturing is also known as the layer by layer process, which procedures a CAD (Computer-Aided Design) model to three-dimensional print desire objects (Kholgh et al., 2020) (Barua et al., 2019) (Bigham et al., 2020). Figure 1 shows the application of additive manufacturing in health care and biomedical engineering field. Additive manufacturing is appropriate to manufacture complex designs with tough to desire geometrical considerations from different materials like metals, ceramics, and polymers (Salmi et al., 2013) (Youssef et al., 2017). Furthermore, additive manufacturing products are comparatively cost-efficient and appropriate for customizing variation (Zhu et al., 2016) (Fantino et al., 2016).

Figure 1. Application of additive manufacturing in biomedical field

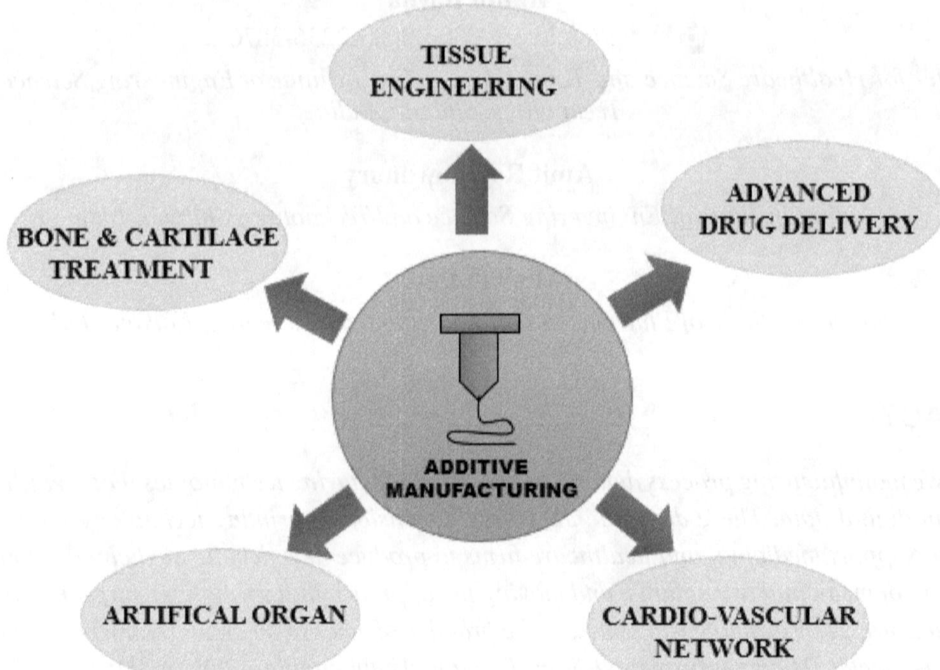

The additive manufacturing technology was primarily familiarized in the 1980s and had been verified in miscellaneous medical and healthcare or therapeutic uses, which contains in different biomedical areas, the scaffolds for tissue engineering of 3D bio-printing methods, research laboratory amenities, and treatment as well as drug delivery methods (Youssef et al., 2017) (Rastogi et al., 2019) (Simpson et al., 2008) (Diment et al., 2017). Additionally, additive manufacturing is engaged to desire healthcare strategies and medicines and to spread them to organic tissues to distinguish beneficial reactions (Whitley et al., 2017). Additive manufacturing practices imaging file arrangements as the geometric descriptions to the model of three-dimensional profiles by several software for accurate uses and allows surgeons to make functioning models for design and operating replications (Dahake et al., 2020). Furthermore, the replicas can be applied to clarify precarious and difficult surgical techniques to patients and their relatives. Figure 2 shows the workflow of additive manufacturing in modern biomedical area. A different therapeutic presentation of additive manufacturing is anatomic customization for splint, orthosis, and prostheses (Moreno et al., 2019) (Yuan et al., 2017).

Figure 2. Workflow of additive manufacturing

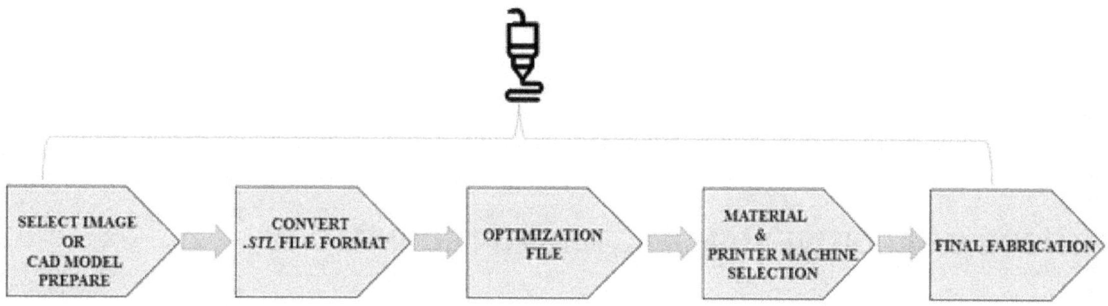

These uses are typically externally applied and non-invasive. Another use of additive manufacturing technologies is for manufacturing hardware and devices for healthcare instruments with the aim of enhancing new abilities or refining their proficiency. These kinds of uses are someway in interaction with bio-fluids or some tissues because of their precise uses (Yahya et al., 2020). The other additive manufacturing application is in bio-engineering implants. As implants are planned for long time usage, they are implanted in the patients' body and are formed by either straight like hip implant, dental implant or indirect additive manufacturing techniques like tissue scaffold. The latest additive manufacturing techniques in the medicinal or biomedical field are recognized as bio-manufacturing which allows additive manufacturing techniques of culture media (Jammalamadaka et al., 2018) (Gopinathan et al., 2018) (Nam et al., 2018) (He et al., 2016). The uses at this time are biocompatible configurations for example tissue scaffolds are applied for cell culture-oriented with a defensive act in contrast to outer forces (Derakhshanfar et al., 2018). Through an extensive variety of investigations that has been issued in the additive manufacturing area, the absence of application-based perceptions in the categorizing of additive manufacturing technologies in the biomedical and tissue engineering field is noticed. Therefore, the current chapter objectives to arrange for a summary of the present additive manufacturing developments along with the upcoming trend of these manufacturing techniques in healthcare and biomedical engineering. Plans and milestones of additive manufacturing technology especially three-dimensional (3D) printing in medical and healthcare above the former two periods are taken in Figure 3. One appropriate feature of additive manufacturing technology is healthcare modeling for the research and teaching of medical practitioners for instance pre-planning and post-planning (Tuomi et al., 2014) (Cui et al., 2020).

Figure 3. Milestones of additive manufacturing technology till date

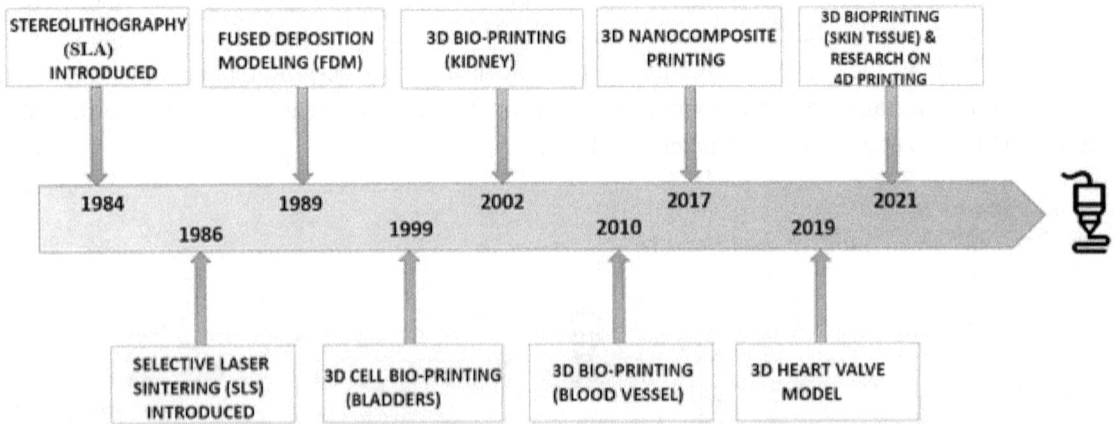

METHODS OF ADDITIVE MANUFACTURING TECHNOLOGY

Additive manufacturing is a modern technique as it thrusts the frontline of engineering especially in the manufacturing sector to an innovative strategy outlook, for example, the aptitude to profile geometries or structures that cannot be designed with any further conventional method (Lv et al., 2021) (Pantermehl et al., 2021). Additive manufacturing has nowadays exposed fruitful uses in numerous areas, for instance, the biomedical area in which it delivers a comparatively rapid and real technique to explain even difficult medicinal cases (Xie et al., 2020). From this viewpoint, the intention of this chapter is to demonstrate the additive manufacturing techniques presently applied in the biomedical area and their aids in consort with existing. The analysis highlights changes in procedures, resources, and strategy of this manufacturing technology applied in biomedical uses (Sigaux et al., 2019) (Jessop et al., 2017). Positive case reports are offered to highlight the potentiality of additive manufacturing methods. Currently, additive manufacturing methods that are now applied for medical uses can be categorized into two groups allowing to the raw material, i.e. metal-based like EMB (Electron Beam Melting), LPBF (Laser Powder Bed Fusion) and non-metal like SLA (Stereolithography), SLS (Selective Laser Sintering) FDM (Fused deposition modeling), 3D Bio-printing (Ingole et al., 2009) (Frazier et al., 2014) (Ataee et al., 2017).

3D (Three-Dimensional) Printing

Among the dissimilar manufacturing approaches that are currently recognized by the manufacturing companies, the 3D (Three-dimensional) printing process is also an additive manufacturing technique (Kessler et al., 2020). It is a procedure from side to side in which a 3D solid item, almost of several forms, is created beginning from digital data. Biomedical three-dimensional printing was formerly a motivating technique (Tarassoli et al., 2018) (Manita et al., 2021). This knowledge has numerous uses, and the fastest increasing advance in the biomedical area has been indicted by the beginning of the three-dimensional printing techniques itself (Tan et al., 2021). The advance of the three-dimensional geometry through the treating of the data images coming from an MRI/CT scan (Calusi et al., 2021), the optimization of the data for the practical printing (Paštyková et al., 2018), and the suitable selection of the three-dimensional printer and associate materials also. This data signifies the direction for the following fabrication, "slicing" that basically a computer design object into cross-sections. This design is then referred to as a three-dimensional printer machine, which fabrications the objective by beginning at the initial layer and structure a sequence of layers on the upper till the item is fabricated by the materials that are required for its arrangement (Kessler et al., 2020). Finally, the desired model (patient-specific) is fabricated from MRI/CT imaging data (Parr et al., 2019). In this technique, additive manufacturing has the prospective to considerably advance the investigation information and the abilities of the upcoming surgeons, the connection between surgeon and patient, growing the level of acceptance of the disease and treatment procedures (Brito etal., 2016). These days, various printing materials and methods are available so as to superior mimic the patient's anatomy (do Ó Silva et al., 2017). Several printing materials are inflexible and for that reason not optimal for elasticity and flexibility, nothing like living tissue. Therefore, there are currently resources capable of solving the gap between the actual anatomy and the replicated one, particularly making an allowance for the soft tissue. The additive manufacturing in the healthcare and biomedical area and strategy requirements to consider external the standard for altering the modern medical system (Marti et al., 2019) (Tam et al., 2018).

FDM (Fused Deposition Modeling)

Fused deposition modeling is an additive manufacturing procedure that is appropriate to the material extrusion group (Mwema et al., 2020) (Dey et al., 2019). Fused deposition modeling printers construct the objects layer-by-layer applying a polymer filament (basically thermoplastic) which is heated by a hot extruder and makes them a liquid material (Figure 4), after then the liquid material is deposited on the surface of the printing bed in consort with a computer-controlled route (Minetota et al., 2018). The fused deposition modeling filaments are divided into two regular dimensions with a diameter of 2.85 mm and 1.75 mm (León-Cabezas et al., 2017) (Podroužek et al., 2019) (León-Cabezas et al., 2017).

As per the extrusion nozzle size, the layer width can differ from nearby 50-500 µm (Podroužek et al., 2019). Though for all additive manufacturing technology, the appropriate accuracy of printing always depends on the smaller thickness of the layer (Pérez et al., 2018) (Espalin et al., 2010). Currently, multiple extruders-based 3D printers are used widely; the extruders are combined with dissimilar materials like flexible or rigid material; or multiple colors in the different layers or single layer (Placone et al., 2018). At present, ABS or acrylonitrile butadiene styrene and PLA or Polylactic acid are the most common two thermoplastic materials used in the fused deposition modeling process (Wickramasinghe et al., 2020). All of these materials have good biodegradable and biocompatibility characteristics, and therefore, all these materials are broadly applied for biomedical and medicinal applications (Yang et al., 2020). These biomaterials allow manufacturers of modern biomedical devices to trust the fused deposition modeling procedure to make the devices that can be used securely for medical trials (Ritzen et al., 2021) (Tan et al., 2018) (Nashed et al., 2021).

SLS (Selective Laser Sintering)

This additive manufacturing process is such a procedure where the powder-type materials (polymer) are applied as a raw material for forming to fabricate the new elements (Lahtinen et al., 2019) (Harada et al., 2020). In the selective laser sintering procedure, basically, a laser is applied on the powder bed to make the structure of the object (Oyar et al., 2017) (Chang et al., 2017) (Goguta et al., 2020). Then another powder layer is applied and this same process is repeated again and this continues till the final structure is formed (Figure 5). The raw materials used in this process are ceramic, polymer, (Zhao et al., 2018) (Fina et al., 2018) etc. As the SLS method have some restriction due to the accuracy of the laser and the grain quality of powder material, therefore it is possible to make mostly in complexity and fragile objects with this type of fabrication method (Lahtinen et al., 2019) (Mazzoli et al., 2013).

Figure 4. Fused deposition modeling (FDM) process

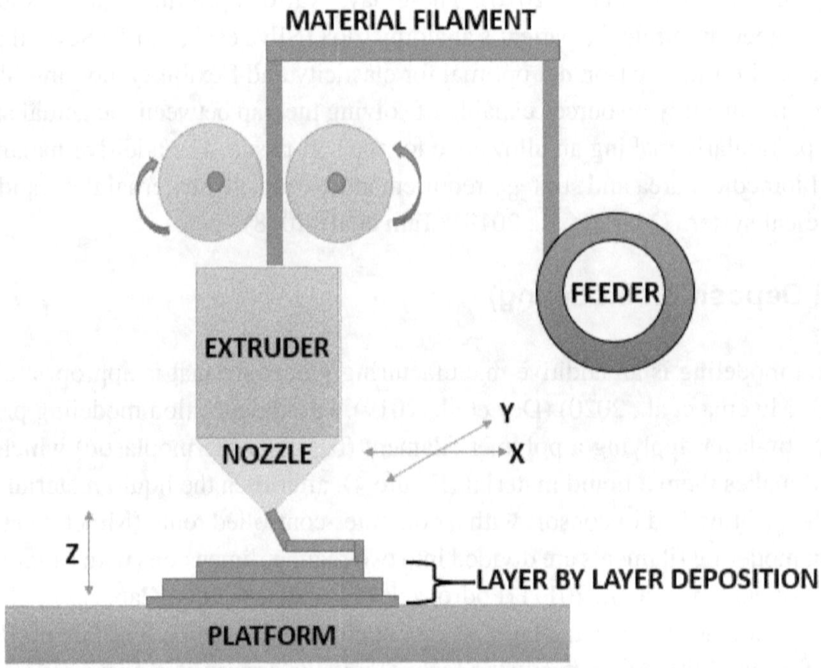

Bio-Printing Process

3D (Three-dimensional) bio-printing technology is one of the most advanced technologies which is applied in a biomedical application specifically in tissue engineering and regenerative medicine (Vijayavenkataraman et al., 2018) (Mandrycky et al., 2016). This technology has an enormous future for the chance in forthcoming medicinal and surgical treatment. Figure 6 shows the bio-printing process. Initially, particular data of organs as well as tissue is gathered for creating the particular replica. With help of the server, the desired model transfers the data into the electrical signal for monitoring the 3D printer to print the model, though, the 3D printer must have the capability to withstand the cell sustainability throughout the process of fabrication period. Usually, normal tissue is a grouping of numerous varieties of cells and similarly, the cells are varied with several ingredients for modified fabrication (Heinrich et al., 2019). Presently, the hard tissues are printed with different types of biomaterial with cells and applied for medicinal trials. In the future, bio-printer also be capable to create vital organs for restructuring and repairing the damaged body area, also to make some valuable medicine screening, tissue engineering, and biomedical investigation. Furthermore, bio-printing can be applied for adapted treatment which will support to decrease the budget of treatment. Biocompatible and biodegradable materials are joined with this bio-printing technology to reduce the irreconcilability affected by materials (Dey et al., 2020).

Figure 5. Selective laser sintering (SLS) process

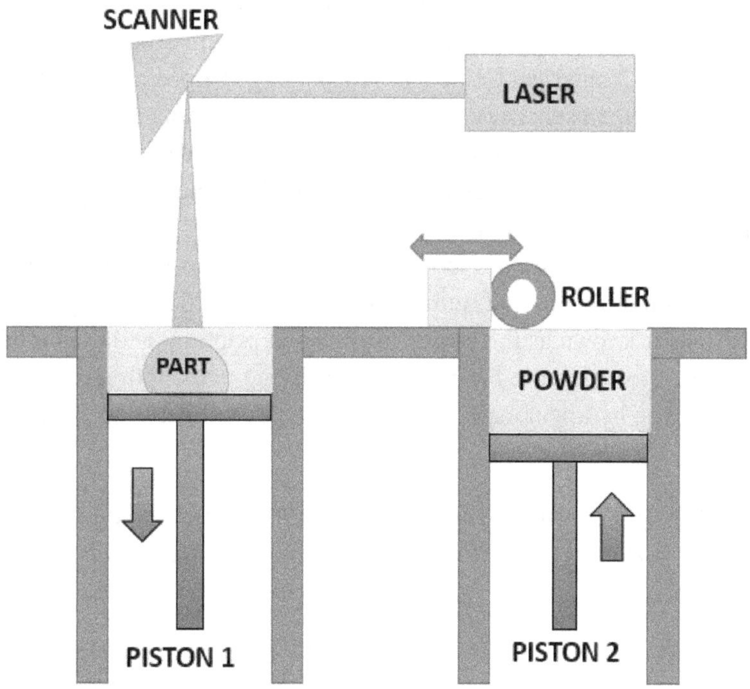

Figure 6. Cell bio-printing process

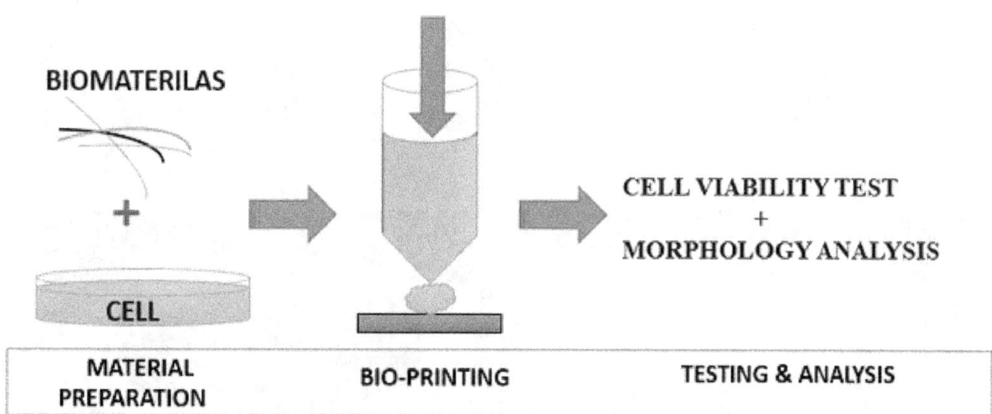

HEALTHCARE AND BIOMEDICAL APPLICATIONS

3D (Three-dimensional) printing process has been considered by specialists and surgeons to make an inclusive range of medicinal uses (Jovic et al., 2020). The patient-specific prosthesis or anatomical parts, and implantations can be exactly fabricated by a 3D printing technique. The high acceptance shows the use of this manufacturing technology in the biomedical and healthcare field (Cui et al., 2018). Associating to traditional manufacturing approaches of biomedical or healthcare procedures, for example, injection molding or CNC (Computer Numerical Controlled) machining (Kim et al., 2019), additive manufacturing technologies are deliberate in specific cases (Lee et al., 2018), though skillful of making substances by CAD (Computer-Aided Design) software with difficult geometric object and multi-functionality, for example, bio-printing (Mandrycky et al., 2016). It was stated that nearly every day twenty patients expire worldwide whereas waiting for the organs transplantation (Peng et al., 2017). Improvements in this type of additive manufacturing techniques led to the increase of bio-printers in such a way, which has the capability of fabrication of organs and stem cells (Barua et al., 2019). The procedure of bio-printing combining with live cells is known as three-dimensional cell printing, and the bioink is used as a feedstock (Tam et al., 2018). Basically, the bio-inks contain living cells with a carrier, which is generally biomaterial or hydrogel. The hydrogel materials can be natural, synthetic, or synthetic-natural mixtures.

Figure 7. 3D printed objects: a. 3D printed nose (alginate–chitosan hydrogel) (Liu et al., 2018); b. native anatomic and axisymmetric aortic valve (PEG-diacrylate hydrogels) (Hockaday et al., 2012); c. vascular structures (alginate) (Tabriz et al., 2015); d. artificial trachea scaffolds (PCL) (Park et al., 2018)

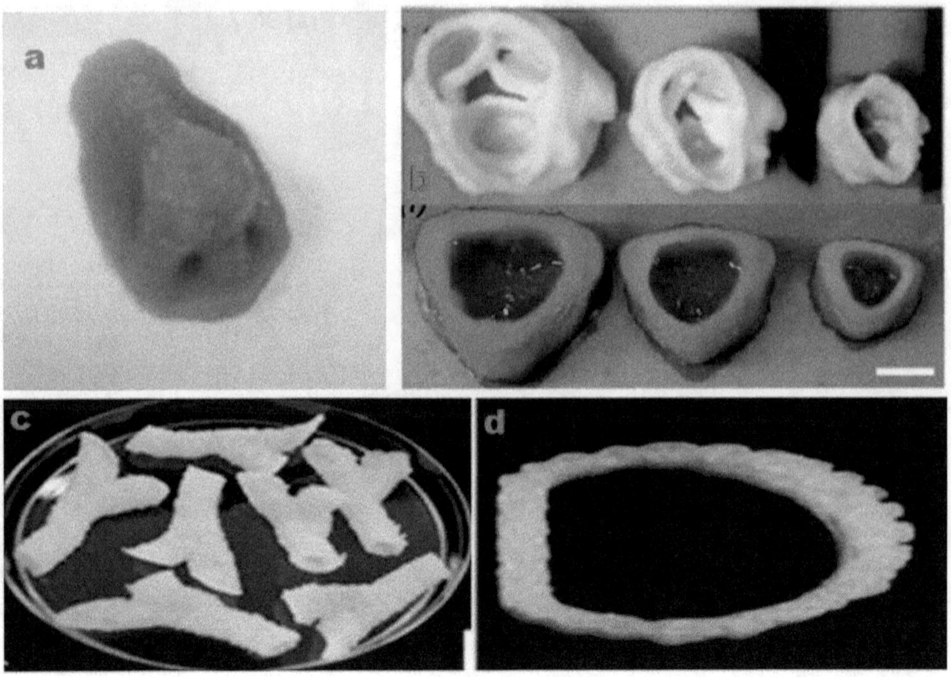

Though the bioink contain living cells, it is assumed that the bio-fabricated structures will be capable to make the growth factors, vasculature, and enhanced incorporate with the host tissues (Mandrycky et al., 2016). Though, due to the existence of live cells in the injected objects, more trials must be required in the bioprinting associating to conventional additive manufacturing techniques (Gao et al., 2018). The area of tissue engineering displays an important character in biocompatible implant fabrication to replace injured or damaged tissues. The main principle of tissue engineering turns around the linking with living cells, biocompatible materials, and different growth factors to make implants that support typical tissue development during the planned hypothesis (Gul et al., 2019) (Rider et al., 2018). Figure 7 shows the 3D printed items fabricated by different additive manufacturing methods. Bio-printing has an appropriate fit into biomedical and tissue engineering, permitting for fabrication of Three-dimensional printed replicas which can imitate the connective tissues' tiny network (Hann et al., 2019).

Table 1. Additive manufacturing in different biomedical fields

Name of the Additive Manufacturing Process	Category	Material	Biomedical Applications	References
Fused filament Fabrication, FDM or Fused deposition modeling, 3D bio-printing	Extrusion	PLA, ABS, Hydrogel with cell	Tissue scaffolds, Surgical plans, Advance drug delivery, Orthosis, Regenerative medicine, Implant etc.	(Parr et al., 2019), (do Ó Silva et al., 2017), (Bandyopadhyay et al., 2015), (Ligon et al., 2017), (Liaw et al., 2017),
DLP (Digital light Processing), SLA (Stereolithography)	Vat polymerization	Photopolymers, Acrylic	Implant, Dentures, Orthosis, Surgical plans etc.	(Liaw et al., 2017), (Bandyopadhyay et al., 2015), (Trenfield et al., 2019), (Mulford et al., 2016)
DLD (Direct laser Deposition)	Focused energy deposition	Metals like steel, alloys etc. Ceramics	Dental model, Orthosis, Implants	(Trenfield et al., 2019), (Li et al., 2020)
Ink-jet printing, Binder jetting	Droplet-based printing	Ceramics, Polymers	Surgical plans, Advance drug delivery, Dental model, Implants	(Bandyopadhyay et al., 2015), (Li et al., 2020)

In-vitro uses of tissue engineering have approached for the advance of functional physical models on behalf of tissue-like humanoid and physical microenvironments for further clinically valid analysis (Liaw et al., 2017). Using good functional, compositional and practical match to the host material, bio-printed three-dimensional replicas can be applied to conduct trials in an In-vivo like functional environment (Kim et al, 2018). Table 1 shows the furthermost active additive manufacturing techniques. Modern research determinations have been effective in replicating purposes of many dissimilar structures through variation and stem cells (Skeldon et al., 2018). Through the additive manufacturing process, patient-derived cells, cell lines, and stem cells can be dropped with appropriate 3-D control to attain any preferred essential preparation (Rey et al., 2020) (Joung et al., 2018). SLA or stereolithography (86) was used to fabricate 4 mm three-dimensional liver lobule replicas, and the materials consist of methacrylate gelatin (GelMA) with biodegradable polyethylene glycol (PEG), replicating the special In-vivo design of the liver. Stereolithography or SLA additive manufacturing technique assists the acceptable fabricating of microscopic particulars inside this object (Elomaa et al., 2020). Table 2 shows the list of some 3D CAD modelling software which helps in additive manufacturing. Several surgical instruments are customized to deal with maximum patients. Collagen bioink material was used to make a vascularized thyroid gland (Baba et al., 2017) by extrusion-based 3D bio-printer. Bio-printing has developed over the last few years as a proficient technique for generating extremely difficult bio-fabrication. So far, numerous laboratories have used this method for making biologically working replicas fabricated of an extensive variation of tissues for example kidney (Lawlor et al., 2021), myocardium (Zhang et al., 2016), mammary epithelium (Swaminathan et al., 2020), liver (Kryou et al., 2019), skin (Knowlton et al., 2018), neurons (Meng et al., 2019), and malignant tumors (Ngan et al., 2020), skeletal muscle (Jamróz et al., 2018) etc. These replicas are fetching beneficial drug selection tools contained by both the pharmaceutical manufacturing and biomedical research laboratories (Lukin et al., 2019) (Mukherjee et al., 2020). Additive manufacturing has been recognized as active tools for the manufacture of implants that encourage regeneration of bone both In-vivo and In-vitro (Sivandzade et al., 2018). Presently, bio-printing offers the high-resolution fabrication of porous tissue scaffolds from an extensive range of materials containing metals, biopolymers, and ceramics (Ng et al., 2018).

Table 2. Different types of CAD modelling software used in additive manufacturing

No.	Name of the software	Solid Modelling
1	Thinker-CAD	Yes
2	Free-CAD	Yes
3	Opens-CAD	Yes
4	Wings3D	No
5	Sketch-UP	No
6	Mol-3D	Yes
7	Solid Works	Yes
8	3DS Max	No
9	Inventor	Yes
10	Rhino 3D	No

The area of surgical treatments, specially prostheses and implants is constantly emerging to permit for an additional cohesive combination of these external objects inside their adjoining tissue, in addition, to raise the functionality (Knowlton et al., 2018). The design of prostheses and implants needs an interdisciplinary method that associates with engineering (mechanical, material, electrical, etc.), biological, and medical science. Additive manufacturing can help connect the between engineering and medical science by making difficult bioactive and biocompatible objects that take benefit of exclusive material characteristics, for example, osteoinductivity and osteoconductivity, to stimulate tissue integration and regeneration of the implantation with the adjacent tissue (Rider et al., 2018) (Liaw et al., 2017). Modern healthcare technology, for example, CT image (Computed Tomography), can permit for the formation of actual precise CAD simulations of a flaw that can help as replicas for printing to confirm an accurate fit into the preferred tissue (Rey et al., 2020). It was confirmed that three-dimensional printed spinal implants (titanium) by EBM (Electron Beam Melting) were greater experienced for making the osteoblast proliferation (Ataee et al., 2018). Nanofabrication procedures have also been applied to make porous coatings for three-dimensional fabricated implants. High-resolution additive manufacturing methods are well matched for making extremely porous implants that permit an intertwined bone mesh for protecting the implant to the neighboring tissue (Tam et al., 2018) (Dey et al., 2020). Moreover, bio-printing of osteoinductive and bio-absorbable materials, for example, calcium phosphate cement, has been presented as active at encouraging bone growth (Dhwan et al., 2019). As radiological imaging has modified a lot, it has permitted the reconstructions of a patient anatomy with the help of CAD software, which allows the prospective designing and construction of patient-modified and precise surgical tools (AlAli et al., 2015). Though, specific anatomical structures or difficult processes could advantage from modified instruments that permit for an extra secured and basic operative practice, in turn reducing the threat of difficulties (Chen et al., 2020). The additive manufacturing method is a very useful tool in quickly converting the computer-aided design into functional tools that could help the continuing desires at the hospital rooms.

CONCLUSION

The important quantity of investigates in current year highlights the significance of additive manufacturing in permitting the upcoming of modern medical and healthcare. Specifically, the additive manufacturing process has previously recognized a different method in the engineering and personalized design of medical and healthcare devices that absolutely influence the efficiency of medical products. Furthermore, the treating of efficient and dedicated biomaterials through additive manufacturing permits superior useful combination, the flexibility of strategy, and optimization of the medical invention. Though, some challenges and problems in terms of insufficient mechanical characteristics in rapid prototyping and 3D (Three-dimensional) printing, scrambling up the mass production of the additive manufacturing fabrication, research on vascularization by bio-printing process, and progressing the modern printable material. It is sensible to predict that in the upcoming the healthcare and medical area can believe a new-fangled example in the additive manufacturing process.

CONFLICTS OF INTEREST

The authors declare that there are no conflicts of interest regarding the publication of this chapter.

FUNDING

No funding has been provided for this work.

ETHICAL APPROVAL

Not required.

ADDITIONAL INFORMATION

Additive manufacturing technology has occurred as a novel disrupting technology that may address the ever-growing demand for medical and healthcare especially organ and tissue transplants. Three-dimensional bio-printing deals with many methodological features which permit for building useful living tissue paradigms by allotting the specific or cluster of cells into exact positions along with numerous types of scaffold ingredients and extracellular matrices, and therefore, may deliver the flexibility required for on-demand adapted manufacture of living organs. Numerous key programs of bio-printing methods are planned, with possible medical and manufacturing uses. The difficult nature of the human tissues, in addition to the permissible and ethical foods for harmless grafting into the living body, would need important investigation and expansion to fabricate marketable bio-printed tissues or organs. This also recommends the probability for additional patenting and authorizing chances from dissimilar divisions of the economy.

ACKNOWLEDGMENT

The authors would like to thank IIEST-Shibpur, Centre for Healthcare Science and Technology lab, and thanks to Mrs. Nibedita Bardhan for language proof reading.

REFERENCES

AlAli, A. B., Griffin, M. F., & Butler, P. E. (2015). Three-Dimensional Printing Surgical Applications. *Eplasty*, *15*, e37.

Ataee, A., Li, Y., Fraser, D., Song, G., & Wen, C. (2018). Anisotropic Ti-6Al-4V gyroid scaffolds manufactured by electron beam melting (EBM) for bone implant applications. *Materials & Design*, *137*, 345–354. doi:10.1016/j.matdes.2017.10.040

Baba, M., Matsumoto, K., Yamasaki, N., Shindo, H., Yano, H., Matsumoto, M., Otsubo, R., John Lawn, M., Matsuo, N., Yamamoto, I., Hidaka, S., & Nagayasu, T. (2017). Development of a Tailored Thyroid Gland Phantom for Fine-Needle Aspiration Cytology by Three-Dimensional Printing. *Journal of Surgical Education*, *74*(6), 1039–1046.

Bandyopadhyay, A., Bose, S., & Das, S. (2015). 3D printing of biomaterials. *MRS Bulletin*, *40*(2), 108–115. doi:10.1557/mrs.2015.3

Barua, R., Datta, S., Datta, P., & Chowdhury, A. R. (2019). Scaffolds and Tissue Engineering Applications by 3D Bio-Printing Process: A New Approach. In K. Kumar & J. Davim (Eds.), *Design, Development, and Optimization of Bio-Mechatronic Engineering Products* (pp. 78–99). IGI Global.

Barua, R., Datta, S., Roychowdhury, A., & Datta, P. (2019). Importance of 3D Printing Technology in Medical Fields. In Additive Manufacturing Technologies From an Optimization Perspective (pp. 21-40). IGI Global.

Bigham, A., Foroughi, F., Rezvani Ghomi, E., Rafienia, M., Neisiany, R. E., & Ramakrishna, S. (2020). The journey of multifunctional bone scaffolds fabricated from traditional toward modern techniques. *Bio-Design and Manufacturing*, *2020*(3), 281–306.

Brito, N. M., Soares, R. S., & Monteiro, E. L. (2016). Additive Manufacturing for Surgical Planning of Mandibular Fracture. *Acta Stomatologica Croatica*, *50*(4), 348–353. doi:10.15644/asc50/4/8

Calusi, S., Arilli, C., Mussi, E., Puggelli, L., Farnesi, D., Casati, M., Compagnucci, A., Marrazzo, L., Talamonti, C., Zani, M., & Pallotta, S. (2021). In phantom evaluation of targeting accuracy in MRI-based brain radiosurgery. *Physica Medica, 85*, 158–164.

Chang, S., Li, L., Lu, L., & Fuh, J. (2017). Selective Laser Sintering of Porous Silica Enabled by Carbon Additive. *Materials (Basel, Switzerland)*, *10*(11), 1313.

Chen, G., Jiang, M., Coles-Black, J., Mansour, K., Chuen, J., & Amott, D. (2020). Three-dimensional printing as a tool in otolaryngology training: A systematic review. *The Journal of Laryngology and Otology*, *134*(1), 14–19.

Cui, H., Miao, S., Esworthy, T., Zhou, X., Lee, S. J., Liu, C., Yu, Z. X., Fisher, J. P., Mohiuddin, M., & Zhang, L. G. (2018). 3D bioprinting for cardiovascular regeneration and pharmacology. *Advanced Drug Delivery Reviews*, *132*, 252–269.

Cui, X., Li, J., Hartanto, Y., Durham, M., Tang, J., Zhang, H., Hooper, G., Lim, K., & Woodfield, T. (2020). Advances in Extrusion 3D Bioprinting: A Focus on Multicomponent Hydrogel-Based Bioinks. *Advanced Healthcare Materials*, *9*(15), e1901648.

Dahake, S., Kuthe, A., Mawale, M., Sapkal, P., Bagde, A., Daronde, S., Kamble, M., & Sarode, B. (2020). Development of customized implant and customized surgical osteotomy guide in ablative tumor surgery for accurate mandibular reconstruction. *The international journal of medical robotics + computer assisted surgery. MRCAS*, *16*(1), e2046.

Derakhshanfar, S., Mbeleck, R., Xu, K., Zhang, X., Zhong, W., & Xing, M. (2018). 3D bioprinting for biomedical devices and tissue engineering: A review of recent trends and advances. *Bioactive Materials*, *3*(2), 144–156.

Dey, A., & Yodo, N. A. (2019). Systematic Survey of FDM Process Parameter Optimization and Their Influence on Part Characteristics. *J. Manuf. Mater. Process.*, *3*, 64.

Dey, M., & Ozbolat, I. T. (2020). 3D bioprinting of cells, tissues and organs. *Scientific Reports*, *10*(1), 14023.

Dhawan, A., Kennedy, P. M., Rizk, E. B., & Ozbolat, I. T. (2019). Three-dimensional Bioprinting for Bone and Cartilage Restoration in Orthopaedic Surgery. *The Journal of the American Academy of Orthopaedic Surgeons*, *27*(5), e215–e226.

Diment, L. E., Thompson, M. S., & Bergmann, J. (2017). Clinical efficacy and effectiveness of 3D printing: A systematic review. *BMJ Open*, *7*(12), e016891.

do Ó Silva, A. L., Dias, I. J., Grempel, R. G., de Morais, H., da Silva Oliveira Brito, N. M., & de Castro Gomes, D. Q. (2017). Prototyping for the treatment of late zygomatic-orbital fracture: A case report. *International Journal of Surgery Case Reports*, *38*, 91–94.

Elomaa, L., Keshi, E., Sauer, I. M., & Weinhart, M. (2020). Development of GelMA/PCL and dECM/PCL resins for 3D printing of acellular in vitro tissue scaffolds by stereolithography. *Materials Science and Engineering C*, *112*, 110958.

Espalin, D., Arcaute, K., Rodriguez, D., Medina, F., Posner, M., & Wicker, R. (2010). Fused deposition modeling of patient-specific polymethylmethacrylate implants. *Rapid Prototyping Journal*, *16*(3), 164–173.

Fan, D., Li, Y., Wang, X., Zhu, T., Wang, Q., Cai, H., Li, W., Tian, Y., & Liu, Z. (2020). Progressive 3D Printing Technology and Its Application in Medical Materials. *Frontiers in Pharmacology*, *11*, 122.

Fantino, E., Chiappone, A., Roppolo, I., Manfredi, D., Bongiovanni, R., Pirri, C. F., & Calignano, F. (2016, May). 3D Printing of Conductive Complex Structures with In Situ Generation of Silver Nanoparticles. *Advanced Materials*, *28*(19), 3712–3717. doi:10.1002/adma.201505109

Fina, F., Madla, C. M., Goyanes, A., Zhang, J., Gaisford, S., & Basit, A. W. (2018). Fabricating 3D printed orally disintegrating printlets using selective laser sintering. *International Journal of Pharmaceutics*, *541*(1-2), 101–107.

Frazier, W. E. (2014). Metal Additive Manufacturing: A Review. *Journal of Materials Engineering and Performance*, *23*, 1917–1928.

Gao, G., Huang, Y., Schilling, A. F., Hubbell, K., & Cui, X. (2018). Organ Bioprinting: Are We There Yet? *Advanced Healthcare Materials*, *7*(1). Advance online publication. doi:10.1002/adhm.201701018

Goguta, L., Lungeanu, D., Negru, R., Birdeanu, M., Jivanescu, A., & Sinescu, C. (2020). Selective Laser Sintering versus Selective Laser Melting and Computer Aided Design - Computer Aided Manufacturing in Double Crowns Retention. *Journal of Prosthodontic Research*. Advance online publication. doi:10.2186/jpr.JPOR_2019_556

Gopinathan, J., & Noh, I. (2018). Recent trends in bioinks for 3D printing. *Biomaterials Research*, *22*, 11.

Gul, M., Arif, A., & Ghafoor, R. (2019). Role of three-dimensional printing in periodontal regeneration and repair: Literature review. *Journal of Indian Society of Periodontology*, *23*(6), 504–510.

Hann, S. Y., Cui, H., Esworthy, T., Miao, S., Zhou, X., Lee, S. J., Fisher, J. P., & Zhang, L. G. (2019). Recent advances in 3D printing: Vascular network for tissue and organ regeneration. *Translational Research; the Journal of Laboratory and Clinical Medicine*, *211*, 46–63.

Harada, Y., Ishida, Y., Miura, D., Watanabe, S., Aoki, H., Miyasaka, T., & Shinya, A. (2020). Mechanical Properties of Selective Laser Sintering Pure Titanium and Ti-6Al-4V, and Its Anisotropy. *Materials (Basel, Switzerland)*, *13*(22), 5081.

He, Y., Yang, F., Zhao, H., Gao, Q., Xia, B., & Fu, J. (2016). Research on the printability of hydrogels in 3D bioprinting. *Scientific Reports*, *6*, 29977.

Heinrich, M. A., Liu, W., Jimenez, A., Yang, J., Akpek, A., Liu, X., Pi, Q., Mu, X., Hu, N., Schiffelers, R. M., Prakash, J., Xie, J., & Zhang, Y. S. (2019). 3D Bioprinting: from Benches to Translational Applications. *Small, 15*(23), e1805510.

Hockaday, L., Kang, K., Colangelo, N., Cheung, P., Duan, B., & Malone, E. (2012). Rapid 3D printing of anatomically accurate and mechanically heterogeneous aortic valve hydrogel scaffolds. *Biofabrication*, *4*(3), 035005.

Ingole, D. S., Kuthe, A. M., Thakare, S. B., & Talankar, A. S. (2009). Rapid prototyping – a technology transfer approach for development of rapid tooling. *Rapid Prototyping Journal*, *15*(4), 280–290.

Jammalamadaka, U., & Tappa, K. (2018). Recent Advances in Biomaterials for 3D Printing and Tissue Engineering. *Journal of Functional Biomaterials*, *9*(1), 22.

Jamróz, W., Szafraniec, J., Kurek, M., & Jachowicz, R. (2018). 3D Printing in Pharmaceutical and Medical Applications - Recent Achievements and Challenges. *Pharmaceutical Research*, *35*(9), 176.

Jessop, Z. M., Al-Sabah, A., Gardiner, M. D., Combellack, E., Hawkins, K., & Whitaker, I. S. (2017). 3D bioprinting for reconstructive surgery: Principles, applications and challenges. *Journal of Plastic, Reconstructive & Aesthetic Surgery; JPRAS*, *70*(9), 1155–1170.

Joung, D., Truong, V., Neitzke, C. C., Guo, S. Z., Walsh, P. J., Monat, J. R., Meng, F., Park, S. H., Dutton, J. R., Parr, A. M., & McAlpine, M. C. (2018). 3D Printed Stem-Cell Derived Neural Progenitors Generate Spinal Cord Scaffolds. *Advanced Functional Materials*, *28*(39), 1801850.

Jovic, T. H., Combellack, E. J., Jessop, Z. M., & Whitaker, I. S. (2020). 3D Bioprinting and the Future of Surgery. *Frontiers in Surgery*, *7*, 609836.

Kessler, A., Hickel, R., & Reymus, M. (2020). 3D Printing in Dentistry-State of the Art. *Operative Dentistry*, *45*(1), 30–40.

Kholgh Eshkalak, S., Rezvani Ghomi, E., Dai, Y., Choudhury, D., & Ramakrishna, S. (2020). The role of three-dimensional printing in healthcare and medicine. *Materials & Design*, *2020*(194), 108940. doi:10.1016/j.matdes.2020.108940

Kim, B. S., Kwon, Y. W., Kong, J. S., Park, G. T., Gao, G., Han, W., Kim, M. B., Lee, H., Kim, J. H., & Cho, D. W. (2018). 3D cell printing of in vitro stabilized skin model and in vivo pre-vascularized skin patch using tissue-specific extracellular matrix bioink: A step towards advanced skin tissue engineering. *Biomaterials*, *168*, 38–53.

Kim, H., Jung, W. K., Choi, I. G., & Ahn, S. H. (2019). A Low-Cost Vision-Based Monitoring of Computer Numerical Control (CNC) Machine Tools for Small and Medium-Sized Enterprises (SMEs). *Sensors (Basel)*, *19*(20), 4506.

Knowlton, S., Anand, S., Shah, T., & Tasoglu, S. (2018). Bioprinting for Neural Tissue Engineering. *Trends in Neurosciences*, *41*(1), 31–46.

Kryou, C., Leva, V., Chatzipetrou, M., & Zergioti, I. (2019). Bioprinting for Liver Transplantation. *Bioengineering (Basel, Switzerland)*, *6*(4), 95.

Lahtinen, E., Precker, R., Lahtinen, M., Hey-Hawkins, E., & Haukka, M. (2019). Selective Laser Sintering of Metal-Organic Frameworks: Production of Highly Porous Filters by 3D Printing onto a Polymeric Matrix. *ChemPlusChem*, *84*(2), 222–225.

Lahtinen, E., Turunen, L., Hänninen, M. M., Kolari, K., Tuononen, H. M., & Haukka, M. (2019). Fabrication of Porous Hydrogenation Catalysts by a Selective Laser Sintering 3D Printing Technique. *ACS Omega*, *4*(7), 12012–12017.

Lawlor, K. T., Vanslambrouck, J. M., Higgins, J. W., Chambon, A., Bishard, K., Arndt, D., Er, P. X., Wilson, S. B., Howden, S. E., Tan, K. S., Li, F., Hale, L. J., Shepherd, B., Pentoney, S., Presnell, S. C., Chen, A. E., & Little, M. H. (2021). Cellular extrusion bioprinting improves kidney organoid reproducibility and conformation. *Nature Materials*, *20*(2), 260–271.

Lee, S. J., Lee, J. B., Park, Y. W., & Lee, D. Y. (2018). 3D Bioprinting for Artificial Pancreas Organ. *Advances in Experimental Medicine and Biology*, *1064*, 355–374.

León-Cabezas, M. A., Martínez-García, A., & Varela-Gandía, F. J. (2017). Innovative functionalized monofilaments for 3D printing using fused deposition modeling for the toy industry. *Proc. Manuf.*, *13*, 738–745.

Li, Y., Li, L., Ma, Y., Zhang, K., Li, G., Lu, B., Lu, C., Chen, C., Wang, L., Wang, H., & Cui, X. (2020). 3D-Printed Titanium Cage with PVA-Vancomycin Coating Prevents Surgical Site Infections (SSIs). *Macromolecular Bioscience*, *20*(3), e1900394.

Liaw, C. Y., & Guvendiren, M. (2017). Current and emerging applications of 3D printing in medicine. *Biofabrication*, *9*(2), 024102.

Ligon, S. C., Liska, R., Stampfl, J., Gurr, M., & Mülhaupt, R. (2017). Polymers for 3D Printing and Customized Additive Manufacturing. *Chemical Reviews*, *117*(15), 10212–10290.

Liu, Q., Li, Q., Xu, S., Zheng, Q., & Cao, X. (2018). Preparation and properties of 3D printed alginate–chitosan polyion complex hydrogels for tissue engineering. *Polymers*, *10*(6), 664.

Lukin, I., Musquiz, S., Erezuma, I., Al-Tel, T. H., Golafshan, N., Dolatshahi-Pirouz, A., & Orive, G. (2019). Can 4D bioprinting revolutionize drug development? *Expert Opinion on Drug Discovery*, *14*(10), 953–956.

Lv, Y., Wang, B., Liu, G., Tang, Y., Lu, E., Xie, K., Lan, C., Liu, J., Qin, Z., & Wang, L. (2021). Metal Material, Properties and Design Methods of Porous Biomedical Scaffolds for Additive Manufacturing: A Review. *Frontiers in Bioengineering and Biotechnology*, *9*, 641130.

Mandrycky, C., Wang, Z., Kim, K., & Kim, D. H. (2016). 3D bioprinting for engineering complex tissues. *Biotechnology Advances*, *34*(4), 422–434.

Manita, P. G., Garcia-Orue, I., Santos-Vizcaino, E., Hernandez, R. M., & Igartua, M. (2021). 3D Bioprinting of Functional Skin Substitutes: From Current Achievements to Future Goals. *Pharmaceuticals (Basel, Switzerland)*, *14*(4), 362.

Marti, P., Lampus, F., Benevento, D., & Setacci, C. (2019). Trends in use of 3D printing in vascular surgery: a survey. *International Angiology: A Journal of the International Union of Angiology, 38*(5), 418–424.

Mazzoli, A. (2013). Selective laser sintering in biomedical engineering. *Medical & Biological Engineering & Computing, 51*(3), 245–256.

Meng, F., Meyer, C. M., Joung, D., Vallera, D. A., McAlpine, M. C., & Panoskaltsis-Mortari, A. (2019). 3D Bioprinted In Vitro Metastatic Models via Reconstruction of Tumor Microenvironments. *Advanced materials (Deerfield Beach, Fla.), 31*(10), e1806899.

Minetola, P., & Galati, M. (2018). A challenge for enhancing the dimensional accuracy of a low-cost 3D printer by means of self-replicated parts. *Additive Manufacturing, 22*, 256–264.

Mondschein, R. J., Kanitkar, A., Williams, C. B., Verbridge, S. S., & Long, T. E. (2017). Polymer structure-property requirements for stereolithographic 3D printing of soft tissue engineering scaffolds. *Biomaterials, 140*, 170–188. doi:10.1016/j.biomaterials.2017.06.005 PMID:28651145

Moreno Madrid, A. P., Vrech, S. M., Sanchez, M. A., & Rodriguez, A. P. (2019). Advances in additive manufacturing for bone tissue engineering scaffolds. *Materials Science and Engineering C, 100*, 631–644.

Mukherjee, P., Chung, J., Cheng, K., Gupta, R., Haag, H., Williams, Z., & Wallace, G. (2021). Invitro and Invivo Study of PCL-Hydrogel Scaffold to Advance Bioprinting Translation in Microtia Reconstruction. *The Journal of Craniofacial Surgery, 32*(5), 1931–1936.

Mulford, J. S., Babazadeh, S., & Mackay, N. (2016). Three-dimensional printing in orthopaedic surgery: Review of current and future applications. *ANZ Journal of Surgery, 86*(9), 648–653.

Mwema, F. M., & Akinlabi, E. T. (2020). Basics of Fused Deposition Modelling (FDM). *Fused Deposition Modeling: Strategies for Quality Enhancement*, 1–15.

Nam, S. Y., & Park, S. H. (2018). ECM Based Bioink for Tissue Mimetic 3D Bioprinting. *Advances in Experimental Medicine and Biology, 1064*, 335–353.

Nashed, N., Lam, M., & Nokhodchi, A. (2021). A comprehensive overview of extended release oral dosage forms manufactured through hot melt extrusion and its combination with 3D printing. *International Journal of Pharmaceutics, 596*, 120237.

Neiman, J. A., Raman, R., Chan, V., Rhoads, M. G., Raredon, M. S., Velazquez, J. J., Dyer, R. L., Bashir, R., Hammond, P. T., & Griffith, L. G. (2015). Photopatterning of hydrogel scaffolds coupled to filter materials using stereolithography for perfused 3D culture of hepatocytes. *Biotechnology and Bioengineering, 112*(4), 777–787.

Ng, W. L., Goh, M. H., Yeong, W. Y., & Naing, M. W. (2018). Applying macromolecular crowding to 3D bioprinting: Fabrication of 3D hierarchical porous collagen-based hydrogel constructs. *Biomaterials Science, 6*(3), 562–574.

Ngan, C., Quigley, A., O'Connell, C., Kita, M., Bourke, J., Wallace, G. G., Choong, P., & Kapsa, R. (2020). 3D Bioprinting and Differentiation of Primary Skeletal Muscle Progenitor Cells. *Methods in Molecular Biology (Clifton, N.J.), 2140*, 229–242.

Oyar, P. (2018). Laser Sintering Technology and Balling Phenomenon. *Photomedicine and Laser Surgery, 36*(2), 72–77.

Pantermehl, S., Emmert, S., Foth, A., Grabow, N., Alkildani, S., Bader, R., Barbeck, M., & Jung, O. (2021). 3D Printing for Soft Tissue Regeneration and Applications in Medicine. *Biomedicines, 9*(4), 336.

Park, H., Lee, J., Jung, H., Kim, D., Kim, S., & Sultan, M. (2018). An omentum-cultured 3D-printed artificial trachea: In vivo bioreactor. *Artif Cell Nanomed B, 46*(3), S1131–S1140.

Parr, W., Burnard, J. L., Wilson, P. J., & Mobbs, R. J. (2019). 3D printed anatomical (bio)models in spine surgery: Clinical benefits and value to health care providers. *Journal of spine surgery (Hong Kong), 5*(4), 549–560.

Paštyková, V., Novotný, J., Veselský, T., Urgošík, D., Liščák, R., & Vymazal, J. (2018). Assessment of MR stereotactic imaging and image co-registration accuracy for 3 different MR scanners by 3 different methods/phantoms: Phantom and patient study. *Journal of Neurosurgery, 129*(Suppl1), 125–132.

Peng, W., Datta, P., Ayan, B., Ozbolat, V., Sosnoski, D., & Ozbolat, I. T. (2017). 3D bioprinting for drug discovery and development in pharmaceutics. *Acta Biomaterialia, 57*, 26–46.

Pérez, M., Medina-Sánchez, G., García-Collado, A., Gupta, M., & Carou, D. (2018). Surface Quality Enhancement of Fused Deposition Modeling (FDM) Printed Samples Based on the Selection of Critical Printing Parameters. *Materials (Basel, Switzerland), 11*(8), 1382.

Phan, K., Sgro, A., Maharaj, M. M., D'Urso, P., & Mobbs, R. J. (2016). Application of a 3D custom printed patient specific spinal implant for C1/2 arthrodesis. *Journal of Spine Surgery, 2*(4), 314–318. doi:10.21037/jss.2016.12.06 PMID:28097249

Placone, J. K., & Engler, A. J. (2018). Recent Advances in Extrusion-Based 3D Printing for Biomedical Applications. *Advanced Healthcare Materials, 7*(8), e1701161.

Podroužek, J., Marcon, M., Ninčević, K., & Wan-Wendner, R. (2019). Bio-Inspired 3D Infill Patterns for Additive Manufacturing and Structural Applications. *Materials (Basel, Switzerland), 12*(3), 499.

Rastogi, P., & Kandasubramanian, B. (2019). Review of alginate-based hydrogel bioprinting for application in tissue engineering. *Biofabrication, 11*(4), 042001.

Rey, F., Barzaghini, B., Nardini, A., Bordoni, M., Zuccotti, G. V., Cereda, C., Raimondi, M. T., & Carelli, S. (2020). Advances in Tissue Engineering and Innovative Fabrication Techniques for 3-D-Structures: Translational Applications in Neurodegenerative Diseases. *Cells, 9*(7), 1636.

Rider, P., Kačarević, Ž. P., Alkildani, S., Retnasingh, S., Schnettler, R., & Barbeck, M. (2018). Additive Manufacturing for Guided Bone Regeneration: A Perspective for Alveolar Ridge Augmentation. *International Journal of Molecular Sciences, 19*(11), 3308.

Ritzen, L., Montano, V., & Garcia, S. J. (2021). 3D Printing of a Self-Healing Thermoplastic Polyurethane through FDM: From Polymer Slab to Mechanical Assessment. *Polymers, 13*(2), 305.

Rodríguez-Panes, A., Claver, J., & Camacho, A. M. (2018). The Influence of Manufacturing Parameters on the Mechanical Behaviour of PLA and ABS Pieces Manufactured by FDM: A Comparative Analysis. *Materials (Basel, Switzerland)*, *11*(8), 1333.

Salmi, M., Paloheimo, K. S., Tuomi, J., Wolff, J., & Mäkitie, A. (2013). Accuracy of medical models made by additive manufacturing (rapid manufacturing). *Journal of Cranio-Maxillo-Facial Surgery*, *41*(7), 603–609. doi:10.1016/j.jcms.2012.11.041 PMID:23333490

Sigaux, N., Pourchet, L., Breton, P., Brosset, S., Louvrier, A., & Marquette, C. A. (2019). 3D Bioprinting:principles, fantasies and prospects. *Journal of Stomatology, Oral and Maxillofacial Surgery*, *120*(2), 128–132.

Simpson, R. L., Wiria, F. E., Amis, A. A., Chua, C. K., Leong, K. F., Hansen, U. N., Chandrasekaran, M., & Lee, M. W. (2008). Development of a 95/5 poly(L-lactide-co-glycolide)/hydroxylapatite and beta-tricalcium phosphate scaffold as bone replacement material via selective laser sintering. *Journal of Biomedical Materials Research. Part B, Applied Biomaterials*, *84*(1), 17–25.

Sivandzade, F., & Cucullo, L. (2018). In-vitro blood-brain barrier modeling: A review of modern and fast-advancing technologies. *Journal of Cerebral Blood Flow & Metabolism (Nihongoban)*, *38*(10), 1667–1681.

Skeldon, G., Lucendo-Villarin, B., & Shu, W. (2018). Three-dimensional bioprinting of stem-cell derived tissues for human regenerative medicine. *Philosophical Transactions of the Royal Society of London. Series B, Biological Sciences*, *373*(1750), 20170224.

Swaminathan, S., & Clyne, A. M. (2020). Direct Bioprinting of 3D Multicellular Breast Spheroids onto Endothelial Networks. *Journal of Visualized Experiments*, (165). Advance online publication. doi:10.3791/61791

Tabriz, A., Hermida, M., Leslie, N., & Shu, W. (2015). Three-dimensional bioprinting of complex cell laden alginate hydrogel structures. *Biofabrication*, *7*(4), 045012.

Tam, C., Chan, Y. C., Law, Y., & Cheng, S. (2018). The Role of Three-Dimensional Printing in Contemporary Vascular and Endovascular Surgery: A Systematic Review. *Annals of Vascular Surgery*, *53*, 243–254.

Tan, D. K., Maniruzzaman, M., & Nokhodchi, A. (2018). Advanced Pharmaceutical Applications of Hot-Melt Extrusion Coupled with Fused Deposition Modelling (FDM) 3D Printing for Personalised Drug Delivery. *Pharmaceutics*, *10*(4), 203.

Tan, S. H., Ngo, Z. H., Leavesley, D., & Liang, K. (2021). Recent Advances in the Design of Three-Dimensional and Bioprinted Scaffolds for Full-Thickness Wound Healing. *Tissue Engineering. Part B, Reviews*. Advance online publication. doi:10.1089/ten.TEB.2020.0339

Tarassoli, S. P., Jessop, Z. M., Al-Sabah, A., Gao, N., Whitaker, S., Doak, S., & Whitaker, I. S. (2018). Skin tissue engineering using 3D bioprinting: An evolving research field. *Journal of Plastic, Reconstructive & Aesthetic Surgery; JPRAS*, *71*(5), 615–623.

Trenfield, S. J., Awad, A., Madla, C. M., Hatton, G. B., Firth, J., Goyanes, A., Gaisford, S., & Basit, A. W. (2019). Shaping the future: Recent advances of 3D printing in drug delivery and healthcare. *Expert Opinion on Drug Delivery*, *16*(10), 1081–1094.

Tuomi, J., Paloheimo, K. S., Vehviläinen, J., Björkstrand, R., Salmi, M., Huotilainen, E., Kontio, R., Rouse, S., Gibson, I., & Mäkitie, A. A. (2014). A novel classification and online platform for planning and documentation of medical applications of additive manufacturing. *Surgical Innovation*, *21*(6), 553–559.

Vijayavenkataraman, S., Yan, W. C., Lu, W. F., Wang, C. H., & Fuh, J. (2018). 3D bioprinting of tissues and organs for regenerative medicine. *Advanced Drug Delivery Reviews*, *132*, 296–332.

Whitley, D. III, Eidson, R. S., Rudek, I., & Bencharit, S. (2017). In-office fabrication of dental implant surgical guides using desktop stereolithographic printing and implant treatment planning software: A clinical report. *The Journal of Prosthetic Dentistry*, *118*(3), 256–263.

Wickramasinghe, S., Do, T., & Tran, P. (2020). FDM-Based 3D Printing of Polymer and Associated Composite: A Review on Mechanical Properties, Defects and Treatments. *Polymers*, *12*(7), 1529.

Xie, Z., Gao, M., Lobo, A. O., & Webster, T. J. (2020). 3D Bioprinting in Tissue Engineering for Medical Applications: The Classic and the Hybrid. *Polymers*, *12*(8), 1717.

Yahya, E. B., Jummaat, F., Amirul, A. A., Adnan, A. S., Olaiya, N. G., Abdullah, C. K., Rizal, S., Mohamad Haafiz, M. K., & Khalil, H. (2020). A Review on Revolutionary Natural Biopolymer-Based Aerogels for Antibacterial Delivery. *Antibiotics (Basel, Switzerland)*, *9*(10), 648.

Yang, T. C., & Yeh, C. H. (2020). Morphology and Mechanical Properties of 3D Printed Wood Fiber/Polylactic Acid Composite Parts Using Fused Deposition Modeling (FDM): The Effects of Printing Speed. *Polymers*, *12*(6), 1334.

Youssef, A., Hollister, S. J., & Dalton, P. D. (2017). Additive manufacturing of polymer melts for implantable medical devices and scaffolds. *Biofabrication*, *9*(1), 012002.

Yuan, B., Zhou, S. Y., & Chen, X. S. (2017). Rapid prototyping technology and its application in bone tissue engineering. *Journal of Zhejiang University. Science. B.*, *18*(4), 303–315.

Zhang, Y. S., Arneri, A., Bersini, S., Shin, S. R., Zhu, K., Goli-Malekabadi, Z., Aleman, J., Colosi, C., Busignani, F., Dell'Erba, V., Bishop, C., Shupe, T., Demarchi, D., Moretti, M., Rasponi, M., Dokmeci, M. R., Atala, A., & Khademhosseini, A. (2016). Bioprinting 3D microfibrous scaffolds for engineering endothelialized myocardium and heart-on-a-chip. *Biomaterials*, *110*, 45–59.

Zhao, M., Wudy, K., & Drummer, D. (2018). Crystallization Kinetics of Polyamide 12 during Selective Laser Sintering. *Polymers*, *10*(2), 168.

Zhu, W., Ma, X., Gou, M., Mei, D., Zhang, K., & Chen, S. (2016). 3D printing of functional biomaterials for tissue engineering. *Current Opinion in Biotechnology*, *40*, 103–112.

KEY TERMS AND DEFINITIONS

3D Printing: 3D printing is basically an additive manufacturing process, which is the manufacture of a 3D objective from digital CAD data.

Additive Manufacturing: Additive manufacturing (AM) is a transformative method to manufacturing products that allow the formation of lighter, tougher quantities and structures.

Bio-Printing: Bio-printing is a 3D printing process; it practices a computational file as a design to fabricate an objective layer by layer. However, different three-dimensional printing, bio-printers print with living cells and biomaterials, constructing organ-like configurations that let living cells reproduce.

Biomaterials: A biomaterial is an element which has been planned to cooperate with living systems for a curative purpose, whichever a beneficial or an investigative one.

Bone-Tissue: Bone tissue is preserved by bone-making cells named osteoblasts and cells which discontinuity bone called osteoclasts.

Healthcare: Health care is the progress or maintenance of healthiness through the anticipation, treatment, diagnosis, recovery, or therapy of illness, disease, wound, and additional bodily and psychological injuries in persons. Health care is provided by health specialists and associated health areas. Prescription, dentistry, dispensary, optometry, care, nursing, audiology, surgery, psychology, work-related treatment, corporeal treatment, athletic exercise, and supplementary health occupations are all together part of health care.

Surgery: Surgery is a remedial or dental department that practices operational manual and instrumental performances on a patient to treat and inspect a pathological situation for example an injury or disease, to assistance recover physical function, advent, or to healing annoying shattered areas. Surgery can contain cutting, grazing, stitching, or else changing living organs and tissues.

Chapter 15
Flammability Index of Nigeria Agro Waste Processed With Polymer and Gum Arabic Hybridized Flame Retardants for Building Applications

Timine Suoware
https://orcid.org/0000-0001-9618-3363
Federal Polytechnic, Ekowe, Nigeria

ABSTRACT

Agro waste polymer composite (AWPC) comprising agro waste as reinforcement and polymer as the matrix is known to contribute additional fuel load when exposed to heat sources owing to their constituent make up. This has restricted their use in areas where strict compliance to fire regulations does exist. GA was mixed with flame retardants (FR) such as aluminium tri-hydroxide, ammonium polyphosphate, and carbon black at 0%, 12%, 15%, and 18% loading proportion, and these were then used to fabricate the AWPC using hand lay-up and compression moulding. The AWPC were tested for flammability properties in the cone calorimeter apparatus at 50kW/m2. Flammability properties such as time to ignition, heat released rates, mass loss rates, and smoke production rates obtained showed the AWPC processed with GA hybridized FR were greatly improved when compared to those without FR. Therefore, GA addition in AWPC proved to be a good FR that could be used in building applications to meet required fire safety standards.

INTRODUCTION

The activity of modern day agricultural processes has caused the availability of more agro waste predicating a threat to the environment.In Nigeria agro-waste is abundant and un-utilized; left to decay constituting a nuisance to the environment especially at huge dump sites. This has raised serious concerns

DOI: 10.4018/978-1-7998-9574-9.ch015

by environmentalist and the general public while attracting researchers all over the globe in finding ways of converting agro waste into useful products. In Nigeria, reinforcement derived from agro waste such as jatrophacurcas seedcake and corn cobs and cassava stalk have been used to produce ceiling boards and particle boards respectively for building applications (Olorunmaiye & Ohijeagbon,2015 and Amenaghawon et al 2016). Fig 1 shows various products developed in the building industry by other researchers using agro waste. It is clear that converting the agro waste in Nigeria to develop value-added products will not only improve its economic growth but create jobs as well.

Figure 1. (a) Coir fibre roofing sheet (b) sugarcane waste based panel (c) oil palm waste roofing tiles (d) coir waste door panel
(Suoware et al 2020)

Fibres or particulates as reinforcement to polymers can be derived from agro waste through various extraction processes (Chauhan & Arya, 2018 and Patil & Kolambe, 2011). The strength and rigidity of the reinforcement are obtained from stem plant and are broadly used in the world of composites due to their wide availability and renewability within a short time as well as their good mechanical properties (Nair et al 2000). The fibres or particulates consist of organic constituents comprising cellulose, hemicellulose, lignin and pectin thus; they are also known as lignocellulosic or cellulosic. Higher cellulosic content in agro waste indicates that higher flammability properties may be obtained which implies more release of flammable volatiles. On the other hand, lower lignin content would mean that a lower concentration of flammability leading to trapping of flammable volatiles may be obtained. Reports from researchers also confirm that the addition of agro waste in processing polymer composite for the design of components can increase the release of flammable or combustible volatiles during a fire scenario; thus making it not safe especially in areas where stringent fire regulations does exist (Alhuthali et al 2012. Ezeh et al 2020, Sebastian Rabe et al 2019).

The flammability of materials is a critical design consideration if agro waste must be used for various component designs. The major challenge is that they respond rapidly to high heat atmosphere, releasing heat energy that can, in some circumstances, contribute additional fuel load thus; making it very destructive. Most agro waste components has failed to pass required fire safety standards for use in aircraft, ships, buildings, land transport, oil and gas facilities, home appliances and other applications requiring high fire resistance (Racier et al 2015). Hence to sustain their viability the high susceptibility to flame needs to be overcome. Agro waste generates heat and combustible products due to the addition of polymers consisting of a complex mixture of gases and solid particulates from incomplete combustion. The amount of heat release from agro waste and polymer is controlled by the combustion of flammable gas products resulting from the decomposition of the organic constituents. Although the polymer matrix is the primary source of flammable volatiles, the decomposition of organic materials when exposed to heat source can also produce volatiles. Therefore, to reduce the flammability risk of agro waste components to meet required fire safety standards for various building applications, flame retardants (FR) are usually processed with them.

Agro waste and polymer form a unified material structure with distinct interface and can be processed with either halogenated based or halogenated free FR (Mallick 1999 &Ishai et al,2006). Recent studies shows that halogenated free FR such as aluminum tri-hydroxide and ammonium polyphosphate are considered the most favourable flame retardant in polymers because they are greener, highly effective and of low toxicity when compared to over 150 chemical compounds from which FR can be derived (Subastinghe et al. 2014, Norzali et al.2011, Redwan et al.2015, Arjmandi et al.2017).The mechanism of the FR when decomposed could delay the start and spread of fire better as well as a reduction in the fuel load in an advent of fire outbreak. The effect of ammonium phosphate in combination with graphite FR in wod-polyprolyene using the cone calorimeter has been studied (Nikolaeva & Karki,2016). The study shows that the fire properties measured in a cone calorimeter apparatus decreased due to the introduction of fire retardants in the shell layer of the composites.

There are different ways to assess the flammability of materials for purpose of estimating the measure of their fire response. In most cases, it requires an observation or measurement of some aspect or combination of flammability, which includes ignition, heat release rates, mass loss and smoke production. Ignition time and heat release rates are the most important parameters in determining the ease a material response to fire. In this chapter, Gum Arabic was used to modify existing FR; aluminium tri-hydroxide and ammonium polyphosphate and then hybridized while adding carbon black. The Gum Arabic hybridized FR was then fabricated with agro waste and polymer. It is intended that when the composite is exposed to fire the Gum Arabic will form a stable char structure to prevent flammable volatiles and improve on flammability properties to meet required safety fire standards. Thus, the main objective of this chapter is to measure the flammability properties using the cone calorimeter apparatus; the agro waste and polymer composite and compare the data obtained in similar conditions in literature.

A Brief Review of Common Agro Waste in Nigeria

Agro-waste can be defined as any material that is un-utilized i.e. left over from a production process. The waste generated from the agricultural sector in Nigeria can be a good alternative to the shortage of raw materials to meet required needs in various industries in Nigeria. The quantity of agro-waste is influenced by increasing agricultural production and productivity. Agro waste such as coconut coir, oil palm, banana, sugarcane and pineapple are easy to access and abundant in Nigeria (Suoware et al 2020). The agro-waste and their fibre extraction as well as their physiochemical compositions and flammability properties are shown respectively in Table 1. It can be observed that pineapple and bagasse waste has the highest percentage of cellulosic content at 85% and 81% respectively followed by oil palm at 65%. This would mean that pineapple, bagasse and oil palm wastes in polymer will generate additional fuel load during a fire scenario. On the other hand, the high lignin content for coir and oil palm signifies a great chance of char formation and consequently acts as a barrier against the flow of flammable volatiles. In Table 1 it also shows that the flammability properties obtained for agro waste reinforced polymer processed with flame retardants is scare in literature. Infact a compilation of reinforcements from natural sources such as the agro waste without flame retardants

Table 1. Flammability properties of agro-waste polymer composite and their physiochemical composition

Agro Waste	Origin (%)	Cellulose (%)	Hemicellulose (%)	Lignin (%)	Pectin (%)	Fibre Extraction	Moisture Content (Wt. %)	Tensile Strength (MPa)	Young Modulus (GPa)	Flammability properties with flame retardants: Tig (s)	p HRR (kW/m²)	MLR (g/s)	SPR (m²/s)	Ref
(a) Coconut waste 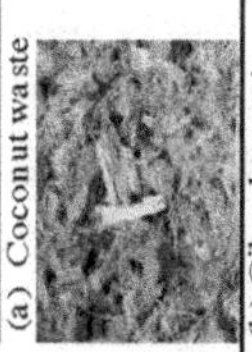	Stem	32-34	0.15-0.25	40-45	3-4	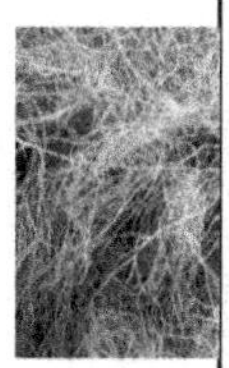	8	220	6	68	571			[20]
(b) Oil palm waste 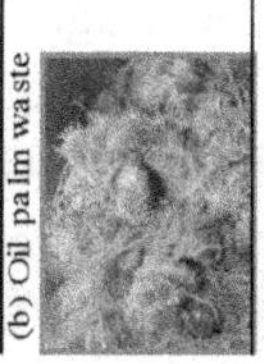	Fruit	65	22.1	29		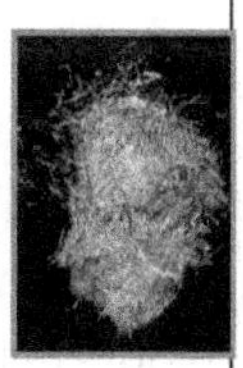	40-46	248	3.2					
(c) Banana waste 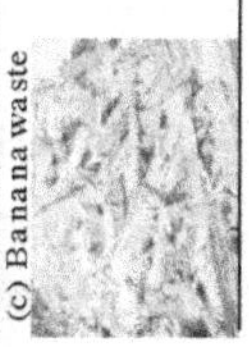	Stem	50-60	25-30	12-18			10-12	355	33.8	17	192.8			[21]
(d) Sugarcane waste 	Stem	81	9.45	25.3		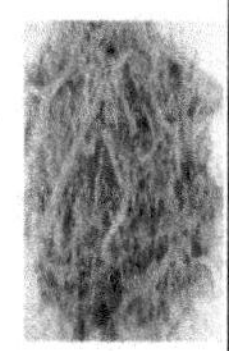	10-40	20-290	19.7-27.1	51	270			[22]
(e) Pineapple Fibre 	Leaf	85	5	12.7			10-15	170-1627	82					

Tig: time to Ignition, pHRR: peak Heat Release Rate, MLR: Mass Loss Rate, SPR: Smoke Production Rate reached 8000 published articles in 2014 while those with flame retardant reached 200 published articles in Elsevier (Prabhakar 2015). The concept of Gum Arabic hybridized in flame retardants such aluminium tri-hydroxide, ammonium polyphosphate and carbon black processed with agro waste polymers and their flammability properties has not been reported.

Research and development has shown that agro waste reinforced in polymers are competing with synthetic based materials especially in areas of non-structural applications despite their inherent challenges. The continuous research work and progress achieved so far in this field of study indicates that there is room for more improvements in a broader applicability. Agro waste reinforced polymers respond to flame rapidly and this is unacceptable to many end users, regulators and government authorities as both materials combust when exposed to heat. This has limited the depth of their applications in many areas where the knowledge of their fire behaviour is unknown. Thus, this chapter is necessary to explore the potentials of some selected Nigeria agro wastes processed with flame retardants and contribute to the scientific knowledge on their reaction to fire behavior when in use. Also, the use of agro-waste that is cheap, easily re-grown and infinite as reinforcing fibres in polymer composites will not only solve the environmental nuisance caused after cultivation but also the need where easily processed lightweight composite materials of complex shape with ease is being sought. The choice of agro-waste for this chapter (i.e. oil palm and wood) has similar chemical, physical and mechanical properties suitable as raw materials for wood based panels.

Experimental Determination of Flammability Properties

In determining flammability properties, engineers and designers rely on fire testing data obtained from bench scale apparatus to predict the safety of materials. The 1998 edition of the compilation of fire test by American Society for Testing and Materials (ASTM) alone listed some bench scale tests (Foley et al.1994). Most of these methods are well elaborated and documented in various literatures (Babrauskas.1986, Scudamore et al.1991, Greyson et al.1994, Babrauskas,1995) Among the fire test methods, the cone calorimeter at 50kW/m^{2}provides a broad range of ignition and flammability properties that can be correlated to real fire scenario. The cone calorimeter is the most versatile in determining a large number of fire properties in a single test using with a small specimen. Cone calorimeter simulation does not simulate post flashover or fully developed conditions but can predict accurately flammability properties up to the point of flashover.

Preparation of Agro Waste as Reinforcements

There are different methods in which agro waste can be prepared as reinforcements to polymers. The study of the flammability properties of oil palm waste (OPW) and wood wastes (WW) were selected in this chapter. The method of obtaining the OPW and WW agro waste reinforcements is shown in Fig 2 (Vijaya et al 2013). The choice of OPW and WW was based on their abundant availability and easy accessibility in Nigeria. OPW as received and shown in Fig 2a was extracted first by washing with hot water to remove the remaining residual oil retained during oil extraction and then soaked in n-Hexane as shown on Fig 2c overnight to complete leaching and further remove impurities. To enhance compatibility with the polymer, the OPW were treated with 5% (NaOH) solution for about 6hrs to avoid their damage. Afterwards, the OPW were then later washed with distilled water until blue litmus paper turned red which indicates that excess concentration of NaOH have been neutralized and then sun dried for 3 days to remove moisture content as shown in Fig 2d. The ready to use fibres as reinforcement are shown in Fig 2f. For the wood waste (WW) as shown in Fig 2b, they were obtained by first grinding and sieving with a 300μm sieve to obtain a fine powder. The particulates were treated with similar treatment approach to the OPW. The particulates were then oven dried at 80°C for 3hrs to remove moisture content as shown in Fig 2e and the ready to use particulate reinforcement shown in Fig 2g.

Figure 2. Process of extracting reinforcements (a) untreated OPW (b) untreated WW (c) treatment with 5% NaOH (d) Sun drying process of OPW (e) oven drying process of WW (f) ready to use OPW reinforcement (g) ready to use WW reinforcement

Processing of Agro Waste with Flame Retardant and Polymer

There is various techniques agro waste and polymers can be processed to form a unified structure with distinct interface mostly referred to as composites (Gokul et al.2017, Dong et al 2014, Ramli et al.2011, Ayrilmis et al.2011, Mahesh et al.2020, Munawar et al 2015, Indira et al 2013 and Messiry 2013). In this study the OPWC and WWC processed without and with the Gum Arabic hybridized flame retardant were fabricated using hand lay-up and compression moulding as shown in Fig. 3. A rectangular wooden frame of 10mm thickness, with internal area of 470,375 mm^2 was positioned on one side of an aluminium sheet and then firmly secured. A releasing agent was then applied within the internal area and the underside cover also made of aluminum sheet. The release agent is to ensure easy removal of the OPWC and WWC after pressing. The required quantities of the OPW and WW, and polyester resin used to form the OPWC and WWC in the mould were obtained using mass fraction model as shown in equations Eq. 1 (Kim et al.2016) The ratio of fibre or particulate to polymer matrix ratio was 20/80 by weight percent and predetermined by the internal area of the mould. Table 2 illustrates the composition of the various OPWC and WWC produced including the percentages of the added Gum Arabic hybridized FR additives. The flame retardants shown in Table 3 were obtained based on suggested range in literature in order not to affect the mechanical properties as well as increase in material cost.

$$V_f = \frac{M_f / \rho_m}{\left(M_f / \rho_m \right) + \left(1 - M_f\right) \rho_m} \tag{1}$$

where, V_f = volume fraction, M_f = mass fraction, ρ_m = density of the matrix (g/cm^3)

Figure 3. Hand lay-up and compression moulding process of the OPWC and WWC in to different panels of gum Arabic hybridized FR

Flammability Index with the Cone Calorimeter Apparatus

The cone calorimetry apparatus (CCA) located at the Rushbrook Fire Laboratory University of Edinburgh, Scotland United Kingdom (Fig 4a) was used to study the flammability (fire reaction properties) of the fabricated agro wastepanels according to ASTM E 1354 standard procedure described by Babrauskas(2002). The specimens (100mm x 100mm x10mm) cut from the OPWC and WWC panels were wrapped in aluminium foil; along the side and bottom to reduce heat losses as specified in the standard. The specimens were inserted in a sample holder and then placed on the load cell as shown in Fig 4b. The load cell measures the specimen mass loss in real time as combustion takes place. The specimens were exposed in the horizontal orientation at irradiance of 50kW/m^2. The magnitude of the applied irradiance can be correlated to the scale and intensity of a fire scenario. An electric truncated cone shape heater is provided to cause thermal decomposition and releasing pyrolysis products from the specimen surface as shown in Fig 4c. The height between the coneheater and the surface of the specimen is maintained at 25mm as specified by the standard. Ignition is produced by an intermediate piloted spark igniter located above the specimen and removed when the pyrolysis products are ignited (Fig 4d). When testing was being conducted, the (CCA) and gas analyzer connected to a data acquisition system and a computer system, records at a fixed interval of 1s the flammability properties. The gasanalyzer needed to perform fundamental cone calorimeter experiments is the oxygen analyzer. Other analyzers such as carbon monoxide and carbon dioxide are usually fitted to give a better understanding of the burning process and to decrease the uncertainties in the results (Babrauskas 1986). The test specimens were repeated severally and a sensible result onaverage was adopted for this study. During the test, the following parameters were obtained: time to ignition (Tig), heat release rates (HRR), mass loss rate (MLR) and smoke production rates

Table 2. Experimental design of the gum arabic hybridized flame retardant composition

Specimen ID	Flame Retardant Composition			
	GA	ATH	APP	CB
OPWC	-	-	-	-
WWC	-	-	-	-
OPWCx	4	-	8	-
WWCx	4	-	8	-
OPWCy	3	-	6	6
WWCy	3	-	6	6
OPWCz	3	9	6	-
WWCz	3	9	6	-

GA: Gum Arabic, **ATH**: Aluminium tri-hydroxide, **APP**: Ammonium polyphosphate, **CB**: Carbon black

RESULTS AND DISCUSSION

In this chapter, flammability properties of the Gum Arabic hybridized FR used in processing the OPWC and WWC at different loading proportion is studied. The flammability results obtained in the cone calorimeter represent all-inclusive fire performance behaviour of the studied OPWC and WWC and a requirement for its safety measure at a bench scale. The obtained flammability properties are significant in predicting full/real scale fire behaviour and provide information for critical modeling of thermal decomposition process. The focus in this section is on the effect of the Gum Arabic with ATH, APP and CB on the reduction of the OPWC and WWC fire risk.

Time to Ignition

One critical property that has been used to quantify a materials performance in fire is the time the material takes to ignite denoted as Tig and it tells how a flame can rapidly spread. Tig occurs when the local concentration of flammable volatiles diffusing through the OPWC and WWC char reaches the flammability lower limit (Kim et al 2016). Among the panels studied, the WWCz containing 3%GA/9%ATH/6%APP and WWCx containing 3%GA/6%APP/6%CB are considered a good fire risk panels for building applications. The flame retardant delayed the Tig respectively by 73.9% and 71.4% compared to those without FR. The mechanism of the FR when thermally decomposed trapped the release of flammable volatiles by releasing constituents that extended the flammable limit thus delaying the start and spread of fire. On the other hand, the FR composition at lower concentration and fibre panels ignited too soon which may have been caused by factors such fibres exposure, high cellulose content of the fibre (Helwig et al. 2000).

Figure 4. Cone Calorimeter set-up at the Rushbrook Fire laboratory, University of Edinburgh, UK (a) Cone calorimeter apparatus (b) Loading of samples by the researcher (c) ignition process taking place (d) Burning of samples after ignition and data recording

Peak Heat Release Rate

The HRR is identified as the key property in evaluating a material's limit to fire safety. High level of HRR indicates a rapid ignition and flame spread. The peak value of the HRR i.e. HRRp gives an indication of the intensity of fire and how it grows as well as the flashover phenomena in real fire situations

(Nikolaeva & Karki 2016). The flashover point is the rapid transition between the primary fire which is essentially localized around a material first ignition and the general conflagration when all surfaces within the enclosed space are burning. High HRR translates to rapid temperature changes, faster flash-over and an increase in the production of products of combustion while lower values are an indication of good flame retardancy.

Figure 5. Time to ignition of the OPWC and WWC at different Gum Arabic hybridized FR loading proportion

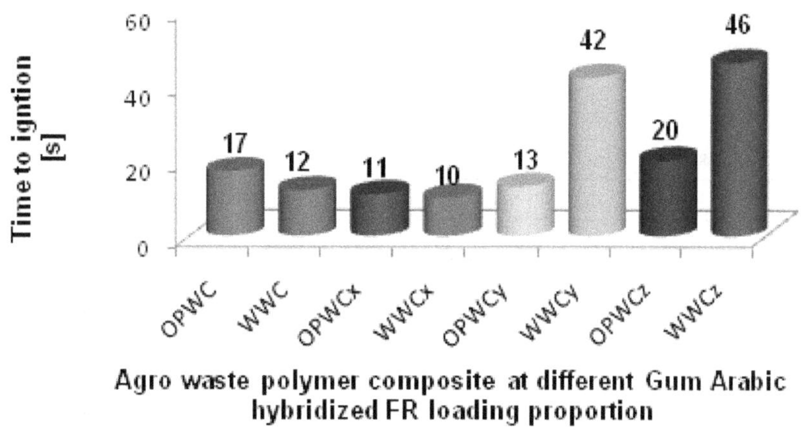

From the HRR curves displayed in Fig. 6 it is clearly seen that shortly after ignition, a sudden rise in the HRR for the OPWC and WWC panels reached a maximum indicating the release of combustible volatiles and then declined slowly. From the curves, it shows that the presence of FR compared to those without FR exhibited a broader appearance which stayed at low level throughout and suggests that the FR composition is interacting with the flammable volatiles by oxygen starvation. This is also a typical characteristic of a char residue-forming material and confirms the role the fibre/particulate constituents that is the lignocellulosic content played in the formation of char during combustion. Besides, the last part of the HRR curves represents glowing combustion of the char residue at solid state interface that is where volatiles has been burnt out. Contrarily, the second sharp peak noticeable on the OPWC and WWC is attributed to a failure in the char structure leading to the release of high rate of combustibles (Schartel and Braun 2003). The underlying composite substrate for OPWC and WWC with double peak HRR is as shown in Fig 7. It is observed that minor pockets of aluminium foil used in wrapping the panels and unfilled char spread indicating a weak char structure that lead to more release of combustibles.

In Table 3 the FR composition in the OPWC and WWC shows that the OPWCz and WWCz were significantly reduced by 67.4% and 71.3% respectively compared to those without FR. This suggests that both panels could be suitable to meet building requirements such as doors, ceilings and floors. The mechanism of GA/ATH/APP acts by diluting the volatiles formed during the burning process, releasing water during dehydration and restricting access of oxygen to the panel's substrate which caused peak

HRR reduction (Alhuthali et al 2012). In addition, the reaction of GA/APP inhibits the mass and volatile transfer between the condense phase and gas phase once a protective carbon layer is formed contributes to the reduction in the HRR.APP in composites creates less reactive radicals in the condense phase and a blanketing effect in the gas phase after initial ignition. The presence of APP dehydrates the carbohydrate in GAP to form stable char structure. The formation of char slows down the rate of combustion of the panels.

Table 3. Summary of the heat release rate of the OPWC and WWC at different Gum Arabic hybridized FR loading proportion

Specimen I.D	Peak Heat Release Rate (kW/m^2)	% Reduction
OPWC	265.5	-
WWC	195.5	-
OPWCx	158.3	40.4
WWCx	68.7	64.9
OPWCy	113.1	57.4
WWCy	78	60.1
OPWCz	86.6	67.4
WWCz	56.1	71.3

Figure 6. Heat release rate of the OPWC and WWC at different Gum Arabic hybridized FR loading proportion

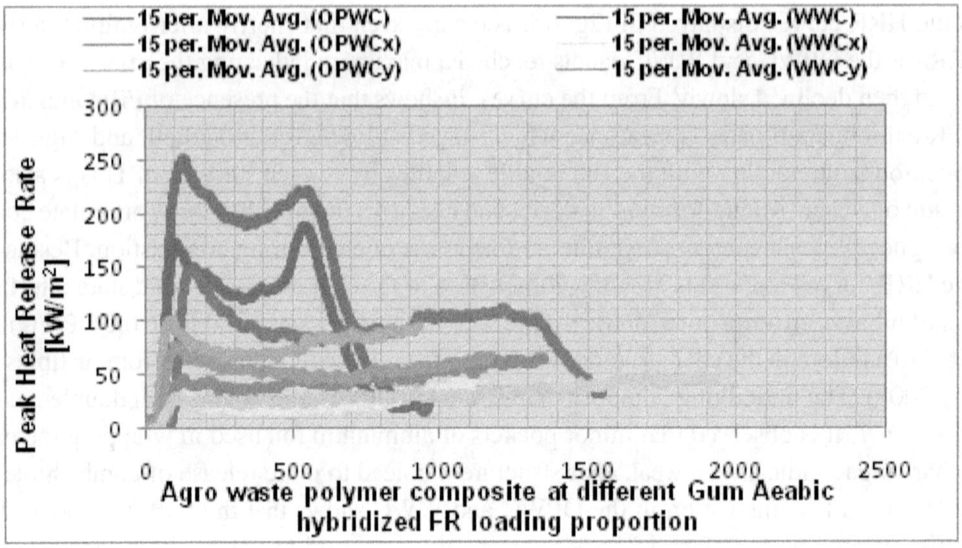

Figure 7. Photographs showing macroscopic images of OPWC and WWC residue surfaces obtained at 50kWm^{-2}in the cone calorimeter (a) OPWC (b) WWC

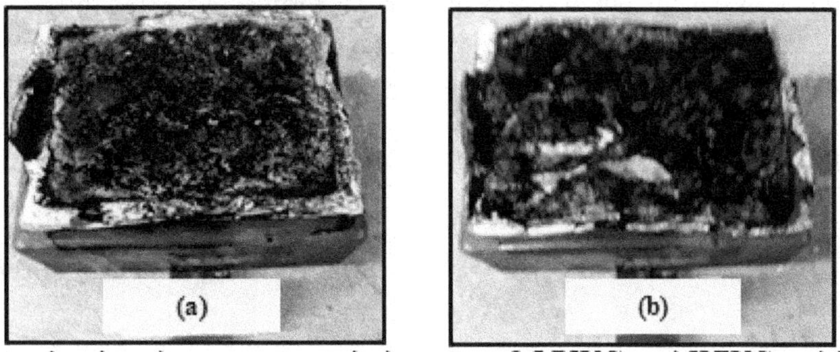

Mass Loss Rate

Mass loss rate (MLR) is a direct measure of dehydration reactions and pyrolysis of materials. The MLR provides the means to evaluate the burning behaviour of the studied OPWC and WWC panels as it plays an important role in heat release and smoke emissions. The MLR as a function of time for the FR compositions in the OPWC and WWC was analyzed for peak mass loss rate. In Fig 8 the MLR curve showed similar characteristics with the HRR curves indicating material loss as heat is evolved. The FR composition; 3%GA/9%ATH/3%APP was more effective in the decrease of the MLR profile which is believed to be due to the stable char formation by the combine FR effect that led to an increased surface area of the char. The cohesive char forming mechanism for both APP/GAP and the engineering ceramic from ATH effectively slowed down the decomposition process of the panel which led to low MLR values as shown in Table 4. The char formation caused a protective layer of compact black char which prevented gas penetration and incomplete combustion. Table 4 also shows that FR composition present in WWCy and WWCz maintained a better stable char structure as the peak MLR was reduced by 62.6% and 66.2% respectively compared to those without FR indicating a slower decomposition rate, yielding more char and suggests good flame retardancy. It was also observed that there was prolong MLR peak time which is thought to be due to accumulation of thermal energy within the studied panels or probably that was when the underlying substrate became involved in the combustion process after being shielded so far by the Gum Arabic hybridized FR (Helwig and Paukszta 2000).

Smoke Production Rate

Smoke production of materials depends on the chemical nature of the organic constituents, oxygen availability and temperature of the fire. When a material burn, heat is released accompanied by very toxic and acrid smoke. The ester bonds present in unsaturated polyester resin which is the polymer used in this study breakdowns at higher pyrolysis temperatures and generates carbon monoxide and carbon oxides (Lee et al. 2011 and Rena et al.2015). The smoke production rate (SPR) of a material is a good parameter in describing fire hazard of materials and a measure of life survivability. The knowledge of SPR is imperative to survival during a fire, as inhalation of smoke is one of the greatest hazards to life. The SPR is a function of time and exhibits similar characteristics with those of the HRR. It is an indication of the release of smoke alongside toxicity from the panel as heat is evolved. In Fig 9 WWCz suppressed the SPR better compared to those without FR. The decomposition of ATH in the 3%GA/9%ATH/6%APP composition which lower the amount of pyrolysis products formed prevented the release of toxic gases (Schartel and Hull 2007). On the other hand, the presence of GA/APP bubble char structure during combustion in the cone calorimeter may have catalyzed the FR components in the panels which played a significant role in the suppression of the smoke and therefore led to an outstanding performance of the WWCz panel.

Table 4. Summary of the mass loss rate of the OPWC and WWC at different Gum Arabic hybridized FR loading proportion

Specimen I.D	Peak Mass Loss Rate (g/s)	% Reduction
OPWC	0.267	-
WWC	0.305	-
OPWCx	0.178	33.3
WWCx	0.162	41.6
OPWCy	0.151	43.4
WWCy	0.108	62.6
OPWCz	0.131	50.9
WWCz	0.103	66.2

Figure 8. Mass loss rate of the OPWC and WWC at different Gum Arabic hybridized FR loading proportion

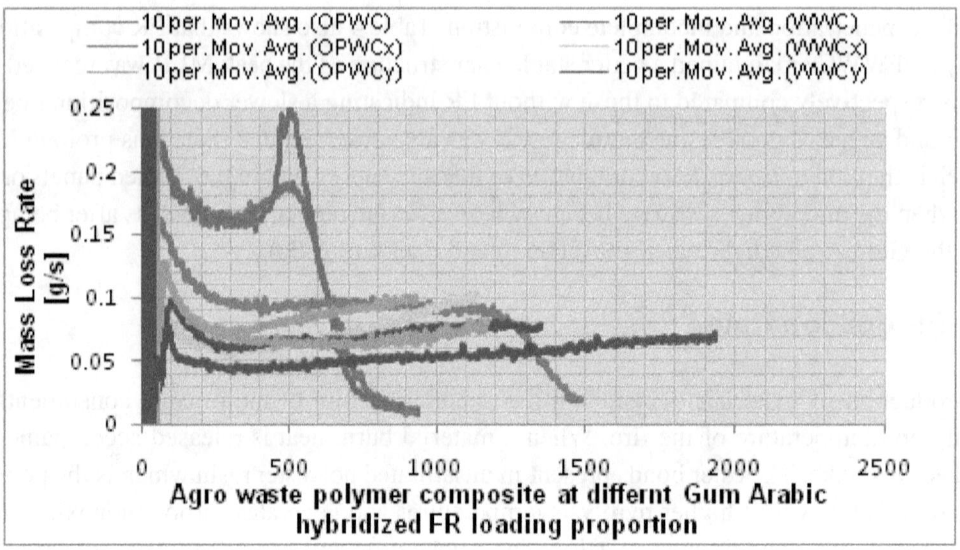

Figure 9. Smoke production rate of the OPWC and WWC at different Gum Arabic hybridized FR loading proportion

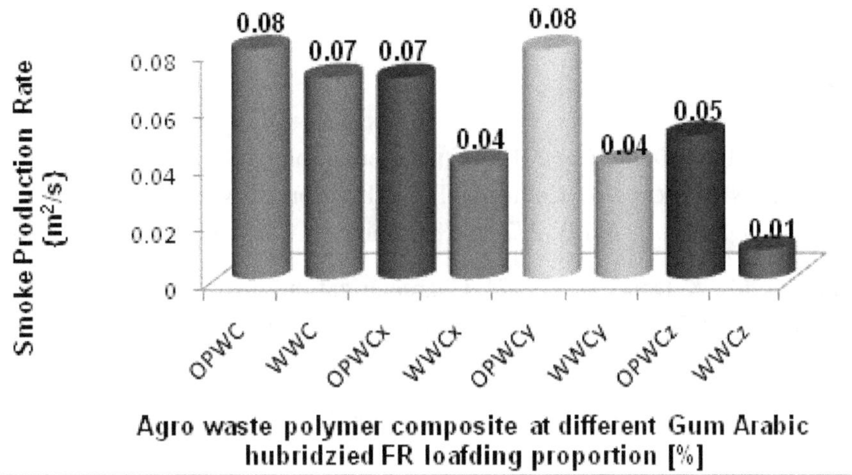

CONCLUSION

The aim of this chapter was to determine the flammability properties of two panels obtained from oil palm waste and wood waste processed with Gum Arabic hybridized aluminium tri-hydroxide, ammonium polyphosphate and carbon. The percentage variation of the flame retardant in the panels is to understand at what optimum loading proportion the required safety measures can be obtained for building applications. On this basis formed the specific objectives of this chapter, the effect of the flame retardant on the panels was drawn and concludes as follows:

1. The OPWC and WWC panels showed high fire risk at lower FR percentage loading proportion but increased in fire safety at a flame retardant composition of 3%GA/9%ATH/6%APP. The OPWCz, WWCy and WWCz represent good flame retardant materials for building applications as the Tig delayed up to 20s from 17s and 46s and 42s from 12s respectively.
2. The intensity of fire and how it grows was significantly reduced with increase in flame retardant. The composition of 3%GA/9%ATH/6%APP in OPWCz and WWCz panels exhibited superior flame retardancy that could meet safety fire conditions in building applications. OPWCz and WWCz were respectively reduced from 265.5kW/m^2 to 86.6kW/m^2 and 195.5kW/m^2 to 56.1kW/m^2 representing 67.4% and 71.3% reduction.
3. The dehydration of the panels were reduced as a result of stable char structure formed in the presence of 3%GA. WWCy and WWCz with the highest percentage reduction of 62.6% and 66.2% followed by OPWCz at 50.9% indicating good flame retardancy that could meet fire safety requirements in building applications.
4. The FR compositions are not good smoke suppressant as there was no significant reduction in the smoke production rate except for WWCz at 85.7% reduction from 0.07m^2/s to 0.01m^2/s compared to those without FR composition.

5. The OPWCz and WWCz at 3%GA/9%ATH/6%APP composition should be incorporated in building applications to meet required fire safety standards.

REFERENCES

Alhuthali, A., Low, I. M., & Dong, C. (2012). Characterization of the Water Absorption, Mechanical and Thermal Properties of Recycled Cellulose Fibre Reinforced Vinyl-Ester Eco-Nanocomposites. *Composites. Part B, Engineering*, *43*(7), 2772–2781. doi:10.1016/j.compositesb.2012.04.038

Alhuthali, A., Low, I. M., & Dong, C. (2012). Characterization of the Water Absorption, Mechanical and Thermal Properties of Recycled Cellulose Fibre Reinforced Vinyl-Ester Eco- Nanocomposites. *Composites. Part B, Engineering*, *43*(7), 2772–2781.

Amenaghawon, N. A., Osayuki-Aguebor, W., & Okiemen, C. (2016). O Production of Particle Boards from Corn Cobs and Cassava Stalks: Optimization of mechanical properties using Response Surface Methodology. *Journal of Materials Environmental Science*, *7*(4), 1236–1244.

Arjmandi, R., Ismail, A., & Hassan, A. (2017). Abu Bakar, A Effects of Ammonium Polyphosphate Content on Mechanical, Thermal and Flammability Properties of Kenaf/Polypropylene and Rice Husk/ Polypropylene Composites. *Construction & Building Materials*, *152*, 484–493.

Ayrilmis, N., Jarusombuti, S., Fueangvivat, V., Bauchongkol, P., & White, R. (2011). Coir Fiber Reinforced Polypropylene Composite Panel for Automotive Interior Applications. *Fibers and Polymers*, *12*(7), 919–926.

Babrauskas, V. (1986). The cone calorimeter – A Versatile Bench-Scale Tool for the Evaluation for Fire Properties. New Technology to Reduce Fire Loses and Cost, 78-87.

Babrauskas, V. (1995). Designing Products for Fire Performance: The State of the Art of Test Methods and Fire Models. *Fire Safety Journal*, *24*, 299–312.

Babrauskas, V. (1986). The cone calorimeter – A Versatile Bench-Scale Tool for the Evaluation for Fire Properties. New Technology to Reduce Fire Loses and Cost, 78-87.

Babrauskas, V. (2002). The Cone Calorimeter. In *SFPE Handbook of Fire Protection Engineering*. Society of FireProduction Engineers.

Chauhan, N., & Arya, N. (2018). Coconut fiber: A natural versatile material. International of Chemical *Studies*, *6*(6), 555-561.

Dong, Y., Ghataura, A., Takagi, H., Haroosh, H. J., Nakagaito, A. N., & Lau, K. T. (2014). Polylactic Acid (PLA) Biocomposites Reinforced with Coir Fibres: Evaluation of Mechanical Performance and Multifunctional Properties. *Compos Part A*, *63*, 76–84.

Dorez, G., Taguet, A., Ferry, L., & Lopez-Cuesta, J. M. (2013). Thermal and Fire Behavior of Natural Fibers/PBS Biocomposites Polymer. Degradation and Stability, 98, 87-95.

Ezeh, E. M., Onukwuli, O. D., Odera, R. S., Ugonabo, V. I., & Okeke, B. (2020). *O Thermal Decomposition and Flame retardancy of Functional Polyester/Banana Peduncle fibre and Aluminum Hydroxide particle Hybrid Composites Chemical and Process* (Vol. 62). Engineering Research.

Foley, M., & Drysdale, D. (1994). Note: Smoke Measurements and the Cone Calorimeter. *Fire and Materials*, *18*, 385–387.

Gokul, K., Prabhu, T. R., & Rajasekaran, T. (2017). Processing and Evaluation of Mechanical Properties of Sugarcane Fiber Reinforced Natural Composites. *Transactions of the Indian Institute of Metals*, *70*(10), 2537–2546. doi:10.100712666-017-1116-8

Greyson, S. J., Babrauskas, V., & Hirschler, M. M. (1994). A New International Standard for Flammability Testing. *Plastics Engineering*, 29–31.

Helwig, M., & Paukszta, D. (2000). Flammability of Composites Based on Polypropylene and Flax Fibres. *Mol Cryst Liquid Cryst*, *354*(1), 373–380.

Indira, K. N. Parameswaranpillai, J., & Thomas, S. (2013). Mechanical Properties and Failure Topography of Banana Fiber PF Macrocomposites Fabricated by RTM and CM Techniques. *International Scholarly Research Notices*. doi:10.1155/2013/936048

Ishai, O., & Daniel, I. (2006). *M Engineering Mechanics of Composite Materials* (2nd ed.). Oxford University Press.

Kim, Y. T., & Eng, H. Y. (2016). Tang LWThe effects of weight fraction on mechanical behaviour of Thermoset Palm EFB composite. *Int. J. Mater. Mech.Manuf*, *4*(4), 232–236.

Lee, B. H., Kim, H. S., Kim, S., Kim, H. J., Lee, B., Deng, Y., Feng, Q., & Luo, J. (2011). Evaluating of the Flammability of Wood-Based Panels and Gypsum Particleboard using Cone Calorimeter. *Construction & Building Materials*, *25*, 3044–3050.

Mahesh, D., Kowshigha, K. R., Raju, N. V., & Aggarwai, P. K. (2020). Characterization of Banana Fibre-Reinforced Polypropylene Composite. *Journal of the Indian Academy of Wood Science*, *17*, 1–8.

Mallick, P. K. (1999). Compression Moulding, Composite materials Technology. Hanser Publishers.

Messiry, M. E. (2013). Theoretical Analysis of Natural fiber Volume Fraction of Reinforced Composites. *Alexandria Engineering Journal*, *52*, 301–306.

Monti, M., & Camino, G. (2013). Thermal and Combustion Behavior of Polyether-sulfone-Boehmite Nanocomposites. *Polymer Degradation & Stability*, *98*(9), 1838–1846.

Mouritz, A. P., Mathys, Z., & Gibson, A. G. (2006). Heat Release of Polymer Composites in Fire Composite Part A. *Appl Sci Manuf*, *37*(7), pp1040–pp1054. doi:10.1016/j.compositesa.2005.01.030

Munawar, R. F., Jamil, N. H., Shahril, M. K., Abdul Rahim, M., Abidin, M. Z. Z., Azam, M. A., & Lau, K. (2015). Development of Green Composite: Pineapple Leaf Fibers (PALF) Reinforced Polylactide (PLA). *Applied Mechanics and Materials*, *761*, 520–525.

Nair, K. C. M., Kumar, R. P., Thomas, S., Schit, S. C., & Ramamurthy, K. (2000). Rheological Behavior of Short Sisal Fiber-Reinforced Polystyrene Composites. *Composites. Part A, Applied Science and Manufacturing*, *31*(11), 1231–1240. doi:10.1016/S1359-835X(00)00083-X

Nikolaeva, M., & Karki, T. (2016). Influence of Fire Retardants on the Reaction to Fire Properties of Coextruded Wood-Polypropylene Composites. *Fire and Materials*, *40*, 535–543.

Norzali, N. A. R., Badri, K. H., & Nuawi, M. (2011). Z Loading Effect of Aluminum Hydroxide onto the Mechanical, Thermal Conductivity, Acoustical and Burning Properties of the Palm-based Polyurethane Composites. *Sains Malaysiana*, *40*(7), 737–742.

Olorunmaiye, J. A., & Ohijeagbon, I. (2015). O Retrofitting Composite Ceiling Boards with Jatropha Curcas Seedcake Material*J. Production Engineering*, *18*(2), 96–102.

Patil, R. G., & Kolambe, B. N. (2011). Development of value added products from Banana Pseudo-stem. An Overview of Progress. National Agricultural Innovation Project.

Prabhakar. (2015). A Review on the Flammability and Flame Retardant Properties of Natural Fibres and Polymer Matrix Based Composites. *Composite Research*, *26*(2), 29–39.

Racier, S., Tomkova, B., Milady, J., Jabbar, J., & Kale, B.M. (2015). Development of a Bio Composite Based on Green Epoxy Polymer and NaturalCellulose Fabric (Bark Cloth) for Automotive Instrument Panel Applications. *Compos Part B, 81*, 149-157.

Rajini, N., Jappes, J.T.W., Siva, I., Rajulu, A.V., & Rajakarunakaran, S. (2017). Fire and Thermal ResistanceProperties of Chemically Treated Ligno-Cellulosic Coconut Fabric–Reinforced PolymerEco-Nanocomposites. *Journal of Industrial Fabric, 47*(1), 104–124.

Rakotomalala, M., Wagner, S., & Doring, M. (2010). Recent Development in Halogen Free Flame Retardants for Epoxy Resins for Electrical and Electronic Applications. *Materials (Basel)*, *3*(8), 4300–4343.

Ramli, R., Yunus, R.M., & Beg, M.D.H., & Prasad, D.M.R. (2011). Oil Palm Fiber Reinforced Polypropylene Composites: Effects of Fiber Loading and Coupling Agents on Mechanical, Thermal, and Interfacial Properties. *Journal of Composite Materials*, *46*(11), 1275–1284.

Redwan, A. M., Badri, K. H., & Tarawneh, M. (2015). A The effect of Aluminium Hydroxide on the Mechanical Properties of Fire Resistivity of Palm-Based Fiberborad Prepared by Pre-Polymerization Method. *Advanced Materials Research*, *1087*, 287–292.

Rena, Y., Wang, Y., Wang, L., & Liu, T. (2015). Evaluation of Intumescent Fire Retardants and synergistic Agents for use in wood Flour/Recycled Polypropylene Composites. *Construction & Building Materials*, *76*, 273–278.

Schartel, B. (Ed.). (2003). *Fourteenth Annual BCC Conference on Flame Retardancy, Fire Retarded Polypropylene/Flax Biocomposites*. Academic Press.

Schartel, B., & Hull, T. (2007). R Development of Fire Retardant Materials-Interpretation of Cone Calorimeter Date. *Fire and Materials*, *31*, 327–354.

Scudamore, M. J., Briggs, P. J., & Prager, F. H. (1991). Cone calorimetry – A Review of Tests Carried out on Plastics for the Association of Plastics Manufacturers in Europe. *Fire and Materials*, *15*, 65–84.

Sebastian Rabe, S., Sanchez-Olivares, G., & Ricardo Pérez-Chávez, R., & Schartel, B. (2019). Natural Keratin and Coconut Fibres from Industrial Wastes in Flame Retarded Thermoplastic Starch Biocomposite, *Materials* , *12,* 344. doi:10.3390/ma12030344

Subasinghe, A., Das, R., & Bhattacharyya, D. (2016). Study of Thermal, Flammability and Mechanical Properties of Intumescent Flame Retardant PP/Kenaf Nanocomposite. *International Journal of Smart and Nano Materials*, *7*(3), 202–220. doi:10.1080/19475411.2016.1239315

Subastinghe, A. (2014). Bhattacharya, D Performance of Different Intumescent Ammonium Polyphosphate Flame Retardants in PP/Kenaf Fibre Composites. *Composites. Part A, Applied Science and Manufacturing*, *65*, 91–99.

Suoware, T.O., Amgbari, C.O, Sorgbara, F., & Kigigha, A.W. (2020). A Review of the Fire Behaviour of Agro-Waste Fibre Composite Material for Industry Utilization in Nigeria. *International Interdisciplinary Conference of Federal Polytechnic Ekowe and Konkan Geographers Association of India.*

Vijaya, S., Ravi, N. M., Helmi, S., & Choo, Y. M. (2013). The Development of a Residual Oil Recovery System to Increase the Revenue of Palm Oil Mill. *Journal of Oil Palm Research*, *25*(1), 116–122.

Zuhudi, N. Z. M. (2014). *Flammability and Durability of Bamboo Fabric Reinforced Polypropylene Composites and their Hybrids* [Ph.D. thesis]. University of Auckland.

Chapter 16
A Review of the Fire Behaviour of Agro-Waste Fibre Composite Material for Industrial Utilization in Nigeria

Timine Suoware
https://orcid.org/0000-0001-9618-3363
Federal Polytechnic, Ekowe, Nigeria

Charles O. Amgbari
Federal Polytechnic, Ekowe, Nigeria

ABSTRACT

Agro-waste fibres (AWFs) are of very reasonable economic value and have attracted a lot of research interest globally. AWFs have significant potential in composites due to low cost, high strength, environmental benign nature, availability, and sustainability. Nevertheless, dozens of the investigations have failed to give the desired attention of the fire behaviour of AWF composite materials when exposed to heat atmosphere. In this chapter, a detailed account of AWF composite material combustion process was discussed to understand the reason for the high heat release rates (HRR) caused by cellulosic content. HRR is a critical component of the flammability properties of AWF composite materials as the fire behaviour largely depends on it. The prospective examination of AWF composite material fire behaviour gives the required direction about the limits of their utilization and developmental trends for future industrial applications in Nigeria.

INTRODUCTION

The activity of modern day agricultural processes has caused the availability of more agro-waste predicating a threat to the environment. Agro-waste are abundant and they of low-cost with the advantage of low density, high toughness, relatively high specific strength/stiffness properties, low abrasiveness, low

DOI: 10.4018/978-1-7998-9574-9.ch016

energy consumption in composites build-up and CO_2 neutrality as reported by Ismail et al (Islam et al., 2011) and Sukyai et al (Sukyai et al., 2012). The waste generated from the agricultural sector in Nigeria can be a good alternative to the shortage of raw materials to meet the needs in the transportation and building industries. Besides, environmental concerns for the use of eco-friendly materials from natural sources have attracted more research attention globally. Currently, research is focused on examining and exploring agro-waste materials that are compatible with the environment and their development to meet required fire safety conditions.

Agro-waste fibres (AWF) such as oil palm fibre (OPF), coconut fibre (CF), banana fibre (BF), sugarcane fibre (SF), pineapple fibre (PF) and wood fibre (WF) are abundant fibres that can be extracted from agro-waste in Nigeria. These fibres consist of organic constituents comprising cellulose, hemicellulose, lignin and pectin and derive their strength and rigidity from cellulose that is semi-crystalline polysaccharide in nature (Sorieul et al., 2016). Cellulose-based AWF obtained from plants (i.e. stem, fruit, seed, leaf etc.) are broadly used in the manufacturing of composite materials due to their wide availability and renewability within a short time. Lignocellulosic fibres are the most abundant and renewable bio-based materials source in nature (Majeed et al., 2013). Among AWF the bastfibres, extracted from the stems of plants have been accepted as the best candidates for reinforcements of polymer composites due to their good mechanical properties (Nair et al., 2000).

Composite materials are a unified materials structure with a distinct interface comprising fibresas reinforcements embedded in polymers referred to as a binder or matrix. The matrix is what positions the fibres in orderly pattern and improve the mechanical properties as well. AWF have attracted considerable interest for the reinforcement in various synthetic polymers such as epoxy, unsaturated polyester, polyurethane, phenolic, polypropylene, polyethylene, chitin, chitosan, soybean, wheat gluten, poly lactic acid etc. Several researchers (Arrakhiz et al., 2013; Eng et al., 2014; Jayamani et al., 2014; John & Thomas, 2008; Khalili et al., 2017; Ku et al., 2011; Shinoj et al., 2011) have shown the utilization and applications of various AWF reinforced polymer composite materials. Notwithstanding, there is a delicate issue that needs to be solved in other to further improve the product performance and general acceptability; their poor inherent fire behaviour limits their usewhere fire hazard is an important design consideration. The fire behaviour of AWF reinforced composite materials is not quite known. It is imperative to give important information of their fire behaviour to aid in industrial applications especially where stringent fire safety regulations do exist.

In this review paper, a detailed examination of AWF composite materials fire behaviour for utilization in the transportation and buildingindustries is presented. A discussion of their combustion process to further identity research gaps for possible future development is examined. The paper is organized into five sections. The first examines the chemical and physical composition of AWF. The second outlines the common agro waste and their fibre extraction. Thirdly this section discusses the novelty research on the fire behaviour of AWF composite materials. Several review reports are examined critically in order to justify the purpose of the study. The fourth section further explores the utilization and prospects of the AWF composite for transportation and building as well as other industries in Nigeria. Finally the fifth section concludes on the findings of these review paper.

COMPOSITION AND PROPERTIES OF AGRO-WASTE FIBRES

Agro-waste can be defined as any material that is un-utilized i.e. left over from a production process. Nigeria agro-waste is being influenced by increasing agricultural production and productivity. The lack of information on the quantity of agro-waste has constrained industrial utilization of agro-waste in Nigeria. This review paper identifies the common agro-waste in abundance in Nigeria and their correspondingfibre extractions as shown in Fig.1. This agro-waste includes but not limited to coconut coir waste, oil palm waste, banana waste, sugarcane waste and pineapple waste.

Figure 1. Common Agro-waste and their processed fibresby extraction

Agro-Waste and Their Fibre Extraction

Coconut Coir Fibre

Coconut palm *(cocos nucifera, family Arecaceae)*is agricultural waste products obtained in the processing of coconut oil, and is obtainable in large quantities in the tropical regions of the world, most especially in Asia, Africa and southern America.Coconut coir fibre is the extract of the outer shell of a coconut husk and this is carried out by a laborious process which is by decorticated and beating done by hand before a bleaching or dyeing process depending on the effectiveness of the wetting process (Chauhan & Arya, 2018). In the shell is covered by the endocarp while exocarp (the smooth outer skin) and mesocarp (the fibrous covering). There are generally two types of coir fibres namely brown fibre which are thick, strong and have high abrasion resistance which are gotten from mature coconuts while the other a white fibre is smoother, finer and also weaker gotten from immature green coconuts. A fully matured coconut tree can produce 50-100 coconuts per year and each fruit takes one year to ripen (Harris, 2012). World production of coconut fibre is in tons from coconut palm residues.

Table 1. Chemical and physical composition of common AWF

AWF	Origin	Cellulose (%)	Hemicellulose (%)	Lignin (%)	Pectin Content	Moisture Angle (Wt. %)	Microfibrillar	Fibre Length (mm)	Fibre Width (mm)	Tensile Strength (MPa	Young Modulus (GPa)	Reference
Coconut Fibre	Stem	32-34	0.15-0.25	40-45	3-4	8	30-49	220	-	-	-	[13]
Oil palm Fibre	Fruit	65	22.1	29	40-46	248	3.2	-	-	-	-	[14]
Banana Fibre		50-60	25-30	12-18	10-12	9-13	355	33.8	-	-	-	[15]
Sugarcane Fibre	Stem	81	9.45	10-40	0.8-2.8	10-3420	290	19.7-27.1	-	-	-	[16]
Pineapple Fibre	Leaf	85	5	12.7	10-15	170-1627	82	-	-	-	-	[17]

Oil Palm Fibre

Oil palm (*Elaeisguineensis,family Arecaceae*) is regarded as huge amount of lignocellulosic waste and unutilized. These products normally cause major environmental pollution that can be readily turned into useful value-added products (Shiner et al., 2011). Thefibres also referred to as a lignocellulosic fibre is extracted from oil palm trunk, oil palm frond, fruit mesocarp and empty fruit bunch by a water retting process. Oil palm fibre (OPF) is hard and tough. The female bunch carries about 2500–3000 fruits borne on 100–120 spikelets attached to a peduncle from the axil of a frond. The fruits produce two main products, palm oil from the outer mesocarp and palm kernel oil from the kernel within the nut. OPF is reported to originate from tropical forests in West Africa and Nigeria is the fifth largest producer nation of oil palm accounting for about 930, 000 metric tons of global output (Izah & Ohimain, 2016).

Banana Fibre

Banana *(Musa paradisiaca, family musaceae)* is one of the oldest fruit after citrus, a central fruit crop of the tropical and subtropical regions of the world grown on about 8.8 million hectares as reported by Mohapatra et al. (Mohapatra et al., 2010). Banana waste is in huge amount as a result of its cultivation which amounts to about 102 metric tons causing environmental hazard and un-utilized, can be converted to value added products (Adeniyi et al., 2019). Banana has one of the tallest herbaceous plants with a pseudo-stem. Its tough treelike pliable stem is composed of the sheathing twisting leaf bases, which contains fibres of sufficient strengths to keep the tree upright. Banana fibre (BF) can be extracted either manually or by a cutting machine. The manual process is quite cumbersome which requires cutting the pseudo-stem into the required length and width before scrapping with a flat blunt blade while the cutting machine is the most preferred process as it cuts the pseudo-stem into two halves from which sheaths are separated as reported by Patil and Kolambe(Patil & Kolambe, 2011).

Sugarcane Fibre

Sugarcane *(saccharum officinarum, family poaceae)* is usually cultivated for its juice from which sugar is processed. The waste i.e. an abundant agricultural lignocellulosic by-products, is a solid fibrous sugarcane Salk residue after crushed sugarcane sucrose extraction which produces about 32% of bagasse (Essays, 2014). Thebagasse is the main residue of the sugarcane industry representing, by weight, almost 30% (hundreds of millions of tons per year world-wide) of the sugarcane agricultural product, therefore bagasse utilization is important for both economic and environmental considerations. The sugarcane fibre from bagasse was extracted by removing the soft-core part pith after juice extraction manually to get outer hard rind and cut across length avoiding the nodes.

Pineapple Leaf Fibre

Pineapple *(Ananascomosusfamily Bromeliaceae)* is an important tropical and subtropical plant widely cultivated in many places including the Philippines, Thailand, Malaysia, Mexico, South Africa, Costa Rica, China, Brazil, and Nigeria. Nigeria is the 7th largest producer of fresh pineapple with 1.4 million metric tons capacity (Baruwa, 2013). The pineapple waste generated consists of residual skin, stem and leaves. The pineapple waste can be converted into highly valuable product.It is a by-product of a pineapple processing industry.Improper handling of this waste would result in the deterioration of environment quality which can be attributed mainly to degradation of the sugar rich contents. The fresh leaves yield about 2% to 3% of fibres (Das et al., 1978). The leaf fibres are extracted either by a mechanical roller and bladder machine system or by a chemical retting process as reported by Sarah et al (Sarah et al., 2018).

Chemical and Physical Composition of Agro-Waste Fibres

Chemical Composition

The agro-waste fibres identified are generally lignocellulosic in nature with high variability in properties. AWF has a complex nature consisting of helically wound cellulose micro-fibrils in matrix of lignin and hemicellulose as reported by Young (Young et al., 1986), Mohanty et al (Mohanty et al., 2001) and Zafeiropoulos et al (Zafeiropoulos et al., 2002).The chemical composition of AWF as reported by Azwa et al (Azwa et al., 2013) and Squire et al (Siqueira et al., 2009) composed mainly of cellulose, hemicellulose, and lignin with minor amount of pectincontents and varies according to fibre type as shown in Table 1.The chemical substances are scattered around the cell wall which is affected by fibre origin, fibre climate condition during cultivation and extraction method (Frollini et al., 2000). The composition of these fibres determines their utilization as well as based on other factors such as: fibre structure, fibre cell dimensions, microfibril angle and chemical composition (Osorio et al., 2010; Frollini et al., 2000). Among the constituents of AWF, cellulose is the major component which occurs largely in a natural crystalline form comprising partially aligned or oriented linear polymer chains. The elementary unit of cellulose macromolecule is anhydrous D-glucose, which contains three hydroxyls (OH). These hydroxyls form hydrogen bonds inside the macromolecule itself and also with hydroxyl groups which is up to 3-13% of moist air (Bledzki et al., 1996). The large amount of hydroxyl group in cellulose gives AWF hydrophilic properties when used to reinforce hydrophobic matrix; the result is a very poor interface and poor resistance to moisture absorption (Alvarez et al., 2003). The chemical composition play a significant role in AWF as it influences many factors from their fire behaviour to their applications and long life behaviour of the composite materials. It is reported by Alhuthali et al (Alhuthali et al., 2012), that high cellulose in fibres obtained from natural sources such as AWF indicates a higher flammability i.e. fire response whereas lower concentration can be found with lower lignin concentration.

Physical Composition

The physical composition of agro-waste fibres is very important to know as there are several parameters to consider which can affect the fibre from reaching its highest potential. The parameters in consideration of AWF composition are fibre dimensions, defects; strength, variability, crystallinity, and structure. Fibre-dimension i.e. its length and width as shown in Table 1 is an important parameter in comparing different types of AWF. High aspect ratio (length/width) in AWF composite materials gives an indication of the strength properties as reported by (Bhal & Singh, 1998; Mallick, 1993). In AWF appropriate selection, the fibre strength plays a significant role for specific applications. Table 1 shows the tensile strength of the AWF and it can be seen that the range of pineapple fibre is high than others and suggests a better application. AWF fibre structure can be obtained using scanning electron microscope (SEM) to show the fibres morphology. The vast array of fibre structures can be found in different literatures (Kavitha, 2016; Ramesha et al., 2013; Senthiil & Sirsshti, 2014; Wang et al., 2019). The intent in this paper is to show the structure of the AWF and not go into details of the cell wall architecture. The physical composition differs from one AWF to another by density, cell wall thickness, and tracheid length and diameter etc. Fig 2 shows the fibre structures of the AWF, the fine structure may account, in part, for their very high equilibrium moisture content as reported by Rowell et al (Frollini et al., 2000).

Figure 2. Samples of scanning electron microscope for the agro-waste fibresstructure

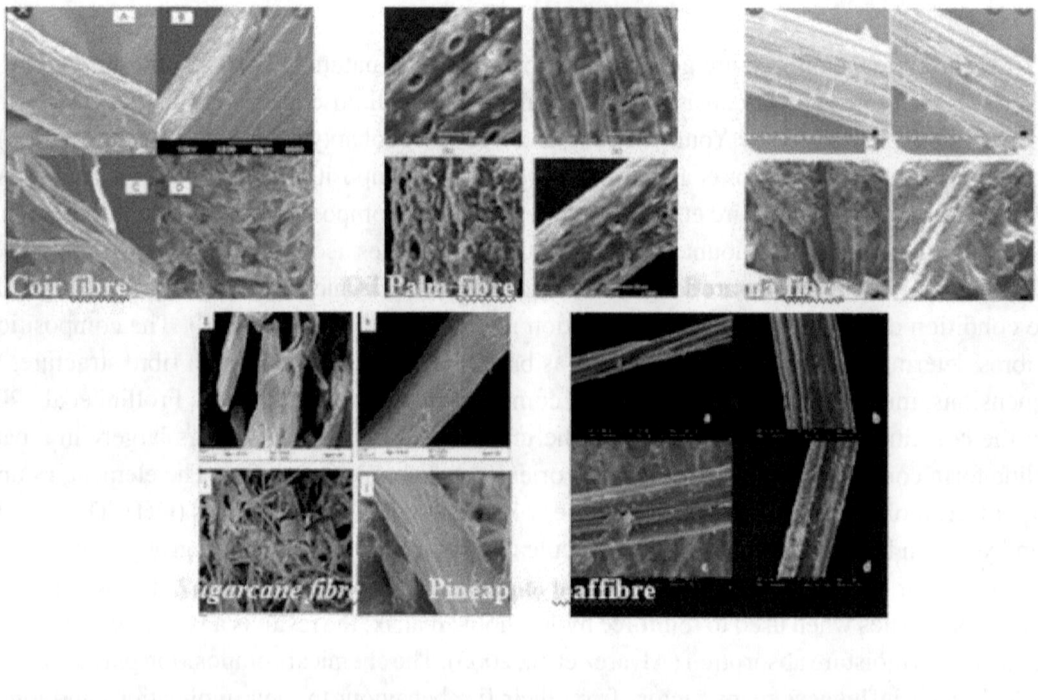

Fabrication Techniques of Agro-Waste Fibre Composite Materials

There are several ways AWF and polymer can be fabricated to form a unified composite material structure (fibre-matrix), fibre - "the reinforcement" and matrix - "the polymer".Polymers are formed by chemical reactions in which a large number of molecules called monomers are joined sequentially, forming a chain. In many polymers, only one monomer is used. In others, two or three different monomers may be combined. Polymers are produced commercially on a very large scale and have a wide range of properties and uses. They also have unique, molecularly definable properties and process advantages unattainable in any other kind of material (Gao et al., 2012). There are a number of categories which polymers can be classified. The simplest of classifications is whether they are synthetic or biodegradable polymers.

Most of the fabrications techniques commonly use for fabricating glass fibre composites are applicable for fabricating AWF composites and they have been fully discussed by various authors (Aveston et al., 1980; Gutowski, 1997; Ishai & Daniel, 2006; Mallick et al., 1990). Among the techniques, hand lay-up as shown in Fig. 3 is the oldest, cheapest, simplest and most common method for making fibre composite products (or laminates). In general the process of fabricating a composite using the hand lay-up method begins with the preparation of a mould according to ASTM standards. The prepared mould is applied with a release agent preferably a polyvinyl chloride (PVC) and waxed with a gel coat (for resistance to weathering and corrosion) with a brush then allow to cure. Composite products are then obtained by impregnating fibre reinforcement's layer by layer with accelerated catalyst and resin matrix which are then rolled with special rollers to consolidate the reinforcement and to drive any entrapped air. When the resin has gelled, but not fully cured, the surplus laminates is trimmed off, to the edges of the mould, the laminate is allowed to cure fully after which it can be freed and lifted from the mould. A summary of the various fabrication techniques by researchers on AWF with polymermatrixes to form various composite materials is presented in Table 2.

Table 2. Reported fabrication techniquesby various researchersfor the selected agro-waste fibres composite materials

Type of Composite Material			
Agro-waste Fibre	**Polymer Matrix**	**Fabrication Technique**	**Reference**
Coconut sheathfibre	Epoxy Resin (ER)	Hand Lay-up	[46]
Oil palm fibre	Unsaturated polyester (UP)	Hand Lay-up/Compression moulding	[47]
Coir fibre	Epoxy Resin (ER)	Hand Lay-up	[45]
Sugarcane fibre	Unsaturated polyester (UP)	Hand Lay-up	[48]
Pineapple leaf fibre	Epoxy Resin (ER)	Hand Lay-up	[49]
Coir fibre	Poly Lactic Acid (PLA)	Laminate fabrication	[50]
Banana fibre	Epoxy Resin(ER)	Hand Lay-up	[51]
Oil palmfibre	Polypropylene (PP)	Injection moulding	[52]
Coir fibre	Polypropylene (PP)	Manual hot pressing	[53]
Sugarcane fibre	Epoxy Resin (ER)	Hand Lay-up	[54]
Pineapple leaf fibre	Polyester	Hand Lay-up	[55]
Oil palm fibre	Epoxy Resin (ER)	Hand Lay-up	[56]
Oil palm fibre	Poly-vinyl Alcohol (PVA)	Hand Lay-up	[57]
Banana fibre	Polypropylene (PP)	Twin screw extruder	[58]
Coir fibre	Polyester	Hand Lay-up	[59]
Banana fibre	Polyester	Hand Lay-up	[60]
Sugarcane fibre	Polystyrene (PS)	Hot pressing	[61]
Sugarcane fibre	Polypropylene (PP)	Screw extruder	[62]
Oil palm fibre	Polystyrene (PS)	Hot press, compression moulding	[63]
Pineapple leaf fibre	Poly-lactic acid (PLA)	Hot pressing laminating	[64]
Pineapple leaf fibre	Polypropylene (PP)	Compression moulding	[65]
Sugarcane fibre	Phenol Formaldehyde Resin	Hot pressing	[66]
Banana fibre	Phenol Formaldehyde Resin	Compression moulding	[67]
Banana fibre	Polyethylene	Compression moulding	[68]
Pineapple leaf fibre	Starch	Hot pressing	[69]

Figure 3. Hand lay-up manufacturing process
(Santos et al., 2018)

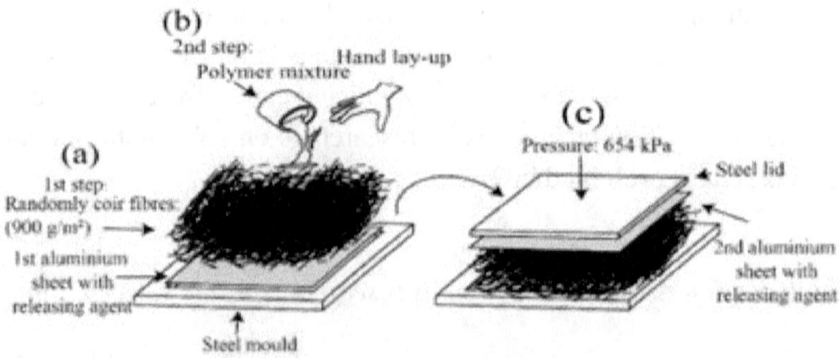

Description of the Fire Behaviour of Agro-Waste Fibre Composites

The fire behaviour of AWF composites from literature reportsshows it is quite complex. To understand their reaction to fire, Baillie and Jayasinghe (Baillie & Jayasinghe, 2017) described in three simplified stages the combustion process which includes the preliminary stage, flame stage of combustion (main stage) and the flameless stage (final stage). The focus in this paper is how the AWF cellulosic and lignin contents contribute or reduce additional fuel load during various stages.

The Preliminary Stage (Thermal Decomposition)

This is the stage where slow thermal decomposition occur releasing gaseous products, weakening the bonds in the composite structure and discoloring of the AWF (Baillie & Jayasinghe, 2017). The thermal decomposition of compositesis an important phase change phenomenon in understanding the behaviour of AWF composites in fire. When solids such as the AWF composite materials are exposed to a heat source, they usually undergo both chemical decomposition and physical degradation. The beginning of thermal decomposition of AWF composites starts with lignin at 160°C where it contributes to the formation of char residue due to ether phenol formation and carbon-carbon links separation (Baillie & Jayasinghe, 2017; Horrocks & Price, 2001). Hemicellulose is next to decompose between 180°C to 260°C which produces high non-combustible gases containing about 70% incombustible CO_2 and 30% combustible CO (Baillie & Jayasinghe, 2017). Finally, cellulose decomposes between 260°C to 360°C forming combustible volatiles and gases, non-combustible gases, tar and char (Baillie & Jayasinghe, 2017; Horrocks & Price, 2001).

The decomposition process of solid state materials is generally conducted through thermogravimetric analysis (TGA), differential scanning calorimeter (DSC) and differential thermal Analysis (DTA). Among these techniques TGA is the most common technique used for kinetic analysis of the decomposition process and it provides the possibility of evaluating the mass loss of a sample with temperature and time (Steinhaus, 2010). A typical thermogravimetric analysis (TGA) curve is shown in Fig 4 for lignin and cellulose. Lignin decomposed at a wider range of 215°C-585°C whereas cellulose decomposed between 286°C-426°C. In table 1 it shows that sugarcane and pineapple AWF exhibited higher percentage of cellulosic content while coir fibre and sugarcane has the highest lignin content. A compilation of the thermal decomposition of the agro-waste fibre composite materials at initial and fully degraded temperatures by various researchers is shown in Table 3. The common and most widely used polymers; epoxy, polyester and polypropylene were considered here because of their availability and cost. Further reading on the polymers decomposition can be found elsewhere (Basnayake et al., 2018). A corresponding plot for the fully degraded temperature in Fig 5 shows the AWF composite materials fall within the range of 330°C-524°C which is within the range reported by Zhao et al (Zhao et al., 2016). It can be seen in Fig 5 that the decomposition of banana AWF in epoxy was fully decomposed at a higher temperature of 524°C containing 50-60% of cellulose against the highest cellulose of 85% for pineapple AWF which was fully degraded at 395°C.For polyester and polypropylene it was observed that sugarcane AWF with 81% cellulose content exhibited the highest decomposed temperature respectively at 480°C and 465°C. On the other hand, coconut coir fibre with the highest lignin content of 40-45% began degradation at 230°C followed by oil palm fibre with 29% lignin content at 150°C less than coir fibre. However, the variability of the temperature valuesmay depend on many factors not limited to sample preparation method, volume fraction of AWF, environmental conditions and manufacturing of the polymer.

Figure 4. TGA curves for cellulose and lignin
(Cordeiro et al., 2017)

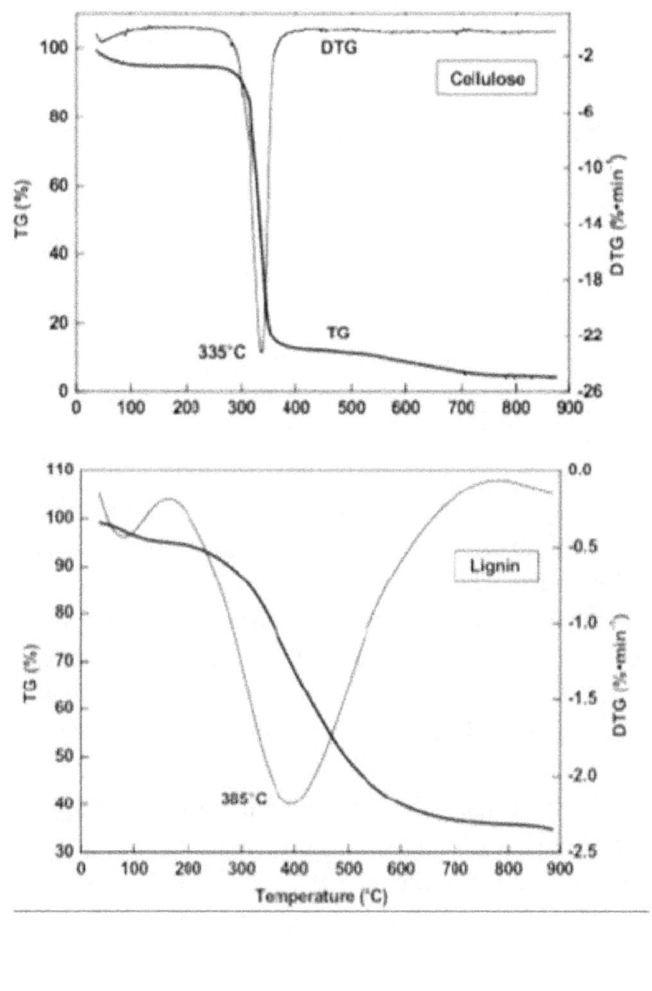

Table 3. Thermal stability of agro-waste fibre composites materials

AWF Composite	Beginning of thermal Degradation (°)	Fully Degraded Tempt. (°)	Ref
Coir Sheath Fibre/Epoxy	230	400	[46]
Peach Palm Tree Fibre/Epoxy	150	330	[73]
Banana Fibre/Epoxy	239	524	[74]
Sugarcane Fibre/Epoxy	200	440	[75]
Pineapple Leaf Fibre/Epoxy	255	395	[76]
Coir Fibre/Polyester	220	380	[77]
Oil Palm Fibre/Polyester	150	442	[47]
Banana Fibre/Polyester	290	420	[78]
Sugarcane Fibre/Polyester	270	480	[79]
Pineapple Leaf Fibre/Polyester	250	460	[80]
Coir Fibre/Polypropylene	250	420	[81]
Oil Palm Fibre/Polypropylene	256	423	[82]
Banana Fibre/Polypropylene	252	405	[83]
Sugarcane Fibre/Polypropylene	260	469	[84]
Pineapple Fibre/Polypropylene	250	455	[85]

The Main Stage (Combustion Properties)

In the second stage classified as the flame stage of combustion and is an active process of decomposition where the product of thermal decomposition of the AWF composite material ignites with a substantial increase in the amount of heat released and mass lost. The combustion mechanism can be clearly understood through the simple concept of the stages of combustion of composite as described by Mouritz and Gibson (Mouritz et al., 2007) and shown in Fig.6.When the AWF composite substrates are exposed to a heat source, they pyrolyze (i.e. gas off), generating combustible fueland non-combustible fuel, charand smoke. In the presence of oxygen, the combustible gases are ignited and a flame occur sustaining combustion. The fire grows by consumption of the fuel source at temperatures exceeding 350°C to 500°C igniting the AWF composite materials. Flashover occurs when the fire temperature exceeds 600°C and at this stage all the combustibles contribute additional fuel load. The flashover point is the rapid transition between the primary fire which is essentially localized around the AWF composite material first ignition and the general conflagration when all surfaces within the enclosed space (compartment) are burning. The fire behaviour before flashover fire is significant in predicting the agro-waste fibre composite materials safety risk margin.

The fibre behaviour of composites is described by some critical combustion properties which determine their safe industrial utilization. These include but not limited to time-to-ignition (Tig), heat release rates (HRR), mass loss rates (MLR) and smoke production rates (SPR). In determining the combustion properties, engineers and designers rely on fire testing data obtained from bench scale apparatus to predict the safety of materials. The 1998 edition of the compilation of fire test by American Society for Testing and Materials (ASTM) alone listed some bench scale tests (Foley & Drysdale, 1994). Most of these methods are well elaborated and documented in various literatures (Babrauskas, 1986; Babrauskas, 1995; Foley & Drysdale, 1994; Greyson et al., 1994; Scudamore et al., 1991). Among the fire test methods, the cone calorimeter at 50kW/m^{2}provides a broad range of ignition and combustion properties that can be correlated to real fire scenario. The cone calorimeter is the most versatile in determining a large number of fire properties in a single test using with a small specimen. Cone calorimeter simulation does not simulate post flashover or fully developed conditions but can predict accurately combustion properties up to the point of flashover.

Among the combustion properties, the HRR defined as the driving force of a fire producing thermal energy per unit area of surface, when flammable decomposition products ignite and burn in the vicinity of the material in fire. The HRR is the most single important parameter in characterizing the firebehaviour of composites and their consequent fire hazard. All other variables used as a parameter in quantifying materials behaviour in fire are correlated to HRR. The generation of other undesirable fire products tends to increase with increasing HRR. Smoke, toxic gases, room temperature and other fire hazard variables generally progress to increase with HRR, as HRR intensifies (Babrauskas, 1986). Furthermore, a high HRR indicates a high threat to life and property. The HRR is characterized by peak HRR which occurs over a very short period of time and often shortly after ignition, and is usually a good indication of the maximum fire behaviour a material can attain while the average HRR is the total heat released averaged over the combustion period, and is considered the most reliable measure of the heat contribution to a sustained fire. Therefore, the average HRR must be taken into consideration in predicting the flashover time inside a compartment as reported by Ramanaiah, et al (Ramanaiah et al., 2013).

Figure 5. Comparison of fully degraded temperature of agro-waste fibre composites

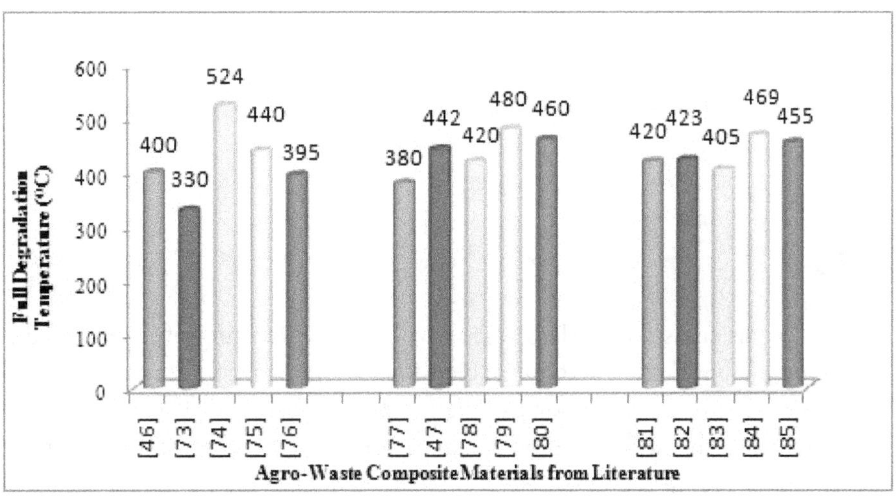

In Fig 7 the HRR of cellulose and lignin burning behaviour compared to the TGA graph in Fig 4 shows that lignin thermally degraded early corresponding to the low HRR curve when compared to cellulose of 2013 and 2014 investigations. This agrees with the reports of Alhuthali et al (Alhuthali et al., 2012). Cellulose and lignin both increased after initial peak HRR but decreased significantly for lignin due to char formation residue preventing the release of combustible volatiles. In Table 1 it shows that the AWF composites with the highest lignin that is coconut coir fibre at 40-45% would form char early followed by oil palm fibre, sugarcane fibre, and banana and pineapple fibres. To further elucidate the effect of the cellulose and lignin content on the fire behaviour of composites,the critical combustion properties (i.e. Tig and peak HRR) of the AWF composite materials found in literature and compared with other sources of natural fibre obtained during a cone calorimeter test by various researchers are shown in Fig 8. It shows that lignin content in coir fibre and oil palm fibre could have been responsible in the reduction of HRR. However, the fire behaviour of composite materials would depend largely on many factors comprising its composition, fibre content, component dimensions, intensity of the fire source, and ventilationas well as the heat intensity as reported by Basnayake et al (Basnayake et al., 2018). There is scare literature on the study of agro-waste fibre composite fire behaviour in different fire conditions and their possible fire reductions to increase industrial utilization.

Figure 6. Stages of fire growth
(Mouritz et al., 2007)

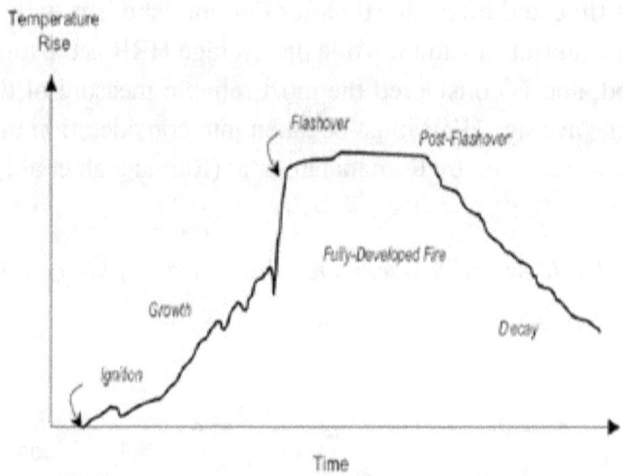

The Final Stage

The final stage also regarded as a flameless stage involves slow burning of residue and ashing of left overs with significant increase in the amount of char formation (Baillie & Jayasinghe, 2017). The significance of char formation is dependent on the chemical structure of the fibre and the polymer matrix as well as combustion atmosphere and combustion temperature. Char is a porous material that contains crystalline and amorphous regions and more carbon than its underlying material (Mouritz et al., 2007). During combustion of AWF composite the formation of char protects the underlying composite from further decomposition from heat. The higher the HRR supposedly the less char is formed. In Fig 5 it shows that the AWF composite materials can form char fully at a range of 330-534°C leaving a substantial amount of char residue in percentage. The char residue for the studied AWF composite can be found in various reports of TGA analysis (Suoware & Edelugo, 2018). Significantly the greater char is formed with increase in lignin content and the polymer matrix structure.

Figure 7. Heat release rate during combustion of cellulose and lignin
(Dorez et al., 2014)

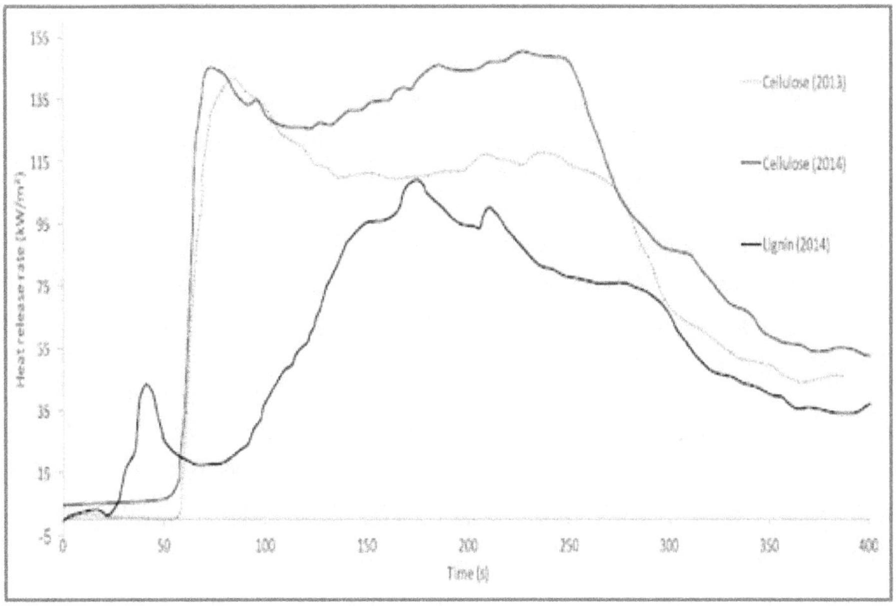

Figure 8. Reported critical combustion properties for some selected agro-waste fibre composites with other sources of fibre composites in literature

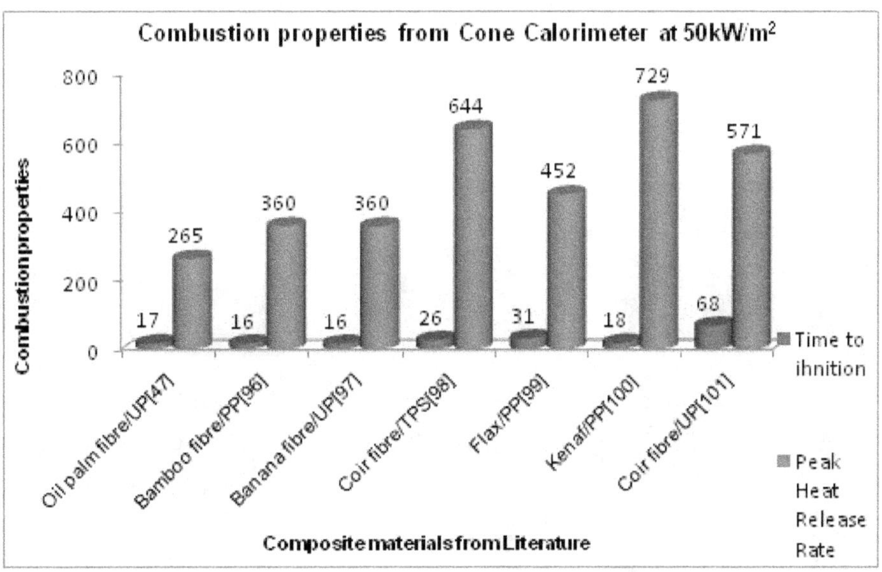

UTILIZATION AND PROSPECTS OF AGRO-WASTE FIBRES TO THE NIGERIA INDUSTRY.

Nigerian Agricultural industries have the potential in contributingmeaningfully to the growth of our economyif AWF is harnessed to produce components by various industries such as automotive, building, sports, food industries etc. The prospects of AWF composite in Nigeria depends on the availability of AWF and the nation is blessed with abundant AWF waste such as coconut, oil palm, banana, sugarcane, pineapple and many others that can be processed into fibres. The process of obtaining the fibres for industrial use can create employment for our timid youths especially in areas where the agro-waste are largely located. It will also create business opportunities to thousands of people in the region. It will also save foreign exchange and reduce importation of raw materials. The AWF composites processed with flame retardants is a way to improve on the fire behaviour as well as increase the potential use of the AWF composites where their fire behaviour is a critical design consideration. AWF flame retardant composite could be used in areas where fire is imminent to reduce additional fuel load and survivability. The utilization of agro-waste reinforced in either synthetic or biopolymers for various industrial applications are shown in Fig 9. Currently research and development has improved component design to meet various industrial needs as landfill of agro-waste is no longer obtainable. These industries includes but not limited to the transportation, building, furniture, food packaging, sports, medical industries as well as many other industries that requires lightweight, low cost, easy to process etc. Materials for component design.

Figure 9. Some selected utilization of agro-waste fibres for potential use in various Nigeria industries (Godavari, n.d.; Treehugger Cannabis in Your Car Doors, n.d.)

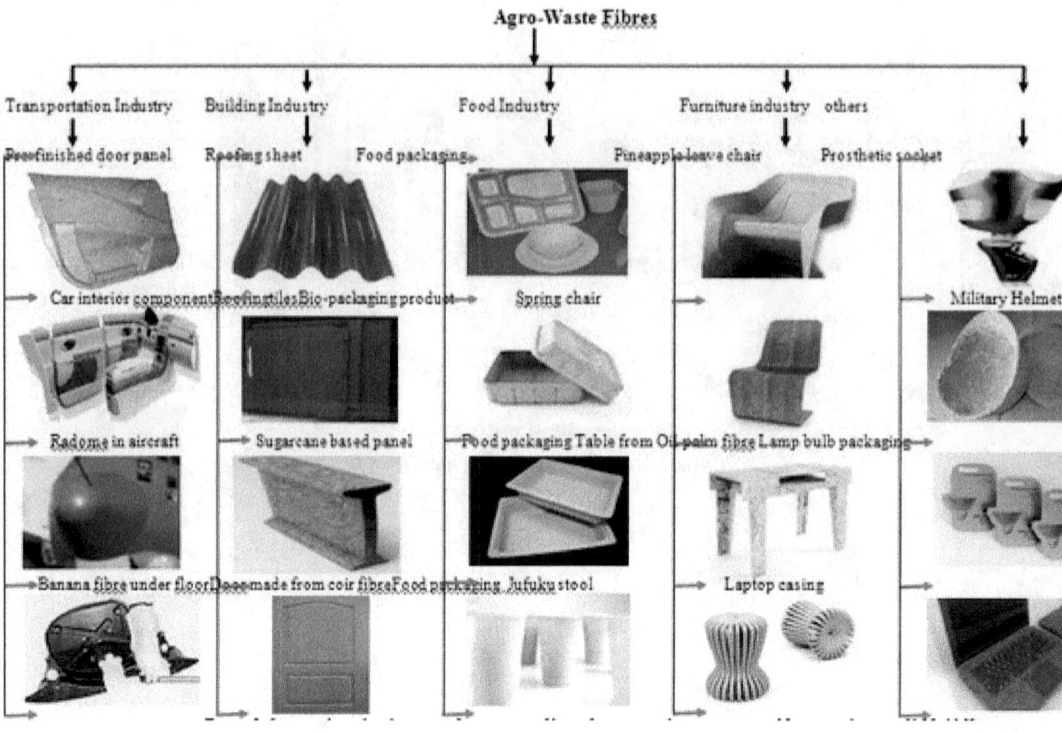

CONCLUSION

The main aim of this review paper is to provide an understandingof the fire behaviour of agro-waste fibrecomposites materials andconcludes as follows:

1. Fibres can be derived from the abundant production of agro-waste in Nigeria such as coconut, oil palm, banana, sugarcane bagasse and pineapple.
2. The agro-waste fibres cellulosic and lignin constituents are primarily responsible for the high and low fire behaviour of composites materials respectively.
3. Thermogravimetric analysis of the AWF composites with polyester, epoxy and polypropylene resins showed thermal decomposition between 150°C to 469°C.
4. The peak heat released rate (pHRR) is identified as the single most important parameter in determining the fire behaviour of composites and it shows that the presence of high lignin content could reduce the pHRR.
5. The AWF composite for coconut coir fibre and oil palm fibre with the highest lignin could reduce additional fuel load during a fire scenario to meet various industrial applications in Nigeria.

REFERENCES

Abdulah, A. M., & Azizi, N. (2018). Thermogravimetric Analysis of Saccharum/Unsaturated Polyester Composites. *Al-qadisiyah J for Engineering Sciences., 11*(4), 411–416.

Abraham, D. P. R., Joseph, A., Jasheelad, A., Binuraj, P. R., & Sarmaa, J. (2016). *Development of Coir-Fibre Cement Composite Roofing Tiles. Procedia Technology, 24*, 169–178.

Adeniyi, A.G., Ighalo, J.O., & Onifade, D.V. (2019). Banana and Plantain Fiber-Reinforced Polymer Composites. *J of Polym Eng.* doi:10.1515/polyeng-20190085

Alhuthali, A., Low, I. M., & Dong, C. (2012). Characterization of the Water Absorption, Mechanical and Thermal Properties of Recycled Cellulose Fibre Reinforced Vinyl-Ester Eco-Nanocomposites. *Composites. Part B, Engineering, 43*(7), 2772–2781. doi:10.1016/j.compositesb.2012.04.038

Alvarez, V. A., Ruscekaite, R. A., & Vazquez, A. (2003). Mechanical properties and water absorption behavior of composites made from a biodegradable matrix and alkaline-treated Sisal Fibers. *Journal of Composite Materials, 37*(17), 1575–1588. doi:10.1177/0021998303035180

Architonic. (2018). *Biocomposites Experimental Pavilion, Stuttgart, Germany*. Available online:https://www.architonic.com/en/project/biomat-group-at-itke-biocompositesexperimental-pavilion/20013658

Arrakhiz, F. Z., Achaby, M., & El. (2013). Mechanical and Thermal Properties of Natural Fibers Reinforced Polymer Composites: Doum/low Density Polyethylene. *Materials & Design, 43*, 200–205. doi:10.1016/j.matdes.2012.06.056

Arun, A., Sathyaseelan, R., Tamilselvan, M., Gowtham, M., & Karthikeyan, A. (2016). Influence of Weight Fractions on Mechanical, Water Absorption and Corrosion Resistance Behaviors of Untreated Hybrid (Coir/Banana) Fiber Reinforced Epoxy Composites. *International Journal of Chemtech Research*, *9*, 932–940.

Asasutjarit, C., Hirunlabh, J., Khedari, J., Charoenvai, S., Zeghmati, B., & Shin, U. C. (2007). Development of Coconut Coir-Based Lightweight Cement Board. *Construction & Building Materials*, *21*(2), 277–288. doi:10.1016/j.conbuildmat.2005.08.028

Asgekar, S.D., Joshi, V.K., Futane, P.S., & Joshi, P.S. (2013). Characteristics of Sugarcane/Coir Fibres Reinforced Composites in Phenol Formaldehyde Resin. *International Journal of Composite Materials, 3*(6), 156-162.

Asim, M., Abdan, K., Jawaid, M., Nasir, M., Dashtizadeh, Z., Ishak, M. R., & Hoque, M. E. (2015). Review on Pineapple Leaves Fibre and Its Composites. *International Journal of Polymer Science*, *2015*, 1–16. Advance online publication. doi:10.1155/2015/950567

Aveston, J., Kelly, A., & Sillwood, J. M. (1980). *Advances in Composite Materials.* Pergamon.

Ayrilmis, N., Jarusombuti, S., Fueangvivat, V., Bauchongkol, P., & White, R. H. (2011). Coir Fiber Reinforced Polypropylene Composite Panel for Automotive Interior Applications. *Fibers and Polymers*, *12*(7), 919–926. doi:10.100712221-011-0919-1

Azwa, Z. N., Yousif, B. F., Manalo, A. C., & Karunasena, W. (2013). A Review on the Degradability of Polymeric Composites Based on Natural Fibres. *Materials & Design*, *47*, 424–442. doi:10.1016/j.matdes.2012.11.025

Babrauskas, V. (1986). The cone calorimeter – A Versatile Bench-Scale Tool for the Evaluation for Fire Properties. In New Technology to Reduce Fire Loses and Cost. Academic Press.

Babrauskas, V. (1995). Designing Products for Fire Performance: The State of the Art of Test Methods and Fire Models. *Fire Safety Journal*, *24*(3), 299–312. doi:10.1016/0379-7112(95)00025-O

Baillie, C., & Jayasinghe, R. (2017). *Green Composites: Waste and Nature-based Materials for a Sustainable Future. Cambridge*. Elsevier Science. doi:10.1016/B978-0-08-100783-9.00001-0

Baruwa, O. I. (2013). Profitability and Constraints of Pineapple Production in Osun State, Nigeria. *Journal of Horticultural Research*, *21*(2), 59–64. doi:10.2478/johr-2013-0022

Basnayake, A.J., Vandi, L., & Heitzmann, M. (2018). *A Literature Review of Natural Fibre Composite Fire Properties.* CIC Fibre-Resin Interface Project.

Bhal, N. S., & Singh, B. (1998). Potential of Natural Fiber Reinforced Polymer Composites for Civil Engineering Applications in India. *Proceedings of the 2nd International Conference in Infrastructure, Department of Civil Engineering and Engineering Mechanics*, 661–673.

Bledzki, A. K., Reihmane, S., & Gassan, J. (1996). Properties and Modification Methods for Vegetable Fibers for Natural Fiber Composites. *Journal of Applied Polymer Science*, *59*(8), 1329–1336. doi:10.1002/(SICI)1097-4628(19960222)59:8<1329::AID-APP17>3.0.CO;2-0

Chaitanya, A. K., Babu, D. K., Narendra, K. B., & Sumanth, K. (2016). Investigation on The Mechanical Properties of Banana Fiber Reinforced Polyester Composites. *International Journal of Engineering Research and Applications*, *6*(9), 76–82.

Chauhan, N., & Arya, N. (2018). Coconut fiber: A natural versatile material. *International of Chemical Studies*, *6*(6), 555–561.

Cordeiro, E. P., Pita, V. J. R. R., & Soares, B. G. (2017, September). Epoxy–Fiber of Peach Palm Trees Composites: The Effect of Composition and Fiber Modification on Mechanical and Dynamic Mechanical Properties. *Journal of Polymers and the Environment*, *25*(3), 913–924. Advance online publication. doi:10.100710924-016-0841-0

Daramola, O. O., Adediran, A. A., Adewuyi, B. O., & Adewole, O. (2017). Mechanical Properties and Water Absorption Behaviour of Treated Pineapple Leaf Fibre Reinforced Polyester Matrix Composites. *Leonardo Journal of Sciences*, (30), 15–30.

Das, B. K., Chakravarty, A. C., Sinha, M. K., & Ghosh, S. K. (1978). Use of polyethylene monofilament in spinning core yarn of jute. *Man Made Textile in India*, *21*(9), 494–498.

De Leao, A. G., Pereira, P. S. C., & Bastos, D. C. (2018). Characterization of Polypropylene Post-Consume/Banana Fibre Composites. *Acta Scientiae and Technicae*, *6*(1), 19–22. doi:10.17648/uezo-ast-v6i1.129

Dong, Y., Ghataura, A., Takagi, H., Haroosh, H. J., Nakagaito, A. N., & Lau, K. T. (2014). Polylactic Acid (PLA)Biocomposites Reinforced with Coir Fibres: Evaluation of Mechanical Performance and Multifunctional Properties. *Compos Part A*, *63*, 76–84. doi:10.1016/j.compositesa.2014.04.003

Dorez, G., Ferry, L., Sonnier, R., Taguet, A., & Lopez-Cuesta, J. M. (2014). Effect of Cellulose, Hemicellulose and Lignin Contents on Pyrolysis and Combustion of Natural Fibers. *Journal of Analytical and Applied Pyrolysis*, *107*, 323–331. doi:10.1016/j.jaap.2014.03.017

Eng, C. C., Ibrahim, N. A., Zainuddin, N., Ariffin, H., & Yunus, W. M. Z. W. (2014). Impact Strength and Flexural Properties Enhancement of Methacrylate Silane Treated Oil Palm Mesocarp Fiber Reinforced Biodegradable Hybrid Composites. The Scientific World Journal. doi:10.1155/2014/213180

Essays, U. K. (2014). *Tensile and impact properties of Sugarcane Bagasse Pith and rind filled Poly(vinyl-chloride) Composites*. Available: https://www.ukessays.co.uk/essays/sciences/sugarcane-bagasse.php

Foley, M., & Drysdale, D. (1994). Note: Smoke Measurements and the Cone Calorimeter. *Fire and Materials*, *18*(6), 385–387. doi:10.1002/fam.810180606

Gao, C. C., Yu, L., Liu, H. S., & Chen, L. (2012). Development of Self-Reinforced Polymer Composites. *Progress in Polymer Science*, *37*(6), 767–780. doi:10.1016/j.progpolymsci.2011.09.005

Godavari. (n.d.). *Natural Fiber Biocomposite*. Available online https://www.somaiya.com/naturomer-%C2%AEnatural-fiber-biocomposite

Gokul, K., Prabhu, T. R., & Rajasekaran, T. (2017). Processing and Evaluation of Mechanical Properties of Sugarcane Fiber Reinforced Natural Composites. *Transactions of the Indian Institute of Metals*, *70*(10), 2537–2546. doi:10.100712666-017-1116-8

Greyson, S. J., Babrauskas, V., & Hirschler, M. M. (1994). A New International Standard for Flammability Testing. *Plastics Engineering*, 29–31.

Gutowski, T. G. (1997). *Advanced Composites Manufacturing*. John Wiley and Sons.

Harris, H.C. (2012). Germination rate is the Significant Characteristic Determining Coconut Palm Diversity. *AoB PANTS.* doi:10.1093/aobpla/pls045

Hassan, M.Z., Sapuan, S.M., & Rasid, Z.A. (2019). Thermal Degradation and Mechanical Behavior of Banana Pseudo-Stem Reinforced Composites. *International Journal of Recent Technology and Engineering, 8*(4), 5599-5902.

Horrocks, A. R., & Price, D. (2001). Fire retardant materials. CRC Press.

Indira, K. N., Parameswaranpillai, J., & Thomas, S. (2013, August 19). Mechanical Properties and Failure Topography of Banana Fiber PF Macrocomposites Fabricated by RTM and CM Techniques. *International Scholarly Research Notices, 2013*, 1–8. Advance online publication. doi:10.1155/2013/936048

Ishai, O., & Daniel, I. M. (2006). *Engineering Mechanics of Composite Materials* (2nd ed.). Oxford University Press.

Islam, M. S., Church, J. S., & Miao, M. (2011). Effect of Removing Polypropylene Fibre Surface Finishing on Mechanical Performance of Kenaf/Polypropylene Composites. *Composites. Part A, Applied Science and Manufacturing, 42*(11), 1687–1693. doi:10.1016/j.compositesa.2011.07.023

Izah, S. C., & Ohimain, E. I. (2016). The opportunities and weakness of Nigerian Oil Palm Industry. *Biotechnological Research*, 2(1), 33–43.

Jain, J., Jain, S., & Sinha, S. (2018). Characterization and Thermal Kinetic Analysis of Pineapple Leaf Fibers and their Reinforcement in Epoxy. *Journal of Elastomers and Plastics*. Advance online publication. doi:10.1177/0095244318783024

Jayamani, E., Hamdan, S., Rahman, M. R., & Bakri, M. K. B. (2014). Comparative Study of Dielectric Properties of Hybrid natural Fiber Composites. *Procedia Engineering, 97*, 536–544. doi:10.1016/j.proeng.2014.12.280

John, M. J., & Thomas, S. (2008). Biofibres and Biocomposites. *Carbohydrate Polymers, 71*(3), 343–364. doi:10.1016/j.carbpol.2007.05.040

Kasim, A. N., Selamat, M. Z., Daud, M. A. M., Yaakob, M. Y., Putra, A., & Sivakumar, D. (2016). Mechanical Properties of Polypropylene Composites reinforced with Alkaline Treated Pineapple Leaf Fibre from Josapine Cultivar. *International Journal of Automotive and Mechanical Engineering, 13*(1), 3157–3167. doi:10.15282/ijame.13.1.2016.3.0263

Kavitha, S. (2016). Bamboo Fibre Analysis by Scanning Electron Microscope Study. *International Journal of Civil Engineering and Technology*, 7(4), 234–241.

Khalili, P., Tshai, K. Y., Hui, D., & Kong, I. (2017). Synergistic of Ammonium Polyphosphate and Alumina Trihydrate as Fire Retardants for Natural Fiber Reinforced Epoxy Composite. *Composites. Part B, Engineering, 114*, 101–110. doi:10.1016/j.compositesb.2017.01.049

Ku, H., Wang, H., Pattarachaiyakoop, N., & Trada, M. (2011). A review on the tensile Properties of natural fiber reinforced polymer composites. *Composites. Part B, Engineering*, *42*(4), 856–873. doi:10.1016/j.compositesb.2011.01.010

Kumar, S. M. S., Duraibabu, D., & Subramanian, K. (2014). Studies on Mechanical, Thermal and Dynamic Mechanical Properties of Untreated (raw) and Treated coconut Sheath Fiber Reinforced Epoxy Composites. *Materials & Design*, *59*, 63–66. doi:10.1016/j.matdes.2014.02.013

Less by Design Coco Form-a Sustainable Packaging Alternative. (2015). Available online: https://lessbydesign.org/2015/02/06/cocoform-a-sustainable-packaging-alternative

Luz, D., da Costa, S. M., Gonçalves, A. R., Del'ArcoJuniord, A. P., & da Costa, S. A. (2016). Polypropylene Composites Reinforced with Biodegraded Sugarcane Bagasse Fibers: Static and Dynamic Mechanical Properties. *Materials Research*, *19*(1), 75–83. doi:10.1590/1980-5373-MR-2015-0410

Mahesh, D., Kowshigha, K. R., Raju, N. V., & Aggarwai, P. K. (2020). Characterization of Banana Fibre-Reinforced Polypropylene Composite. *Journal of the Indian Academy of Wood Science*, *17*(1), 1–8. doi:10.100713196-019-00244-x

Majeed, K., Jawaid, M., Hassan, A., Abu Bakar, A., Abdul Khalil, H.P.S., Salami, A.A., & Inuwa, I. (2013). Potential Materials for Food Packaging from Nanoclay/Natural Fibres Filled Hybrid Composites. *Mater. Des, 46*, 391–410.

Mallick, P. K. (1990). *Compression Moulding, Composite Materials Technology* (P. K. Mallick & S. Newman, Eds.). Hanser Publishers.

Mallick, P. K. (1993) Fiber reinforced composites: materials, manufacturing and design (2nd ed.). Boca Raton, FL: CRC Press.

Mittal, V., & Sinha, S. (2017). Effect of chemical Treatment on Thermal Properties of Bagasse Fiber-Reinforced Epoxy Composite. *Science and Engineering of Composite Materials*, *24*(2), 237–243. doi:10.1515ecm-2014-0434

Mohanty, A. K., Misra, M., & Drzal, L. T. (2001). Surface modifications of Natural Fibers and Performance of the Resulting Biocomposites: An overview. *Composite Interfaces*, *8*(5), 313–343. doi:10.1163/156855401753255422

Mohapatra, D., Mishra, S., & Sutar, N. (2010). Banana and its by-product utilization: An overview. *Journal of Scientific and Industrial Research*, *69*, 323–329.

Mothe, C.G., de Araujo, C.R., & Maradela, F.S. (2003). Mechanical and Thermal Properties of Coir Fibre Reject/Polypropylene Composite. *7thCongressoBrasileiro de Polimeros,* 973-974.

Mouritz, A. P., Gibson, A. G., & Gladwell, G. M. L. (Eds.). (2007). Fire Properties of Polymer Composite Materials (Solid Mechanics & Its Applications S., v. 143). Springer Netherlands.

Mulinari, D. R., Baptista, C. A. R. P., Souza, J. V. C., & Voorwald, H. V. C. (2011). Mechanical Properties of Coconut Fibre Reinforced Polyester Composites. *Procedia Engineering*, *10*, 2074–2079. doi:10.1016/j.proeng.2011.04.343

Munawar, R. F., Jamil, N. H., Shahril, M. K., Abdul Rahim, M., Abidin, M. Z. Z., Azam, M. A., & Lau, K. (2015). Development of Green Composite: Pineapple Leaf Fibers (PALF) Reinforced Polylactide (PLA). *Applied Mechanics and Materials*, *761*, 520–525. doi:10.4028/www.scientific.net/AMM.761.520

Muragan, M.D., Zakaria, Z., & Hassan, A. (2019). Mechanical and Thermal Properties of Graphene Oxide reinforced Polypropylene/Pineapple Leaves Fibre Composites. *PERINTIS eJournal*, *9*(2), 11-20.

Mustafa Al Bakri, A. M. M., Liyana, J., Norazian, M. N., Kamarudin, H., & Ruzaidi, C. M. (2013). Mechanical Properties of Polymer Composites with Sugarcane Bagasse Filler. *Advanced Materials Research*, *740*, 739–744. doi:10.4028/www.scientific.net/AMR.740.739

Nair, K. C. M., Kumar, R. P., Thomas, S., Schit, S. C., & Ramamurthy, K. (2000). Rheological behavior of short sisal fiber-reinforced polystyrene composites. *Composites. Part A, Applied Science and Manufacturing*, *31*(11), 1231–1240. doi:10.1016/S1359-835X(00)00083-X

Natsa, S., Akindapo, J. O., & Garba, D. K. (2015). Development of a Military Helmet Using Coconut Fiber Reinforced Polymer Matrix Composite. *European Journal of Engineering Technology*, *3*(7), 55–65.

Obasi, H. C., Iheaturu, N. C., Onuoha, F., Chike-Onyegbula, C., Akanbi, M. N., & Ezeh, V. O. (2014). Influence of Alkali Treatment and Fibre Content on the Properties of Oil Palm Press Fibre Reinforced Epoxy Biocomposites. *American Journal of Engineering Research*, *3*(2), 117–123.

Odusote, J. K., & Oyewo, A. T. (2016). Mechanical Properties of Pineapple Leaf Fiber Reinforced Polymer Composites for Application as a Prosthetic Socket. *Journal of Engineering Technology*, *7*(1), 125–139.

Osita, O., Ignatius, O., & Udeani Henry, U. (2016). Study on the Mechanical Properties of Palm Kernel Fibre Reinforced Epoxy and Poly-Vinyl Alcohol (PVA) Composite Material. *International Journal of Engineering and Technologies*, *7*, 68–77. doi:10.18052/www.scipress.com/IJET.7.68

Osorio, L., Trujillo, E., van Vuure, A. W., Lens, F., Ivens, J., & Verpoest, I. (2010). The Relationship between the Bamboo Fibre Microstructure and Mechanical Properties. *Proceedings of the 14th European Conference on Composite Materials*, 467-473.

Patil, R. G., & Kolambe, B. N. (2011). Development of value added products from Banana Pseudo-stem. An Overview of Progress. National Agricultural Innovation Project.

Ramanaiah, K., Prasad, A. V. R., & Reddy, K. H. C. (2013). Thermopyhsical and Fire Properties of Vakka Natural Fiber Reinforced Polyester Composites. *Journal of Reinforced Plastics and Composites*, *32*(15), 1092–1098. doi:10.1177/0731684413486366

Ramesha, M., Atreyaa, T. S. A. S., Aswina, U. S., Eashwara, H., & Deepab, C. (2014). Processing and Mechanical Property Evaluation of Banana Fiber Reinforced Polymer Composites. *Procedia Engineering*, *97*, 563–572. doi:10.1016/j.proeng.2014.12.284

Ramesha, M., Palanikumarb, K., & Hemachandra Reddy, K. H. (2013). Comparative Evaluation on Properties of Hybrid Glass Fiber- Sisal/Jute Reinforced Epoxy Composites. *Procedia Engineering*, *51*, 745–750. doi:10.1016/j.proeng.2013.01.106

Ramli, R., Yunus, R. M., Beg, M. D. H., & Prasad, D. M. R. (2011). Oil Palm Fiber Reinforced Polypropylene Composites: Effects of Fiber Loading and Coupling Agents on Mechanical, Thermal, and Interfacial Properties. *Journal of Composite Materials*, *46*(11), 1275–1284. doi:10.1177/0021998311417647

Ridzuan, R., & Beg, M.D.HRosli, M.YRohaya, M.HAstimar, A.ASamahani, SZawawi, I. (2013). Effects of Coupling Agent and Flame Retardant on the Performance s of Oil Palm Empty Fruit Bunch Fibre Reinforced Polypropylene Composites. *World Academy of Science, Engineering and Technology*, *7*(6), 445–453.

Rowell, R. M., Han, J. S., & Rowell, J. S. (2000). Characterization and Factors Effecting Fiber Properties. In E. Frollini, A. L. Leao, & L. H. C. Mattoso (Eds.), Natural Polymers and Agrofibers Composites (pp. 115–134). Academic Press.

Santafé Júnior, H. P. G., Lopes, F. P. D., Costa, L. L., & Monteiro, S. N. (2010). Mechanical Properties of Tensile Tested Coir Fiber Reinforced Polyester Composites. *Revista Matéria*, *15*(2), 113–118. doi:10.1590/S1517-70762010000200017

Santos, J. C. D., Siqueira, R. L., Gomes, L. M., Freier, R. T. S., Mano, V., & Panzera, T. H. (2018). Effects of Sodium Carbonate on the Performance of Epoxy and Polyester-Coir Reinforced Composite. *Polymer Testing*, *67*, 533–544. doi:10.1016/j.polymertesting.2018.03.043

Santosha, P.., Gowda, A., & Manikanth, V. (2018). Effect of Fibre Loading on Thermal Properties of Banana and Pineapple Leaf. *Fibre Reinforced Polyester Composite, 5*(2), 5631-5635.

Sapuan, S. M., & Bachtiara, D. (2012). Mechanical Properties of Sugar Palm Fibre Reinforced High Impact Polystyrene Composites. *Procedia Chemistry*, *4*, 101–106. doi:10.1016/j.proche.2012.06.015

Sarah, A. S., Rahman, W. A., Majid, R. A., Yahya, W. J., Adrus, N., Hasannuddin, A. K., & Low, J. H. (2018). Optimization of Pineapple Leaf Fibre Extraction Methods and Their Biodegradabilities for Soil Cover Application. *Journal of Polymers and the Environment*, *26*(1), 319–329. doi:10.100710924-017-0942-4

Scudamore, M. J., Briggs, P. J., & Prager, F. H. (1991). Cone calorimetry – A Review of Tests Carried out on Plastics for the Association of Plastics Manufacturers in Europe. *Fire and Materials*, *15*(2), 65–84. doi:10.1002/fam.810150205

Selamat, M. Z., Zhafri Tahir, M. S., Kasim, A. N., Dharmalingam, S., Azma Putra, A., Yaakob, M., & Yand Daud, M. A. M. (2018). Effect of Starch Sizes Particle as Binder on Short Pineapple Leaf Fiber Composite Mechanical Properties. *MATEC Web of Conferences*, *150*, 04008. Advance online publication. doi:10.1051/matecconf/201815004008

Senthiil, P. V., & Sirsshti, A. (2014). Studies on Material and Mechanical Properties of Natural Fiber Reinforced Composites. *The International Journal of Engineering And Science*, *3*(11), 18–27.

Shah, H., Srinivasulu, B., & Shit, S. C. (2013). Influence of Banana Fibre Chemical Modification on the Mechanical and Morphological Properties of Woven Banana Fabric/Unsaturated Polyester Resin Composites. Polymers from Renewable Resources, 4(2), 51-84.

Shiner, S., Visvanathan, R., Panigrahi, S., & Kochubabu, M. (2011). Oil Palm Fiber (OPF) and its Composites: A Review. *Industrial Crops and Products*, *33*(1), 7–22. doi:10.1016/j.indcrop.2010.09.009

Shinoj, S., Visvanathan, R., Panigrahi, S., & Kochubabu, M. (2011). Oil palm fiber (OPF) and its Composites: A Review. *Industrial Crops and Products*, *33*(1), 7–22. doi:10.1016/j.indcrop.2010.09.009

Simdo, J. A., Carmona, V. B., Marconcini, J. M., Mattoso, L. H. C., Arsberg, S. T., & Sanadi, A. R. (2016). Effect of Fibre Treatment Condition and Coupling Agent on the Mechanical and Thermal Properties in Highly Filled Composites of Sugarcane Bagasse Fibre/PP. *Materials Research*, *19*(4), 7446–7751.

Siqueira, G., Bras, J., & Dufresne, A. (2009). *New Process of Chemical Grafting of Cellulose Nanoparticles with a Long Chain Isocyanate Langmuir*. PMID:19921797

Sorieul, M., Dickson, A., Hill, S. J., & Pearson, H. (2016). Plant Fibre: Molecular Structure and Biomechanical Properties, of a Complex Living Material, Influencing Its Deconstruction towards a Bio based Composite. *Materials (Basel)*, *9*(12), 6–18. doi:10.3390/ma9080618 PMID:28773739

Steinhaus, T. (2010). *Determination of Intrinsic Material Flammability Properties from Material Tests assisted by Numerical Modelling*. University of Edinburgh.

Sukyai, P., Sriroth, K., Lee, B. K., & Kim, H. J. (2012). The Effect of Bacterial Cellulose on the Mechanical and Thermal Expansion Properties of Kenaf/Poly Lactic Acid Composites. Shenzhan, 1343-1351.

Suoware, T. O., & Edelugo, S. O. (2018). Thermal, Mechanical and Flame Spread Properties of Hybridized FlameRetardant in oil palm fibre-polyester panel. *Springer Nature Applied Science*, *1*(1), 42. doi:10.100742452-018-0048-1

Treehugger Cannabis in Your Car Doors, but Not to Smuggle It. (n.d.). Available online: https://www.treehugger.com/sustainable-agriculture/cannabis-your-car-doors-not-smuggle-it.html

Vidyashri, V., Lewis, H., Narayanasamy, P., Mahesha, G. T., & Bhat, K. S. (2019). Preparation of Chemically Treated Sugarcane Bagasse Fiber Reinforced Epoxy Composites and their Characterization. *Cogent Engineering*, *6*(1), 1708644. doi:10.1080/23311916.2019.1708644

Wang, H., Memon, H., & Hassan, E.A.MMiah, MArshad Ali, M. A. (2019). Effect of Jute Fiber Modification on Mechanical Properties of Jute Fiber Composite. *Materials (Basel)*, *12*(8), 1226. doi:10.3390/ma12081226 PMID:30991643

Young, R. A. (1986). Structure, Swelling and Bonding of Cellulose Fibers. In R. A. Young & R. M. Rowell (Eds.), *Cellulose: structure, modification and hydrolysis* (pp. 91–128). John Wiley & Sons.

Zafeiropoulos, N. E., Williams, D. R., Baillie, C. A., & Matthews, F. L. (2002). Engineering and Characterization of the interface in flax fibre/polypropylene composite materials. Part I. Development and investigation of surface treatments. *Composites. Part A, Applied Science and Manufacturing*, *33*(8), 1083–1093. doi:10.1016/S1359-835X(02)00082-9

Zhao, C., Jiang, E., & Chen, A. (2016). Volatile production from pyrolysis of cellulose, hemicellulose and lignin. *Journal of the Energy Institute*.

Compilation of References

Less by Design Coco Form-a Sustainable Packaging Alternative. (2015). Available online: https://lessbydesign.org/2015/02/06/cocoform-a-sustainable-packaging-alternative

Architonic. (2018). *Biocomposites Experimental Pavilion, Stuttgart, Germany*. Available online:https://www.architonic.com/en/project/biomat-group-at-itke-biocompositesexperimental-pavilion/20013658

Treehugger Cannabis in Your Car Doors, but Not to Smuggle It. (n.d.). Available online: https://www.treehugger.com/sustainable-agriculture/cannabis-your-car-doors-not-smuggle-it.html

Godavari. (n.d.). *Natural Fiber Biocomposite*. Available online https://www.somaiya.com/naturomer-%C2%AEnatural-fiber-biocomposite

Jayamani, E., Hamdan, S., Rahman, M. R., & Bakri, M. K. B. (2014). Comparative Study of Dielectric Properties of Hybrid natural Fiber Composites. *Procedia Engineering*, *97*, 536–544. doi:10.1016/j.proeng.2014.12.280

Oreper, G. M., & Szekely, J. (1984). Heat-and fluid-flow phenomena in weld pools. *Journal of Fluid Mechanics*, *147*(-1), 53–79. doi:10.1017/S0022112084001981

Sahu, A. K., Sahu, N. K., & Sahu, A. K. (2014). Appraisal of CNC machine tool by integrated MULTI-MOORA-IVGN circumferences: An empirical study. *Grey Systems: Theory and Application*.

Turley, D. M., & Doyle, E. D. (1982). Microstructural behavior--Its influence on machining. *On the Art of Cutting Metals—75 Years Later. A Tribute to FW Taylor*, 14-19.

Zhao, L., Cui, C., Wang, Q., & Bu, S. (2010). Growth characteristics and corrosion resistance of micro-arc oxidation coating on pure magnesium for biomedical applications. *Corrosion Science*, *52*(7), 2228–2234. doi:10.1016/j.corsci.2010.03.008

Bosman, L., & Fernhaber, S. (2018). Example Engineering Curriculum: Online Discussion Prompts. In *Teaching the Entrepreneurial Mindset to Engineers* (pp. 65–90). Springer. doi:10.1007/978-3-319-61412-0_9

Darmanin, T., Taffin de Givenchy, E., Amigoni, S., & Guittard, F. (2010). Hydrocarbon versus fluorocarbon in the electrodeposition of superhydrophobic polymer films. *Langmuir*, *26*(22), 17596–17602. doi:10.1021/la103310m PMID:20879773

Eng, C. C., Ibrahim, N. A., Zainuddin, N., Ariffin, H., & Yunus, W. M. Z. W. (2014). Impact Strength and Flexural Properties Enhancement of Methacrylate Silane Treated Oil Palm Mesocarp Fiber Reinforced Biodegradable Hybrid Composites. The Scientific World Journal. doi:10.1155/2014/213180

Li, D., Lu, S., Dong, W., Li, D., & Li, Y. (2012). Study of the law between the weld pool shape variations with the welding parameters under two TIG processes. *Journal of Materials Processing Technology*, *212*(1), 128–136. doi:10.1016/j.jmatprotec.2011.08.015

Rahman, M. A., Rahman, M., Kumar, A. S., & Lim, H. S. (2005). CNC microturning: An application to miniaturization. *International Journal of Machine Tools & Manufacture*, *45*(6), 631–639. doi:10.1016/j.ijmachtools.2004.10.003

Ku, H., Wang, H., Pattarachaiyakoop, N., & Trada, M. (2011). A review on the tensile Properties of natural fiber reinforced polymer composites. *Composites. Part B, Engineering*, *42*(4), 856–873. doi:10.1016/j.compositesb.2011.01.010

Kumar, R., Thanigaivelan, R., Rajanikant, G. K., Jagadeesha, T., & Das, J. (2021). Evaluation of hydroxyapatite-and zinc-coated Ti-6Al-4V surface for biomedical application using electrochemical process. *Journal of the Australian Ceramic Society*, *57*(1), 107–116. doi:10.100741779-020-00517-6

Rojko, A. (2017). Industry 4.0 concept: Background and overview. *International Journal of Interactive Mobile Technologies*, *11*(5), 77. doi:10.3991/ijim.v11i5.7072

Selvi, S., Vishvaksenan, A., & Rajasekar, E. (2018). Cold metal transfer (CMT) technology-An overview. *Defence Technology, 14*(1), 28-44.

Vyas, A., & Shaw, M. C. (1999). *Mechanics of saw-tooth chip formation in metal cutting*. Academic Press.

Chauhan, N., & Arya, N. (2018). Coconut fiber: A natural versatile material. *International of Chemical Studies*, *6*(6), 555–561.

Dzhurinskiy, D. V., Dautov, S. S., Shornikov, P. G., & Akhatov, I. S. (2021). Surface Modification of Aluminum 6061-O Alloy by Plasma Electrolytic Oxidation to Improve Corrosion Resistance Properties. *Coatings*, *11*(1), 4. doi:10.3390/coatings11010004

Kumar, N. P., Vendan, S. A., & Shanmugam, N. S. (2016). Investigations on the parametric effects of cold metal transfer process on the microstructural aspects in AA6061. *Journal of Alloys and Compounds*, *658*, 255–264. doi:10.1016/j.jallcom.2015.10.166

Manyindo, B. M., & Oxley, P. L. B. (1986). Modelling the catastrophic shear type of chip when machining stainless steel. *Proceedings of the Institution of Mechanical Engineers. Part C, Journal of Mechanical Engineering Science*, *200*(5), 349–358. doi:10.1243/PIME_PROC_1986_200_138_02

Pereira, A. C., & Romero, F. (2017). A review of the meanings and the implications of the Industry 4.0 concept. *Procedia Manufacturing*, *13*, 1206–1214. doi:10.1016/j.promfg.2017.09.032

Auer, M. E., & May, D. (Eds.). (2020). *Cross Reality and Data Science in Engineering: Proceedings of the 17th International Conference on Remote Engineering and Virtual Instrumentation* (*Vol. 1231*). Springer Nature.

Prieto, J. R., Carbonell, J. M., Cante, J. C., Oliver, J., & Jonsen, P. (2018). Generation of segmental chips in metal cutting modeled with the PFEM. *Computational Mechanics*, *61*(6), 639–655. doi:10.100700466-017-1442-z PMID:31007328

Qi, B. J., Yang, M. X., Cong, B. Q., & Liu, F. J. (2013). The effect of arc behavior on weld geometry by high-frequency pulse GTAW process with 0Cr18Ni9Ti stainless steel. *International Journal of Advanced Manufacturing Technology*, *66*(9-12), 1545–1553. doi:10.100700170-012-4438-z

Saha, S. K., Park, Y. J., & Cho, S. O. (2021). Fabrication of highly ordered nanoporous oxide layer on Ti6Al4V surfaces for improved corrosion resistance property. *Journal of Molecular Structure*, *1223*, 129244. doi:10.1016/j.molstruc.2020.129244

Shiner, S., Visvanathan, R., Panigrahi, S., & Kochubabu, M. (2011). Oil Palm Fiber (OPF) and its Composites: A Review. *Industrial Crops and Products*, *33*(1), 7–22. doi:10.1016/j.indcrop.2010.09.009

Kovacic, M., Brezocnik, M., Pahole, I., Balic, J., & Kecelj, B. (2005). Evolutionary programming of CNC machines. *Journal of Materials Processing Technology*, *164*, 1379–1387. doi:10.1016/j.jmatprotec.2005.02.047

Kutelu, B. J., Seidu, S. O., Eghabor, G. I., & Ibitoye, A. I. (2018). Review of GTAW welding parameters. *Journal of Minerals & Materials Characterization & Engineering*, *6*(05), 541–554. doi:10.4236/jmmce.2018.65039

Mohapatra, D., Mishra, S., & Sutar, N. (2010). Banana and its by-product utilization: An overview. *Journal of Scientific and Industrial Research*, *69*, 323–329.

Monostori, L., Kádár, B., Bauernhansl, T., Kondoh, S., Kumara, S., Reinhart, G., Sauer, O., Schuh, G., Sihn, W., & Ueda, K. (2016). Cyber-physical systems in manufacturing. *CIRP Annals*, *65*(2), 621–641. doi:10.1016/j.cirp.2016.06.005

Tekdir, H., & Yetim, A. F. (2021). Additive manufacturing of multiple layered materials (Ti6Al4V/316L) and improving their tribological properties with glow discharge surface modification. *Vacuum*, *184*, 109893. doi:10.1016/j.vacuum.2020.109893

Essays, U. K. (2014). *Tensile and impact properties of Sugarcane Bagasse Pith and rind filled Poly(vinyl-chloride) Composites*. Available: https://www.ukessays.co.uk/essays/sciences/sugarcane-bagasse.php

Lu, Y., Chen, S., Shi, Y., Li, X., Chen, J., Kvidahl, L., & Zhang, Y. M. (2014). Double-electrode arc welding process: Principle, variants, control and developments. *Journal of Manufacturing Processes*, *16*(1), 93–108. doi:10.1016/j.jmapro.2013.08.003

Teti, R. (2002). Machining of composite materials. *CIRP Annals*, *51*(2), 611–634. doi:10.1016/S0007-8506(07)61703-X

Yang, L., Yang, X., Zhang, T., & Sun, R. (2021). Optimization of microstructure and properties of composite coatings by laser cladding on titanium alloy. *Ceramics International*, *47*(2), 2230–2243. doi:10.1016/j.ceramint.2020.09.063

Baruwa, O. I. (2013). Profitability and Constraints of Pineapple Production in Osun State, Nigeria. *Journal of Horticultural Research*, *21*(2), 59–64. doi:10.2478/johr-2013-0022

Karademir, I., Celik, M. B., Husem, F., Maleki, E., Amanov, A., & Unal, O. (2021). Effects of constrained groove pressing, severe shot peening and ultrasonic nanocrystal surface modification on microstructure and mechanical behavior of S500MC high strength low alloy automotive steel. *Applied Surface Science*, *538*, 147935. doi:10.1016/j.apsusc.2020.147935

Li, K., & Zhang, Y. (2007). *Metal transfer in double-electrode gas metal arc welding*. Academic Press.

Thandra, S. K., & Choudhury, S. K. (2010). Effect of cutting parameters on cutting force, surface finish and tool wear in hot machining. *International Journal of Machining and Machinability of Materials*, *7*(3-4), 260–273. doi:10.1504/IJMMM.2010.033070

Harris, H.C. (2012). Germination rate is the Significant Characteristic Determining Coconut Palm Diversity. *AoB PANTS*. doi:10.1093/aobpla/pls045

Traidia, A., Roger, F., Chidley, A., Schroeder, J., & Marlaud, T. (2011). Effect of helium-argon mixtures on the heat transfer and fluid flow in gas tungsten arc welding. *Journal of Chemistry and Chemical Engineering*, *5*(9), 854–861.

Valiev, R. (2004). Nanostructuring of metals by severe plastic deformation for advanced properties. *Nature Materials*, *3*(8), 511–516. doi:10.1038/nmat1180 PMID:15286754

Ye, C., Telang, A., Gill, A. S., Suslov, S., Idell, Y., Zweiacker, K., Wiezorek, J. M. K., Zhou, Z., Qian, D., Mannava, S. R., & Vasudevan, V. K. (2014). Gradient nanostructure and residual stresses induced by Ultrasonic Nano-crystal Surface Modification in 304 austenitic stainless steel for high strength and high ductility. *Materials Science and Engineering A*, *613*, 274–288. doi:10.1016/j.msea.2014.06.114

Chen, J. H., Chen, P. N., Hua, P. H., Chen, M. C., Chang, Y. Y., & Wu, W. (2009). Deposition of multicomponent alloys on low-carbon steel using gas tungsten arc welding (GTAW) cladding process. *Materials Transactions*, *50*(3), 689–694. doi:10.2320/matertrans.MRP2008276

Izah, S. C., & Ohimain, E. I. (2016). The opportunities and weakness of Nigerian Oil Palm Industry. *Biotechnological Research*, *2*(1), 33–43.

Jiang, F., Liu, Z., Wan, Y., & Shi, Z. (2013). Analytical modeling and experimental investigation of tool and workpiece temperatures for interrupted cutting 1045 steel by inverse heat conduction method. *Journal of Materials Processing Technology*, *213*(6), 887–894. doi:10.1016/j.jmatprotec.2013.01.004

Zhao, W., Liu, D., Yang, J., Zhang, H., Ma, A., Zhang, X., Ren, Z., Zhang, R., Dong, Y., & Ye, C. (2021). Improving plain and fretting fatigue resistance of A100 steel using ultrasonic nanocrystal surface modification. *International Journal of Fatigue*, *148*, 106204. doi:10.1016/j.ijfatigue.2021.106204

Allen-Bradley PLCs. (2013). *An Emphasis on Design and Application, Kelvin T. Erickson*. Dogwood Valley Press, LLC.

De Weck, O. (2014). Trends in advanced manufacturing technology innovation. *Production in the Innovation Economy*, 235.

Grum, J., & Šturm, R. (1997). Laser surface melt-hardening of gray and nodular irons. *Applied Surface Science*, *109*, 128–132. doi:10.1016/S0169-4332(96)00648-4

Islam, M. S., Church, J. S., & Miao, M. (2011). Effect of Removing Polypropylene Fibre Surface Finishing on Mechanical Performance of Kenaf/Polypropylene Composites. *Composites. Part A, Applied Science and Manufacturing*, *42*(11), 1687–1693. doi:10.1016/j.compositesa.2011.07.023

Vidyarthy, R., & Dwivedi, D. (2016). Activating flux tungsten inert gas welding for enhanced weld penetration. *J Manuf Process, 22*, 211–228.

Adeniyi, A.G., Ighalo, J.O., & Onifade, D.V. (2019). Banana and Plantain Fiber-Reinforced Polymer Composites. *J of Polym Eng*. doi:10.1515/polyeng-20190085

Lawal, S. A., Choudhury, I. A., & Nukman, Y. (2012). Application of vegetable oil-based metalworking fluids in machining ferrous metals—A review. *International Journal of Machine Tools & Manufacture*, *52*(1), 1–12. doi:10.1016/j.ijmachtools.2011.09.003

Sanchez, A. G., You, C., Leering, M., Glaser, D., Furfari, D., Fitzpatrick, M. E., Wharton, J., & Reed, P. A. S. (2021). Effects of laser shock peening on the mechanisms of fatigue short crack initiation and propagation of AA7075-T651. *International Journal of Fatigue*, *143*, 106025. doi:10.1016/j.ijfatigue.2020.106025

Wu, C. S., Wang, L., Ren, W. J., & Zhang, X. Y. (2014). Plasma arc welding: Process, sensing, control and modeling. *Journal of Manufacturing Processes*, *16*(1), 74–85. doi:10.1016/j.jmapro.2013.06.004

Chang, K. C., Zhao, J. R., & Hung, F. Y. (2021). Microstructure, Mechanical Properties, and Fatigue Fracture Characteristics of High-Fracture-Resistance Selective Laser Melting Al-Ni-Cu Alloys. *Metals*, *11*(1), 87. doi:10.3390/met11010087

M'Saoubi, R., Axinte, D., Soo, S. L., Nobel, C., Attia, H., Kappmeyer, G., Engin, S., & Sim, W. M. (2015). High performance cutting of advanced aerospace alloys and composite materials. *CIRP Annals*, *64*(2), 557–580. doi:10.1016/j.cirp.2015.05.002

Patil, R. G., & Kolambe, B. N. (2011). Development of value added products from Banana Pseudo-stem. An Overview of Progress. National Agricultural Innovation Project.

Yin, X., He, G., Meng, W., Xu, Z., Hu, L., & Ma, Q. (2020). Comparison study of low-heat-input wire arc-fabricated nickel-based alloy by cold metal transfer and plasma arc. *Journal of Materials Engineering and Performance*, *29*(7), 4222–4232. doi:10.100711665-020-04942-3

Das, B. K., Chakravarty, A. C., Sinha, M. K., & Ghosh, S. K. (1978). Use of polyethylene monofilament in spinning core yarn of jute. *Man Made Textile in India*, *21*(9), 494–498.

Li, K., Wu, Z., Zhu, Y., & Liu, C. (2017). Metal transfer in submerged arc welding. *Journal of Materials Processing Technology*, *244*, 314–319. doi:10.1016/j.jmatprotec.2017.02.004

O'Toole, L., Kang, C., & Fang, F. (2020). Advances in rotary ultrasonic-assisted machining. *Nanomanufacturing and Metrology*, *3*(1), 1–25. doi:10.100741871-019-00053-3

Sarah, A. S., Rahman, W. A., Majid, R. A., Yahya, W. J., Adrus, N., Hasannuddin, A. K., & Low, J. H. (2018). Optimization of Pineapple Leaf Fibre Extraction Methods and Their Biodegradabilities for Soil Cover Application. *Journal of Polymers and the Environment*, *26*(1), 319–329. doi:10.100710924-017-0942-4

Thanigaivelan, R., & Deepa, D. (2018). Heat transfer enhancement by coated fins in the microscale domain. *Thermal Science*, *22*(6 Part B), 2783–2789. doi:10.2298/TSCI160901089T

Wada, T., & Cox, T. B. (1983). *Preparation and evaluation of submerged-arc weld in 4 inch thick 3Cr-1. 5Mo-0. 1V steel plate.* Final report (No. DOE/NBM-4005684). AMAX Metals Group.

Xu, W. X., & Zhang, L. C. (2015). Ultrasonic vibration-assisted machining: Principle, design and application. *Advances in Manufacturing*, *3*(3), 173–192. doi:10.100740436-015-0115-4

Eshkabilov, K., Berdiyev, S., & Kamolov, B. (2021, January). Hardening of cutting tools by combined gas nitriding method. *IOP Conference Series. Materials Science and Engineering*, *1030*(1), 012019. doi:10.1088/1757-899X/1030/1/012019

Gillner, A., Petring, D., & Olowinsky, A. (2021). Laser joining. In *Advanced Joining Processes* (pp. 3–34). Elsevier. doi:10.1016/B978-0-12-820787-1.00001-2

Jeon, Y., & Lee, C. M. (2012). Current research trend on laser assisted machining. *International Journal of Precision Engineering and Manufacturing*, *13*(2), 311–317. doi:10.100712541-012-0040-4

Young, R. A. (1986). Structure, Swelling and Bonding of Cellulose Fibers. In R. A. Young & R. M. Rowell (Eds.), *Cellulose: structure, modification and hydrolysis* (pp. 91–128). John Wiley & Sons.

Kibria, G., Doloi, B., & Bhattacharyya, B. (2013). Experimental investigation and multi-objective optimization of Nd: YAG laser micro-turning process of alumina ceramic using orthogonal array and grey relational analysis. *Optics & Laser Technology*, *48*, 16–27. doi:10.1016/j.optlastec.2012.09.036

Mashrabov, N., Erofeev, V. V., & Korotkov, V. A. (2021). Plasma hardening of elements for soil cultivating instruments. *Welding International*, 1-5.

Mohanty, A. K., Misra, M., & Drzal, L. T. (2001). Surface modifications of Natural Fibers and Performance of the Resulting Biocomposites: An overview. *Composite Interfaces*, *8*(5), 313–343. doi:10.1163/156855401753255422

Schultz, H. (1993). *Electron beam welding* (Vol. 93). Woodhead Publishing.

Azar, P., Li, P., Patnaik, P. C., Thamburaj, R., & Immarigeon, J. P. (2001). Electron beam weld repair and qualification of titanium fan blades for military gas turbine engines. *RTO AVT Specialsits' Meeting on Cost Effective Application of Titanium Alloys in Military Platforms.*

Chaus, A. S., Sahul, M., Moravčík, R., & Sobota, R. (2021). Role of microstructural factor in wear resistance and cutting performance of high-speed steel end mills. *Wear*, *474*, 203865. doi:10.1016/j.wear.2021.203865

Sun, S., Brandt, M., Barnes, J. E., & Dargusch, M. S. (2011). Experimental investigation of cutting forces and tool wear during laser-assisted milling of Ti-6Al-4V alloy. *Proceedings of the Institution of Mechanical Engineers. Part B, Journal of Engineering Manufacture*, *225*(9), 1512–1527. doi:10.1177/0954405411411608

Zafeiropoulos, N. E., Williams, D. R., Baillie, C. A., & Matthews, F. L. (2002). Engineering and Characterization of the interface in flax fibre/polypropylene composite materials. Part I. Development and investigation of surface treatments. *Composites. Part A, Applied Science and Manufacturing*, *33*(8), 1083–1093. doi:10.1016/S1359-835X(02)00082-9

Azwa, Z. N., Yousif, B. F., Manalo, A. C., & Karunasena, W. (2013). A Review on the Degradability of Polymeric Composites Based on Natural Fibres. *Materials & Design*, *47*, 424–442. doi:10.1016/j.matdes.2012.11.025

Lopez de Lacalle, L. N., Sanchez, J. A., Lamikiz, A., & Celaya, A. (2004). Plasma assisted milling of heat-resistant superalloys. *J. Manuf. Sci. Eng.*, *126*(2), 274-285.

Rao, C. M., Sachin, B., Rao, S. S., & Herbert, M. A. (2021). Minimum Quantity Lubrication through the micro-hole textured PCD and PCBN inserts in the machining of the Ti–6Al–4V alloy. *Tribology International*, *153*, 106619. doi:10.1016/j.triboint.2020.106619

Thomas, W. M., Threadgill, P. L., & Nicholas, E. D. (1999). Feasibility of friction stir welding steel. *Science and Technology of Welding and Joining*, *4*(6), 365–372. doi:10.1179/136217199101538012

Cui, L., Fujii, H., Tsuji, N., & Nogi, K. (2007). Friction stir welding of a high carbon steel. *Scripta Materialia*, *56*(7), 637–640. doi:10.1016/j.scriptamat.2006.12.004

Sanchez, L. E., Mello, H. J., Neto, R. R. I., & Davim, J. P. (2014). Hot turning of a difficult-to-machine steel (saexev-f) aided by infrared radiation. *International Journal of Advanced Manufacturing Technology*, *73*(5-8), 887–898. doi:10.100700170-014-5879-3

Sasi, R., Subbu, S. K., & Palani, I. A. (2017). Performance of laser surface textured high speed steel cutting tool in machining of Al7075-T6 aerospace alloy. *Surface and Coatings Technology*, *313*, 337–346. doi:10.1016/j.surfcoat.2017.01.118

Siqueira, G., Bras, J., & Dufresne, A. (2009). *New Process of Chemical Grafting of Cellulose Nanoparticles with a Long Chain Isocyanate Langmuir*. PMID:19921797

Mishra, R. S., & Ma, Z. Y. (2005). Friction stir welding and processing. *Materials Science and Engineering R Reports*, *50*(1-2), 1–78. doi:10.1016/j.mser.2005.07.001

Pandey, K., & Datta, S. (2021). Hot machining of difficult-to-cut materials: A review. *Materials Today: Proceedings*, *44*, 2710–2715. doi:10.1016/j.matpr.2020.12.687

Rowell, R. M., Han, J. S., & Rowell, J. S. (2000). Characterization and Factors Effecting Fiber Properties. In E. Frollini, A. L. Leao, & L. H. C. Mattoso (Eds.), Natural Polymers and Agrofibers Composites (pp. 115–134). Academic Press.

Singh, B., Sasi, R., Subbu, S. K., & Muralidharan, B. (2019). Electric discharge texturing of HSS cutting tool and its performance in dry machining of aerospace alloy. *Journal of the Brazilian Society of Mechanical Sciences and Engineering*, *41*(3), 1–11. doi:10.100740430-019-1654-6

Bruzzo, F., Catalano, G., Demir, A. G., & Previtali, B. (2021). Surface finishing by laser re-melting applied to robotized laser metal deposition. *Optics and Lasers in Engineering*, *137*, 106391. doi:10.1016/j.optlaseng.2020.106391

Chai, G., & Zhu, Y. (2010). Spectra and thermal analysis of the arc in activating flux plasma arc welding. *Guangpuxueyuguangpu fen xi = Guangpu*, *30*(4), 1141–1145.

Gawali, D., & Sharma, V. K. (2009, December). FPGA based micro-PLC design approach. In *2009 International conference on advances in computing, control, and telecommunication technologies* (pp. 660-663). IEEE.

Schiller, G. J. (2015, March). *Additive manufacturing for Aerospace. In 2015 IEEE Aerospace Conference.* IEEE.

Sukyai, P., Sriroth, K., Lee, B. K., & Kim, H. J. (2012). The Effect of Bacterial Cellulose on the Mechanical and Thermal Expansion Properties of Kenaf/Poly Lactic Acid Composites. Shenzhan, 1343-1351.

Besharati-Givi, M. K., & Asadi, P. (2014). *Advances in friction-stir welding and processing.* Elsevier.

Gürgen, S., & Sofuoğlu, M. A. (2021). Advancements in conventional machining: a case of vibration and heat-assisted machining of aerospace alloys. In *Advanced Machining and Finishing* (pp. 143–175). Elsevier. doi:10.1016/B978-0-12-817452-4.00015-4

Meng, X., Zhang, K., Guo, X., Wang, C., & Sun, L. (2021). Preparation of micro-textures on cemented carbide substrate surface by plasma-assisted laser machining to enhance the PVD tool coatings adhesion. *Journal of Materials Processing Technology*, *288*, 116870. doi:10.1016/j.jmatprotec.2020.116870

Osorio, L., Trujillo, E., van Vuure, A. W., Lens, F., Ivens, J., & Verpoest, I. (2010). The Relationship between the Bamboo Fibre Microstructure and Mechanical Properties. *Proceedings of the 14th European Conference on Composite Materials*, 467-473.

Bledzki, A. K., Reihmane, S., & Gassan, J. (1996). Properties and Modification Methods for Vegetable Fibers for Natural Fiber Composites. *Journal of Applied Polymer Science*, *59*(8), 1329–1336. doi:10.1002/(SICI)1097-4628(19960222)59:8<1329::AID-APP17>3.0.CO;2-0

Skoczypiec, S., Grabowski, M., & Ruszaj, A. (2016). The impact of electrochemical assistance on the microturning process. *International Journal of Advanced Manufacturing Technology*, *86*(5), 1873–1880. doi:10.100700170-015-8320-7

Suresh, S., Venkatesan, K., Natarajan, E., & Rajesh, S. (2020a). Performance Analysis of Nano Silicon Carbide Reinforced Swept Friction Stir Spot Weld Joint in AA6061-T6 Alloy. *Silicon*, 1-14.

Vieira, J. M., Machado, A. R., & Ezugwu, E. O. (2001). Performance of cutting fluids during face milling of steels. *Journal of Materials Processing Technology*, *116*(2-3), 244–251. doi:10.1016/S0924-0136(01)01010-X

Alvarez, V. A., Ruscekaite, R. A., & Vazquez, A. (2003). Mechanical properties and water absorption behavior of composites made from a biodegradable matrix and alkaline-treated Sisal Fibers. *Journal of Composite Materials*, *37*(17), 1575–1588. doi:10.1177/0021998303035180

Anderson, M., Patwa, R., & Shin, Y. C. (2006). Laser-assisted machining of Inconel 718 with an economic analysis. *International Journal of Machine Tools & Manufacture*, *46*(14), 1879–1891. doi:10.1016/j.ijmachtools.2005.11.005

Suresh, S., Venkatesan, K., Natarajan, E., & Rajesh, S. (2020b). Influence of tool rotational speed on the properties of friction stir spot welded AA7075-T6/Al2O3 composite joint. *Materials Today: Proceedings*, *27*, 62–67. doi:10.1016/j.matpr.2019.08.220

Zubkov, N., Poptsov, V., & Vasiliev, S. (2017). Surface Hardening by Turning without Chip Formation. *Jordan Journal of Mechanical & Industrial Engineering*, *11*(1).

Gunasekaran, K., Kumar, P. G., Thanigaivelan, R., Arunachalam, R., & Shanmugam, V. (2021). Optimization of Turning Parameters of Cryogenic Soaked AZ91 Magnesium Alloy using TOPSIS coupled Taguchi Technique. *Journal of New Materials for Electrochemical Systems*, *24*(1), 49–54. doi:10.14447/jnmes.v24i1.a09

Suresh, S., Venkatesan, K., Natarajan, E., Rajesh, S., & Lim, W. H. (2019). Evaluating weld properties of conventional and swept friction stir spot welded 6061-T6 aluminium alloy. *Materials Express*, *9*(8), 851–860. doi:10.1166/mex.2019.1584

Witczak, M., Mrugalski, M., & Korbicz, J. (2015). Towards robust neural-network-based sensor and actuator fault diagnosis: Application to a tunnel furnace. *Neural Processing Letters*, *42*(1), 71–87. doi:10.100711063-014-9387-0

Mallick, P. K. (1993) Fiber reinforced composites: materials, manufacturing and design (2nd ed.). Boca Raton, FL: CRC Press.

Parida, A. K., & Maity, K. (2021). Study of machinability in heat-assisted machining of nickel-base alloy. *Measurement*, *170*, 108682. doi:10.1016/j.measurement.2020.108682

Tolcha, M. A., Gutu Jiru, M., & Lemu, H. G. (2021). Experimental Investigation of Al-CuMgMn Alloy Developed by Laser Beam Surface Modification for Wear Resistance. *Metals*, *11*(5), 712. doi:10.3390/met11050712

Yeh, H. J. (2013). Ultrasonic welding of medical plastics. In *Joining and assembly of medical materials and devices* (pp. 296–323e). Woodhead Publishing. doi:10.1533/9780857096425.3.296

Bhal, N. S., & Singh, B. (1998). Potential of Natural Fiber Reinforced Polymer Composites for Civil Engineering Applications in India. *Proceedings of the 2nd International Conference in Infrastructure, Department of Civil Engineering and Engineering Mechanics*, 661–673.

Guan, P., Zhou, L., Yu, Z., Sun, Y., Liu, Y., Wu, F., & Chu, D. (2020). Recent progress of surface coating on cathode materials for high-performance lithium-ion batteries. *Journal of Energy Chemistry*, *43*, 220–235. doi:10.1016/j.jechem.2019.08.022

Komanduri, R., Flom, D. G., & Lee, M. (1985). *Highlights of the DARPA advanced machining research program*. Academic Press.

Zhang, Y., Li, Y., Luo, Z., Yuan, T., Bi, J., Wang, Z. M., Wang, Z. P., & Chao, Y. J. (2016). Feasibility study of dissimilar joining of aluminum alloy 5052 to pure copper via thermo-compensated resistance spot welding. *Materials & Design*, *106*, 235–246. doi:10.1016/j.matdes.2016.05.117

Devaraj, S., Malkapuram, R., & Singaravel, B. (2021). Performance analysis of micro textured cutting insert design parameters on machining of Al-MMC in turning process. *International Journal of Lightweight Materials and Manufacture*, *4*(2), 210–217. doi:10.1016/j.ijlmm.2020.11.003

Kavitha, S. (2016). Bamboo Fibre Analysis by Scanning Electron Microscope Study. *International Journal of Civil Engineering and Technology*, *7*(4), 234–241.

Morozow, D., Barlak, M., Werner, Z., Pisarek, M., Konarski, P., Zagórski, J., & Caban, J. (2021). Wear Resistance Improvement of Cemented Tungsten Carbide Deep-Hole Drills after Ion Implantation. *Materials (Basel)*, *14*(2), 239. doi:10.3390/ma14020239 PMID:33418938

Schulz, H. (Ed.). (2001). *Scientific fundamentals of HSC*. Hanser.

Wang, H., Memon, H., & Hassan, E.A.MMiah, MArshad Ali, M. A. (2019). Effect of Jute Fiber Modification on Mechanical Properties of Jute Fiber Composite. *Materials (Basel)*, *12*(8), 1226. doi:10.3390/ma12081226 PMID:30991643

Ramesha, M., Palanikumarb, K., & Hemachandra Reddy, K. H. (2013). Comparative Evaluation on Properties of Hybrid Glass Fiber- Sisal/Jute Reinforced Epoxy Composites. *Procedia Engineering*, *51*, 745–750. doi:10.1016/j.proeng.2013.01.106

Smith, S., & Tlusty, J. (1997). *Current trends in high-speed machining*. Academic Press.

Senthiil, P. V., & Sirsshti, A. (2014). Studies on Material and Mechanical Properties of Natural Fiber Reinforced Composites. *The International Journal of Engineering And Science*, *3*(11), 18–27.

Toenshoff, H. K., &Denkena, B. (2013). *Basics of cutting and abrasive processes*. Academic Press.

Cao, X., Hu, J., Huo, W., Xi, X., & Zhao, W. (2021). Surface microstructure and property modifications in AISI 304 stainless steel induced by pseudospark pulsed electron beam treatments. *Vacuum*, *184*, 109914. doi:10.1016/j.vacuum.2020.109914

Kuo, M., Sun, Z., & Pan, D. (2013). Laser welding with activating flux. *Sci Technol Weld Joining*, *6*(1), 17–22.

Naves, V. T. G., Da Silva, M. B., & Da Silva, F. J. (2013). Evaluation of the effect of application of cutting fluid at high pressure on tool wear during turning operation of AISI 316 austenitic stainless steel. *Wear*, *302*(1-2), 1201–1208. doi:10.1016/j.wear.2013.03.016

Sorieul, M., Dickson, A., Hill, S. J., & Pearson, H. (2016). Plant Fibre: Molecular Structure and Biomechanical Properties, of a Complex Living Material, Influencing Its Deconstruction towards a Bio based Composite. *Materials (Basel)*, *9*(12), 6–18. doi:10.3390/ma9080618 PMID:28773739

Zhang, P. (2010). *Advanced industrial control technology*. William Andrew.

Coromant, S. (1999). Application Guide: Die &Mould Making. *Sandvik Coromant*, 41-42.

Gao, C. C., Yu, L., Liu, H. S., & Chen, L. (2012). Development of Self-Reinforced Polymer Composites. *Progress in Polymer Science*, *37*(6), 767–780. doi:10.1016/j.progpolymsci.2011.09.005

Mallick, P. K. (1990). *Compression Moulding, Composite Materials Technology* (P. K. Mallick & S. Newman, Eds.). Hanser Publishers.

Ye, G. G., Xue, S. F., Ma, W., Jiang, M. Q., Ling, Z., Tong, X. H., & Dai, L. H. (2012). Cutting AISI 1045 steel at very high speeds. *International Journal of Machine Tools & Manufacture*, *56*, 1–9. doi:10.1016/j.ijmachtools.2011.12.009

Aveston, J., Kelly, A., & Sillwood, J. M. (1980). *Advances in Composite Materials*. Pergamon.

Gutowski, T. G. (1997). *Advanced Composites Manufacturing*. John Wiley and Sons.

Ishai, O., & Daniel, I. M. (2006). *Engineering Mechanics of Composite Materials* (2nd ed.). Oxford University Press.

Santos, J. C. D., Siqueira, R. L., Gomes, L. M., Freier, R. T. S., Mano, V., & Panzera, T. H. (2018). Effects of Sodium Carbonate on the Performance of Epoxy and Polyester-Coir Reinforced Composite. *Polymer Testing*, *67*, 533–544. doi:10.1016/j.polymertesting.2018.03.043

Kumar, S. M. S., Duraibabu, D., & Subramanian, K. (2014). Studies on Mechanical, Thermal and Dynamic Mechanical Properties of Untreated (raw) and Treated coconut Sheath Fiber Reinforced Epoxy Composites. *Materials & Design*, *59*, 63–66. doi:10.1016/j.matdes.2014.02.013

Suoware, T. O., & Edelugo, S. O. (2018). Thermal, Mechanical and Flame Spread Properties of Hybridized FlameRetardant in oil palm fibre-polyester panel. *Springer Nature Applied Science*, *1*(1), 42. doi:10.100742452-018-0048-1

Odusote, J. K., & Oyewo, A. T. (2016). Mechanical Properties of Pineapple Leaf Fiber Reinforced Polymer Composites for Application as a Prosthetic Socket. *Journal of Engineering Technology*, *7*(1), 125–139.

Abdel-Malek, L., & Wolf, C. (1991). Evaluating flexibility of alternative FMS designs—A comparative measure. *International Journal of Production Economics*, *23*(1-3), 3–10. doi:10.1016/0925-5273(91)90042-R

Childs, T. H. C., Maekawa, K., Obikawa, T., & Yamane, Y. (2000). *Metal cutting. In Theory and Applications*. Arnold.

Huang, H.-Y. (2010). Effects of activating flux on the welded joint characteristics in gas metal arc welding. *Materials & Design*, *31*(5), 2488–2495. doi:10.1016/j.matdes.2009.11.043

Majeed, K., Jawaid, M., Hassan, A., Abu Bakar, A., Abdul Khalil, H.P.S., Salami, A.A., & Inuwa, I. (2013). Potential Materials for Food Packaging from Nanoclay/Natural Fibres Filled Hybrid Composites. *Mater. Des, 46*, 391–410.

Voznesenskaya, A. A., Kochuev, D. A., Raznoschikov, A. S., Zhdanov, A. V., & Morozov, V. V. (2020, July). Effect of part orientation on temperature stresses in the structure of a material during selective laser melting. *IOP Conference Series. Materials Science and Engineering*, *896*(1), 012131. doi:10.1088/1757-899X/896/1/012131

Dong, Y., Ghataura, A., Takagi, H., Haroosh, H. J., Nakagaito, A. N., & Lau, K. T. (2014). Polylactic Acid (PLA)Biocomposites Reinforced with Coir Fibres: Evaluation of Mechanical Performance and Multifunctional Properties. *Compos Part A*, *63*, 76–84. doi:10.1016/j.compositesa.2014.04.003

Arun, A., Sathyaseelan, R., Tamilselvan, M., Gowtham, M., & Karthikeyan, A. (2016). Influence of Weight Fractions on Mechanical, Water Absorption and Corrosion Resistance Behaviors of Untreated Hybrid (Coir/Banana) Fiber Reinforced Epoxy Composites. *International Journal of Chemtech Research*, *9*, 932–940.

Ramli, R., Yunus, R. M., Beg, M. D. H., & Prasad, D. M. R. (2011). Oil Palm Fiber Reinforced Polypropylene Composites: Effects of Fiber Loading and Coupling Agents on Mechanical, Thermal, and Interfacial Properties. *Journal of Composite Materials*, *46*(11), 1275–1284. doi:10.1177/0021998311417647

Vidyashri, V., Lewis, H., Narayanasamy, P., Mahesha, G. T., & Bhat, K. S. (2019). Preparation of Chemically Treated Sugarcane Bagasse Fiber Reinforced Epoxy Composites and their Characterization. *Cogent Engineering*, *6*(1), 1708644. doi:10.1080/23311916.2019.1708644

Daramola, O. O., Adediran, A. A., Adewuyi, B. O., & Adewole, O. (2017). Mechanical Properties and Water Absorption Behaviour of Treated Pineapple Leaf Fibre Reinforced Polyester Matrix Composites. *Leonardo Journal of Sciences*, (30), 15–30.

Obasi, H. C., Iheaturu, N. C., Onuoha, F., Chike-Onyegbula, C., Akanbi, M. N., & Ezeh, V. O. (2014). Influence of Alkali Treatment and Fibre Content on the Properties of Oil Palm Press Fibre Reinforced Epoxy Biocomposites. *American Journal of Engineering Research*, *3*(2), 117–123.

Osita, O., Ignatius, O., & Udeani Henry, U. (2016). Study on the Mechanical Properties of Palm Kernel Fibre Reinforced Epoxy and Poly-Vinyl Alcohol (PVA) Composite Material. *International Journal of Engineering and Technologies*, *7*, 68–77. doi:10.18052/www.scipress.com/IJET.7.68

Santafé Júnior, H. P. G., Lopes, F. P. D., Costa, L. L., & Monteiro, S. N. (2010). Mechanical Properties of Tensile Tested Coir Fiber Reinforced Polyester Composites. *Revista Matéria*, *15*(2), 113–118. doi:10.1590/S1517-70762010000200017

Byrne, G., Dornfeld, D., & Denkena, B. (2003). Advancing cutting technology. *CIRP Annals*, *52*(2), 483–507. doi:10.1016/S0007-8506(07)60200-5

Tümer, M., Mert, T., & Karahan, T. (2021). Investigation of microstructure, mechanical, and corrosion behavior of nickel-based alloy 625/duplex stainless steel UNS S32205 dissimilar weldments using ERNiCrMo-3 filler metal. *Welding in the World*, *65*(2), 171–182. doi:10.100740194-020-01011-0

You, Y., Zhang, B., Tao, S., Liang, Z., Tang, B., Zhou, R., & Yuan, D. (2021). Effect of Surface Microstructure on the Heat Dissipation Performance of Heat Sinks Used in Electronic Devices. *Micromachines*, *12*(3), 265. doi:10.3390/mi12030265 PMID:33806561

Chaitanya, A. K., Babu, D. K., Narendra, K. B., & Sumanth, K. (2016). Investigation on The Mechanical Properties of Banana Fiber Reinforced Polyester Composites. *International Journal of Engineering Research and Applications*, *6*(9), 76–82.

Mustafa Al Bakri, A. M. M., Liyana, J., Norazian, M. N., Kamarudin, H., & Ruzaidi, C. M. (2013). Mechanical Properties of Polymer Composites with Sugarcane Bagasse Filler. *Advanced Materials Research*, *740*, 739–744. doi:10.4028/www.scientific.net/AMR.740.739

Luz, D., da Costa, S. M., Gonçalves, A. R., Del'ArcoJuniord, A. P., & da Costa, S. A. (2016). Polypropylene Composites Reinforced with Biodegraded Sugarcane Bagasse Fibers: Static and Dynamic Mechanical Properties. *Materials Research*, *19*(1), 75–83. doi:10.1590/1980-5373-MR-2015-0410

Sapuan, S. M., & Bachtiara, D. (2012). Mechanical Properties of Sugar Palm Fibre Reinforced High Impact Polystyrene Composites. *Procedia Chemistry*, *4*, 101–106. doi:10.1016/j.proche.2012.06.015

Kasim, A. N., Selamat, M. Z., Daud, M. A. M., Yaakob, M. Y., Putra, A., & Sivakumar, D. (2016). Mechanical Properties of Polypropylene Composites reinforced with Alkaline Treated Pineapple Leaf Fibre from Josapine Cultivar. *International Journal of Automotive and Mechanical Engineering*, *13*(1), 3157–3167. doi:10.15282/ijame.13.1.2016.3.0263

Asgekar, S.D., Joshi, V.K., Futane, P.S., & Joshi, P.S. (2013). Characteristics of Sugarcane/Coir Fibres Reinforced Composites in Phenol Formaldehyde Resin. *International Journal of Composite Materials, 3*(6), 156-162.

Ramesha, M., Atreyaa, T. S. A. S., Aswina, U. S., Eashwara, H., & Deepab, C. (2014). Processing and Mechanical Property Evaluation of Banana Fiber Reinforced Polymer Composites. *Procedia Engineering*, *97*, 563–572. doi:10.1016/j.proeng.2014.12.284

Selamat, M. Z., Zhafri Tahir, M. S., Kasim, A. N., Dharmalingam, S., Azma Putra, A., Yaakob, M., & Yand Daud, M. A. M. (2018). Effect of Starch Sizes Particle as Binder on Short Pineapple Leaf Fiber Composite Mechanical Properties. *MATEC Web of Conferences*, *150*, 04008. Advance online publication. doi:10.1051/matecconf/201815004008

Arifvianto, B., & Mahardika, M. (2018). Surface morphology, roughness and wettability of the medical grade 316L stainless steel processed with surface mechanical attrition treatment and electropolishing for the preparation of osteosynthesis plate. *The Journal of Physiological Sciences; JPS*, *29*(3), 83–94.

Boothroyd, G. (1988). *Fundamentals of metal machining and machine tools* (Vol. 28). CRC Press.

Khalili, P., Tshai, K. Y., Hui, D., & Kong, I. (2017). Synergistic of Ammonium Polyphosphate and Alumina Trihydrate as Fire Retardants for Natural Fiber Reinforced Epoxy Composite. *Composites. Part B, Engineering*, *114*, 101–110. doi:10.1016/j.compositesb.2017.01.049

Solowjow, E., Ugalde, I., Shahapurkar, Y., Aparicio, J., Mahler, J., Satish, V., . . . Claussen, H. (2020, August). Industrial robot grasping with deep learning using a programmable logic controller (plc). In *2020 IEEE 16th International Conference on Automation Science and Engineering (CASE)* (pp. 97-103). IEEE.

Xu, L. Y., Jing, H. Y., & Han, Y. D. (2018). Effect of welding on the corrosion behavior of X65/Inconel 625 in simulated solution. *Welding in the World*, *62*(2), 363–375. doi:10.100740194-018-0549-y

Baillie, C., & Jayasinghe, R. (2017). *Green Composites: Waste and Nature-based Materials for a Sustainable Future. Cambridge*. Elsevier Science. doi:10.1016/B978-0-08-100783-9.00001-0

Horrocks, A. R., & Price, D. (2001). Fire retardant materials. CRC Press.

Steinhaus, T. (2010). *Determination of Intrinsic Material Flammability Properties from Material Tests assisted by Numerical Modelling*. University of Edinburgh.

Cordeiro, E. P., Pita, V. J. R. R., & Soares, B. G. (2017, September). Epoxy–Fiber of Peach Palm Trees Composites: The Effect of Composition and Fiber Modification on Mechanical and Dynamic Mechanical Properties. *Journal of Polymers and the Environment*, *25*(3), 913–924. Advance online publication. doi:10.100710924-016-0841-0

Hassan, M.Z., Sapuan, S.M., & Rasid, Z.A. (2019). Thermal Degradation and Mechanical Behavior of Banana Pseudo-Stem Reinforced Composites. *International Journal of Recent Technology and Engineering, 8*(4), 5599-5902.

Mittal, V., & Sinha, S. (2017). Effect of chemical Treatment on Thermal Properties of Bagasse Fiber-Reinforced Epoxy Composite. *Science and Engineering of Composite Materials*, *24*(2), 237–243. doi:10.1515ecm-2014-0434

Jain, J., Jain, S., & Sinha, S. (2018). Characterization and Thermal Kinetic Analysis of Pineapple Leaf Fibers and their Reinforcement in Epoxy. *Journal of Elastomers and Plastics*. Advance online publication. doi:10.1177/0095244318783024

Mulinari, D. R., Baptista, C. A. R. P., Souza, J. V. C., & Voorwald, H. V. C. (2011). Mechanical Properties of Coconut Fibre Reinforced Polyester Composites. *Procedia Engineering*, *10*, 2074–2079. doi:10.1016/j.proeng.2011.04.343

Shah, H., Srinivasulu, B., & Shit, S. C. (2013). Influence of Banana Fibre Chemical Modification on the Mechanical and Morphological Properties of Woven Banana Fabric/Unsaturated Polyester Resin Composites. Polymers from Renewable Resources, 4(2), 51-84.

Abdulah, A. M., & Azizi, N. (2018). Thermogravimetric Analysis of Saccharum/Unsaturated Polyester Composites. *Al-qadisiyah J for Engineering Sciences.*, *11*(4), 411–416.

Carlsson, J. O., & Martin, P. M. (2010). Chemical vapor deposition. In *Handbook of Deposition Technologies for films and coatings* (pp. 314–363). William Andrew Publishing. doi:10.1016/B978-0-8155-2031-3.00007-7

König, W. (2008). *Fertigungsverfahren 1: Drehen, Fräsen, Bohren*. Springer-Verlag.

Sivakumar, J., Vasudevan, M., & Korra, N. N. (2021). Effect of activated flux tungsten inert gas (A-TIG) welding on the mechanical properties and the metallurgical and corrosion assessment of Inconel 625. *Welding in the World*, *65*(6), 1061–1077. doi:10.100740194-020-01061-4

Trsek, H. (2016). *Isochronous wireless network for real-time communication in industrial automation*. Springer. doi:10.1007/978-3-662-49158-4

Santosha, P.., Gowda, A., & Manikanth, V. (2018). Effect of Fibre Loading on Thermal Properties of Banana and Pineapple Leaf. *Fibre Reinforced Polyester Composite, 5*(2), 5631-5635.

Mothe, C.G., de Araujo, C.R., & Maradela, F.S. (2003). Mechanical and Thermal Properties of Coir Fibre Reject/Polypropylene Composite. *7thCongressoBrasileiro de Polimeros,* 973-974.

Ridzuan, R., & Beg, M.D.HRosli, M.YRohaya, M.HAstimar, A.ASamahani, SZawawi, I. (2013). Effects of Coupling Agent and Flame Retardant on the Performance s of Oil Palm Empty Fruit Bunch Fibre Reinforced Polypropylene Composites. *World Academy of Science, Engineering and Technology*, *7*(6), 445–453.

De Leao, A. G., Pereira, P. S. C., & Bastos, D. C. (2018). Characterization of Polypropylene Post-Consume/Banana Fibre Composites. *Acta Scientiae and Technicae*, *6*(1), 19–22. doi:10.17648/uezo-ast-v6i1.129

Simdo, J. A., Carmona, V. B., Marconcini, J. M., Mattoso, L. H. C., Arsberg, S. T., & Sanadi, A. R. (2016). Effect of Fibre Treatment Condition and Coupling Agent on the Mechanical and Thermal Properties in Highly Filled Composites of Sugarcane Bagasse Fibre/PP. *Materials Research*, *19*(4), 7446–7751.

Muragan, M.D., Zakaria, Z., & Hassan, A. (2019). Mechanical and Thermal Properties of Graphene Oxide reinforced Polypropylene/Pineapple Leaves Fibre Composites. *PERINTIS eJournal*, *9*(2), 11-20.

Basnayake, A.J., Vandi, L., & Heitzmann, M. (2018). *A Literature Review of Natural Fibre Composite Fire Properties.* CIC Fibre-Resin Interface Project.

Zhao, C., Jiang, E., & Chen, A. (2016). Volatile production from pyrolysis of cellulose, hemicellulose and lignin. *Journal of the Energy Institute.*

Mouritz, A. P., Gibson, A. G., & Gladwell, G. M. L. (Eds.). (2007). Fire Properties of Polymer Composite Materials (Solid Mechanics & Its Applications S., v. 143). Springer Netherlands.

Arrakhiz, F. Z., Achaby, M., & El. (2013). Mechanical and Thermal Properties of Natural Fibers Reinforced Polymer Composites: Doum/low Density Polyethylene. *Materials & Design*, *43*, 200–205. doi:10.1016/j.matdes.2012.06.056

Daniyan, I. A., & Mpofu, K. (2018). Application of Design for Manufacturing and Assembly: Development of a Multi-feedstock Biodiesel Processor. In *Applications of Design for Manufacturing and Assembly*. IntechOpen.

Sun, L., Yuan, G., Gao, L., Yang, J., Chhowalla, M., Gharahcheshmeh, M. H., Gleason, K. K., Choi, Y. S., Hong, B. H., & Liu, Z. (2021). Chemical vapour deposition. *Nature Reviews Methods Primers*, *1*(1), 1–20. doi:10.103843586-020-00005-y

Zhang, Z., & Zhang, F. (2009). Spectral analysis of welding plasma of magnesium alloy using flux coated wire. *Materials Transactions*, *50*(8), 1909–1914. doi:10.2320/matertrans.M2009023

Babrauskas, V. (1986). The cone calorimeter – A Versatile Bench-Scale Tool for the Evaluation for Fire Properties. In New Technology to Reduce Fire Loses and Cost. Academic Press.

Ramanaiah, K., Prasad, A. V. R., & Reddy, K. H. C. (2013). Thermopyhsical and Fire Properties of Vakka Natural Fiber Reinforced Polyester Composites. *Journal of Reinforced Plastics and Composites*, *32*(15), 1092–1098. doi:10.1177/0731684413486366

Dorez, G., Ferry, L., Sonnier, R., Taguet, A., & Lopez-Cuesta, J. M. (2014). Effect of Cellulose, Hemicellulose and Lignin Contents on Pyrolysis and Combustion of Natural Fibers. *Journal of Analytical and Applied Pyrolysis*, *107*, 323–331. doi:10.1016/j.jaap.2014.03.017

Natsa, S., Akindapo, J. O., & Garba, D. K. (2015). Development of a Military Helmet Using Coconut Fiber Reinforced Polymer Matrix Composite. *European Journal of Engineering Technology*, *3*(7), 55–65.

Asasutjarit, C., Hirunlabh, J., Khedari, J., Charoenvai, S., Zeghmati, B., & Shin, U. C. (2007). Development of Coconut Coir-Based Lightweight Cement Board. *Construction & Building Materials*, *21*(2), 277–288. doi:10.1016/j.conbuildmat.2005.08.028

Abraham, D. P. R., Joseph, A., Jasheelad, A., Binuraj, P. R., & Sarmaa, J. (2016). *Development of Coir-Fibre Cement Composite Roofing Tiles. Procedia Technology*, *24*, 169–178.

Asim, M., Abdan, K., Jawaid, M., Nasir, M., Dashtizadeh, Z., Ishak, M. R., & Hoque, M. E. (2015). Review on Pineapple Leaves Fibre and Its Composites. *International Journal of Polymer Science*, *2015*, 1–16. Advance online publication. doi:10.1155/2015/950567

Baptista, A., Silva, F., Porteiro, J., Míguez, J., & Pinto, G. (2018). Sputtering physical vapour deposition (PVD) coatings: A critical review on process improvement and market trend demands. *Coatings*, *8*(11), 402. doi:10.3390/coatings8110402

Berthier, A., Paillard, P., Carin, M., Valensi, F., & Pellerin, S. (2012). TIG and A-TIG welding experimental investigations and comparison to simulation: Part 1: Identification of Marangoni effect. *Science and Technology of Welding and Joining*, *17*(8), 609–615. doi:10.1179/1362171812Y.0000000024

Jawahir, I. S., & Van Luttervelt, C. A. (1993). Recent developments in chip control research and applications. *CIRP Annals*, *42*(2), 659–693. doi:10.1016/S0007-8506(07)62531-1

John, M. J., & Thomas, S. (2008). Biofibres and Biocomposites. *Carbohydrate Polymers*, *71*(3), 343–364. doi:10.1016/j.carbpol.2007.05.040

Zhang, D. T., & Chen, J. G. (1993). An NC lathe simulator for part programming and machine operation training. *Computers in Industry*, *21*(2), 139–148. doi:10.1016/0166-3615(93)90132-K

Abdulateef, O. F., Mustafa, F. F., & Salman, S. A. (2010). Prediction the Effect of Flame Cutting Parameters on the Quality of Metal Surface In CNC Flame Cutting Machine Using Artificial Neural Network. *The Iraqi Journal For Mechanical And Material Engineering, 10*(3).

Abu-Izneid, T., AlHajri, N., Ibrahim, A. M., Javed, M. N., Salem, K. M., Pottoo, F. H., & Kamal, M. A. (2020). MicroRNAs in the regulation of immune response against SARS COV-2 and other viral infections. *Journal of Advanced Research*. PMID:33282419

AlAli, A. B., Griffin, M. F., & Butler, P. E. (2015). Three-Dimensional Printing Surgical Applications. *Eplasty*, *15*, e37.

Alam, M. S., Garg, A., Pottoo, F. H., Saifullah, M. K., Tareq, A. I., Manzoor, O., Mohsin, M., & Javed, M. N. (2017). Gum ghatti mediated, one pot green synthesis of optimized gold nanoparticles: Investigation of process-variables impact using Box-Behnken based statistical design. *International Journal of Biological Macromolecules*, *104*, 758–767. doi:10.1016/j.ijbiomac.2017.05.129 PMID:28601649

Alam, M. S., Javed, M. N., Pottoo, F. H., Waziri, A., Almalki, F. A., Hasnain, M. S., Garg, A., & Saifullah, M. K. (2019). QbD approached comparison of reaction mechanism in microwave synthesized gold nanoparticles and their superior catalytic role against hazardous nirto-dye. *Applied Organometallic Chemistry*, *33*(9), e5071. doi:10.1002/aoc.5071

Alhuthali, A., Low, I. M., & Dong, C. (2012). Characterization of the Water Absorption, Mechanical and Thermal Properties of Recycled Cellulose Fibre Reinforced Vinyl-Ester Eco- Nanocomposites. *Composites. Part B, Engineering*, *43*(7), 2772–2781.

Alhuthali, A., Low, I. M., & Dong, C. (2012). Characterization of the Water Absorption, Mechanical and Thermal Properties of Recycled Cellulose Fibre Reinforced Vinyl-Ester Eco-Nanocomposites. *Composites. Part B, Engineering*, *43*(7), 2772–2781. doi:10.1016/j.compositesb.2012.04.038

Altinkok, N., Özsert, I., & Findik, F. (2013). Dry sliding wear behavior of Al2O3/SiC particle reinforced aluminium based mmcs fabricated by stir casting method. *Acta Physica Polonica A*, *124*(1), 11–19. doi:10.12693/APhysPolA.124.11

Amenaghawon, N. A., Osayuki-Aguebor, W., & Okiemen, C. (2016). O Production of Particle Boards from Corn Cobs and Cassava Stalks: Optimization of mechanical properties using Response Surface Methodology. *Journal of Materials Environmental Science*, *7*(4), 1236–1244.

Ananth, S., Prakash, J. U., Juliyana, S. J., Rubi, C. S., & Sadhana, A. D. (2021). Effect of process parameters on WEDM of Al–Fly ash composites using Taguchi Technique. *Materials Today: Proceedings*, *39*, 1786–1790. doi:10.1016/j.matpr.2020.07.615

Anasane, S. S., & Bhattacharyya, B. (2016). Experimental investigation on suitability of electrolytes for electrochemical micromachining of titanium. *International Journal of Advanced Manufacturing Technology*, *86*(5), 2147–2160. doi:10.100700170-015-8309-2

Anderson. (2006). *Hypersonic and high-temperature gas dynamics*. American Institute of Aeronautics and Astronautics.

Arjmandi, R., Ismail, A., & Hassan, A. (2017). Abu Bakar, A Effects of Ammonium Polyphosphate Content on Mechanical, Thermal and Flammability Properties of Kenaf/Polypropylene and Rice Husk/Polypropylene Composites. *Construction & Building Materials*, *152*, 484–493.

Arumugam, R., & Rajalingam, M. (2020). A review on the effect of electro chemical machining on metal matrix composite materials. *Materials Today: Proceedings*, *33*, 3854–3857. doi:10.1016/j.matpr.2020.06.229

Aslam, M., Javed, M. N., Deeb, H. H., Nicola, M. K., Mirza, A., Akhtar, M. H., & Waziri, A. (2021). Lipid Nanocarriers for Neurotherapeutics: Introduction, Challenges, Blood-brain Barrier, and Promises of Delivery Approaches. *CNS & Neurological Disorders - Drug Targets*, *20*. Advance online publication. doi:10.2174/18715273206662107061042 40 PMID:34967302

Ataee, A., Li, Y., Fraser, D., Song, G., & Wen, C. (2018). Anisotropic Ti-6Al-4V gyroid scaffolds manufactured by electron beam melting (EBM) for bone implant applications. *Materials & Design*, *137*, 345–354. doi:10.1016/j.matdes.2017.10.040

Avery, R. E., & Tuthill, A. H. (1994). *Guidelines for the Welded Fabrication of Nickel Alloys for Corrosion-resistant Service*. Nickel Development Institute.

Ayrilmis, N., Jarusombuti, S., Fueangvivat, V., Bauchongkol, P., & White, R. (2011). Coir Fiber Reinforced Polypropylene Composite Panel for Automotive Interior Applications. *Fibers and Polymers*, *12*(7), 919–926.

Baba, M., Matsumoto, K., Yamasaki, N., Shindo, H., Yano, H., Matsumoto, M., Otsubo, R., John Lawn, M., Matsuo, N., Yamamoto, I., Hidaka, S., & Nagayasu, T. (2017). Development of a Tailored Thyroid Gland Phantom for Fine-Needle Aspiration Cytology by Three-Dimensional Printing. *Journal of Surgical Education*, *74*(6), 1039–1046.

Babrauskas, V. (1986). The cone calorimeter – A Versatile Bench-Scale Tool for the Evaluation for Fire Properties. New Technology to Reduce Fire Loses and Cost, 78-87.

Babrauskas, V. (2002). The Cone Calorimeter. In *SFPE Handbook of Fire Protection Engineering*. Society of FireProduction Engineers.

Babrauskas, V. (1995). Designing Products for Fire Performance: The State of the Art of Test Methods and Fire Models. *Fire Safety Journal*, *24*, 299–312.

Badini, C., La Vecchia, M., Fino, P., & Cobelli, P. (2000). Gravity casting processing of brake discs: Use of partially recycled (Al-Si-Mg)/SiCp composite. *Materials Science and Technology*, *16*(6), 681–686. doi:10.1179/026708300101508270

Bagotsky, V. S. (Ed.). (2005). *Fundamentals of electrochemistry* (Vol. 44). John Wiley & Sons. doi:10.1002/047174199X

Bajaj, G., Kim, M. R., Mohammed, S. I., & Yeo, Y. (2012). Hyaluronic acid-based hydrogel for regional delivery of paclitaxel to intraperitoneal tumors. *Journal of Controlled Release*, *158*(3), 386–392. doi:10.1016/j.jconrel.2011.12.001 PMID:22178261

Bandyopadhyay, A., Bose, S., & Das, S. (2015). 3D printing of biomaterials. *MRS Bulletin*, *40*(2), 108–115. doi:10.1557/mrs.2015.3

Barkallah, R., Taktak, R., Guermazi, N., Elleuch, K., & bouaziz, J. (2021). Mechanical properties and wear behaviour of alumina/tricalcium phosphate/titania ceramics as coating for orthopedic implant. *Engineering Fracture Mechanics*, *241*, 107399. doi:10.1016/j.engfracmech.2020.107399

Barua, R., Datta, S., Roychowdhury, A., & Datta, P. (2019). Importance of 3D Printing Technology in Medical Fields. In Additive Manufacturing Technologies From an Optimization Perspective (pp. 21-40). IGI Global.

Barua, R., Datta, S., Datta, P., & Chowdhury, A. R. (2019). Scaffolds and Tissue Engineering Applications by 3D Bio-Printing Process: A New Approach. In K. Kumar & J. Davim (Eds.), *Design, Development, and Optimization of Bio-Mechatronic Engineering Products* (pp. 78–99). IGI Global.

Behera, M. P., Dougherty, T., Singamneni, S., & de Silva, K. (2020). Selective laser melting of aluminium metal-matrix composites and the challenges. *Materials Today: Proceedings*, *33*, 5729–5733. doi:10.1016/j.matpr.2020.05.086

Bharti, C, Alam, MS, Javed, MN, Khalid, M, Saifullah, FA, & Manchanda, R. Silica Based Nanomaterial for Drug Delivery. (2021). Nanomaterials: Evolution and Advancement towards Therapeutic. *Drug Delivery*, *2*(2), 57.

Bhat, N. N., Kumari, K., Dutta, S., Pal, S. K., & Pal, S. (2015). Friction stir weld classification by applying wavelet analysis and support vector machine on weld surface images. *Journal of Manufacturing Processes*, *20*, 274–281. doi:10.1016/j.jmapro.2015.07.002

Bhattacharyya & Mitra. (2002). Development of Electrochemical Micromachining (ECMM) Setup. *Robotics and Computer Integrated Manufacturing, 18*, 283–289.

Bhattacharyya, B. (2015). *Electrochemical micromachining for nanofabrication, MEMS and nanotechnology*. William Andrew.

Bhattacharyya, B., & Munda, J. (2003). Experimental investigation into electrochemical micromachining (EMM) process. *Journal of Materials Processing Technology*, *140*(1-3), 287–291. doi:10.1016/S0924-0136(03)00722-2

Bigham, A., Foroughi, F., Rezvani Ghomi, E., Rafienia, M., Neisiany, R. E., & Ramakrishna, S. (2020). The journey of multifunctional bone scaffolds fabricated from traditional toward modern techniques. *Bio-Design and Manufacturing*, *2020*(3), 281–306.

Binjawadagi, B., Dwivedi, V., Manickam, C., Ouyang, K., Wu, Y., & Lee, L. J. (2014). Adjuvanted poly (lactic-co-glycolic) acid nanoparticle-entrapped inactivated porcine reproductive and respiratory syndrome virus vaccine elicits cross-protective immune response in pigs. *International Journal of Nanomedicine*, *24*(9), 679–694. PMID:24493925

Bi, Y. G., Lin, Z. T., & Deng, S. T. (2019). Fabrication and characterization of hydroxyapatite/sodium alginate/chitosan composite microspheres for drug delivery and bone tissue engineering. *Materials Science and Engineering C*, *100*, 576–583. doi:10.1016/j.msec.2019.03.040 PMID:30948094

Bobo, D., Robinson, K. J., Islam, J., Thurecht, K. J., & Corrie, S. R. (2016). Nanoparticle-Based Medicines: A Review of FDA-Approved Materials and Clinical Trials to Date. *Pharmaceutical Research*, *33*(10), 2373–2387. doi:10.100711095-016-1958-5 PMID:27299311

Brito, N. M., Soares, R. S., & Monteiro, E. L. (2016). Additive Manufacturing for Surgical Planning of Mandibular Fracture. *Acta Stomatologica Croatica*, *50*(4), 348–353. doi:10.15644/asc50/4/8

Browning, R. J., Reardon, P. J. T., Parhizkar, M., Pedley, R. B., Edirisinghe, M., Jonathan, C., Knowles, J. C., & Stride, E. (2017). Drug Delivery Strategies for Platinum-Based Chemotherapy. *ACS Nano*, *11*(9), 8560–8578. doi:10.1021/acsnano.7b04092 PMID:28829568

Bulei, C., Todor, M. P., & Kiss, I. (2018). Metal matrix composites processing techniques using recycled aluminium alloy. *IOP Conference Series. Materials Science and Engineering*, *393*(1), 012089. Advance online publication. doi:10.1088/1757-899X/393/1/012089

Cai, S., Thati, S., Bagby, T. R., Diab, H. M., Davies, N. M., Cohen, M. S., & Forrest, M. L. (2010). Localized doxorubicin chemotherapy with a biopolymeric nanocarrier improves survival and reduces toxicity in xenografts of human breast cancer. *Journal of Controlled Release*, *146*(2), 212–218. doi:10.1016/j.jconrel.2010.04.006 PMID:20403395

Calusi, S., Arilli, C., Mussi, E., Puggelli, L., Farnesi, D., Casati, M., Compagnucci, A., Marrazzo, L., Talamonti, C., Zani, M., & Pallotta, S. (2021). In phantom evaluation of targeting accuracy in MRI-based brain radiosurgery. *Physica Medica, 85*, 158–164.

Campbell, F. C. (2006). Metal Matrix Composites. In *Manufacturing Technology for Aerospace Structural Materials* (1st ed., pp. 419–458). Elsevier Science. doi:10.1016/B978-185617495-4/50009-3

Canavese, G., Ancona, A., Racca, L., Canta, M., Dumontel, B., Barbaresco, F., Limongi, T., & Cauda, V. (2018). Nanoparticle-assisted ultrasound: A special focus on sonodynamic therapy against cancer. *Chemical Engineering Journal, 340*, 155–172. doi:10.1016/j.cej.2018.01.060 PMID:30881202

Cao, Q., Giustozzi, F., Zanni-Merk, C., de Bertrand de Beuvron, F., & Reich, C. (2019). Smart condition monitoring for industry 4.0 manufacturing processes: An ontology-based approach. *Cybernetics and Systems, 50*(2), 82–96. doi:10.1080/01969722.2019.1565118

Carbolated, J. P. A., da Silva, K. B., Chaves, J. A. M., Netipanyj, M. F. D., Popat, K. C., & Claro, A. P. R. A. (2020). Nanoporous layer formation on the Ti10Mo8Nb alloy surface using anodic oxidation. *Surface and Coatings Technology, 386*, 125467. doi:10.1016/j.surfcoat.2020.125467

Carvalho, A., Grenho, L., Fernandes, M. H., Daskalova, A., Trifonov, A., Buchvarov, I., & Monteiro, F. J. (2020). Femtosecond laser microstructuring of alumina toughened zirconia for surface functionalization of dental implants. *Ceramics International, 46*(2), 1383–1389. doi:10.1016/j.ceramint.2019.09.101

Chakraborty, A., Mulroney, A. T., & Gupta, M. C. (2021). Superhydrophobic Surfaces by Microtexturing: A Critical Review. *Reviews of Adhesion and Adhesives, 9*(1), 35–64.

Chalisgaonkar, R. (2020). Insight in applications, manufacturing and corrosion behaviour of magnesium and its alloys–A review. *Materials Today: Proceedings, 26*, 1060–1071. doi:10.1016/j.matpr.2020.02.211

Chandrasekhar, S., & Prasad, N. B. V. (2020). Multi-response optimization of electrochemical machining parameters in the micro-drilling of AA6061-TiB2in situ composites using the Entropy–VIKOR method. *Proceedings of the Institution of Mechanical Engineers. Part B, Journal of Engineering Manufacture, 234*(10), 1311–1322. doi:10.1177/0954405420911539

Chang, S., Li, L., Lu, L., & Fuh, J. (2017). Selective Laser Sintering of Porous Silica Enabled by Carbon Additive. *Materials (Basel, Switzerland), 10*(11), 1313.

Chaubaroux, C., Vrana, E., Debry, C., Schaaf, P., Senger, B., Voegel, J. C., Haikel, Y., Ringwald, C., Hemmerlé, J., Lavalle, P., & Boulmedais, F. (2012). Collagen-based fibrillar multilayer films cross-linked by a natural agent. *Biomacromolecules, 13*(7), 2128–2135. doi:10.1021/bm300529a PMID:22662909

Chauhan, N., & Arya, N. (2018). Coconut fiber: A natural versatile material. International of Chemical *Studies, 6*(6), 555-561.

Chen, B., Zhan, H., & Zhou, W. (2015). Aerodynamic design of a re-entry capsule for high-speed manned re-entry. *Acta Astronautica, 106*, 160–169. doi:10.1016/j.actaastro.2014.10.036

Chen, G., Jiang, M., Coles-Black, J., Mansour, K., Chuen, J., & Amott, D. (2020). Three-dimensional printing as a tool in otolaryngology training: A systematic review. *The Journal of Laryngology and Otology, 134*(1), 14–19.

Chen, G., Svirskis, D., Lu, W., Ying, M., Huang, Y., & Wen, J. (2018). N-trimethyl chitosan nanoparticles and CSKSSDYQC peptide: N-trimethyl chitosan conjugates enhance the oral bioavailability of gemcitabine to treat breast cancer. *Journal of Controlled Release, 277*, 142–153. doi:10.1016/j.jconrel.2018.03.013 PMID:29548985

Cheng, Y. F., Mei, Y. H., Sathishkumar, G., Lu, Z. S., Li, C. M., Wang, F., Xia, Q. Y., & Xu, L. Q. (2020). Tannic acid-assisted deposition of silk sericin on the titanium surfaces for antifouling application. *Colloid and Interface Science Communications*, *35*, 100241. doi:10.1016/j.colcom.2020.100241

Chen, X., Xu, Z., Zhu, D., Fang, Z., & Zhu, D. (2016). Experimental research on electrochemical machining of titanium alloy Ti60 for a blisk. *Chinese Journal of Aeronautics*, *29*(1), 274–282. doi:10.1016/j.cja.2015.09.010

Chen, Y., Han, Z., Cao, K., Zheng, X., & Xu, X. (2020). Manufacturing upgrading in industry 4.0 era. *Systems Research and Behavioral Science*, *37*(4), 766–771. doi:10.1002res.2717

Choi, K. Y., Jeon, E. J., Yoon, H. Y., Lee, B. S., Na, J. H., Min, K. H., Kim, S. Y., Myung, S.-J., Lee, S., Chen, X., Kwon, I. C., Choi, K., Jeong, S. Y., Kim, K., & Park, J. H. (2012). Theranostic nanoparticles based on PEGylated hyaluronic acid for the diagnosis, therapy and monitoring of colon cancer. *Biomaterials*, *33*(26), 6186–6193. doi:10.1016/j.biomaterials.2012.05.029 PMID:22687759

Choudhary, S. K. (2006). *Study of the Effects of Heat Input on Microstructure and Grain Size of Submerged Arc Weilding (Saw)*. Welds by Digital Image Analysis Technique.

Christofi, T., Baritaki, S., Falzone, L., Libra, M., & Zaravinos, A. (2019). Current Perspectives in Cancer Immunotherapy. *Cancers (Basel)*, *11*(10), 1472. doi:10.3390/cancers11101472 PMID:31575023

Cui, H., Miao, S., Esworthy, T., Zhou, X., Lee, S. J., Liu, C., Yu, Z. X., Fisher, J. P., Mohiuddin, M., & Zhang, L. G. (2018). 3D bioprinting for cardiovascular regeneration and pharmacology. *Advanced Drug Delivery Reviews*, *132*, 252–269.

Cui, X., Li, J., Hartanto, Y., Durham, M., Tang, J., Zhang, H., Hooper, G., Lim, K., & Woodfield, T. (2020). Advances in Extrusion 3D Bioprinting: A Focus on Multicomponent Hydrogel-Based Bioinks. *Advanced Healthcare Materials*, *9*(15), e1901648.

D'Emilia, G., di Ilio, A., Gaspari, A., Natale, E., & Stamopoulos, A. G. (2020). Uncertainty assessment for measurement and simulation in selective laser melting: A case study of an aerospace part. *Acta IMEKO*, *9*(4), 96–105. doi:10.21014/acta_imeko.v9i4.720

Dahake, S., Kuthe, A., Mawale, M., Sapkal, P., Bagde, A., Daronde, S., Kamble, M., & Sarode, B. (2020). Development of customized implant and customized surgical osteotomy guide in ablative tumor surgery for accurate mandibular reconstruction. *The international journal of medical robotics + computer assisted surgery. MRCAS*, *16*(1), e2046.

Das, B., Bag, S., & Pal, S. (2017). Probing weld quality monitoring in friction stir welding through characterization of signals by fractal theory. *Journal of Mechanical Science and Technology*, *31*(5), 2459–2465. doi:10.100712206-017-0444-2

Das, B., Pal, S., & Bag, S. (2016). A combined wavelet packet and Hilbert-Huang transform for defect detection and modelling of weld strength in friction stir welding process. *Journal of Manufacturing Processes*, *22*, 260–268. doi:10.1016/j.jmapro.2016.04.002

Das, B., Pal, S., & Bag, S. (2016). Monitoring of friction stir welding process using weld image information. *Science and Technology of Welding and Joining*, *21*(4), 317–324. doi:10.1080/13621718.2015.1109805

Das, B., Pal, S., & Bag, S. (2017a). Design and development of force and torque measurement setup for real time monitoring of friction stir welding process. *Measurement: Journal of the International Measurement Confederation*, *103*, 186–198. doi:10.1016/j.measurement.2017.02.034

Das, B., Pal, S., & Bag, S. (2017b). Weld quality prediction in friction stir welding using wavelet analysis. *International Journal of Advanced Manufacturing Technology*, *89*(1–4), 711–725. doi:10.100700170-016-9140-0 de la Torre, P., Pérez-Lorenzo, M. J., Alcázar-Garrido, Á., & Flores, A. I. (2020). Cell-Based Nanoparticles Delivery Systems for Targeted Cancer Therapy: Lessons from Anti-Angiogenesis Treatments. *Molecules (Basel, Switzerland)*, *25*(3), 715. doi:10.3390/molecules25030715 PMID:32046010

de Viteri, V. S., Bayón, R., Igartua, A., Barandika, G., Moreno, J. E., Peremarch, C. P. J., & Pérez, M. M. (2016). Structure, tribocorrosion and biocide characterization of Ca, P and I containing TiO2 coatings developed by plasma electrolytic oxidation. *Applied Surface Science*, *367*, 1–10. doi:10.1016/j.apsusc.2016.01.145

Delfi, M., Ghomi, M., Zarrabi, A., Mohammadinejad, R., Taraghdari, Z. B., Ashrafizadeh, M., Zare, E. N., Agarwal, T., Padil, V. V. T., Mokhtari, B., Rossi, F., Perale, G., Sillanpaa, M., Borzacchiello, A., Kumar Maiti, T., & Makvandi, P. (2020). Functionalization of polymers and nanomaterials for biomedical applications: Antimicrobial platforms and drug carriers. *Prosthesis*, *2*(2), 117–139. doi:10.3390/prosthesis2020012

Derakhshanfar, S., Mbeleck, R., Xu, K., Zhang, X., Zhong, W., & Xing, M. (2018). 3D bioprinting for biomedical devices and tissue engineering: A review of recent trends and advances. *Bioactive Materials*, *3*(2), 144–156.

Dev, S., Ramkumar, K. D., Arivazhagan, N., & Rajendran, R. (2018). Investigations on the microstructure and mechanical properties of dissimilar welds of inconel 718 and sulphur rich martensitic stainless steel, AISI 416. *Journal of Manufacturing Processes*, *32*, 685–698. doi:10.1016/j.jmapro.2018.03.035

Dey, A., & Yodo, N. A. (2019). Systematic Survey of FDM Process Parameter Optimization and Their Influence on Part Characteristics. *J. Manuf. Mater. Process.*, *3*, 64.

Dey, H. C., Albert, S. K., Bhaduri, A. K., & Mudali, U. K. (2013). Activated flux TIG welding of titanium. *Welding in the World*, *57*(6), 903–912.

Dey, M., & Ozbolat, I. T. (2020). 3D bioprinting of cells, tissues and organs. *Scientific Reports*, *10*(1), 14023.

Dhawan, A., Kennedy, P. M., Rizk, E. B., & Ozbolat, I. T. (2019). Three-dimensional Bioprinting for Bone and Cartilage Restoration in Orthopaedic Surgery. *The Journal of the American Academy of Orthopaedic Surgeons*, *27*(5), e215–e226.

Diment, L. E., Thompson, M. S., & Bergmann, J. (2017). Clinical efficacy and effectiveness of 3D printing: A systematic review. *BMJ Open*, *7*(12), e016891.

Dinardo, G., Fabbiano, L., & Vacca, G. (2018). A smart and intuitive machine condition monitoring in the Industry 4.0 scenario. *Measurement*, *126*, 1–12. doi:10.1016/j.measurement.2018.05.041

Din, F. U., Aman, W., Ullah, I., Qureshi, O. S., Mustapha, O., Shafique, S., & Zeb, A. (2017). Effective use of nanocarriers as drug delivery systems for the treatment of selected tumors. *International Journal of Nanomedicine*, *12*, 7291–7309. doi:10.2147/IJN.S146315 PMID:29042776

do Ó Silva, A. L., Dias, I. J., Grempel, R. G., de Morais, H., da Silva Oliveira Brito, N. M., & de Castro Gomes, D. Q. (2017). Prototyping for the treatment of late zygomatic-orbital fracture: A case report. *International Journal of Surgery Case Reports*, *38*, 91–94.

Dong, Y., Ghataura, A., Takagi, H., Haroosh, H. J., Nakagaito, A. N., & Lau, K. T. (2014). Polylactic Acid (PLA) Biocomposites Reinforced with Coir Fibres: Evaluation of Mechanical Performance and Multifunctional Properties. *Compos Part A*, *63*, 76–84.

Dorez, G., Taguet, A., Ferry, L., & Lopez-Cuesta, J. M. (2013). Thermal and Fire Behavior of Natural Fibers/PBS Biocomposites Polymer. Degradation and Stability, 98, 87-95.

Dragan, E.S., & Dinu, M.V. (2019). Polysaccharides constructed hydrogels as vehicles for proteins and peptides. A review. *CarbohydrPolym., 1*(225).

El-Hofy, H. (2018). *Fundamentals of machining processes: conventional and nonconventional processes*. CRC Press. doi:10.1201/9780429443329

Elomaa, L., Keshi, E., Sauer, I. M., & Weinhart, M. (2020). Development of GelMA/PCL and dECM/PCL resins for 3D printing of acellular in vitro tissue scaffolds by stereolithography. *Materials Science and Engineering C*, *112*, 110958.

Elzoghby, A.O., Saad, N.I., Helmy, M.W., Samy, W.M., & Elgindy, N.A. (2013). Ionically-crosslinked milk protein nanoparticles as flutamide carriers for effective anticancer activity in prostate cancer-bearing rats. *Eur J Pharm Biopharm.*, *85*(3 Pt A), 444-51.

Espalin, D., Arcaute, K., Rodriguez, D., Medina, F., Posner, M., & Wicker, R. (2010). Fused deposition modeling of patient-specific polymethylmethacrylate implants. *Rapid Prototyping Journal*, *16*(3), 164–173.

Ezeh, E. M., Onukwuli, O. D., Odera, R. S., Ugonabo, V. I., & Okeke, B. (2020). *O Thermal Decomposition and Flame retardancy of Functional Polyester/Banana Peduncle fibre and Aluminum Hydroxide particle Hybrid Composites Chemical and Process* (Vol. 62). Engineering Research.

Fadl-allah, S. A., Quahtany, M., & El-Shenawy, N. S. (2013). *Surface modification of titanium plate with anodic oxidation and its application in bone growth*. Academic Press.

Fakhar, N., & Sabbaghian, M. (2021). A good combination of ductility, strength, and corrosion resistance of fine-grained ZK60 magnesium alloy produced by repeated upsetting process for biodegradable applications. *Journal of Alloys and Compounds*, *862*, 158334. doi:10.1016/j.jallcom.2020.158334

Fan, D., Li, Y., Wang, X., Zhu, T., Wang, Q., Cai, H., Li, W., Tian, Y., & Liu, Z. (2020). Progressive 3D Printing Technology and Its Application in Medical Materials. *Frontiers in Pharmacology*, *11*, 122.

Fantino, E., Chiappone, A., Roppolo, I., Manfredi, D., Bongiovanni, R., Pirri, C. F., & Calignano, F. (2016, May). 3D Printing of Conductive Complex Structures with In Situ Generation of Silver Nanoparticles. *Advanced Materials*, *28*(19), 3712–3717. doi:10.1002/adma.201505109

Farahani, S., Brown, N., Loftis, J., Krick, C., Pichl, F., Vaculik, R., & Pilla, S. (2019). Evaluation of in-mold sensors and machine data towards enhancing product quality and process monitoring via Industry 4.0. *International Journal of Advanced Manufacturing Technology*, *105*(1), 1371–1389. doi:10.100700170-019-04323-8

Feldman, D. (2019). Polymers and Polymer Nanocomposites for Cancer Therapy. *Applied Sciences (Basel, Switzerland)*, *9*(18), 3899. doi:10.3390/app9183899

Fereiduni, E., Ghasemi, A., & Elbestawi, M. (2020). Selective laser melting of aluminum and titanium matrix composites: Recent progress and potential applications in the aerospace industry. *Aerospace (Basel, Switzerland)*, *7*(6), 77. doi:10.3390/aerospace7060077

Fina, F., Madla, C. M., Goyanes, A., Zhang, J., Gaisford, S., & Basit, A. W. (2018). Fabricating 3D printed orally disintegrating printlets using selective laser sintering. *International Journal of Pharmaceutics*, *541*(1-2), 101–107.

Fintová, S., Kuběna, I., Palán, J., Mertová, K., Duchek, M., Hutař, P., Pastorek, F., & Kunz, L. (2020). Influence of sandblasting and acid etching on fatigue properties of ultra-fine grained Ti grade 4 for dental implants. *Journal of the Mechanical Behavior of Biomedical Materials*, *111*, 104016. doi:10.1016/j.jmbbm.2020.104016 PMID:32799132

Florence, T., Dharmalingam, S., Ragupathy, V., & Satishkumar, P. (2015). Machinablity study on Electrochemical Machining–A Review. *International Journal of Chemtech Research*, *7*(6), 2596.

Foley, M., & Drysdale, D. (1994). Note: Smoke Measurements and the Cone Calorimeter. *Fire and Materials*, *18*, 385–387.

Frazier, W. E. (2014). Metal Additive Manufacturing: A Review. *Journal of Materials Engineering and Performance*, *23*, 1917–1928.

Fu, C., Yang, R.M., Wang, L., Li, N.N., Qi, M., & Xu, X.D. (2017). Surface functionalization of superparamagnetic nanoparticles by an acid-liable polysaccharide-based prodrug for combinatorial monitoring and chemotherapy of hepatocellular carcinoma. *RSC Adv., 66*.

Fujimoto, K., Negishi, H., Nakamura, R., & Nakakita, K. (2018). *Aero-Acoustics CFD Prediction for Re-entry Capsule at Subsonic to Supersonic Regine*. AIAA. doi:10.2514/6.2018-0520

Funderburk, R. S. (1999). Key concepts in welding engineering. *Welding Innovation*, *16*(1), 1–4.

Gao, G., Huang, Y., Schilling, A. F., Hubbell, K., & Cui, X. (2018). Organ Bioprinting: Are We There Yet? *Advanced Healthcare Materials*, *7*(1). Advance online publication. doi:10.1002/adhm.201701018

Gao, L., Yu, S., Yu, R., He, T., & Shi, Y. (2020). Study on Arc additive manufacturing process and properties of 5356 aluminum alloy rocket booster module transition end frame. *Jixie Gongcheng Xuebao*, *56*(8), 28–36. doi:10.3901/JME.2020.08.028

Gharibshahiyan, E., Raouf, A. H., Parvin, N., & Rahimian, M. (2011). The effect of microstructure on hardness and toughness of low carbon welded steel using inert gas welding. *Materials & Design*, *32*(4), 2042–2048. doi:10.1016/j.matdes.2010.11.056

Gnanavelbabu, A., & Rajkumar, K. (2020). Experimental characterization of dimensional and surface alternation of straight and angular cutting on self-lubricating composite: A wire EDM approach. *Arabian Journal for Science and Engineering*, *45*(7), 5859–5872. doi:10.100713369-020-04596-2

Godec, M., Malej, S., Feizpour, D., Donik, Č., Balažic, M., Klobčar, D., Pambaguian, L., Conradi, M., & Kocijan, A. (2021). Hybrid additive manufacturing of Inconel 718 for future space applications. *Materials Characterization*, *172*, 172. doi:10.1016/j.matchar.2020.110842

Goguta, L., Lungeanu, D., Negru, R., Birdeanu, M., Jivanescu, A., & Sinescu, C. (2020). Selective Laser Sintering versus Selective Laser Melting and Computer Aided Design - Computer Aided Manufacturing in Double Crowns Retention. *Journal of Prosthodontic Research*. Advance online publication. doi:10.2186/jpr.JPOR_2019_556

Gokul, K., Prabhu, T. R., & Rajasekaran, T. (2017). Processing and Evaluation of Mechanical Properties of Sugarcane Fiber Reinforced Natural Composites. *Transactions of the Indian Institute of Metals*, *70*(10), 2537–2546. doi:10.100712666-017-1116-8

Golasiński, K. M., Detsch, R., Szklarska, M., Łosiewicz, B., Zubko, M., Mackiewicz, S., Pieczyska, E. A., & Boccaccini, A. R. (2021). Evaluation of mechanical properties, in vitro corrosion resistance and biocompatibility of Gum Metal in the context of implant applications. *Journal of the Mechanical Behavior of Biomedical Materials*, *115*, 104289. doi:10.1016/j.jmbbm.2020.104289 PMID:33388535

Golla, K., Bhaskar, C., Ahmed, F., & Kondapi, A. K. (2013). A target-specific oral formulation of Doxorubicin-protein nanoparticles: Efficacy and safety in hepatocellular cancer. *Journal of Cancer*, *4*(8), 644–652. doi:10.7150/jca.7093 PMID:24155776

Gopinathan, J., & Noh, I. (2018). Recent trends in bioinks for 3D printing. *Biomaterials Research*, *22*, 11.

Govindarao, R., Ganguly, R. I., Dash, R. R., Rao, P. S. P., Reddy, G. S., & Singh, S. K. (2015). Development of a Novel Aluminium Based Metal Matrix Composite using Insitu Ternary Mixture (Al 2 O 3 -SiC-C) Prepared by Thermal Treatment of Fly-ash. *Transactions of the Indian Institute of Metals*, *68*(5), 951–958. doi:10.100712666-015-0532-x

Greyson, S. J., Babrauskas, V., & Hirschler, M. M. (1994). A New International Standard for Flammability Testing. *Plastics Engineering*, 29–31.

Gu, D., Zhang, H., Chen, H., Zhang, H., & Xi, L. (2020). Laser additive manufacturing of high-performance metallic aerospace components. *Chinese Journal of Lasers*, *47*(5).

Gulfam, M., Kim, J. E., Lee, J. M., Ku, B., Chung, B. H., & Chung, B. G. (2012). Anticancer drug-loaded gliadin nanoparticles induce apoptosis in breast cancer cells. *Langmuir*, *28*(21), 8216–8223. doi:10.1021/la300691n PMID:22568862

Gul, M., Arif, A., & Ghafoor, R. (2019). Role of three-dimensional printing in periodontal regeneration and repair: Literature review. *Journal of Indian Society of Periodontology*, *23*(6), 504–510.

Habib, S., Zavahir, S., Abusrafa, A. E., Abdulkareem, A., Sobolčiak, P., Lehocky, M., Vesela, D., Humpolíček, P., & Popelka, A. (2021). Slippery Liquid-Infused Porous Polymeric Surfaces Based on Natural Oil with Antimicrobial Effect. *Polymers*, *13*(2), 206. doi:10.3390/polym13020206 PMID:33430082

Han, J., Wang, Q., Zhang, Z., Gong, T., & Sun, X. (2014). Cationic bovine serum albumin based self-assembled nanoparticles as siRNA delivery vector for treating lung metastatic cancer. *Small*, *10*(3), 524–535. doi:10.1002mll.201301992 PMID:24106138

Hann, S. Y., Cui, H., Esworthy, T., Miao, S., Zhou, X., Lee, S. J., Fisher, J. P., & Zhang, L. G. (2019). Recent advances in 3D printing: Vascular network for tissue and organ regeneration. *Translational Research; the Journal of Laboratory and Clinical Medicine*, *211*, 46–63.

Harada, Y., Ishida, Y., Miura, D., Watanabe, S., Aoki, H., Miyasaka, T., & Shinya, A. (2020). Mechanical Properties of Selective Laser Sintering Pure Titanium and Ti-6Al-4V, and Its Anisotropy. *Materials (Basel, Switzerland)*, *13*(22), 5081.

Harris, L. G., & Richards, R. G. (2006). Staphylococci and implant surfaces: A review. *Injury*, *37*(2), S3–S14. doi:10.1016/j.injury.2006.04.003 PMID:16651069

Hasnain, M. S., Javed, M. N., Alam, M. S., Rishishwar, P., Rishishwar, S., Ali, S., Nayak, A. K., & Beg, S. (2019). Purple heart plant leaves extract-mediated silver nanoparticle synthesis: Optimization by Box-Behnken design. *Materials Science and Engineering C*, *99*, 1105–1114. doi:10.1016/j.msec.2019.02.061 PMID:30889643

Hehr, A., Norfolk, M., Wenning, J., Sheridan, J., Ahlfors, M., Mastropietro, A. J., Maghsoudi, E., & Roberts, S. N. (2020). Hot isostatic pressing of ultrasonic additive manufacturing liquid cold plate heat exchangers. *Journal of Spacecraft and Rockets*, 1–5.

Heinrich, M. A., Liu, W., Jimenez, A., Yang, J., Akpek, A., Liu, X., Pi, Q., Mu, X., Hu, N., Schiffelers, R. M., Prakash, J., Xie, J., & Zhang, Y. S. (2019). 3D Bioprinting: from Benches to Translational Applications. *Small*, *15*(23), e1805510.

Helwig, M., & Paukszta, D. (2000). Flammability of Composites Based on Polypropylene and Flax Fibres. *Mol Cryst Liquid Cryst*, *354*(1), 373–380.

He, Y., Xiao, H., Gan, W., Yu, Q., & Yin, F. (2018). The Electrochemical Dissolution Behavior Research of Titanium Alloy under Different Electrolyte. *Procedia CIRP*, *68*, 751–756. doi:10.1016/j.procir.2017.12.132

He, Y., Yang, F., Zhao, H., Gao, Q., Xia, B., & Fu, J. (2016). Research on the printability of hydrogels in 3D bioprinting. *Scientific Reports*, *6*, 29977.

Hockaday, L., Kang, K., Colangelo, N., Cheung, P., Duan, B., & Malone, E. (2012). Rapid 3D printing of anatomically accurate and mechanically heterogeneous aortic valve hydrogel scaffolds. *Biofabrication*, *4*(3), 035005.

Ho, K. H., Newman, S. T., Rahimifard, S., & Allen, R. D. (2004). State of the art in wire electrical discharge machining (WEDM). *International Journal of Machine Tools & Manufacture*, *44*(12-13), 1247–1259. doi:10.1016/j.ijmachtools.2004.04.017

Hollis, B. R., & Borrelli, S. (2012). Aerothermodynamics of blunt body entry vehicles. *Progress in Aerospace Sciences*, *48–49*, 42–56. doi:10.1016/j.paerosci.2011.09.005

Hong, S. H., Li, Y., Eom, J. B., & Choi, Y. (2018). Responsive alginate-cisplatin nanogels for selective imaging and combined chemo/radio therapy of proliferating macrophages. *Quantitative Imaging in Medicine and Surgery*, *8*(8), 733–742. doi:10.21037/qims.2018.09.01 PMID:30306054

Hong, S., Choi, D. W., Kim, H. N., Park, C. G., Lee, W., & Park, H. H. (2020). Protein-Based Nanoparticles as Drug Delivery Systems. *Pharmaceutics*, *12*(7), 604. doi:10.3390/pharmaceutics12070604 PMID:32610448

Hosseinifar, T., Sheybani, S., Abdouss, M., Hassani Najafabadi, S. A., & Shafiee Ardestani, M. (2018). Pressure responsive nanogel base on Alginate-Cyclodextrin with enhanced apoptosis mechanism for colon cancer delivery. *Journal of Biomedical Materials Research. Part A*, *06*(2), 349–359. doi:10.1002/jbm.a.36242 PMID:28940736

Huang, S. F., & Liu, Y. (2014). Electrochemical micromachining of complex shapes on nickel and nickel-based superalloys. *Materials and Manufacturing Processes*, *29*(11-12), 1483–1487. doi:10.1080/10426914.2014.930897

Huang, S., & Huang, G. (2019). Preparation and drug delivery of dextran-drug complex. *Drug Delivery*, *26*(1), 252–261. doi:10.1080/10717544.2019.1580322 PMID:30857442

Huo, M., Wang, H., Zhang, Y., Cai, H., Zhang, P., Li, L., Zhou, J., & Yin, T. (2020). Co-delivery of silybin and paclitaxel by dextran-based nanoparticles for effective anti-tumor treatment through chemotherapy sensitization and microenvironment modulation. *Journal of Controlled Release*, *321*, 198–210. doi:10.1016/j.jconrel.2020.02.017 PMID:32044390

Hynes, N. R. J., & Kumar, R. (2018). Electrochemical Machining of Aluminium Metal Matrix Composites. *Surface Engineering and Applied Electrochemistry*, *54*(4), 367–373. doi:10.3103/S1068375518040087

Ige, Umoru, & Aribo. (2012). Natural Products: A Minefield of Biomaterials. *International Scholarly Research Notices*.

Indira, K. N. Parameswaranpillai, J., & Thomas, S. (2013). Mechanical Properties and Failure Topography of Banana Fiber PF Macrocomposites Fabricated by RTM and CM Techniques. *International Scholarly Research Notices*. doi:10.1155/2013/936048

Ingole, D. S., Kuthe, A. M., Thakare, S. B., & Talankar, A. S. (2009). Rapid prototyping – a technology transfer approach for development of rapid tooling. *Rapid Prototyping Journal*, *15*(4), 280–290.

Ishai, O., & Daniel, I. (2006). *M Engineering Mechanics of Composite Materials* (2nd ed.). Oxford University Press.

Jaishankar, M., Tseten, T., Anbalagan, N., Mathew, B. B., & Beeregowda, K. N. (2014). Toxicity, mechanism and health effects of some heavy metals. *Interdisciplinary Toxicology*, *7*(2), 60–72. doi:10.2478/intox-2014-0009 PMID:26109881

Jammalamadaka, U., & Tappa, K. (2018). Recent Advances in Biomaterials for 3D Printing and Tissue Engineering. *Journal of Functional Biomaterials*, *9*(1), 22.

Jamróz, W., Szafraniec, J., Kurek, M., & Jachowicz, R. (2018). 3D Printing in Pharmaceutical and Medical Applications - Recent Achievements and Challenges. *Pharmaceutical Research*, *35*(9), 176.

Jang, Y., Choi, W. T., Johnson, C. T., García, A. J., Singh, P. M., Breedveld, V., & Champion, J. A. (2018). Inhibition of bacterial adhesion on nanotextured stainless steel 316L by electrochemical etching. *ACS Biomaterials Science & Engineering*, *4*(1), 90–97. doi:10.1021/acsbiomaterials.7b00544 PMID:29333490

Javed, M.N., Dahiya, E.S., Ibrahim, A.M., Alam, M.S., Khan, F.A., & Pottoo, F.H. (2020a). Recent Advancement in Clinical Application of Nanotechnological Approached Targeted Delivery of Herbal Drugs. *Nanophytomedicine*, 151-172.

Javed, M.N., Pottoo, F.H., Shamim, A., Hasnain, M.S., & Alam, M.S. (2021). Design of Experiments for the Development of Nanoparticles, Nanomaterials, and Nanocomposites. *Design of Experiments for Pharmaceutical Product Development*, 151-169.

Javed, M. N., Ashraf, G. M., Barreto, G. E., & Naim, M. (2020b). Current Trends for Rationalizing Brain Targeting Nanoparticles in Neurological Disorders (Part-I). *Current Drug Metabolism*, *21*(9), 648. doi:10.2174/138920022108201113094659 PMID:33327901

Javed, M. N., Kohli, K., & Amin, S. (2018). Risk assessment integrated QbD approach for development of optimized bicontinuous mucoadhesive limicubes for oral delivery of rosuvastatin. *AAPS PharmSciTech*, *19*(3), 1377–1391. doi:10.120812249-018-0951-1 PMID:29388027

Jeffus, L. (2020). *Welding: Principles and applications*. Cengage Learning.

Jessop, Z. M., Al-Sabah, A., Gardiner, M. D., Combellack, E., Hawkins, K., & Whitaker, I. S. (2017). 3D bioprinting for reconstructive surgery: Principles, applications and challenges. *Journal of Plastic, Reconstructive & Aesthetic Surgery; JPRAS*, *70*(9), 1155–1170.

Jiang, H., Liang, G., Dai, M., Dong, Y., Wu, Y., Zhang, L., Xi, Q., & Qi, L. (2020). Preparation of doxorubicin-loaded collagen-PAPBA nanoparticles and their anticancer efficacy in ovarian cancer. *Annals of Translational Medicine*, *8*(14), 880. doi:10.21037/atm-20-5028 PMID:32793724

Jiang, J. Y., Xu, J. L., Liu, Z. H., Deng, L., Sun, B., Liu, S. D., Wang, L., & Liu, H. Y. (2015). Preparation, corrosion resistance and hemocompatibility of the superhydrophobic TiO_2 coatings on biomedical Ti-6Al-4V alloys. *Applied Surface Science*, *347*, 591–595. doi:10.1016/j.apsusc.2015.04.075

Joung, D., Truong, V., Neitzke, C. C., Guo, S. Z., Walsh, P. J., Monat, J. R., Meng, F., Park, S. H., Dutton, J. R., Parr, A. M., & McAlpine, M. C. (2018). 3D Printed Stem-Cell Derived Neural Progenitors Generate Spinal Cord Scaffolds. *Advanced Functional Materials*, *28*(39), 1801850.

Jovic, T. H., Combellack, E. J., Jessop, Z. M., & Whitaker, I. S. (2020). 3D Bioprinting and the Future of Surgery. *Frontiers in Surgery*, *7*, 609836.

Kah, P., & Martikainen, J. (2013). Influence of shielding gases in the welding of metals. *International Journal of Advanced Manufacturing Technology*, *64*(9-12), 1411–1421. doi:10.100700170-012-4111-6

Karimi, M., Eslami, M., Sahandi-Zangabad, P., Mirab, F., Farajisafiloo, N., Shafaei, Z., Ghosh, D., Bozorgomid, M., Dashkhaneh, F., & Hamblin, M. R. (2016). pH-Sensitive stimulus-responsive nanocarriers for targeted delivery of therapeutic agents. *Wiley Interdisciplinary Reviews. Nanomedicine and Nanobiotechnology*, *8*(5), 696–716. doi:10.1002/wnan.1389 PMID:26762467

Karmakar, R., & Maji, P. (2021). Research advancements in machining of metal matrix composites through wire electro discharge machining technique. *Materials Today: Proceedings*, *44*, 4457–4461. doi:10.1016/j.matpr.2020.10.717

Kaushikh, K., Arunvinthan, S., & Pillai, S. N. (2018). Aerodynamics and Aerothermodynamics of undulated re-entry vehicles. *Acta Astronautica*, *142*, 95–102. doi:10.1016/j.actaastro.2017.10.024

Kavimani, V., Prakash, K. S., Thankachan, T., Nagaraja, S., Jeevanantham, A. K., & Jhon, J. P. (2020). WEDM parameter optimization for silicon@ r-GO/magneisum composite using taguchi based GRA coupled PCA. *Silicon*, *12*(5), 1161–1175. doi:10.100712633-019-00205-6

Kessler, A., Hickel, R., & Reymus, M. (2020). 3D Printing in Dentistry-State of the Art. *Operative Dentistry*, *45*(1), 30–40.

Kholgh Eshkalak, S., Rezvani Ghomi, E., Dai, Y., Choudhury, D., & Ramakrishna, S. (2020). The role of three-dimensional printing in healthcare and medicine. *Materials & Design*, *2020*(194), 108940. doi:10.1016/j.matdes.2020.108940

Kim, B. S., Kwon, Y. W., Kong, J. S., Park, G. T., Gao, G., Han, W., Kim, M. B., Lee, H., Kim, J. H., & Cho, D. W. (2018). 3D cell printing of in vitro stabilized skin model and in vivo pre-vascularized skin patch using tissue-specific extracellular matrix bioink: A step towards advanced skin tissue engineering. *Biomaterials*, *168*, 38–53.

Kim, H., Jung, W. K., Choi, I. G., & Ahn, S. H. (2019). A Low-Cost Vision-Based Monitoring of Computer Numerical Control (CNC) Machine Tools for Small and Medium-Sized Enterprises (SMEs). *Sensors (Basel)*, *19*(20), 4506.

Kim, Y. T., & Eng, H. Y. (2016). Tang LWThe effects of weight fraction on mechanical behaviour of Thermoset Palm EFB composite. *Int. J. Mater. Mech.Manuf*, *4*(4), 232–236.

Knowlton, S., Anand, S., Shah, T., & Tasoglu, S. (2018). Bioprinting for Neural Tissue Engineering. *Trends in Neurosciences*, *41*(1), 31–46.

Kobayashi, D. Y., & Wolynec, S. (1999). Evaluation of the low corrosion resistant phase formed during the sigma phase precipitation in duplex stainless steels. *Materials Research*, *2*(4), 239–247.

Koli, D. K., Agnihotri, G., & Purohit, R. (2015). Advanced aluminium matrix composites: The critical need of automotive and aerospace engineering fields. *Materials Today: Proceedings*, *2*(4-5), 3032–3041. doi:10.1016/j.matpr.2015.07.290

Kolodziejczak, G. C. (1987). *Pulsed GMAW parameter variation to minimize interference form tack welds* (Doctoral dissertation). Massachusetts Institute of Technology.

Koodaryan, R., & Hafezeqoran, A. (2016). Surface modification of dental polymers by plasma treatment: A review. *Biomedical & Pharmacology Journal*, *9*(1), 317–321. doi:10.13005/bpj/942

Koopaie, M., Bordbar-Khiabani, A., Kolahdooz, S., Darbandsari, A. K., & Mozafari, M. (2020). Advanced surface treatment techniques counteract biofilm-associated infections on dental implants. *Materials Research Express*, *7*(1), 015417. doi:10.1088/2053-1591/ab6a57

Kou, S. (2003). *Welding metallurgy*. Academic Press.

Kozak, J., Rajurkar, K. P., & Makkar, Y. (2004). Selected problems of micro-electrochemical machining. *Journal of Materials Processing Technology*, *149*(1-3), 426–431. doi:10.1016/j.jmatprotec.2004.02.031

Kruger, G. H., Miller, S. F., Shih, A. J., & van Niekerk, T. I. (2015). Robustness of 802.15 Wireless Interface for Real-Time Thermal Feedback Control of the Friction Stir Welding Process. *ASME 2015 International Manufacturing Science and Engineering Conference*, V001T02A085-V001T02A085. American Society of Mechanical Engineers.

Kryou, C., Leva, V., Chatzipetrou, M., & Zergioti, I. (2019). Bioprinting for Liver Transplantation. *Bioengineering (Basel, Switzerland)*, *6*(4), 95.

Kumar, A., Walia, H., Pottoo, F.H., & Javed, M.N. (2020). Insights of Nanophytomedicines as a Combinatorial Therapy in Disease Diagnosis and Treatment. *Nanophytomedicine*, 113-132.

Kumarasamy, G., Lakshmanan, P., & Thangamani, G. (2021). Electrochemical micromachining of hastelloy C276 by different electrolyte solutions. *Arabian Journal for Science and Engineering*, *46*(3), 2243–2259. doi:10.100713369-020-05032-1

Kumar, D., Pandey, J., Raj, V., & Kumar, P. (2017). A Review on the Modification of Polysaccharide Through Graft Copolymerization for Various Potential Applications. *The Open Medicinal Chemistry Journal*, *11*(1), 109–126. doi:10.2174/1874104501711010109 PMID:29151987

Kumar, H., Manna, A., & Kumar, R. (2018). Modeling and desirability approach-based multi-response optimization of WEDM parameters in machining of aluminum metal matrix composite. *Journal of the Brazilian Society of Mechanical Sciences and Engineering*, *40*(9), 1–19. doi:10.100740430-018-1368-1

Kumar, R., Dilthey, U., Dwivedi, D. K., & Ghosh, P. K. (2009). Thin sheet welding of Al 6082 alloy by AC pulse-GMA and AC wave pulse-GMA welding. *Materials & Design*, *30*(2), 306–313. doi:10.1016/j.matdes.2008.04.073

Kumar, S. S., Erdemir, F., Varol, T., Kumaran, S. T., Uthayakumar, M., & Canakci, A. (2020). Investigation of WEDM process parameters of Al–SiC–B4C composites using response surface methodology. *International Journal of Lightweight Materials and Manufacture*, *3*(2), 127–135. doi:10.1016/j.ijlmm.2019.09.003

Kumar, U., Yadav, I., Kumari, S., Kumari, K., Ranjan, N., Kesharwani, R. K., ... Chakravarty, D. (2015). Defect identification in friction stir welding using discrete wavelet analysis. *Advances in Engineering Software*, *85*, 43–50. doi:10.1016/j.advengsoft.2015.02.001

Kuntanapreeda, S., & Hess, D. (2020). Opening access to space by maximizing utilization of 3D printing in launch vehicle design and production. *Appl. Sci. Eng. Prog.*, *14*(2), 1–3. doi:10.14416/j.asep.2020.12.002

Kyzioł, K., Kaczmarek, Ł., Brzezinka, G., & Kyzioł, A. (2014). Structure, characterization and cytotoxicity study on plasma surface modified Ti–6Al–4V and γ-TiAl alloys. *Chemical Engineering Journal*, *240*, 516–526. doi:10.1016/j.cej.2013.10.091

Lachowicz, D., Karabasz, A., Bzowska, M., Szuwarzyński, M., Karewicz, A., & Nowakowska, M. (2019). Blood-compatible, stable micelles of sodium alginate – Curcumin bioconjugate for anti-cancer applications. *European Polymer Journal*, *113*, 208–219. doi:10.1016/j.eurpolymj.2019.01.058

Lahtinen, E., Precker, R., Lahtinen, M., Hey-Hawkins, E., & Haukka, M. (2019). Selective Laser Sintering of Metal-Organic Frameworks: Production of Highly Porous Filters by 3D Printing onto a Polymeric Matrix. *ChemPlusChem*, *84*(2), 222–225.

Lahtinen, E., Turunen, L., Hänninen, M. M., Kolari, K., Tuononen, H. M., & Haukka, M. (2019). Fabrication of Porous Hydrogenation Catalysts by a Selective Laser Sintering 3D Printing Technique. *ACS Omega*, *4*(7), 12012–12017.

Lathabai, S. (2018). Additive Manufacturing of Aluminium-Based Alloys and Composites. In *Fundamentals of Aluminium Metallurgy*. Elsevier Ltd., doi:10.1016/B978-0-08-102063-0.00002-3

Lawlor, K. T., Vanslambrouck, J. M., Higgins, J. W., Chambon, A., Bishard, K., Arndt, D., Er, P. X., Wilson, S. B., Howden, S. E., Tan, K. S., Li, F., Hale, L. J., Shepherd, B., Pentoney, S., Presnell, S. C., Chen, A. E., & Little, M. H. (2021). Cellular extrusion bioprinting improves kidney organoid reproducibility and conformation. *Nature Materials*, *20*(2), 260–271.

Lebaudy, E., Fournel, S., Lavalle, P., Vrana, N. E., & Gribova, V. (2021). Recent Advances in Antiinflammatory Material Design. *Advanced Healthcare Materials*, *10*(1), 2001373. doi:10.1002/adhm.202001373 PMID:33052031

Lee, B. H., Kim, H. S., Kim, S., Kim, H. J., Lee, B., Deng, Y., Feng, Q., & Luo, J. (2011). Evaluating of the Flammability of Wood-Based Panels and Gypsum Particleboard using Cone Calorimeter. *Construction & Building Materials*, *25*, 3044–3050.

Lee, K. Y., & Mooney, D. J. (2012). Alginate: Properties and biomedical applications. *Progress in Polymer Science*, *37*(1), 106–126. doi:10.1016/j.progpolymsci.2011.06.003 PMID:22125349

Lee, S. J., Lee, J. B., Park, Y. W., & Lee, D. Y. (2018). 3D Bioprinting for Artificial Pancreas Organ. *Advances in Experimental Medicine and Biology*, *1064*, 355–374.

León-Cabezas, M. A., Martínez-García, A., & Varela-Gandía, F. J. (2017). Innovative functionalized monofilaments for 3D printing using fused deposition modeling for the toy industry. *Proc. Manuf.*, *13*, 738–745.

Leśniak-Ziółkowska, K., Kazek-Kęsik, A., Rokosz, K., Raaen, S., Stolarczyk, A., Krok-Borkowicz, M., Pamuła, E., & Simka, W. (2021). Plasma electrolytic oxidation as an effective tool for production of copper incorporated bacteriostatic coatings on Ti-15Mo alloy. *Applied Surface Science*, *563*, 150284. doi:10.1016/j.apsusc.2021.150284

Liaw, C. Y., & Guvendiren, M. (2017). Current and emerging applications of 3D printing in medicine. *Biofabrication*, *9*(2), 024102.

Ligon, S. C., Liska, R., Stampfl, J., Gurr, M., & Mülhaupt, R. (2017). Polymers for 3D Printing and Customized Additive Manufacturing. *Chemical Reviews*, *117*(15), 10212–10290.

Li, J., Cai, C., Li, J., Li, J., Li, J., Sun, T., Wang, L., Wu, H., & Yu, G. (2018). Chitosan-Based Nanomaterials for Drug Delivery. *Molecules (Basel, Switzerland)*, *23*(10), 2661. doi:10.3390/molecules23102661 PMID:30332830

Li, J., Huo, M., Wang, J., Zhou, J., Mohammad, J. M., Zhang, Y., Zhu, Q., Waddad, A. Y., & Zhang, Q. (2012). Redox-sensitive micelles self-assembled from amphiphilic hyaluronic acid-deoxycholic acid conjugates for targeted intracellular delivery of paclitaxel. *Biomaterials*, *33*(7), 2310–2320. doi:10.1016/j.biomaterials.2011.11.022 PMID:22166223

Li, J., & Laghari, R. A. (2019). A review on machining and optimization of particle-reinforced metal matrix composites. *International Journal of Advanced Manufacturing Technology*, *100*(9), 2929–2943. doi:10.100700170-018-2837-5

Li, J., Yang, X., Yang, P., & Gao, F. (2017). Hyaluronic acid–conjugated silica nanoparticles for breast cancer therapy. *Inorganic and Nano-Metal Chemistry*, *47*(5), 777–782. doi:10.1080/15533174.2016.1218509

Li, M., Neoh, K. G., Xu, L. Q., Wang, R., Kang, E. T., Lau, T., Olszyna, D. P., & Chiong, E. (2012). Surface modification of silicone for biomedical applications requiring long-term antibacterial, antifouling, and hemocompatible properties. *Langmuir*, *28*(47), 16408–16422. doi:10.1021/la303438t PMID:23121175

Lin, T., Zhao, P., Jiang, Y., Tang, Y., Jin, H., Pan, Z., He, H., Yang, V. C., & Huang, Y. (2016). Blood-Brain-Barrier-Penetrating Albumin Nanoparticles for Biomimetic Drug Delivery via Albumin-Binding Protein Pathways for Antiglioma Therapy. *ACS Nano*, *10*(11), 9999–10012. doi:10.1021/acsnano.6b04268 PMID:27934069

Lippold, J. C., & Kotecki, D. J. (2005). Welding metallurgy and weldability of stainless steels. Academic Press.

Li, Q., Bao, X., Cai, S., Xie, Y., Liu, Y., Liu, J., & Xu, G. (2021). Fabrication of superhydrophobic composite coating of hydroxyapatite/stearic acid on magnesium alloy and its corrosion resistance, antibacterial adhesion. *Journal of Materials Science*, *56*(8), 5233–5249. doi:10.100710853-020-05592-5

Liu, G., Li, D., Xiang, J., & Luo, W. (2015). Design, Modeling and Analysis of a Sharp-edge Hypersonic Stealthy Reentry Vehicle. *Procedia Engineering*, *99*, 163–167. doi:10.1016/j.proeng.2014.12.521

Liu, Q., Li, Q., Xu, S., Zheng, Q., & Cao, X. (2018). Preparation and properties of 3D printed alginate–chitosan polyion complex hydrogels for tissue engineering. *Polymers*, *10*(6), 664.

Liu, W., Luo, Z., Yuan, T., Li, Y., Zhang, H., & Ao, S. (2018). The multi-physics analysis for a novel tool structure to improve the accuracy in electrochemical micro-machining. *International Journal of Advanced Manufacturing Technology, 94*(5), 1991–2001. doi:10.100700170-017-0902-0

Liu, Y., Rath, B., Tingart, M., & Eschweiler, J. (2020). Role of implants surface modification in osseointegration: A systematic review. *Journal of Biomedical Materials Research, 108A*(3), 470–484. doi:10.1002/jbm.a.36829 PMID:31664764

Li, Y., Li, L., Ma, Y., Zhang, K., Li, G., Lu, B., Lu, C., Chen, C., Wang, L., Wang, H., & Cui, X. (2020). 3D-Printed Titanium Cage with PVA-Vancomycin Coating Prevents Surgical Site Infections (SSIs). *Macromolecular Bioscience, 20*(3), e1900394.

Lohcharoenkal, W., Wang, L., Chen, Y. C., & Rojanasakul, Y. (2014). Protein nanoparticles as drug delivery carriers for cancer therapy. *BioMed Research International, 2014*, 180549. doi:10.1155/2014/180549 PMID:24772414

Longhurst, W. R., Wilbur, I. C., Osborne, B. E., & Gaither, B. W. (2018). Process monitoring of friction stir welding via the frequency of the spindle motor current. *Proceedings of the Institution of Mechanical Engineers. Part B, Journal of Engineering Manufacture, 232*(4), 720–730. doi:10.1177/0954405416654089

Łukaszewska-Kuska, M., Krawczyk, P., Martyla, A., Hędzelek, W., & Dorocka-Bobkowska, B. (2019). Effects of a hydroxyapatite coating on the stability of endosseous implants in rabbit tibiae. *Dental and Medical Problems, 56*(2), 123–129. doi:10.17219/dmp/103851 PMID:31274249

Lukin, I., Musquiz, S., Erezuma, I., Al-Tel, T. H., Golafshan, N., Dolatshahi-Pirouz, A., & Orive, G. (2019). Can 4D bioprinting revolutionize drug development? *Expert Opinion on Drug Discovery, 14*(10), 953–956.

Lv, Y., Wang, B., Liu, G., Tang, Y., Lu, E., Xie, K., Lan, C., Liu, J., Qin, Z., & Wang, L. (2021). Metal Material, Properties and Design Methods of Porous Biomedical Scaffolds for Additive Manufacturing: A Review. *Frontiers in Bioengineering and Biotechnology, 9*, 641130.

Maclin John Vasanth, K., Lokendar Ram, P. S., Pon Anand, V., Prabu, M., & Rahul, S. (2020). Experimental investigation of mechanical and tribological properties of Aluminium metal matrix composites fabricated by powder metallurgy route - A review. *Materials Today: Proceedings, 33*, 1058–1072. doi:10.1016/j.matpr.2020.07.057

Mahesh, D., Kowshigha, K. R., Raju, N. V., & Aggarwai, P. K. (2020). Characterization of Banana Fibre-Reinforced Polypropylene Composite. *Journal of the Indian Academy of Wood Science, 17*, 1–8.

Makino, A., & Kimura, S. (2014). "Solid Tumor-Targeting Theranostic Polymer Nanoparticle in Nuclear Medicinal Fields", The Scientific World Journal, vol. 2014. *Article ID, 424513*, 12.

Mallick, P. K. (1999). Compression Moulding, Composite materials Technology. Hanser Publishers.

Mandrycky, C., Wang, Z., Kim, K., & Kim, D. H. (2016). 3D bioprinting for engineering complex tissues. *Biotechnology Advances, 34*(4), 422–434.

Maniraj, S., Thanigaivelan, R., Viswanathan, R., & Elumalai, P. (2021). Experimental investigation of MRR and ROC in aluminium metal matrix composites. *Materials Today: Proceedings, 45*, 1102–1106. doi:10.1016/j.matpr.2020.03.190

Manita, P. G., Garcia-Orue, I., Santos-Vizcaino, E., Hernandez, R. M., & Igartua, M. (2021). 3D Bioprinting of Functional Skin Substitutes: From Current Achievements to Future Goals. *Pharmaceuticals (Basel, Switzerland), 14*(4), 362.

Manivasagam, V. K., & Popat, K. C. (2020). In Vitro Investigation of Hemocompatibility of Hydrothermally Treated Titanium and Titanium Alloy Surfaces. *ACS Omega, 5*(14), 8108–8120. doi:10.1021/acsomega.0c00281 PMID:32309720

Marques, A. B. L. R. D. S. (2017). *Antifouling surfaces as a strategy to prevent microorganism's adhesion on medical devices* (Doctoral dissertation).

Marti, P., Lampus, F., Benevento, D., & Setacci, C. (2019). Trends in use of 3D printing in vascular surgery: a survey. *International Angiology: A Journal of the International Union of Angiology, 38*(5), 418–424.

Matthews, C. (2005). *ASME engineer's data book.* Amer Society of Mechanical.

Matusiak, J., & Pfeifer, T. (2013). The research of technological and environmental conditions during low-energetic gas-shielded metal arc welding of aluminium alloys. *Welding International, 27*(5), 338–344. doi:10.1080/09507116.2011.600040

Ma, Z. Y., Liu, Z. Y., Zhang, Q., Ni, D. R., & Xiao, B. L. (2013). Fabrication of metal matrix composites via friction stir processing. *Proceedings of the 1st International Joint Symposium on Joining and Welding*, 395–399. 10.1533/978-1-78242-164-1.395

Mazzoli, A. (2013). Selective laser sintering in biomedical engineering. *Medical & Biological Engineering & Computing*, *51*(3), 245–256.

McAtee, C. O., Barycki, J. J., & Simpson, M. A. (2014). Emerging roles for hyaluronidase in cancer metastasis and therapy. *Advances in Cancer Research*, *123*, 1–34. doi:10.1016/B978-0-12-800092-2.00001-0 PMID:25081524

Mehta, M. G., & Vadher, J. A. (2020). Mechanical characterization of aluminum LM25 based metal matrix composite made with conventional process. *Materials Today: Proceedings*, *38*, 3091–3096. doi:10.1016/j.matpr.2020.09.490

Mehta, M., Chatterjee, K., & De, A. (2013). Monitoring torque and traverse force in friction stir welding from input electrical signatures of driving motors. *Science and Technology of Welding and Joining*, *18*(3), 191–197. doi:10.1179/1362171812Y.0000000084

Meng, F., Meyer, C. M., Joung, D., Vallera, D. A., McAlpine, M. C., & Panoskaltsis-Mortari, A. (2019). 3D Bioprinted In Vitro Metastatic Models via Reconstruction of Tumor Microenvironments. *Advanced materials (Deerfield Beach, Fla.)*, *31*(10), e1806899.

Meng, X., Cheng, Y., Wang, P., Chen, K., Chen, Z., Liu, X., Fu, X., Wang, K., Liu, K., Liu, Z., & Duan, X. (2021). Enhanced Hemocompatibility of a Direct Chemical Vapor Deposition-Derived Graphene Film. *ACS Applied Materials & Interfaces*, *13*(4), 4835–4843. doi:10.1021/acsami.0c19790 PMID:33474941

Messiry, M. E. (2013). Theoretical Analysis of Natural fiber Volume Fraction of Reinforced Composites. *Alexandria Engineering Journal*, *52*, 301–306.

Millett, J. C. F., Whiteman, G., Park, N. T., Case, S., & Appleby-Thomas, G. (2019). The effects of heat treatment upon the shock response of a copper-beryllium alloy. *Acta Materialia*, *165*, 678–685. doi:10.1016/j.actamat.2013.01.021

Minetola, P., & Galati, M. (2018). A challenge for enhancing the dimensional accuracy of a low-cost 3D printer by means of self-replicated parts. *Additive Manufacturing*, *22*, 256–264.

Mishra, D., Gupta, A., Raj, P., Kumar, A., Anwer, S., Pal, S. K., ... Pal, A. (2020). Real time monitoring and control of friction stir welding process using multiple sensors. *CIRP Journal of Manufacturing Science and Technology*, *30*, 1–11. doi:10.1016/j.cirpj.2020.03.004

Mishra, S., Sharma, S., Javed, M. N., Pottoo, F. H., Barkat, M. A., Alam, M. S., Amir, M., & Sarafroz, M. (2019). Bioinspired nanocomposites: Applications in disease diagnosis and treatment. *Pharmaceutical Nanotechnology*, *7*(3), 206–219. doi:10.2174/2211738507666190425121509 PMID:31030662

Mohandas, T., Reddy, G. M., & Kumar, B. S. (1999). Heat-affected zone softening in high-strength low-alloy steels. *Journal of Materials Processing Technology*, *88*(1-3), 284–294. doi:10.1016/S0924-0136(98)00404-X

Mohandoss, S., Ganeasan, M. S., Manovasuki, J., & Sridhar, T. M. (2020). Surface Modification of 316L Stainless Steel dental implants by Novel Bioinert Nano Zirconia coating. *Trends in Biomaterials & Artificial Organs*, *34*(3).

Mohanta, B.C., Javed, M.N., Hasnain, M.S., & Nayak, A.K. (2020). Polyelectrolyte complexes of alginate for controlling drug release. *Alginates in Drug Delivery*, 297-321.

Mondschein, R. J., Kanitkar, A., Williams, C. B., Verbridge, S. S., & Long, T. E. (2017). Polymer structure-property requirements for stereolithographic 3D printing of soft tissue engineering scaffolds. *Biomaterials*, *140*, 170–188. doi:10.1016/j.biomaterials.2017.06.005 PMID:28651145

Montgomerie, Z., & Popat, K. C. (2021). Improved hemocompatibility and reduced bacterial adhesion on superhydrophobic titania nanoflower surfaces. *Materials Science and Engineering C*, *119*, 111503. doi:10.1016/j.msec.2020.111503 PMID:33321602

Monti, M., & Camino, G. (2013). Thermal and Combustion Behavior of Polyether-sulfone-Boehmite Nanocomposites. *Polymer Degradation & Stability*, *98*(9), 1838–1846.

Moon, Y. H., Lee, E. S., & Park, J. W. (2002). A study on electrochemical micromachining for fabrication of microgrooves in an air-lubricated hydrodynamic bearing. *International Journal of Advanced Manufacturing Technology*, *20*(10), 720–726. doi:10.1007001700200229

Moreno Madrid, A. P., Vrech, S. M., Sanchez, M. A., & Rodriguez, A. P. (2019). Advances in additive manufacturing for bone tissue engineering scaffolds. *Materials Science and Engineering C*, *100*, 631–644.

Motiei, M., Kashanian, S., Lucia, L. A., & Khazaei, M. (2017). Intrinsic parameters for the synthesis and tuned properties of amphiphilic chitosan drug delivery nanocarriers. *Journal of Controlled Release*, *260*, 213–225. doi:10.1016/j.jconrel.2017.06.010 PMID:28625671

Moungomo, J. B. M., Kouya, D. N., & Songmene, V. (2016). Aluminium machining chips formation, treatment & recycling: A review. *Key Engineering Materials*, *710*(1977), 71–76. doi:10.4028/www.scientific.net/KEM.710.71

Mouritz, A. P., Mathys, Z., & Gibson, A. G. (2006). Heat Release of Polymer Composites in Fire Composite Part A. *Appl Sci Manuf*, *37*(7), pp1040–pp1054. doi:10.1016/j.compositesa.2005.01.030

Mughal, M. P., Farooq, M. U., Mumtaz, J., Mia, M., Shareef, M., Javed, M., ... Pruncu, C. I. (2021). Surface modification for osseointegration of Ti6Al4V ELI using powder mixed sinking EDM. *Journal of the Mechanical Behavior of Biomedical Materials*, *113*, 104145.

Mukherjee, P., Chung, J., Cheng, K., Gupta, R., Haag, H., Williams, Z., & Wallace, G. (2021). Invitro and Invivo Study of PCL-Hydrogel Scaffold to Advance Bioprinting Translation in Microtia Reconstruction. *The Journal of Craniofacial Surgery*, *32*(5), 1931–1936.

Mulford, J. S., Babazadeh, S., & Mackay, N. (2016). Three-dimensional printing in orthopaedic surgery: Review of current and future applications. *ANZ Journal of Surgery*, *86*(9), 648–653.

Munawar, R. F., Jamil, N. H., Shahril, M. K., Abdul Rahim, M., Abidin, M. Z. Z., Azam, M. A., & Lau, K. (2015). Development of Green Composite: Pineapple Leaf Fibers (PALF) Reinforced Polylactide (PLA). *Applied Mechanics and Materials*, *761*, 520–525.

Murugan, C., Rayappan, K., Thangam, R., Bhanumathi, R., Shanthi, K., Vivek, R., Thirumurugan, R., Bhattacharyya, A., Sivasubramanian, S., Gunasekaran, P., & Kannan, S. (2016). Combinatorial nanocarrier based drug delivery approach for amalgamation of anti-tumor agents in breast cancer cells: An improved nanomedicine strategy. *Scientific Reports*, *6*(1), 34053. doi:10.1038rep34053 PMID:27725731

Murugan, K., & Guruprasad, B. (1964). Varatharaju Perumal, and Arul T G,(2021). Renewable Energy Based Grid System Using UPQC for PQ Improvement. *Journal of Physics: Conference Series*, 062094.

Mwema, F. M., & Akinlabi, E. T. (2020). Basics of Fused Deposition Modelling (FDM). *Fused Deposition Modeling: Strategies for Quality Enhancement*, 1–15.

Nair, K. C. M., Kumar, R. P., Thomas, S., Schit, S. C., & Ramamurthy, K. (2000). Rheological Behavior of Short Sisal Fiber-Reinforced Polystyrene Composites. *Composites. Part A, Applied Science and Manufacturing*, *31*(11), 1231–1240. doi:10.1016/S1359-835X(00)00083-X

Nam, S. Y., & Park, S. H. (2018). ECM Based Bioink for Tissue Mimetic 3D Bioprinting. *Advances in Experimental Medicine and Biology*, *1064*, 335–353.

Nashed, N., Lam, M., & Nokhodchi, A. (2021). A comprehensive overview of extended release oral dosage forms manufactured through hot melt extrusion and its combination with 3D printing. *International Journal of Pharmaceutics*, *596*, 120237.

Natarajan, N., Arunachalam, R. M., & Thanigaivelan, R. (2013). Experimental study and analysis of micro holes machining in EDM of SS 304. *International Journal of Machining and Machinability of Materials*, *13*(1), 1–16. doi:10.1504/IJMMM.2013.051905

Nayak, K. C., Pandey, A. K., & Date, P. P. (2020). Mechanical and physical characterization of powder metallurgy based aluminium metal matrix hybrid composite. *Materials Today: Proceedings*, *33*, 5408–5413. doi:10.1016/j.matpr.2020.03.134

Neiman, J. A., Raman, R., Chan, V., Rhoads, M. G., Raredon, M. S., Velazquez, J. J., Dyer, R. L., Bashir, R., Hammond, P. T., & Griffith, L. G. (2015). Photopatterning of hydrogel scaffolds coupled to filter materials using stereolithography for perfused 3D culture of hepatocytes. *Biotechnology and Bioengineering*, *112*(4), 777–787.

Ngan, C., Quigley, A., O'Connell, C., Kita, M., Bourke, J., Wallace, G. G., Choong, P., & Kapsa, R. (2020). 3D Bioprinting and Differentiation of Primary Skeletal Muscle Progenitor Cells. *Methods in Molecular Biology (Clifton, N.J.)*, *2140*, 229–242.

Ng, W. L., Goh, M. H., Yeong, W. Y., & Naing, M. W. (2018). Applying macromolecular crowding to 3D bioprinting: Fabrication of 3D hierarchical porous collagen-based hydrogel constructs. *Biomaterials Science*, *6*(3), 562–574.

Nikolaeva, M., & Karki, T. (2016). Influence of Fire Retardants on the Reaction to Fire Properties of Coextruded Wood-Polypropylene Composites. *Fire and Materials*, *40*, 535–543.

Norrish, J. (2006). *Advanced welding processes*. Elsevier. doi:10.1533/9781845691707

Norzali, N. A. R., Badri, K. H., & Nuawi, M. (2011). Z Loading Effect of Aluminum Hydroxide onto the Mechanical, Thermal Conductivity, Acoustical and Burning Properties of the Palm-based Polyurethane Composites. *Sains Malaysiana*, *40*(7), 737–742.

Nurgali, K., Jagoe, R. T., & Abalo, R. (2018). Editorial: Adverse Effects of Cancer Chemotherapy: Anything New to Improve Tolerance and Reduce Sequelae? *Frontiers in Pharmacology*, *9*, 245. doi:10.3389/fphar.2018.00245 PMID:29623040

Nwanji, E. C., & Babalola, P. O. (2018). Synthesizing a Metal Matrix Composite (Al-Al2O3) from Aluminium (AA1237) Reinforced with Alumina (Al2O3) Particulates. *IOP Conference Series. Materials Science and Engineering*, *413*(1), 012043. Advance online publication. doi:10.1088/1757-899X/413/1/012043

O'Brien, R. L. (2017). Jefferson's Welding Encyclopedia (18th ed.). American Welding Society.

Okamoto, Y., Taguchi, K., Sakuragi, M., Imoto, S., Yamasaki, K., & Otagiri, M. (2019). Preparation, Characterization, and in Vitro/in Vivo Evaluation of Paclitaxel-Bound Albumin-Encapsulated Liposomes for the Treatment of Pancreatic-Cancer. *ACS Omega*, *4*(5), 8693–8700. doi:10.1021/acsomega.9b00537 PMID:31459959

Olorunmaiye, J. A., & Ohijeagbon, I. (2015). O Retrofitting Composite Ceiling Boards with Jatropha Curcas Seedcake Material*J. Production Engineering*, *18*(2), 96–102.

Oyar, P. (2018). Laser Sintering Technology and Balling Phenomenon. *Photomedicine and Laser Surgery*, *36*(2), 72–77.

Oyesola, M. O., Mpofu, K., Mathe, N. R., & Daniyan, I. A. (2020). Hybrid-additive manufacturing cost model: A sustainable through-life engineering support for maintenance repair overhaul in the aerospace. *Procedia Manufacturing*, *49*, 199–205. doi:10.1016/j.promfg.2020.07.019

P, R, S,, N. R., A. K., & Raj. (2016). Preparation, characterization and biological evaluation of curcumin loaded alginate aldehyde-gelatin nanogels. *Materials Science and Engineering C*, *68*, 251–257. doi:10.1016/j.msec.2016.05.046 PMID:27524019

Palani, S., Lakshmanan, P., & Kaliyamurthy, R. (2020). Experimental investigations of electrochemical micromachining of nickel aluminum bronze alloy. *Materials and Manufacturing Processes*, *35*(16), 1860–1869. doi:10.1080/10426914.2020.1813888

Paliwal, S. R., Paliwal, R., Agrawal, G. P., & Vyas, S. P. (2016). Hyaluronic acid modified pH-sensitive liposomes for targeted intracellular delivery of doxorubicin. *Journal of Liposome Research*, *26*(4), 276–287. doi:10.3109/08982104.2015.1117489 PMID:26784587

Panahi, Z., Tamjid, E., & Rezaei, M. (2020). Surface modification of biodegradable AZ91 magnesium alloy by electrospun polymer nanocomposite: Evaluation of in vitro degradation and cytocompatibility. *Surface and Coatings Technology*, *386*, 125461. doi:10.1016/j.surfcoat.2020.125461

Pandey, M., Saleem, S., Nautiyal, H., Pottoo, F.H., & Javed, M.N. (2020). PINK1/Parkin in neurodegenerative disorders: crosstalk between mitochondrial stress and neurodegeneration. *Quality Control of Cellular Protein in Neurodegenerative Disorders*, 282-301.

Pang, G., Xin, K., Bu, F., & Gao, T. (2019). Modeling and experiment study on inter-electrode gap in electrochemical machining with a floating cathode for rotary parts. *International Journal of Advanced Manufacturing Technology*, *101*(5), 2151–2164. doi:10.100700170-018-3051-1

Pantermehl, S., Emmert, S., Foth, A., Grabow, N., Alkildani, S., Bader, R., Barbeck, M., & Jung, O. (2021). 3D Printing for Soft Tissue Regeneration and Applications in Medicine. *Biomedicines*, *9*(4), 336.

Panwar, N., & Chauhan, A. (2018). ScienceDirect Fabrication methods of particulate reinforced Aluminium metal matrix composite-A review. *Materials Today: Proceedings*, *5*(2), 5933–5939. doi:10.1016/j.matpr.2017.12.194

Park, H., Lee, J., Jung, H., Kim, D., Kim, S., & Sultan, M. (2018). An omentum-cultured 3D-printed artificial trachea: In vivo bioreactor. *Artif Cell Nanomed B*, *46*(3), S1131–S1140.

Parr, W., Burnard, J. L., Wilson, P. J., & Mobbs, R. J. (2019). 3D printed anatomical (bio)models in spine surgery: Clinical benefits and value to health care providers. *Journal of spine surgery (Hong Kong)*, *5*(4), 549–560.

Paštyková, V., Novotný, J., Veselský, T., Urgošík, D., Liščák, R., & Vymazal, J. (2018). Assessment of MR stereotactic imaging and image co-registration accuracy for 3 different MR scanners by 3 different methods/phantoms: Phantom and patient study. *Journal of Neurosurgery*, *129*(Suppl1), 125–132.

Patel, J. D., & Maniya, K. D. (2018). A Review on: Wire cut electrical discharge machining process for metal matrix composite. *Procedia Manufacturing*, *20*, 253–258. doi:10.1016/j.promfg.2018.02.037

Patil, R. G., & Kolambe, B. N. (2011). Development of value added products from Banana Pseudo-stem. An Overview of Progress. National Agricultural Innovation Project.

Patnaik, L., Maity, S. R., & Kumar, S. (2021). Mechanical and tribological assessment of composite AlCrN or aC: Ag-based thin films for implant application. *Ceramics International*, *47*(5), 6736–6752. doi:10.1016/j.ceramint.2020.11.016

Peat, T., Galloway, A., Toumpis, A., Mcnutt, P., & Iqbal, N. (2016). The erosion performance of cold spray deposited metal matrix composite coatings with subsequent friction stir processing. *Applied Surface Science*. Advance online publication. doi:10.1016/j.apsusc.2016.10.156

Pei, M., Jia, X., Zhao, X., Li, J., & Liu, P. (2018). Alginate-based cancer-associated, stimuli-driven and turn-on theranostic prodrug nanogel for cancer detection and treatment. *Carbohydrate Polymers*, *183*, 131–139. doi:10.1016/j.carbpol.2017.12.013 PMID:29352868

Peng, W., Datta, P., Ayan, B., Ozbolat, V., Sosnoski, D., & Ozbolat, I. T. (2017). 3D bioprinting for drug discovery and development in pharmaceutics. *Acta Biomaterialia*, *57*, 26–46.

Pérez, M., Medina-Sánchez, G., García-Collado, A., Gupta, M., & Carou, D. (2018). Surface Quality Enhancement of Fused Deposition Modeling (FDM) Printed Samples Based on the Selection of Critical Printing Parameters. *Materials (Basel, Switzerland)*, *11*(8), 1382.

Perla, V. R., Suraparaju, S. K., Thimmarayappa, M., & Pathi, V. R. (2021). Effect of wire electrical discharge machining process parameters of Al-6082 hybrid nano metal matrix composites. *Materials Today: Proceedings*, *43*, 551–556. doi:10.1016/j.matpr.2020.12.047

Phan, K., Sgro, A., Maharaj, M. M., D'Urso, P., & Mobbs, R. J. (2016). Application of a 3D custom printed patient specific spinal implant for C1/2 arthrodesis. *Journal of Spine Surgery*, *2*(4), 314–318. doi:10.21037/jss.2016.12.06 PMID:28097249

Pires, F., Cachada, A., Barbosa, J., Moreira, A. P., & Leitão, P. (2019). Digital twin in industry 4.0: Technologies, applications and challenges. *2019 IEEE 17th International Conference on Industrial Informatics (INDIN)*, *1*, 721–726.

Placone, J. K., & Engler, A. J. (2018). Recent Advances in Extrusion-Based 3D Printing for Biomedical Applications. *Advanced Healthcare Materials*, *7*(8), e1701161.

Podroužek, J., Marcon, M., Ninčević, K., & Wan-Wendner, R. (2019). Bio-Inspired 3D Infill Patterns for Additive Manufacturing and Structural Applications. *Materials (Basel, Switzerland)*, *12*(3), 499.

Pol, N., Verma, G., Pandey, R., & Shanmugasundaram, T. (2018). *Fabrication of AA7005/TiB2-B4C surface composite by friction stir processing: Evaluation of ballistic behaviour*. Defence Technology. doi:10.1016/j.dt.2018.08.002

Potter, J. M. (1990). *Fatigue and fracture testing of weldments* (Vol. 1058). ASTM International.

Pottoo, F.H., Abu-Izneid, T., Ibrahim, A.M., Javed, M.N., AlHajri, N., & Hamrouni, A.M. (2021c). Immune system response during viral Infections: Immunomodulators, cytokine storm (CS) and Immunotherapeutics in COVID-19. *Saudi Pharmaceutical Journal*.

Pottoo, F.H., Barkat, M.A., Ansari, M.A., Javed, M.N., Jamal, Q.M., & Kamal, M.A. (2021b). Nanotechnological based miRNA intervention in the therapeutic management of neuroblastoma. *Seminars in Cancer Biology, 69*, 100-108.

Pottoo, F.H., Javed, M.N., Rahman, J.U., Abu-Izneid, T., & Khan, F.A. (2021a). Targeted delivery of miRNA based therapeuticals in the clinical management of Glioblastoma Multiforme. *Seminars in Cancer Biology, 69*, 391-398.

Pottoo, F. H., Javed, M., Barkat, M., Alam, M., Nowshehri, J. A., Alshayban, D. M., & Ansari, M. A. (2019a). Estrogen and serotonin: Complexity of interactions and implications for epileptic seizures and epileptogenesis. *Current Neuropharmacology, 17*(3), 214–231. doi:10.2174/1570159X16666180628164432 PMID:29956631

Pottoo, F. H., Sharma, S., Javed, M. N., Barkat, M. A., Harshita, Alam, M. S., Naim, M. J., Alam, O., Ansari, M. A., Barreto, G. E., & Ashraf, G. M. (2020). Lipid-based nanoformulations in the treatment of neurological disorders. *Drug Metabolism Reviews, 52*(1), 185–204. doi:10.1080/03602532.2020.1726942 PMID:32116044

Pottoo, F. H., Tabassum, N., Javed, M. N., Nigar, S., Rasheed, R., Khan, A., Barkat, M. A., Alam, M. S., Maqbool, A., Ansari, M. A., Barreto, G. E., & Ashraf, G. M. (2019b). The synergistic effect of raloxifene, fluoxetine, and bromocriptine protects against pilocarpine-induced status epilepticus and temporal lobe epilepsy. *Molecular Neurobiology, 56*(2), 1233–1247. doi:10.100712035-018-1121-x PMID:29881945

Prabhakar. (2015). A Review on the Flammability and Flame Retardant Properties of Natural Fibres and Polymer Matrix Based Composites. *Composite Research, 26*(2), 29–39.

Prabhu, R. H., Patravale, V. B., & Joshi, M. D. (2015). Polymeric nanoparticles for targeted treatment in oncology: Current insights. *International Journal of Nanomedicine, 2*(10), 1001–1018. PMID:25678788

Prakash, J., & Gopalakannan, S. (2021). Teaching—Learning-Based Optimization Coupled with Response Surface Methodology for Micro Electrochemical Machining of ALUMINIUM NANOCOMPOSITE. *Silicon, 13*(2), 409–432. doi:10.100712633-020-00434-0

Prathipati, R. Ch. R., & Dora, S. P. (2019). Corrosion behavior of surface induced by wire EDM on 316L stainless steel: An experimental investigation. *SN Applied Sciences, 1*(9), 1–11. doi:10.100742452-019-0992-4

Praveen, D. V., Raju, D. R., & Raju, M. J. (2020). Optimization of machining parameters of wire-cut EDM on ceramic particles reinforced Al-metal matrix composites–A review. *Materials Today: Proceedings, 23*, 495–498. doi:10.1016/j.matpr.2019.05.392

Racier, S., Tomkova, B., Milady, J., Jabbar, J., & Kale, B.M. (2015). Development of a Bio Composite Based on Green Epoxy Polymer and NaturalCellulose Fabric (Bark Cloth) for Automotive Instrument Panel Applications. *Compos Part B, 81*, 149-157.

Raja, K., Ravikumar, R., Thirunavukkarasu, V., & Kumaravel, S. (2016). Experimental Investigations on Machinability of Beryllium Copper C17200using Electrochemical Micro Machining. *Asian Journal of Research in Social Sciences and Humanities, 6*(11), 367–373. doi:10.5958/2249-7315.2016.01198.9

Rajini, N., Jappes, J.T.W., Siva, I., Rajulu, A.V., & Rajakarunakaran, S. (2017). Fire and Thermal ResistanceProperties of Chemically Treated Ligno-Cellulosic Coconut Fabric–Reinforced PolymerEco-Nanocomposites. *Journal of Industrial Fabric, 47*(1), 104–124.

Rajurkar, K. P., Sundaram, M. M., & Malshe, A. P. (2013). Review of electrochemical and electro discharge machining. *Procedia CIRP, 6*, 13–26. doi:10.1016/j.procir.2013.03.002

Rakotomalala, M., Wagner, S., & Doring, M. (2010). Recent Development in Halogen Free Flame Retardants for Epoxy Resins for Electrical and Electronic Applications. *Materials (Basel), 3*(8), 4300–4343.

Raman, S., Mahmood, S., Hilles, A. R., Javed, M. N., Azmana, M., & Al-Japairai, K. A. (2020). Polymeric nanoparticles for brain drug delivery-a review. *Current Drug Metabolism*, *21*(9), 649–660. doi:10.2174/13892002216662005080748 PMID:32384025

Ramesh, S., & Subburam, V. (2019). Electrochemical micromachining of aluminium alloy composite. In *Advances in Manufacturing Technology* (pp. 309–317). Springer. doi:10.1007/978-981-13-6374-0_36

Ramli, R., Yunus, R.M., & Beg, M.D.H., & Prasad, D.M.R. (2011). Oil Palm Fiber Reinforced Polypropylene Composites: Effects of Fiber Loading and Coupling Agents on Mechanical, Thermal, and Interfacial Properties. *Journal of Composite Materials*, *46*(11), 1275–1284.

Ramya,, A.N., Joseph,, M.M., Maniganda,, S., & Karunakaran,, V., T., T. S., & Maiti, K. K. (2017). Emergence of Gold-Mesoporous Silica Hybrid Nanotheranostics: Dox-Encoded, Folate Targeted Chemotherapy with Modulation of SERS Fingerprinting for Apoptosis Toward Tumor Eradication. *Small*, *13*(31). Advance online publication. doi:10.1002mll.201700819 PMID:28671767

Ranjan, R., Khan, A. R., Parikh, C., Jain, R., Mahto, R. P., Pal, S., Pal, S. K., & Chakravarty, D. (2016). Classification and identification of surface defects in friction stir welding: An image processing approach. *Journal of Manufacturing Processes*, *22*, 237–253. doi:10.1016/j.jmapro.2016.03.009

Rastogi, P., & Kandasubramanian, B. (2019). Review of alginate-based hydrogel bioprinting for application in tissue engineering. *Biofabrication*, *11*(4), 042001.

Redwan, A. M., Badri, K. H., & Tarawneh, M. (2015). A The effect of Aluminium Hydroxide on the Mechanical Properties of Fire Resistivity of Palm-Based Fiberborad Prepared by Pre-Polymerization Method. *Advanced Materials Research*, *1087*, 287–292.

Rena, Y., Wang, Y., Wang, L., & Liu, T. (2015). Evaluation of Intumescent Fire Retardants and synergistic Agents for use in wood Flour/Recycled Polypropylene Composites. *Construction & Building Materials*, *76*, 273–278.

Rey, F., Barzaghini, B., Nardini, A., Bordoni, M., Zuccotti, G. V., Cereda, C., Raimondi, M. T., & Carelli, S. (2020). Advances in Tissue Engineering and Innovative Fabrication Techniques for 3-D-Structures: Translational Applications in Neurodegenerative Diseases. *Cells*, *9*(7), 1636.

Rezaee, A., Najafizadeh, A., Kermanpur, A., & Moallemi, M. (2011). The influence of reversion annealing behavior on the formation of nanograined structure in AISI 201L austenitic stainless steel through martensite treatment. *Materials & Design*, *32*(8-9), 4437–4442. doi:10.1016/j.matdes.2011.03.065

Rider, P., Kačarević, Ž. P., Alkildani, S., Retnasingh, S., Schnettler, R., & Barbeck, M. (2018). Additive Manufacturing for Guided Bone Regeneration: A Perspective for Alveolar Ridge Augmentation. *International Journal of Molecular Sciences*, *19*(11), 3308.

Rippe, M., Cosenza, V., & Auzély-Velty, R. (2019). Design of Soft Nanocarriers Combining Hyaluronic Acid with Another Functional Polymer for Cancer Therapy and Other Biomedical Applications. *Pharmaceutics*, *11*(7), 338. doi:10.3390/pharmaceutics11070338 PMID:31311150

Ritzen, L., Montano, V., & Garcia, S. J. (2021). 3D Printing of a Self-Healing Thermoplastic Polyurethane through FDM: From Polymer Slab to Mechanical Assessment. *Polymers*, *13*(2), 305.

Rizwanullah, M., Alam, M., Harshita, Mir, S. R., Rizvi, M. M. A., & Amin, S. (2020). Polymer-Lipid Hybrid Nanoparticles: A Next-Generation Nanocarrier for Targeted Treatment of Solid Tumors. *Current Pharmaceutical Design*, *26*(11), 1206–1215. doi:10.2174/1381612826666200116150426 PMID:31951163

Rodríguez-Panes, A., Claver, J., & Camacho, A. M. (2018). The Influence of Manufacturing Parameters on the Mechanical Behaviour of PLA and ABS Pieces Manufactured by FDM: A Comparative Analysis. *Materials (Basel, Switzerland)*, *11*(8), 1333.

Rose, J. B. R., Saranya, P., & Benifa, J. V. B. (2018). Investigation of computational flow fields and aeroacoustic characteristics over a re-entry command module. *Proceedings of the Institution of Mechanical Engineers. Part G, Journal of Aerospace Engineering*, *232*(3), 532–544. doi:10.1177/0954410016682272

Rostam, H. M., Singh, S., Salazar, F., Magennis, P., Hook, A., Singh, T., Vrana, N. E., Alexander, M. R., & Ghaemmaghami, A. M. (2016). The impact of surface chemistry modification on macrophage polarisation. *Immunobiology*, *221*(11), 1237–1246. doi:10.1016/j.imbio.2016.06.010 PMID:27349596

Safdar, M. H., Hussain, Z., Abourehab, M. A. S., Hasan, H., Afzal, S., & Ei Thu, H. (2018). New developments and clinical transition of hyaluronic acid-based nanotherapeutics for treatment of cancer: Reversing multidrug resistance, tumour-specific targetability and improved anticancer efficacy. *Artificial Cells, Nanomedicine, and Biotechnology*, *46*, 8. PMID:29082766

Sagnella, S.M., Duong, H., MacMillan, A., Boyer, C., Whan, R., & McCarroll, J.A. (2014). Dextran-Based Doxorubicin Nanocarriers with Improved Tumor Penetration. *Biomacromolecules, 15*(1), 262-275.

Salmi, M., Paloheimo, K. S., Tuomi, J., Wolff, J., & Mäkitie, A. (2013). Accuracy of medical models made by additive manufacturing (rapid manufacturing). *Journal of Cranio-Maxillo-Facial Surgery*, *41*(7), 603–609. doi:10.1016/j.jcms.2012.11.041 PMID:23333490

Samal, P., Vundavilli, P. R., Meher, A., & Mahapatra, M. M. (2020). Recent progress in aluminum metal matrix composites: A review on processing, mechanical and wear properties. *Journal of Manufacturing Processes, 59*(November), 131–152. doi:10.1016/j.jmapro.2020.09.010

Schartel, B. (Ed.). (2003). *Fourteenth Annual BCC Conference on Flame Retardancy, Fire Retarded Polypropylene/Flax Biocomposites*. Academic Press.

Schartel, B., & Hull, T. (2007). R Development of Fire Retardant Materials-Interpretation of Cone Calorimeter Date. *Fire and Materials*, *31*, 327–354.

Schiller, G. J. (2015).Additive manufacturing for Aerospace. *IEEE Aerosp. Conf. Proc.*, 1-8.

Schoenmakers, J. W., Heuker, M., López-Álvarez, M., Nagengast, W. B., van Dam, G. M., van Dijl, J. M., Jutte, P. C., & van Oosten, M. (2021). Image-guided in situ detection of bacterial biofilms in a human prosthetic knee infection model: A feasibility study for clinical diagnosis of prosthetic joint infections. *European Journal of Nuclear Medicine and Molecular Imaging*, *48*(3), 757–767. doi:10.100700259-020-04982-w PMID:32901352

Scudamore, M. J., Briggs, P. J., & Prager, F. H. (1991). Cone calorimetry – A Review of Tests Carried out on Plastics for the Association of Plastics Manufacturers in Europe. *Fire and Materials*, *15*, 65–84.

Sebastian Rabe, S., Sanchez-Olivares, G., & Ricardo Pérez-Chávez, R., & Schartel, B. (2019). Natural Keratin and Coconut Fibres from Industrial Wastes in Flame Retarded Thermoplastic Starch Biocomposite, *Materials* , *12,* 344. doi:10.3390/ma12030344

Senapati, S., Mahanta, A. K., Kumar, S., & Maiti, P. (2018). Controlled drug delivery vehicles for cancer treatment and their performance. *Signal Transduction and Targeted Therapy*, *3*(1), 7. doi:10.103841392-017-0004-3 PMID:29560283

Senthilkumar, C., Ganesan, G., & Karthikeyan, R. (2011). Parametric optimization of electrochemical machining of Al/15% SiCp composites using NSGA-II. *Transactions of Nonferrous Metals Society of China*, *21*(10), 2294–2300. doi:10.1016/S1003-6326(11)61010-8

Senthil, S., Raguraman, M., & Manalan, D. T. (2020). Manufacturing processes & recent applications of aluminium metal matrix composite materials: A review. *Materials Today: Proceedings*, *45*, 5934–5938. doi:10.1016/j.matpr.2020.08.792

Serio, L. M., Palumbo, D., Galietti, U., De Filippis, L. A. C., & Ludovico, A. D. (2016). Monitoring of the friction stir welding process by means of thermography. *Nondestructive Testing and Evaluation*, *31*(4), 371–383. doi:10.1080/10589759.2015.1121266

Serrano-Sevilla, I., Artiga, Á., Mitchell, S. G., De Matteis, L., & de la Fuente, J. M. (2019). Natural Polysaccharides for siRNA Delivery: Nanocarriers Based on Chitosan, Hyaluronic Acid, and Their Derivatives. *Molecules (Basel, Switzerland)*, *24*(14), 2570. doi:10.3390/molecules24142570 PMID:31311176

Sha, J., Lü, Z., Sha, R., Zu, Y., Dai, J., Xian, Y., Zhang, W., Cui, D., & Yan, C. (2021). Improved wettability and mechanical properties of metal coated carbon fiber-reinforced aluminum matrix composites by squeeze melt infiltration technique. *Transactions of Nonferrous Metals Society of China*, *31*(2), 317–330. doi:10.1016/S1003-6326(21)65498-5

Shankar, V., Gill, T. P. S., Mannan, S. L., & Sundaresan, S. (2003). Solidification cracking in austenitic stainless steel welds. *Sadhana*, *28*(3-4), 359–382. doi:10.1007/BF02706438

Shanmughasundaram, P., Subramanian, R., & Prabhu, G. (2011). Some studies on aluminium - Fly ash composites fabricated by two step stir casting method. *European Journal of Scientific Research*, *63*(2), 204–218.

Sharma, A. K., Bhandari, R., Aherwar, A., Rimašauskienė, R., & Pinca-Bretotean, C. (2020). A study of advancement in application opportunities of aluminum metal matrix composites. *Materials Today: Proceedings*, *26*, 2419–2424. doi:10.1016/j.matpr.2020.02.516

Sharma, S., Javed, M. N., Pottoo, F. H., Rabbani, S. A., Barkat, M., Sarafroz, M., & Amir, M. (2019). Bioresponse inspired nanomaterials for targeted drug and gene delivery. *Pharmaceutical Nanotechnology*, *7*(3), 220–233. doi:10.2174/2211738507666190429103814 PMID:31486751

Shi, G., Guan, C., Quan, D., Wu, D., Tang, L., & Gao, T. (2020). An aerospace bracket designed by thermo-elastic topology optimization and manufactured by additive manufacturing. *Chinese Journal of Aeronautics*, *33*(4), 1252–1259. doi:10.1016/j.cja.2019.09.006

Siegel, R. L., Miller, K. D., Goding Sauer, A., Fedewa, S. A., Butterly, L. F., Anderson, J. C., Cercek, A., Smith, R. A., & Jemal, A. (2020). Colorectal cancer statistics, 2020. *CA: a Cancer Journal for Clinicians*, *70*(3), 145–164. doi:10.3322/caac.21601 PMID:32133645

Sigaux, N., Pourchet, L., Breton, P., Brosset, S., Louvrier, A., & Marquette, C. A. (2019). 3D Bioprinting:principles, fantasies and prospects. *Journal of Stomatology, Oral and Maxillofacial Surgery*, *120*(2), 128–132.

Simpson, R. L., Wiria, F. E., Amis, A. A., Chua, C. K., Leong, K. F., Hansen, U. N., Chandrasekaran, M., & Lee, M. W. (2008). Development of a 95/5 poly(L-lactide-co-glycolide)/hydroxylapatite and beta-tricalcium phosphate scaffold as bone replacement material via selective laser sintering. *Journal of Biomedical Materials Research. Part B, Applied Biomaterials*, *84*(1), 17–25.

Singh, N., Singh, R., Ahuja, I. P. S., Farina, I., & Fraternali, F. (2019). Metal matrix composite from recycled materials by using additive manufacturing assisted investment casting. *Composite Structures*, *207*, 129–135. doi:10.1016/j.compstruct.2018.09.072

Sinha, A., Chakraborty, A., & Jana, N. R. (2014). Dextran-gated, multifunctional mesoporous nanoparticle for glucose-responsive and targeted drug delivery. *ACS Applied Materials & Interfaces*, *6*(24), 22183–22191. doi:10.1021/am505848p PMID:25458145

Sivandzade, F., & Cucullo, L. (2018). In-vitro blood-brain barrier modeling: A review of modern and fast-advancing technologies. *Journal of Cerebral Blood Flow & Metabolism (Nihongoban)*, *38*(10), 1667–1681.

Skeldon, G., Lucendo-Villarin, B., & Shu, W. (2018). Three-dimensional bioprinting of stem-cell derived tissues for human regenerative medicine. *Philosophical Transactions of the Royal Society of London. Series B, Biological Sciences*, *373*(1750), 20170224.

Song, R., Murphy, M., Li, C., Ting, K., Soo, C., & Zheng, Z. (2018). Current development of biodegradable polymeric materials for biomedical applications. *Drug Design, Development and Therapy*, *12*, 3117–3145. doi:10.2147/DDDT.S165440 PMID:30288019

Soundarrajan, M., & Thanigaivelan, R. (2020). Effect of coated and geometrically modified tools on performance of electrochemical micromachining. *Materials and Manufacturing Processes*, *35*(7), 775–782. doi:10.1080/10426914.2020.1740252

Stokes, V. K. (1989). Joining methods for plastics and plastic composites: An overview. *Polymer Engineering and Science*, *29*(19), 1310–1324.

Stolt, R., & Elgh, F. (2020). Introducing design for selective laser melting in aerospace industry. *Journal of Computational Design and Engineering*, *7*(4), 489–497. doi:10.1093/jcde/qwaa042

Subasinghe, A., Das, R., & Bhattacharyya, D. (2016). Study of Thermal, Flammability and Mechanical Properties of Intumescent Flame Retardant PP/Kenaf Nanocomposite. *International Journal of Smart and Nano Materials*, *7*(3), 202–220. doi:10.1080/19475411.2016.1239315

Subastinghe, A. (2014). Bhattacharya, D Performance of Different Intumescent Ammonium Polyphosphate Flame Retardants in PP/Kenaf Fibre Composites. *Composites. Part A, Applied Science and Manufacturing*, *65*, 91–99.

Sudhagar, S., Sakthivel, M., Ashok, A., & Ajith Arul Daniel, S. (2017). Investigation of discontinuities in friction stir welding of 2024 aluminum alloy using radiography. *Materials Evaluation*, *75*(8).

Sudhagar, S., Sakthivel, M., & Ganeshkumar, P. (2019). Monitoring of friction stir welding based on vision system coupled with Machine learning algorithm. *Measurement*, *144*, 135–143. doi:10.1016/j.measurement.2019.05.018

Sudhagar, S., Sakthivel, M., Mathew, P. J., & Daniel, S. A. A. (2017). A multi criteria decision making approach for process improvement in friction stir welding of aluminium alloy. *Measurement*, *108*, 1–8. doi:10.1016/j.measurement.2017.05.023

Sun, D., Qu, F., & Yan, C. (2018). An effective flux scheme for hypersonic heating prediction of re-entry vehicles. *Computers & Fluids*, *176*, 109–116. doi:10.1016/j.compfluid.2018.05.028

Sun, F., Wang, L., Li, X. C., Cheng, W., Lin, Z., Ba, D. C., Song, G.-Q., & Sun, C. S. (2020). Effect of surface modification on the long-term stability of dental implant abutment screws by plasma nitriding treatment. *Surface and Coatings Technology*, *399*, 126089. doi:10.1016/j.surfcoat.2020.126089

Sun, M., Li, J., Zhang, C., Xie, Y., Qiao, H., Su, Z., Oupický, D., & Ping, Q. (2017). Arginine-Modified Nanostructured Lipid Carriers with Charge-Reversal and pH-Sensitive Membranolytic Properties for Anticancer Drug Delivery. *Advanced Healthcare Materials*, *6*(8), 1600693. doi:10.1002/adhm.201600693 PMID:28187243

Suoware, T.O., Amgbari, C.O, Sorgbara, F., & Kigigha, A.W. (2020). A Review of the Fire Behaviour of Agro-Waste Fibre Composite Material for Industry Utilization in Nigeria. *International Interdisciplinary Conference of Federal Polytechnic Ekowe and Konkan Geographers Association of India*.

Suresh, V., Praneet, A. D., & Anoop, J. (2020). Ingenious analysis on machining parameters of aluminium alloy (LM25)/graphite (Gr)/boron carbide (B4C) hybrid composites using wire electrical discharge machining (WEDM). *Materials Today: Proceedings*.

Suutala, N., Takalo, T., & Moisio, T. (1980). Ferritic-austenitic solidification mode in austenitic stainless steel welds. *Metallurgical Transactions. A, Physical Metallurgy and Materials Science*, *11*(5), 717–725.

Swaminathan, S., & Clyne, A. M. (2020). Direct Bioprinting of 3D Multicellular Breast Spheroids onto Endothelial Networks. *Journal of Visualized Experiments*, (165). Advance online publication. doi:10.3791/61791

Tabriz, A., Hermida, M., Leslie, N., & Shu, W. (2015). Three-dimensional bioprinting of complex cell laden alginate hydrogel structures. *Biofabrication*, *7*(4), 045012.

Taghavi, S., Ramezani, M., Alibolandi, M., Abnous, K., & Taghdisi, S. M. (2017). Chitosan-modified PLGA nanoparticles tagged with 5TR1 aptamer for in vivo tumor-targeted drug delivery. *Cancer Letters*, *400*, 1–8. doi:10.1016/j.canlet.2017.04.008 PMID:28412238

Tam, C., Chan, Y. C., Law, Y., & Cheng, S. (2018). The Role of Three-Dimensional Printing in Contemporary Vascular and Endovascular Surgery: A Systematic Review. *Annals of Vascular Surgery*, *53*, 243–254.

Tan, D. K., Maniruzzaman, M., & Nokhodchi, A. (2018). Advanced Pharmaceutical Applications of Hot-Melt Extrusion Coupled with Fused Deposition Modelling (FDM) 3D Printing for Personalised Drug Delivery. *Pharmaceutics*, *10*(4), 203.

Tan, S. H., Ngo, Z. H., Leavesley, D., & Liang, K. (2021). Recent Advances in the Design of Three-Dimensional and Bioprinted Scaffolds for Full-Thickness Wound Healing. *Tissue Engineering. Part B, Reviews*. Advance online publication. doi:10.1089/ten.TEB.2020.0339

Tarassoli, S. P., Jessop, Z. M., Al-Sabah, A., Gao, N., Whitaker, S., Doak, S., & Whitaker, I. S. (2018). Skin tissue engineering using 3D bioprinting: An evolving research field. *Journal of Plastic, Reconstructive & Aesthetic Surgery; JPRAS*, *71*(5), 615–623.

Thambi, T., You, D. G., Han, H. S., Deepagan, V. G., Jeon, S. M., Suh, Y. D., Choi, K. Y., Kim, K., Kwon, I. C., Yi, G.-R., Lee, J. Y., Lee, D. S., & Park, J. H. (2014). Bioreduciblecarboxymethyl dextran nanoparticles for tumor-targeted drug delivery. *Advanced Healthcare Materials*, *3*(11), 1829–1838. doi:10.1002/adhm.201300691 PMID:24753360

Thangamani, G., Thangaraj, M., Moiduddin, K., Mian, S. H., Alkhalefah, H., & Umer, U. (2021). Performance Analysis of Electrochemical Micro Machining of Titanium (Ti-6Al-4V) Alloy under Different Electrolytes Concentrations. *Metals*, *11*(2), 247. doi:10.3390/met11020247

Thanigaivelan, R., & Arunachalam, R. M. (2010). Experimental study on the influence of tool electrode tip shape on electrochemical micromachining of 304 stainless steel. *Materials and Manufacturing Processes*, *25*(10), 1181–1185. doi:10.1080/10426914.2010.508806

Thanigaivelan, R., Arunachalam, R. M., Nithish, A., Venkatesh, S., Naveenkumar, P., & Selvaganapathy, S., & Aravind, A. S. (2020). Optimization of Laser and Electrochemical Process Parameters for Surface Modification of Hardness and Hydrophobicity on 316L Steel. *Lasers in Engineering*, 45.

Thomas, R. G., Moon, M., Lee, S., & Jeong, Y. Y. (2015). Paclitaxel loaded hyaluronic acid nanoparticles for targeted cancer therapy: In vitro and in vivo analysis. *International Journal of Biological Macromolecules, 72*, 510–518. doi:10.1016/j.ijbiomac.2014.08.054 PMID:25224289

Trenfield, S. J., Awad, A., Madla, C. M., Hatton, G. B., Firth, J., Goyanes, A., Gaisford, S., & Basit, A. W. (2019). Shaping the future: Recent advances of 3D printing in drug delivery and healthcare. *Expert Opinion on Drug Delivery, 16*(10), 1081–1094.

Tuomi, J., Paloheimo, K. S., Vehviläinen, J., Björkstrand, R., Salmi, M., Huotilainen, E., Kontio, R., Rouse, S., Gibson, I., & Mäkitie, A. A. (2014). A novel classification and online platform for planning and documentation of medical applications of additive manufacturing. *Surgical Innovation, 21*(6), 553–559.

Ulbrich, K., Holá, K., Šubr, V., Bakandritsos, A., Tuček, J., & Zbořil, R. (2016). Targeted Drug Delivery with Polymers and Magnetic Nanoparticles: Covalent and Noncovalent Approaches, Release Control, and Clinical Studies. *Chemical Reviews, 116*(9), 5338–5431. doi:10.1021/acs.chemrev.5b00589 PMID:27109701

Uttarwar, S. S., & Chopde, I. K. (2013). A Study of Influence of Electrochemical Process Parameters on the Material Removal Rate and Surface Roughness of SS AISI. *International Journal of Computational Engineering Research, 3*(3).

Vellingiri, S., Soundararajan, R., Mohankumar, N., Nithyananthakumar, K., & Muthuselvam, K. (2021). Exploration on WEDM process parameters effect on LM13 alloy and LM13/SiC composites using Taguchi method. *Materials Today: Proceedings, 45*, 997–1003. doi:10.1016/j.matpr.2020.03.050

Vijaya, S., Ravi, N. M., Helmi, S., & Choo, Y. M. (2013). The Development of a Residual Oil Recovery System to Increase the Revenue of Palm Oil Mill. *Journal of Oil Palm Research, 25*(1), 116–122.

Vijayavenkataraman, S., Yan, W. C., Lu, W. F., Wang, C. H., & Fuh, J. (2018). 3D bioprinting of tissues and organs for regenerative medicine. *Advanced Drug Delivery Reviews, 132*, 296–332.

Villanueva, J., Trino, L., Thomas, J., Bijukumar, D., Royhman, D., Stack, M. M., & Mathew, M. T. (2017). Corrosion, tribology, and tribocorrosion research in biomedical implants: Progressive trend in the published literature. *Journal of Bio- and Tribo-Corrosion, 3*(1), 1–8. doi:10.100740735-016-0060-1

Walther, B., Schilm, J., Michaelis, A., & Lohrengel, M. M. (2007). Electrochemical dissolution of hard metal alloys. *Electrochimica Acta, 52*(27), 7732–7737. doi:10.1016/j.electacta.2006.12.038

Wang, Z., & Yang, Y. (2016, July). Comparison of welder performance qualification rules between Chinese regulation and ASME BPVS Sec. IX-2015. In *Pressure Vessels and Piping Conference* (*Vol. 50367*, p. V01BT01A025). American Society of Mechanical Engineers.

Wang, B., Wu, Z., Lan, J., Li, Y., Xie, L., Huang, X., Zhang, A., Qiao, H., Chang, X., Lin, H., Zhang, H., Li, T., & Huang, Y. (2021). Surface modification of titanium implants by silk fibroin/Ag co-functionalized strontium titanate nanotubes for inhibition of bacterial-associated infection and enhancement of in vivo osseointegration. *Surface and Coatings Technology, 405*, 126700. doi:10.1016/j.surfcoat.2020.126700

Wang, G., Wang, J. J., Tang, X. J., Du, L., & Li, F. (2016). In vitro and in vivo evaluation of functionalized chitosan-Pluronic micelles loaded with myricetin on glioblastoma cancer. *Nanomedicine (London), 12*(5), 1263–1278. doi:10.1016/j.nano.2016.02.004 PMID:26970027

Wang, L., Wang, W., Zhao, H., Liu, Y., Liu, J., & Bai, N. (2020). Bioactive effects of low-temperature argon–oxygen plasma on a titanium implant surface. *ACS Omega, 5*(8), 3996–4003. doi:10.1021/acsomega.9b03504 PMID:32149226

Wang, M., Chen, X., Tong, W., Wang, J., & Wang, X. (2020). Influences of gap pressure on machining performance in radial ultrasonic rolling electrochemical micromachining. *International Journal of Advanced Manufacturing Technology*, *107*(1), 157–166. doi:10.100700170-020-05012-7

Wang, Q., Zhou, P., Liu, S., Attarilar, S., Ma, R. L. W., Zhong, Y., & Wang, L. (2020). Multi-scale surface treatments of titanium implants for rapid osseointegration: A review. *Nanomaterials (Basel, Switzerland)*, *10*(6), 1244. doi:10.3390/nano10061244 PMID:32604854

Wang, Y., Zhou, J., Qiu, L., Wang, X., Chen, L., Liu, T., & Di, W. (2014). Cisplatin-alginate conjugate liposomes for targeted delivery to EGFR-positive ovarian cancer cells. *Biomaterials*, *35*(14), 4297–4309. doi:10.1016/j.biomaterials.2014.01.035 PMID:24565522

Waziri, A., Bharti, C., Aslam, M., Jamil, P., Mirza, A., Javed, M. N., Pottoo, U., Ahmadi, A., & Alam, M. S. (2021). Probiotics for the Chemoprotective Role Against the Toxic Effect of Cancer Chemotherapy. *Anti-cancer Agents in Medicinal Chemistry*, *21*. Advance online publication. doi:10.2174/1871520621666210514000615 PMID:33992067

Wei, T., Yu, Q., & Chen, H. (2019). Responsive and synergistic antibacterial coatings: Fighting against bacteria in a smart and effective way. *Advanced Healthcare Materials*, *8*(3), 1801381. doi:10.1002/adhm.201801381 PMID:30609261

Weman, K. (2011). *Welding processes handbook*. Elsevier.

Whitley, D. III, Eidson, R. S., Rudek, I., & Bencharit, S. (2017). In-office fabrication of dental implant surgical guides using desktop stereolithographic printing and implant treatment planning software: A clinical report. *The Journal of Prosthetic Dentistry*, *118*(3), 256–263.

Wickramasinghe, S., Do, T., & Tran, P. (2020). FDM-Based 3D Printing of Polymer and Associated Composite: A Review on Mechanical Properties, Defects and Treatments. *Polymers*, *12*(7), 1529.

Wong, K. H., Lu, A., Chen, X., & Yang, Z. (2020). Natural Ingredient-Based Polymeric Nanoparticles for Cancer Treatment. *Molecules (Basel, Switzerland)*, *25*(16), 3620. doi:10.3390/molecules25163620 PMID:32784890

Wu, D., Hua, X., Huang, L., Li, F., & Cai, Y. (2018). Elucidation of keyhole induced bubble formation mechanism in fiber laser welding of low carbon steel. *International Journal of Heat and Mass Transfer*, *127*, 1077–1086. doi:10.1016/j.ijheatmasstransfer.2018.07.107

Wu, J., Tang, C., & Yin, C. (2017). Co-delivery of doxorubicin and interleukin-2 via chitosan based nanoparticles for enhanced antitumor efficacy. *Acta Biomaterialia*, *47*, 81–90. doi:10.1016/j.actbio.2016.10.012 PMID:27729232

Xie, Z., Gao, M., Lobo, A. O., & Webster, T. J. (2020). 3D Bioprinting in Tissue Engineering for Medical Applications: The Classic and the Hybrid. *Polymers*, *12*(8), 1717.

Yahya, E. B., Jummaat, F., Amirul, A. A., Adnan, A. S., Olaiya, N. G., Abdullah, C. K., Rizal, S., Mohamad Haafiz, M. K., & Khalil, H. (2020). A Review on Revolutionary Natural Biopolymer-Based Aerogels for Antibacterial Delivery. *Antibiotics (Basel, Switzerland)*, *9*(10), 648.

Yang, N. J., & Hinner, M. J. (2015). Getting across the cell membrane: An overview for small molecules, peptides, and proteins. *Methods in Molecular Biology (Clifton, N.J.)*, *1266*, 29–53. doi:10.1007/978-1-4939-2272-7_3 PMID:25560066

Yang, T. C., & Yeh, C. H. (2020). Morphology and Mechanical Properties of 3D Printed Wood Fiber/Polylactic Acid Composite Parts Using Fused Deposition Modeling (FDM): The Effects of Printing Speed. *Polymers*, *12*(6), 1334.

Yayla, P. A. Ş. A., Kaluc, E., & Ural, K. (2007). Effects of welding processes on the mechanical properties of HY 80 steel weldments. *Materials & Design*, *28*(6), 1898–1906. doi:10.1016/j.matdes.2006.03.028

Yoshinari, M., Oda, Y., Kato, T., & Okuda, K. (2001). Influence of surface modifications to titanium on antibacterial activity in vitro. *Biomaterials*, *22*(14), 2043–2048. doi:10.1016/S0142-9612(00)00392-6 PMID:11426884

Youssef, A., Hollister, S. J., & Dalton, P. D. (2017). Additive manufacturing of polymer melts for implantable medical devices and scaffolds. *Biofabrication*, *9*(1), 012002.

Yuan, B., Zhou, S. Y., & Chen, X. S. (2017). Rapid prototyping technology and its application in bone tissue engineering. *Journal of Zhejiang University. Science. B.*, *18*(4), 303–315.

Yu, N., Li, G., Gao, Y., Jiang, H., & Tao, Q. (2016). Thermo-sensitive complex micelles from sodium alginate-graft-poly(N-isopropylacrylamide) for drug release. *International Journal of Biological Macromolecules*, *86*, 296–301. doi:10.1016/j.ijbiomac.2016.01.066 PMID:26806647

Yu, X., Trase, I., Ren, M., Duval, K., Guo, X., & Chen, Z. (2016). Design of Nanoparticle-Based Carriers for Targeted Drug Delivery. *Journal of Nanomaterials*, *2016*, 1087250. doi:10.1155/2016/1087250 PMID:27398083

Zhang, Q., Dong, J., Peng, M., Yang, Z., Wan, Y., Yao, F., Zhou, J., Ouyang, C., Deng, X., & Luo, H. (2020). Laser-induced wettability gradient surface on NiTi alloy for improved hemocompatibility and flow resistance. *Materials Science and Engineering C*, *111*, 110847. doi:10.1016/j.msec.2020.110847 PMID:32279801

Zhang, X., Zhang, T., Ma, X., Wang, Y., Lu, Y., Jia, D., Huang, X., Chen, J., Xu, Z., & Wen, F. (2020). The design and synthesis of dextran-doxorubicin prodrug-based pH-sensitive drug delivery system for improving chemotherapy efficacy. *AJPS*, *15*(5), 605–616. doi:10.1016/j.ajps.2019.10.001 PMID:33193863

Zhang, Y. S., Arneri, A., Bersini, S., Shin, S. R., Zhu, K., Goli-Malekabadi, Z., Aleman, J., Colosi, C., Busignani, F., Dell'Erba, V., Bishop, C., Shupe, T., Demarchi, D., Moretti, M., Rasponi, M., Dokmeci, M. R., Atala, A., & Khademhosseini, A. (2016). Bioprinting 3D microfibrous scaffolds for engineering endothelialized myocardium and heart-on-a-chip. *Biomaterials*, *110*, 45–59.

Zhao, M., Wudy, K., & Drummer, D. (2018). Crystallization Kinetics of Polyamide 12 during Selective Laser Sintering. *Polymers*, *10*(2), 168.

Zhen, X., Wang, X., Xie, C., Wu, W., & Jiang, X. (2013). Cellular uptake, antitumor response and tumor penetration of cisplatin-loaded milk protein nanoparticles. *Biomaterials*, *34*(4), 1372–1382. doi:10.1016/j.biomaterials.2012.10.061 PMID:23158934

Zhu, D. C., Ke, T. A. N. G., Song, M. Z., & Tu, M. J. (2006). Effects of annealing process on electrical conductivity and mechanical property of Cu-Te alloys. *Transactions of Nonferrous Metals Society of China*, *16*(2), 459–462. doi:10.1016/S1003-6326(06)60078-2

Zhu, W., Ma, X., Gou, M., Mei, D., Zhang, K., & Chen, S. (2016). 3D printing of functional biomaterials for tissue engineering. *Current Opinion in Biotechnology*, *40*, 103–112.

Zhu, Y., Liu, Y., Jin, K., & Pang, Z. (2019). Polysaccharide nanoparticles for cancer drug targeting. In *Polysacch. Carriers Drug Deliv* (pp. 365–396). Elsevier. doi:10.1016/B978-0-08-102553-6.00013-1

Zielińska, A., Carreiró, F., Oliveira, A. M., Neves, A., Pires, B., Venkatesh, D. N., Durazzo, A., Lucarini, M., Eder, P., Silva, A. M., Santini, A., & Souto, E. B. (2020). Polymeric Nanoparticles: Production, Characterization, Toxicology and Ecotoxicology. *Molecules (Basel, Switzerland)*, *25*(16), 3731. doi:10.3390/molecules25163731 PMID:32824172

Zuhudi, N. Z. M. (2014). *Flammability and Durability of Bamboo Fabric Reinforced Polypropylene Composites and their Hybrids* [Ph.D. thesis]. University of Auckland.

About the Contributors

Mirzada Ahmed is Project Scholar in the Department of Aerospace Engineering, Dayananda Sagar University, Bangalore, India.

Charles Amgbari is a lecturer and currently Ag HOD department of mechanical engineering technology from 2017 to date, Federal Polytechnic Ekowe. He obtained his Bachelor of engineering degree from Niger Delta University in 2006 and a Master of engineering in industrial and production engineering in 2013 from the University of Benin all in Nigeria. He have published in many scientific journals and conference paper.

Aswin B. is a PG CAD/CAM student and his research area includes CAD/CAM, Micro machining and polymer composites. He has attended workshop and seminars and participated in national conference conducted at Muthayammal Engineering College, Rasipuram.

Ranjit Barua is a Senior Researcher at Centre for Healthcare Science and Technology, IIEST-SHIBPUR. Core research area Bio-Mechanical System, 3D Bioprinting, Tissue Engineering.

Vempannan C. is a B.E. student from Muthayammal Engineering College,Rasipuram and his area of interest are ECM, EDM and metal matrix composites. He has participated in various workshops and seminars during his studies.

Clement Christy Deepak Charles completed Bachelor in Mechanical Engineering in 2010, Master in Engineering CAD/CAM in 2013 and Master of Business Administration in 2016 and Teaching Experience of 6 Years and Industrial Experience of 5 Years. He published papers in various National and International Journals. He published two Patents and conducted the many funded Seminar and Workshop for the student's development. He is Editorial Board member and Reviewer for National and International Journals. He published books on "Manufacturing Technology-I" and "A Concept AWD Mechanism in Automobile Applications" for the student's development. He guided many projects for UG students and received best projects award in 2014 and 2015. He is a member in many professional bodies.

Thirumanikandan D. is a PG CAD/CAM student and his area of interest are ECM, EDM and Micromachining. He has participated in various workshop and seminars conducted by reputed institution.

Kavitha G. is presently working as a Professor in the Department of Computer Science & Engineering, Muthayammal Engineering College, TamilNadu, India. She has a good Academics & Research experience in various areas of Computer Science. She is completed Ph.D. from Anna University, Chennai. Her area of research includes Networking, Machine Learning, Data Science and Big Data. She has filed successfully 3 Patents and received one award called "Distinguished Professor". She has published number of research papers in various International/National Journals (including Elsevier, Springer etc.). She is an active member of 05 various Professional International Societies (IEEE, IE, ACM, ISTE, CSI). She has also attended seminars, workshops and short-term courses in Various University. She has published one text book and one book Chapter content in reputed publications.

Prasad G., Assistant Professor (Sr.Gr.), Aerospace Engineering, Dayananda Sagar University, Bangalore, India. Prof. Prasad G. received his Masters degree and Bachelors degree in Aeronautical Engineering from Anna University, Chennai, India. He is pursuing his PhD at Anna University. He has 7 years of teaching experience. Prior to joining Dayananda Sagar University, Bangalore, India he worked at Anna University. He has published 11 articles in reputed Scopus Indexed International Journals with good Impact Factor. He has completed two funded project sponsored by The Institution of Engineers (India) and Tamilnadu State Council for Science and Technology. Awarded Indian National Science Academy (INSA) Visiting Scientist Programme 2019 and Awarded Science Academies' Summer Research Fellowship Programme (SRPF) 2019. Reviewer of Aircraft Engineering and Aerospace Technology, Reviewer International Journal of Engine Research, Reviewer Journal of The Institution of Engineers Series C and Chaired as Jury member-IEEE Project competition. He has attended several International Conferences, workshops and provided guest Lectures. He is a Professional Membership of American Institute of Aeronautics and Astronautics, Institution of Engineers, Life Member in Indian Cryogenic Council and Life Member in Shock Wave Society.

Suganya Priyadharshini G. received her Ph.D (Surface Modification Technique) and M.E (Computer Aided Design) from Anna University, Chennai, India and B.E degree in Mechanical Engineering from Periyar University. She is working as an Assistant Professor in the Department of Mechanical Engineering, Coimbatore Institute of Technology, Coimbatore, since March 2014. She has a flair for teaching, with more than 15 years of experience. She is a lifetime membership in Indian Society for Technical Education (ISTE), Indian Welding Society (IWS) and Indian Society of Systems for Science and Engineering (ISSE). She received good academic performance indicator award for the year 2018. She has published over 30 papers in international journals and conferences to her credit. She is the Co-Author of book chapter in Elsevier publication and is acting as a reviewer of Materials Research Express, iop publication, Measurements, and Materials Today Proceedings an Elsevier publication. In addition she also filed 4 patents and 1 patent got grant (Australian) for Flying vehicle abetted additive manufacturing, others Virtual reality based wireless rescue robot and a method to rescue child fallen to deep well, Gearless slider flapping wing mechanism and 3D Printed PLA sheets embedded bio composite manufacturing method thereof. One of her conference paper has got best paper award among 114 colleges all over India. She had awarded best academic performance indicator award by Coimbatore Institute of Technology for the academic year 2018-19 and 2019-20. She also received her best faculty award for the academic year 2019-20. Her area of interests includes Friction Stir Processing (FSP), Under Water Welding, Material Characterization and Metal Matrix Composites. She is in-charge of Field Emission Scanning Electron Microscope (FE-SEM) laboratory and generated over 25 lakhs in revenue since 2015.

Deepika G. N. is a Project Scholar in the Department of Aerospace Engineering, Dayananda Sagar University, Bangalore, India.

Vinod Kumaar J. R. is currently working as an Assistant Professor in the Department of Mechanical Engineering, Mahendra Engineering College (Autonomous), Namakkal, Tamilnadu, India. He graduated (B.E-Mechanical Engineering) from Mahendra Engineering College in 2008 and obtained his post-graduation (M.E-Engineering Design) from Sona college of technology, Salem in 2011. He is currently pursuing Ph.D. from Anna University, Chennai. He has 10 years of teaching experience and 1 year of industrial experience. He is a life member of ISTE and IEI. He has published papers in International, National, SCI and SCOPUS indexed Journals. His areas of interest are Micro and Nano manufacturing, Non-traditional machining, traditional machining.

Md. Noushad Javed is working as an Assistant Professor at the School of Pharmaceutical Sciences. He had received his Ph.D. degree from the Department of Pharmaceutics, School of Pharmaceutical Education & Research, Jamia Hamdard, India. Earlier he had completed his MPharm (Pharmaceutics) and BPharm from the same institution. His area of interest includes QbD integrated optimization (DoE and Artificial Intelligence) of novel drug delivery systems for neuronal disorders, cancer, and parasitic infections.

Sahana Jawahar is an MSc student in the Department of Aerospace Engineering, Delft University of Technology, The Netherlands. Recipient of the Excellence scholarship at TU Delft which funds her master studies.

Saravanan K. G. is working as a Associate Professor in Department of Mechanical Engineering at Sona College of Technology, Salem, India. He is having total of 14 years of teaching and industrial experience. He research area includes micromachining, metal matrix composites, surface modification techniques and optimization. He has published 10 papers in international and national journals. He has published 5 papers in international conference and published chapters in reputed publication house. He served as a reviewer for many international journals. He is the member of SAE India and ISTE.

Sripad Kulkarni S. has post graduated in field of Advanced Material Technology from University Visvesvaraya College of Engineering, Bengaluru and graduated in the field of Mechanical Engineering from APS college of Engineering. His area of interest is manufacturing science, Rapid prototyping. He has completed NPTEL Certification in Structural analysis of Nanomaterials and Introduction to Research.

Gopal M. received B.E., M.E., and Ph.D. from Anna university in the year of 2012, 2014 and 2017 respectively. Published more than 30 articles in international journals and 5 book chapters in reputed publishers.

Makesh M. is working as a Professor in Mechanical Engineering at Paavai Engineering College, Namakkal. He is have 18 Years of experience in Teaching and Research. His research interest includes Metal matrix composites, machining and surface modification process. He has published 12 international journals and 7 conference papers. He is the professional member of ISTE and IE(I).

Soundarrajan Madesh completed Bachelor in Mechanical Engineering in 2012, Master in Engineering Design in 2014 and Doctorate of philosophy in 2020 and Teaching Experience of 7 Years. He published papers in various National and International Journals. He published two Patents and conducted the many funded Seminar and Workshop for the student's development. He is Editorial Board member and Reviewer for National and International Journals. He guided many projects for UG, PG students and received best projects award in 2019 and 2020. He is a member in many professional bodies such as IE (I), ISTE.

Suresh P. completed his PhD in manufacturing in 2010 and having 18 years of Teaching and research experience.He has published 106 paper in national and international journal.He is a Fellow Member Institution of Engineers (FIE), Senior Member IEEE, Life Member of ISTE, Life Member of ISNT, Life Member of IIIE, Life Member of Indian Society of Ergonomics (MISE), Life Member of IACSIT, Life Member of IAENG.He is a recipient of Best Engineer Award.

Varatharaju Perumal obtained his B.E. degree in Automobile Engineering in the year 1992 from IRTT, Bharathiar University. He did his master degree, M.E. CAD, in Anna University. He completed his Ph.D. in Anna University in the year 2018 by analysing Alternate Fuels in I.C. Engine. He has teaching experience of 25 years in reputed Institutions occupying various positions. During the tenure of his service he had organised Guest Lectures, Workshops, Seminars and Symposium for the benefit of students. He had published in international journals indexed in Web of Science and Scopus. Journal Reviewer and Editor He is reviewer of various reputed journals like Fuels, Applied energy, Energy Conversion and Management, Journal of Nanostructure in Chemistry, Environmental Science and Pollution Research, Renewable energy, Journal of Nanostructure in Chemistry, Environmental Science, Part A: Journal of Power and Energy, Indian Journal of Engineering & Materials Sciences, The Open Ecology Journal and Pollution Research, etc. published by Elsevier, Springer, SAGE, NISCAIR etc. Awards and Recognition His research interests are in the field of Alternate fuels, IC engines, Combustion, Emission and Pollution control and Automobile Engineering. For the contribution made by him in the field of teaching, research and to the society he has been awarded with different titles by various organisations.

Satya Prakash has completed his Bachelor of Engineering in Mechanical Engineering (2015-2019 Batch), from Anna University Chennai and currently Pursuing Master of Engineering in CAD/CAM (Mechanical Engineering-2019-2021 Batch) from Muthayammal Engineering College, Under Anna University Chennai. His Current Research area is Machining, Biomaterials and Optimisation.

Geetha Priyan completed his Bachelor of Engineering in Mechanical Engineering, Master of Engineering in Manufacturing Engineering and Doctor of Philosophy in Mechanical Engineering from Anna University, Tamil Nadu, India in 2011, 2013 and 2019 respectively. He is currently working as assistant professor in Department of Mechanical Engineering, School of Mechanical Engineering, SRM Institute of Science and Technology (Formerly SRM University), Chennai, India. Currently, he has guiding three Ph.D. students. He has a 6 years 3 month academic and research experience in the manufacturing field. He has published more than 32 technical publications, 34 international conference publications, 7 national conference publications, 4 Patent Publication, Reviewer for 8 International Journal and 9 International Conference, 4 Online courses Completed, 9 workshop/FDP/Conference organized and 47 short-term courses (workshop) attended so far. His areas of interest are micro manufacturing, Additive

Manufacturing Processes, Advanced Manufacturing Processes, Micro and Nanomachining, Composite Materials, Design of Experiments and Optimization Techniques.

Amit Roy Chowdhury is a Professor of Aerospace and Applied Mechanics Dept. IIEST SHIBPUR.

Likitha S. is a Project Scholar in the Department of Aerospace Engineering, Dayananda Sagar University, Bangalore, India.

Maniraj S. has completed PhD in Micro machining and published his research findings in SCI Journals. His area of interest are micromachining, metal matrix composites and optimization. He has 7 years of Teaching experience and 5 years of research experience.

Chandika Samynathan has completed her Bachelors in Electronics and Communication and Masters in Mechatronics. She has completed her PhD from Anna University, India and currently working as a Professor in Algonquin College, Canada. Her area of interest are Robotics, Automation, PLC and MEMS. She has authored paper and published in high impact factor journals.

Timine O. Suoware's first and master's degree were in Mechanical Engineering from Rivers State University of Science and Technology, Port Harcourt. A Doctor of Philosophy (Ph.D) in Mechanical Engineering specialty in Design and Production Technology from University of Nigeria, Nsukka. He has spent 18 years of his postgraduation; training and equipping middle level manpower in Engineering. He is currently a Principal Lecturer in the Department of Mechanical Engineering Technology and Acting Director of Academic Planning, Advancement and Linkages at the Federal Polytechnic, Ekowe, Bayelsa State. He has served as Acting Dean Students' Affairs, Coordinator School of Engineering Technology and Head of Mechanical Engineering Technology. His flair for research which has earned him many published articles in reputable journals such as fire and materials a Scopus publisher, springer nature, materials de construccion and many others. His area of research interest is on material characterization with emphasis on design of components with fibre reinforced composites and their flame retardancy; He recently served as an expert evaluator on the technical session and joint secretary at the international multidisciplinary conference organized by Federal Polytechnic Ekowe and Konkan Geographers Association of India as well as a technical committee member at the 4th international conference on advanced composite materials.

Mythili T. is currently working as Assistant professor in Department of Aeronautical Engineering, Mahendra Engineering College (Autonomous), Namakkal, Tamilnadu, India. She graduated (B.E- Mechanical Engineering) from Mahendra Engineering College in 2004 and obtained her post graduation (M.E-Computer Aided Design) from Mahendra Engineering college, Namakkal in 2007.She obtained her Ph.D from Anna University, Chennai in 2021.She has 14 years of teaching experience. She has published papers in international, national and Scopus and indexed journals. Her areas of interest are Composite Materials, Conventional machining and Non-Conventional Machining.

Kanish T. C. obtained his MTech in Mechatronics and Ph.D. in Manufacturing from Vellore Institute of Technology and his area of interest are abrasive finishing and Condition monitoring of Machine tools. He has published more than 25 papers in international journals and conferences. He is a member of the

Institution of Engineers (IE) India, SAE, and Soft computing research society. Currently, he is guiding PhD scholars at VIT University in the field of Condition Monitoring of Machine tools, Machining, and soft computing techniques. Guided more than 30 Post Graduate and more than 50 undergraduate student Research Projects. Visited Various Foreign universities in Thailand, Singapore, Australia, China, France, UK, and USA for Technical Collaborations. Served as the session chair, resource person, organizer, and executive committee member of many conferences held in India, Thailand, China, Australia, and USA. Organized Various workshops, technical symposiums, training programs on behalf of VIT University. Provided a series of lectures on Non-traditional Machining, Low-cost Automation, Industrial automation etc. to the various industrial Executives and students.

Mohamad Taleuzzaman passed his B. Pharm. and M. Pharm. (Pharmaceutical Analysis) from Jamia Hamdard, Delhi in 2005 and 2007 respectively. Ph. D. From Glocal University Saharanpur, 2018. Qualified GATE examination, conducted by IIT Mumbai in 2005 with percentile score 92.49. Completed M. Pharm. research project from Fortis clinical research Ltd. (FCRL), Faridabad in 2007.

Index

E

F

G

H

I

L

M

N

O

P

www.ingramcontent.com/pod-product-compliance
Lightning Source LLC
LaVergne TN
LVHW080634011026
849160LV00014B/1217
* 9 7 8 1 7 9 9 8 9 5 7 5 6 *